Lecture Notes in Networks and Systems

Volume 60

Series editor

Janusz Kacprzyk, Polish Academy of Sciences, Warsaw, Poland
e-mail: kacprzyk@ibspan.waw.pl

The series "Lecture Notes in Networks and Systems" publishes the latest developments in Networks and Systems—quickly, informally and with high quality. Original research reported in proceedings and post-proceedings represents the core of LNNS.

Volumes published in LNNS embrace all aspects and subfields of, as well as new challenges in, Networks and Systems.

The series contains proceedings and edited volumes in systems and networks, spanning the areas of Cyber-Physical Systems, Autonomous Systems, Sensor Networks, Control Systems, Energy Systems, Automotive Systems, Biological Systems, Vehicular Networking and Connected Vehicles, Aerospace Systems, Automation, Manufacturing, Smart Grids, Nonlinear Systems, Power Systems, Robotics, Social Systems, Economic Systems and other. Of particular value to both the contributors and the readership are the short publication timeframe and the world-wide distribution and exposure which enable both a wide and rapid dissemination of research output.

The series covers the theory, applications, and perspectives on the state of the art and future developments relevant to systems and networks, decision making, control, complex processes and related areas, as embedded in the fields of interdisciplinary and applied sciences, engineering, computer science, physics, economics, social, and life sciences, as well as the paradigms and methodologies behind them.

Advisory Board

More information about this series at http://www.springer.com/series/15179

Samir Avdaković

Editor

Advanced Technologies, Systems, and Applications III

Proceedings of the International Symposium on Innovative and Interdisciplinary Applications of Advanced Technologies (IAT), Volume 2

Springer

Editor
Samir Avdaković
Faculty of Electrical Engineering
University of Sarajevo
Sarajevo, Bosnia and Herzegovina

ISSN 2367-3370   ISSN 2367-3389   (electronic)
Lecture Notes in Networks and Systems
ISBN 978-3-030-02576-2   ISBN 978-3-030-02577-9   (eBook)
https://doi.org/10.1007/978-3-030-02577-9

Library of Congress Control Number: 2016954521

This Springer imprint is published by the registered company Springer Nature Switzerland AG
The registered company address is: Gewerbestrasse 11, 6330 Cham, Switzerland

Contents

Electrical Machines and Drives

Computer Science

Mechanical Engineering

Civil Engineering

Volume-Delay Functions: A Review

Ammar Saric[✉], Sanjin Albinovic, Suada Dzebo, and Mirza Pozder

Faculty of Civil Engineering, Department of Roads and Transportation,
University of Sarajevo, Sarajevo, Bosnia and Herzegovina
ammar.saric@hotmail.com, pozder.mirza@hotmail.com,
sanjin.albinovic@gmail.com, suada.dzebo@gf.unsa.ba

Abstract. The last step of four-step process in traffic modelling is assignment. The dependence of traffic flow speed on number of vehicles in traffic stream is described by volume-delay functions (VDF). They represent key element in traffic assignment process. There are several different types of these functions that lead to different speed and travel time estimation. The paper gives a review of most frequently used volume-delay functions with their advantages and limitations in practical use.

Keywords: Volume-delay functions (VDF) · Speed · Traffic flow

1 Introduction

In transportation planning process, four-step demand model is the most popular travel demand forecasting model. Components of this model are trip generation, trip distribution, modal split, and trip assignment. Last step, i.e. traffic assignment, represents process to assign the traffic demand to the links of network. In this step drivers choose the best travel route based on traffic conditions and travel costs. As an indicator of this choice, travel time is used, which is equivalent to the travel cost and represent one of the most important factor in decision making regarding destinations, routes and transport modes. [1] Another quality measure of chosen route is average speed of traffic flow which is in direct correlation with travel time.

It is well known that travel time is increasing with increasing of traffic flow, which results in higher degree of saturation. The relationship between travel time or average speed of traffic flow and number of vehicles in traffic stream is described by volume-delay functions (VDF). [1] *"The correct determination of the volume-delay function is very important due to its strong effect on the results and, as a consequence, on the reliability of the traffic model."* [2] In general, three components are necessary to describe this relation: free-flow speed (or free-flow travel time), link capacity and number of vehicles in traffic flow or volume. Free-flow speed can be measured directly on the field, which is detail described in several research and manuals (i.e. Highway Capacity Manual – HCM). Unlike this, capacity is much harder to measure due its dynamic nature. However, usually both components are taken as constant for one road section.

© Springer Nature Switzerland AG 2019
S. Avdaković (Ed.): IAT 2018, LNNS 60, pp. 3–12, 2019.
https://doi.org/10.1007/978-3-030-02577-9_1

2   Volume-Delay Functions

First research on volume-delay functions are from the 1950s. They describe speed-volume relationship in very simplified form, such as linear by Irwin Dodd and Von Cube. Later, more complex functions were developed by Overgaard (exponential) and Mosher (logarithmic and hyperbolic). [1, 2] In the most popular software package for traffic macrosimulations PTV Visum 17, volume-delay functions are listed as follows: [3]

1. Akçelik

$$t_{cur} = t_0 + \frac{3600}{4} \cdot a \cdot \left[(sat - 1) + \sqrt{(sat - 1)^2 + \frac{8 \cdot b \cdot sat}{d \cdot a}} \right] \tag{1}$$

where are:
a – length of the analysis time slot Tf (h)
d – lane capacity Q (veh/h)
and with default values: a = 1; b = 1; c = 1; d = 1800
In all models *sat* is defined as:

$$sat = \frac{q}{q_{max} \cdot c} \tag{2}$$

2. Akçelik 2

$$t_{cur} = 3600 \cdot length \cdot \left[\frac{1}{v_0} + \frac{a}{4} \cdot \left((sat - 1) + \sqrt{(sat - 1)^2 + \frac{8 \cdot b \cdot sat}{d \cdot a \cdot q_{max}}} \right) \right] \tag{3}$$

where are:
a – length of the analysis time slot Tf (h)
d – 1/number of lanes
and with default values: a = 1; b = 1; c = 1; d = 1

3. BPR

$$t_{cur} = t_0 \cdot \left(1 + a \cdot sat^b \right) \tag{4}$$

with default values: a = 1; b = 3; c = 1

4. BPR 2

$$t_{cur} = \begin{cases} t_0 \cdot \left(1 + a \cdot sat^b \right), & sat \leq sat_{crit} \\ t_0 \cdot \left(1 + a \cdot sat^{b'} \right), & sat > sat_{crit} \end{cases} \tag{5}$$

with default values: a = 1; b = 2; b' = 2; c = 1

5. BPR 3

$$
t_{cur} = \begin{cases} t_0 \cdot \left(1 + a \cdot sat^b\right), \ sat \le sat_{crit} \\ t_0 \cdot \left(1 + a \cdot sat^{b'}\right) + (q - q_{max}) \cdot d, \ sat > sat_{crit} \end{cases} \tag{6}
$$

with default values: a = 1; b = 2; c = 1; d = 0

6. CONICAL

$$
t_{cur} = t_0 \cdot \left(2 + \sqrt{a^2 \cdot (1 - sat)^2 + b^2} - a \cdot (1 - sat) - b\right); b = \frac{2a - 1}{2a - 2} \tag{7}
$$

with default values: a = 4; c = 1

7. CONICAL-MARGINAL

$$
t_{cur} = t_0 \cdot \left(2 + \frac{a^2 \cdot (1 - sat) \cdot (1 - 2 \cdot sat) + b^2}{\sqrt{a^2 \cdot (1 - sat)^2 + b^2}} - a \cdot (1 - 2 \cdot sat) - b\right); b = \frac{2a - 1}{2a - 2} \tag{8}
$$

with default values: a = 4; c = 1

8. Exponential

$$
t_{cur} = \begin{cases} t_0 + \frac{e^{a \cdot sat}}{b}, \ sat \le sat_{crit} \\ t_0 + \frac{e^{a \cdot sat_{crit}}}{b} + d \cdot (sat - sat_{crit}), \ sat > sat_{crit} \end{cases} \tag{9}
$$

with default values: a = 1; b = 1; c = 1; d = 1

9. INRETS

$$
t_{cur} = \begin{cases} t_0 \cdot \frac{1,1 - a \cdot sat}{1,1 - sat}, \ sat \le sat_{crit} \\ t_0 \cdot \frac{1,1}{0,1} \cdot sat^2, \ sat > sat_{crit} \end{cases} \tag{10}
$$

with default values: a = 1; c = 1

10. Logistic

$$
t_{cur} = t_0 + \frac{a}{1 + f \cdot e^{(b - d \cdot sat)}} \tag{11}
$$

with default values: a = 1; b = 1; c = 1; d = 1; f = 1

11. Lohse

$$t_{cur} = \begin{cases} t_0 \cdot (1 + a \cdot sat^b), \ sat \leq sat_{crit} \\ t_0 \cdot \left(1 + a \cdot (sat_{crit})^b\right) + a \cdot b \cdot t_0 \cdot (sat_{crit})^{b-1} \cdot (sat - sat_{crit}), \ sat > sat_{crit} \end{cases}$$

(12)

with default values: a = 1; b = 3; c = 1

12. Quadratic

$$t_{cur} = t_0 + a + b \cdot sat + d \cdot sat^2$$

(13)

with default values: a = 1; b = 1; c = 1; d = 1

13. SIGMOIDAL-MMF

- for links

$$t_{cur} = t_0 \cdot \frac{a \cdot b + d \cdot sat^f}{b + sat^f}$$

(14)

with default values: a = 1; b = 1; c = 1; d = 1; f = 1

- for nodes

$$t_{cur} = t_0 + \frac{a \cdot b + d \cdot sat^f}{b + sat^f}$$

(15)

with default values: a = 1; b = 1; c = 1; d = 1; f = 1

14. TMODEL

- for links

$$t_{cur} = \begin{cases} (t_0 + a) \cdot \left(1 + d \cdot (sat + f)^b\right), \ sat \leq sat_{crit} \\ (t_0 + a') \cdot \left(1 + d' \cdot (sat + f)^{b'}\right), \ sat > sat_{crit} \end{cases}$$

(16)

with default values: a = 0; a' = 0; b = 2; b' = 2; c = 1; d = 1; d' = 1; f = 0; f' = 0

- for nodes

$$t_{cur} = \begin{cases} (t_0 + a) \ + \ d \cdot (sat + f)^b, \ sat \leq sat_{crit} \\ (t_0 + a') \ + \ d' \cdot (sat + f)^{b'}, \ sat > sat_{crit} \end{cases}$$

(17)

with default values: a = 0; a' = 0; b = 2; b' = 2; c = 1; d = 1; d' = 1; f = 0; f' = 0.

It should be noted that these volume-delay functions are expressed in form of travel time. Very simple transformation can make these formulas suitable for calculation of average travel speed.

This list of VDF models is not final. There are more and specific VDF models developed for specific area and calibrated for particular road type. However, reliability of all of these models in practical application must be verified with field investigation.

In the 1989. Spiess [4] defined conditions that VDF models need to satisfy:

- Function must be strictly increasing.
- The value of the function at zero volume must be equal to free-flow speed and the value of the function at critical volume of 1 must be equal to the half of the free-flow speed.
- First derivate of the function must exist and be strictly increasing.
- The value of the first derivate of the function at a flow equaling capacity should be equal α. This parameter is equal to α in BPR formula and explain change of speed or travel time when capacity is reached.
- The first derivate of the function must be less than Mα, where M is a positive constant. This control the steepness of the function in congested conditions.
- The first derivate of the functions for volume equal to 0 must be positive which guarantees the uniqueness of the link volume.
- The calculation of new model should not take more computing time than BPR model [2].

All listed models meet previous requirements, but the most commonly used VDF models are: BPR, Akçelik and Conical.

The BPR function was developed by the US Bureau of Public Roads (BPR) in 1964. One of the most important factor of popularity of this model is its simplicity. Application of this model requires knowledge of two parameters; parameters α and β (or a and b) (Eq. 3). Parameter α refers to *"the ratio of travel time* (or average travel speed) *per unit distance at practical capacity to that at free flow"*, while parameter β determines change of average travel speed from free-flow conditions to congested conditions. [5] This change is moderate with smaller values of β, while the higher values makes it more sudden (Fig. 1). Also, shape of the BPR function depends on β values; with smaller values ($\beta < 1$) function is concave, instead of convex shape for $\beta > 1$ (Fig. 1). According to the FHWA (2014) [6] recommendations, default values for parameters α and β (or a and b) are 0,15 and 4, respectively. However, these values do not represent traffic conditions for all types of roads and different type of traffic control. [7] Therefore, calibration process, which is mathematically very fast, with correct field data is necessary.

Regardless simplicity of this model, it has several limitations. According to Singh [8] one major problem with BPR function is that it overestimates speeds in congested conditions (v/c > 1) and underestimates speeds at v/c < 1. [1] Another problem is that this model does not consider the existence of traffic lights or number of lanes. Spiess [4] also has pointed several shortcomings of BPR model. For these reasons, several planning organizations proposed modified BPR function (like BPR 2 or BPR 3) or entire different VDF model [5].

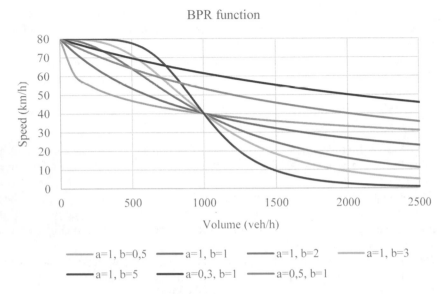

Fig. 1. Different BPR curves

Conical congested functions are the most common substitution for BPR model (Eq. 7). Spiess [4] developed this model in order to overcome problems with high value of β parameter in BPR function. [5] During the first iterations in equilibrium assignment v/c relation can be highly over 1 (very often 3 or 5). In combination with high value of β parameter it can cause some numerical problems and overload links. In addition to this, for volume much below capacity, free-flow speed becomes independent of actual traffic volume when high value of β parameter is used [4].

Difference between BPR and Conical model with the same parameter is negligible because the same parameter is used for specification of congestion behavior of a road link, such as capacity and steepness, so transition from one to another model is very simple [1, 4].

Davidson [9] has proposed function based on queuing theory and Taylor [10] introduce method for estimating the parameters in Davidson's function. However, this definition of its parameter implies equality between flow capacity and reciprocal of the free-flow travel time [11]. Davidson [12] endeavor to modified delay parameter ("b" in Eq. 1) led to another problem; service quality was better with increased free-flow travel time [13]. Akçelik [13] developed a time-dependent form of Davidson's function using coordinate transformations to overcome problems of inconsistent parameter definition and overestimation of travel time around the capacity flow. [14] Akçelik's model improved modeling of link travel speed in conditions when intersection delays prevail. Also, this model has better convergence and more realistic speed estimation under congested conditions [8].

2.1 Comparison of VDF Models

Most discussion about reliability of VDF models is focused on description of volume-speed relationship in congested conditions, i.e. when the capacity is exceeded. There are two main problems with these models in such conditions; the shape of VDF curve and steepness. According to [15] the shape of the VDF for the volume beyond capacity is easy to define and can be done with field data. Of course, there are two possible shapes, convex and concave. Fundamental diagram of traffic flow suggests that relationship volume-speed must be convex. Two most used models, BPR and Conical, also defined this relation as convex (Fig. 2), both with default parameters and calibrated values. However, lower β values which cause concave shape are also very likely. For example, calibrated BPR and Conical models for two-way highways have β < 1 (e.g. [16]). In conditions when v/c > 1 the shape of VDF model is more result of author's intuition and often insufficiently large amount of realistic data. During the peak hours many links of highways in big cities have high volumes over capacity, which correspond to the bottom part of fundamental volume-speed diagram. [15] This part of diagram cannot be represented with VDF models.

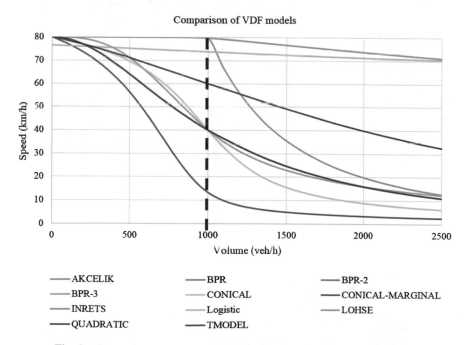

Fig. 2. Comparison of different VDF models (volume-speed relationship)

Another important difference between these models is steepness of the curve in area near or beyond capacity. Some models are tend to dramatically drop at point v/c = 1 on volume-speed diagram (Fig. 2) or rise very fast on volume-travel time diagram (Fig. 3) (e.g. Akçelik, Inrets, Conical-marginal). Other VDF curves have moderate change of behavior at this point, but it still can have big influence on final results.

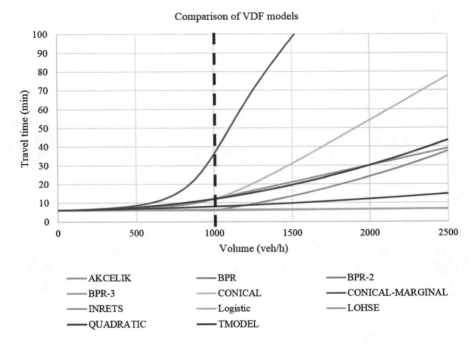

Fig. 3. Comparison of different VDF models (volume-travel time relationship)

In order to make an overview of different VDF models a hypothetical test was made. Eleven volume-delay functions were compared with these inputs:

1. Values of unknown parameters in VDF models are default like in PTV Visum 17.
2. Free-flow speed is $V_0 = 80$ km/h. This speed correspond to the posted speed limit on two-lane highways in Bosnia and Herzegovina.
3. Capacity (q_{max}) is equal to 1000 veh/h. This value is obtained as practical capacity (80%) of one-lane capacity on two-lane highways according to HBS 2010. This is very relevant for conditions in Bosnia and Herzegovina, but may not be correct for other countries or types of roads. This topic is no further discussed here.
4. Length of the hypothetical road is 8 km with travel time at free-flow of 6 min.

On Fig. 2 volume-speed relationship of VDF models is displayed. As it can be seen there is a big difference among volume-delay functions in terms of shape, steepness and predicted speed. Some models (like Akçelik and Inrets) don't show any sensitivity in the area below capacity and then fall fast after capacity is reached. This behavior is more characteristic for road segments with signalized intersections. Other models have similar trend for all volume values. Predicted speed is very similar for all volume-delay functions in low-flow conditions (i.e. up to 400 veh/h). This state could also be viewed as free-flow condition.

Similar observations can also be made for volume-travel time relationship on Fig. 3. For low value of traffic flow, travel time is almost constant, and this trend is obvious for all models except for Conical-marginal. The biggest change is at capacity point where most volume-delay functions tend to rise very fast. Exceptions from this are Akçelik and Logistic models.

All of these models are developed only for homogeneous traffic. As it can be seen from Eqs. (1)–(17), volume-delay functions can calculate speed only for one vehicle class. They do not account for speeds of various classes present in the stream. [17] In reality, traffic flow is mixed with several types of vehicles. Each vehicle category has specific traffic performance and impact on overall traffic stream. The biggest influence on speed of traffic flow have heavy vehicles, which are slower than personal cars, especially in combination with bad horizontal and vertical geometry. In addition, faster vehicles must overpass slower vehicles within segment of road with a sufficient length of passing zone, which is also important factor for correct description of traffic flow behavior.

All stated limitations and drawbacks of existing VDF models indicate need for its improvements. New improvements of volume-delay functions must include the following:

- New calibration parameter for different type of vehicles or develop new model suitable for different type of vehicles.
- Incorporate impact of vertical and horizontal geometry including the length of passing zone sufficient for safe passing manoeuvre.
- Develop new or improve existing model based on stochastic capacity.
- Improve existing models for congested conditions.
- Investigate in detail which models are appropriate for different type of roads and possibility of usage one model for several type of roads.

3 Conclusion

Volume-delay functions have key role in process of traffic assignment. They need to reliably present a relationship between travel speed (or travel time) and traffic flow. There are numerous models in usage, and all of them require at least calibration of unknown parameters in practical application.

However, beside calibration, these models need improvements in several fields for more realistic represent of traffic performance. Those improvements must include: better definition of models in congested conditions, influence of road and vehicle characteristics on speed and travel time and development of different models for different types of road facilities.

References

1. Leong, L.V.: Delay functions in trip assignment for transport planning process. In: Proceedings of the International Conference of Global Network for Innovative Technology and AWAM International Conference in Civil Engineering (IGNITE – AICCE 2017) (2017)
2. Oskarbiski, J., Jamroz, K., Smolarek, L., Zawisza, M., Žarski, K.: Analysis of possibilities for the use of volume-delay functions in the planning module of the tristar system. Transp. Probl. **12**(1), 39–50 (2017)

3. PTV Visum, Manual. PTV AG, Karlsruhe, Germany (2017)
4. Spiess, H.: Technical note – conical volume-delay functions. Transp. Sci. **24**(2), 153–158 (1990)
5. Mtoi, E.T., Moses, R.: Calibration and evaluation of link congestion functions: applying intrinsic sensitivity of link speed as a practical consideration to heterogeneous facility types within urban network. J. Transp. Technol. **4**, 141–149 (2014)
6. U.S. Department of Transportation, Federal Highway Administration, Office of Planning, Environment, and Reality: Travel Model Improvement Program (TMIP), TMIP Email List Technical Synthesis Series 2007–2010 (2014)
7. Marquez, L.: Conical and the BPR Volume-delay functions for multilane roads. Boletín Técnico **54**(3), 14–24 (2016)
8. Singh, R., Dowling, R.: Improved speed-flow relationship: application to transportation planning models. In: Donnelly, R. (ed.) Proceedings of the Seventh TRB Conference on the Application of Transportation Planning Methods, pp. 340–349. Transportation Research Board (1999)
9. Davidson, K.B.: A flow travel time relationship for use in transportation planning. In: 3rd Australian Road Research Board (ARRB) Conference, Sydney, pp. 183–194 (1966)
10. Taylor, M.A.P.: Parameter estimation and sensitivity of parameter values in a flow-rate/travel-time relation. Transp. Sci. **11**, 275–292 (1977)
11. Golding, S.: On Davidson's flow/travel time relationship. Aust. Road Res. **7**, 36–37 (1977)
12. Davidson, K.B.: The theoretical basis of a flow-travel time relationship for use in transportation planning. Aust. Road Res. **8**, 32–35 (1978)
13. Akçelik, R.: Travel time functions for transport planning purposes: Davidson's function, its time dependent form and alternative travel time function. Aust. Road Res. **21**, 44–59 (1991)
14. Wong. W., Wong, S.C.: Network topological effects on the macroscopic bureau of public roads function. Transp. A: Transp. Sci. **12**(2), 272–296 (2015)
15. Jastrzebski, W.: Volume delay functions. In: 15th International EMME/2 Users Group Conference, Vancouver, BC (2000)
16. Lovrić, I.: Modeli brzine prometnog toka izvangradskih dvotračnih cesta (eng. Speed models of traffic flow on two-lane highways). PhD thesis, University of Mostar (2007)
17. Leong, L.V.: Effects of volume-delay function on time, speed and assigned volume in transportation planning process. Int. J. Appl. Eng. Res. **11**(13), 8010–8018 (2016)

Model of Existing Road Using Aerial Photogrammetry

Mirza Pozder[1(⊠)], Sanjin Albinovic[1], Ammar Saric[1],
Dzevad Krdzalic[2], and Marko Savic[3]

[1] Faculty of Civil Engineering, Department of Roads and Transportation,
University of Sarajevo, Sarajevo, Bosnia and Herzegovina
pozder.mirza@hotmail.com, ammar.saric@hotmail.com,
sanjin.albinovic@gmail.com
[2] Faculty of Civil Engineering, Department of Geodesy and Geoinformatics,
University of Sarajevo, Sarajevo, Bosnia and Herzegovina
dzevadkrdzalic@gmail.com
[3] Survey Engineer, Survey Agency "Marcos", Bratunac,
Bosnia and Herzegovina
savic.marko5@yahoo.com

Abstract. This paper presents developing a model of existing road using aerial photogrammetry. Project is consisting of two phases. In the first phase, point cloud was made using aerial photogrammetry. Second phase was building of 3D model of existing road. Primary objective of this project was testing a method and possibility to use it in future for "as built" projects on road network in B&H.

Keywords: Aerial photogrammetry · Point cloud · Model · Road

1 Introduction

Lately, application of point clouds in civil engineering is more popular and even necessary. Usage of point clouds is based on digitalization of existing objects and 3D model making for future objects. This paper presents a continuation of research on application of photogrammetry in road engineering. The first results are presented in [1].

For the past few years, many researchers presented their research results about ability to use photogrammetry technique (terrestrial and aerial) in exploration road surface features.

For example, Knyaz and Chibunichev presented research results about using photogrammetric techniques for road surface analysis. Two photogrammetric techniques for road surface analysis are presented. For accurate measuring of road pavement and for road surface reconstruction based on imagery obtained from unmanned aerial vehicle. The first technique uses photogrammetric system based on structured light for fast and accurate surface 3D reconstruction and it allows analysing the characteristics of road texture and monitoring the pavement behaviour. The second technique provides dense 3D model road suitable for road macro parameters estimation [2].

© Springer Nature Switzerland AG 2019
S. Avdaković (Ed.): IAT 2018, LNNS 60, pp. 13–20, 2019.
https://doi.org/10.1007/978-3-030-02577-9_2

David et al. presented study about using close range photogrammetry and how hand-held laserscanning can be used to derive 3D models of pavement surfaces [3].

Tiong et al. presented paper about using close range digital photogrammetry method for determination road pothole severity. Ten random pothole samples were chosen randomly and their severity levels were assessed through both the conventional visual approach and the photogrammetric method. The study reveals that digitalclose range photogrammetry method can be used as the alternative way in visual assessment of pothole severity [4].

2   Results from Previous Project

The first part of several connected projects related to photogrammetry usage in road engineering is about terrestrial photogrammetry [1].

In order to determine the condition parameters of road pavement surface. Research was conducted on a small test site at Faculty of Civil Engineering, University of Sarajevo. Currently pavement distress data, in Bosnia and Herzegovina, are usually collected with different techniques. On highways and main roads, pavement distress data are collected using falling weight deflectometer (FWD) or road profilometer. On regional and local road, there are no measuring distresses at all in regular periods, only on projects levels if necessary.

Primary objective of previous study was to explore and popularize this technique, and explore possibility to use it in the future for road pavement surface distresses.

Figure 1 represents high density point cloud made from photos using terestrial photogrammetry and consist about 1,5 milion points.

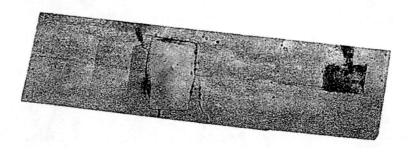

Fig. 1. High density point cloud - top view [1].

Another objective of this project was also creation of the model used for identification road pavement surface distress such as potholes, cracks and damaged patch. These parameters are often used for determination road pavement performance.

Figure 2 shows shaded point cloud model where road surface distresses can be identified. Based on this model, severity and intensity of these distresses can be estimated (especially for potholes and patches) [1].

Fig. 2. Shaded point cloud model with ditress identification [1].

Finally high density point cloud model is developed (Fig. 3). Based on this model, mesh (Fig. 4) and surface model is created and it can be used for determination macro texture or texture depth.

Fig. 3. Point cloud model [1].

Fig. 4. Road surface mesh model [1].

Small scale model was used for modeling of road surface texture model.

In following section, second part of this project, which is related to aerial photogrammetry application for making of DTM (digital terrain model) of roads, is described.

3   Aerial Photogrammetry Survey

In process of generation a test point cloud, aerial photogrammetry was used with application of unmanned aerial vehicle (UAV) and GNSS receiver. The type of UAV was DJI Phantom 4 (Fig. 5).

Fig. 5. UAV Phantom 4 with the table for photosignaling of control points

Unmanned aerial vehicle Phantom 4 is equipped with all necessary sensors to carry out an autonomous flight, as well as with a 13 megapixel camera which is embedded within a gyroscope that allows horizontal position of camera during the flight. The duration of the flight was approximately 10 min and during this time, 103 photographs were taken from which a photogrammetric point cloud was generated. As regards the survey of the constructed road, it was not necessary to filter the data because it is a built-up area with easily recognizable contours.

The flight was executed from a height of 60 m above the take-off point, and 80% of the longitudinal and transverse overlap were used, because it is a narrow and long object. The flight of UAV during the aerial survey was carried out in five parallel flights that are oriented in the northwest-southeast direction as well as the direction of this section of the road. The control points in the field are photosignalized with signs measuring 0.50×0.50 m. The position and altitude of the control points were determined using the Trimble R4 GNSS receiver in the RTK mode of operation, connected to the permanent station system of B&H. There were 9 control points in the field (Figs. 6 and 7).

Data processing resulted in a cloud of 7 million points, of which 460.000 were in the zone of the road. The recorded length of the road is 300 m. The density of points after processing is 100 points/m^2. Classical methods of measurement cannot give the quantity and detail of the model to such an extent, and the application of this model is multiple. When processing data with a given flight height and shooting parameters, it is possible to see objects of dimensions over 3 cm using orthomosaic, and the

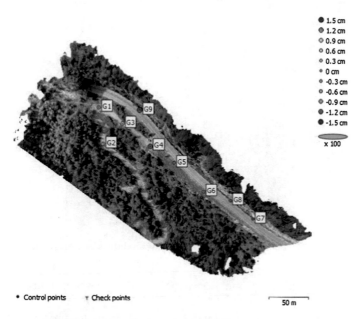

Fig. 6. Position of control points with elllipse of errors

Fig. 7. Shaded point cloud model with ditress identification

determination of the precise position of the existing objects on the ground is very easily feasible.

This way of survey is much faster and more efficient than the traditional approach. All important objects were recorded with high accuracy, with the possibility of survey of inaccessible parts of the terrain.

The obtained model can be used for various purposes, such as: 3D model of terrain for designing new roads and reconstruction of existing ones, analysis of the condition of the pavement construction on a much larger scale, etc.

As the original point cloud contained a large number of points, and the objective of this project was to create a 3D model of existing road, a reduction in the number of

Fig. 8. Shaded point cloud model with ditress identification

points (crop) in the area of the traffic was done. Figure 8 shows the cropped point cloud model of road.

The next step in designing a 3D model is to extract points and feature lines from point cloud. For this purpose, the Autodesk Infraworks software package was used. Selection and shading of points were performed and finally extraction features lines (especially alignment of the road) and cross sections was done. In this case, from the original point cloud with almost 460 thousand points, the tendency is to get points on longitudinal and transversal profiles. Figures 9 and 10 show feature lines and cross sections. The number of cross sections is defined at every 2.5 m along the alignment of the road, producing a model with almost 5000 points.

Fig. 9. Shaded point cloud - perspective view

Finally, based on the points from longitudinal feature lines and cross sections, a 3D model of an existing roadway was created (Fig. 11). As the original point cloud model

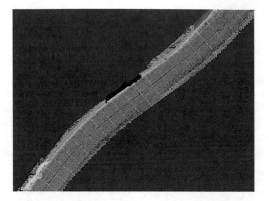

Fig. 10. Shaded point cloud - plan view

contained a large number of points, the accuracy of the 3D model of the road, which was created with a significantly smaller number of points, is quite satisfactory. The transversal and longitudinal slopes, the width of carriageway can be clearly identified, as well as the other characteristics of the road surface, such as larger cracks and rutting.

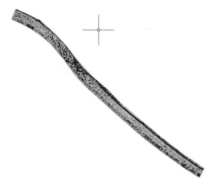

Fig. 11. Shaded point cloud model with ditress identification

4   Conclusion

The use of photogrammetric techniques for making as built projects of a roadway is a technique that could in the future be more and more applied. The research has shown, through these three examples that 3D models can be obtained in a relatively fast and inexpensive way. The use of an aerial photogrammetry is particularly suitable as a short period of time can make models of roads of significant length. Practically, the entire process is limited by flight time and hardware and software support for processing points. The density of point cloud model would ultimately be defined by the investors themselves, and in accordance with the project's goal or the amount of information required from the model.

References

1. Pozder, M., Albnovic, S., Saric, A., Krdzalic, D.: Determination of road surface characteristics using photogrammetry technique. In: 5th International Conference on Road and Rail Infrastructure, Zadar, Croatia (2018)
2. Knyaz, V.A., Chibunichev, A.G.: Photogrammetric techniques for road surface analysis. The International Archives of the Photogrammetry, Remote Sensing and Spatial Information Sciences, vol. XLI-B5, XXIII ISPRS Congress, Prague, Czech Republic (2016)
3. David, W., Millar, P., McQuaid, G.: Use of 3D modelling techniques to better understand road surface textures. In: 5th International SaferRoads Conference, Auckland, New Zealand (2017)
4. Tiong, P.L.Y., Mushairry, M., Hainin, M.R.: Road surface assessment of pothole severity by close range digital photogrammetry method. World Appl. Sci. J. **19**(6), 867–873 (2012)

Importance and Comparison of Factors Influencing Success in Construction Project in Bosnia and Herzegovina and Croatia

Žanesa Ljevo[✉] and Suada Džebo

Faculty of Civil Engineering, Department of Roads and Transportation,
University of Sarajevo, Sarajevo, Bosnia and Herzegovina
zanesalj@gmail.com, suada.dzebo.gf@gmail.com

Abstract. More than 2/3 organization don't fully understands the value of project management. Organizations that undervalue project management as a strategic competency report an average of 50% more of their projects failing outright. Projects are 2.5 times more successful when proven project management practices are used. This paper show the results of research in construction organization in Bosnia and Herzegovina about project management and factors success. In the research conducted in Bosnia and Herzegovina and Croatia, 154 respondents participated. The only successful projects are completed on time, within budget and quality for 16.7% Investors, 33.3% Contractors and 25.0% Project Management. The following success factors were analyzed: project mission, project schedule/plans, top management support, client consultation and acceptance, monitoring and feedback, communication, personnel. Communication and top management support are most important for performance/ execution phase.

Keywords: Factor · Success · Construction projects · B&H · Croatia
Questionnaire · Results

1 Introduction

According to a PMI (Project Management Institute) report from 2018 year only 58% of organizations fully understand the value of project management. The same report shows the trends of success factors throughout the years (Fig. 1). In 2018, 69% projects met original goals, and 2013 that was around 60%. The failed projects budgets lost in 2013 year was over 35%, and 2018 that number is 32%. Report showed that 52% of the experienced scope creep or uncontrolled changes to the project's scope, which is a significant increase from 43% reported five years ago.

Every organization, regardless of the industry, is required today to adjust much more quickly than in the past, due to the speed of change and fierce competition on the market. To do this, organizations start projects and expect them to deliver results. To do this, organizations start projects and expect them to deliver results, and that can metrics with project performance. Only 52% Project Management Practitioners believed that their organization fully understands the value of project management, but 87% Senior Executive believed in this [1].

© Springer Nature Switzerland AG 2019
S. Avdaković (Ed.): IAT 2018, LNNS 60, pp. 21–27, 2019.
https://doi.org/10.1007/978-3-030-02577-9_3

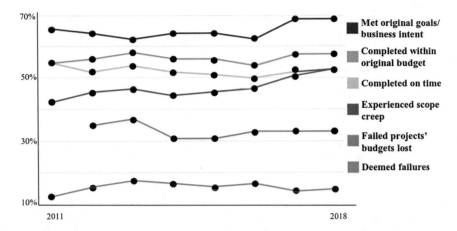

Fig. 1. Project performance metrics

2 Overview of Literature

Traditionally, the project's success measure reflected three aspects of "three constraints" or "iron triangles": cost, time and quality/performance. These dimensions are still considered basic to measure the success of the project.

However, many authors have agreed that the project's performance goes beyond these three criteria regardless of the type of project.

In 1988, Wit [2] showed that these measures were not sufficient to determine the success of the project.

Increasing the scope and complexity of contracts and projects leads to an increase in criteria such as security, quality of set requirements or impact on contractual arrangements [3].

Project success has historically been measured against the triple constraints of cost, time, and quality; often referred to as the iron triangle. Cost and time are easily measurably throughout the life of the project. However, frequently, quality will not be measurable until the final project output is delivered.

The concept of a project success can have a different meaning to different people. Project management literature has not defined unambiguous criteria for successful projects. On the basis of previous research in project management, critical success/failure factors in project phases and conflict situations have been reviewed [4].

During the literature review we have found many studies that were dealing with success factors, and some of them findings that it is possible to identify critical success/failure factors in project management in different organizational conditions and in different project phases [4].

It was observed that authors were building on Pinto and Slevin's (1987) success factors as opposed to creating original factors, which implies that current literature views these factors as adequate without the need for further research. That factors are: project mission, top management support, schedule and plans, client consultation, personnel, technical tasks, client acceptance, monitoring and feedback, communication and trouble-shooting [5–8].

The importance of owner involvement to create four success conditions:

1. "Success criteria should be agreed on with the stakeholders before the start of the project, and repeatedly at configuration review points throughout the project.
2. A collaborative working relationship should be maintained between the project owner (or sponsor) and project manager, with both viewing the project as a partnership.
3. The project manager should be empowered with flexibility to deal with unforeseen circumstances as they see best, and with the owner giving guidance as to how they think the project should be best achieved.
4. The owner should take an interest in the performance of the project" [9].

A strong project manager helps a company to define the vision for a project or initiative. If they can bring robust project management skills it can make the difference between a successful project that helps companies to grow and a failed project that consumes resources with no measurable return.

Factors F1-project mission, F2-project schedule and plans, F3-personnel and technical task, F4-communication, F5-client acceptance and consultation, F6-monitoring and feedback, F7-top management support have been identified as success factors of the project, which were selected and amended (after analysis and consultation with university professors).

3 Research Methodology and Results

After reviewing the literature, analysis and consultation with university professors, we selected key factors, and they are used in further research. There was based on a survey conducted in Bosnia and Herzegovina and Croatia, followed by the analysis of the results, and making conclusions. In the questionnaires the phases of concept, defining and planning, execution, monitoring and control (which runs in parallel with the performance/executions phase) as well as phases that are present in the construction projects were taken into account, but were used only first three phases.

The paper assumes that the different participants perceive (Investor, Contractor/Subcontractor and Project manager/Consultant/Architect/Designer) differently the importance of the key success factors in different phases of the construction project, and ranking them differently.

The questionnaire contained questions that were related to the key success factors, whose importance was evaluated according to the Likert scale of assessment (1 - not at all important; ... 6 - most important), after which each key success factors was allocated a phase of the project (concept, defining and planning, execution, monitoring and control) which was considered to be noteworthy.

The ranking of importance (Relative Importance Index - RII) the key quality factors from the perception of investors, contractors, project managers [10].

RII is in the interval $0 \div 1$ - when used in ordinal grading scale in research many researchers advocate this method of ranking; as RII is higher the factor is considered to be more important (1 being the most important and 7 being the least important). The method was propagated by many authors in similar cases, but not the same, because in literature we did not find something like this case [10].

In the survey participated investors of construction project, civil engineers, architects. From 154 filled questionnaires, 79 were from Bosnia and Herzegovina and 75 from Croatia.

The following will show the results of the ranking of quality factors (Table 1) and the analysis of the importance of certain factors that affect the success of the project by project phases and by groups investor, contractor, project manager. In this case the ranking of importance - RII just measure perceptions (strength of feeling).

Table 1. RII and ranking of success factors from the perspective of different respondents

		Success factors						
		F1	F2	F3	F4	F5	F6	F7
All respondents	B&H	0.757 / 7	0.795 / 5	0.806 / 4	0.816 / 3	0.821 / 2	0.827 / 1	0.762 / 6
	CRO	0.744 / 7	0.820 / 3	0.791 / 5	0.847 / 1	0.798 / 4	0.847 / 1	0.764 / 6
Investor	B&H	0.854 / 1	0.778 / 6	0.806 / 4	0.826 / 2	0.826 / 2	0.799 / 5	0.771 / 7
	CRO	0.806 / 4	0.796 / 5	0.750 / 6	0.815 / 2	0.750 / 6	0.833 / 1	0.815 / 2
Contractor	B&H	0.722 / 7	0.796 / 5	0.821 / 4	0.827 / 3	0.840 / 2	0.858 / 1	0.747 / 6
	CRO	0.638 / 7	0.790 / 5	0.790 / 5	0.877 / 1	0.812 / 4	0.862 / 2	0.843 / 3
Project manager	B&H	0.708 / 7	0.810 / 2	0.792 / 5	0.798 / 3	0.798 / 3	0.821 / 1	0.768 / 6
	CRO	0.784 / 6	0.853 / 1	0.814 / 4	0.843 / 2	0.814 / 4	0.843 / 2	0.755 / 7

Davis [8] in the study ranking success factors on this way: communication, client consultation, client acceptance, top management support, project schedule/plans, project mission, technical task, trouble-shooting, personnel and monitoring and feedback, which is different compared to the ranking factors of this research (for all a respondents).

The results of the research showed that all participants in B&H and Croatia consider (Table 1) that the most important success factor is monitoring and feedback (RII = 0.827 - B&H; RII = 0.847 - CRO). Other rankings are different from one another, so in Bosnia and Herzegovina investors believe that the most important factors are the project mission (RII = 0.854), communication, and client acceptance and consultation (RII = 0.826), personnel and technical task (RII = 0.806), monitoring and feedback (RII = 0.799). Constructors believe the most important are monitoring and feedback (RII = 0.858), client acceptance and consultation (RII = 0.840), communication (RII = 0.827). Project managers believe that the most important are: monitoring and feedback (RII = 0.821), project schedule and plans (RII = 0.810), communication, and client acceptance and consultation (RII = 0.798). In Croatia investors believe that the most important factors are: monitoring and feedback (RII = 0.833), communication, and top management support (RII = 0.815), project mission (RII = 0.806). Constructors believe the most important are communication (RII = 0.877), monitoring and feedback (RII = 0.862), client acceptance and consultation (RII = 0.812). Project managers believe that the most important are: project schedule and plans (RII = 0.853), communication, and monitoring and feedback (RII = 0.843).

Based on the analysis it was concluded that there are significant differences in understanding the importance of each individual success factor at certain phases of the project by different participants in the project (Figs. 2, 3 and 4).

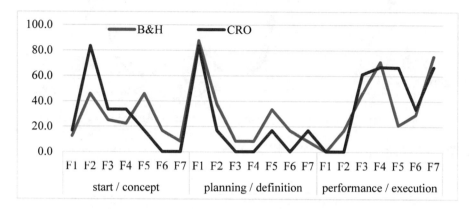

Fig. 2. Importance of key success factors in project phases from investor in B&H and Croatia

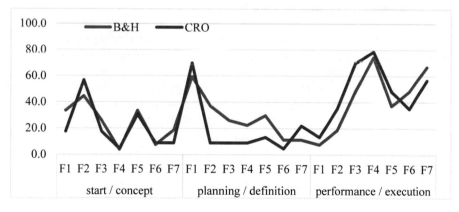

Fig. 3. Importance of key success factors in project phases from contractor in B&H and Croatia

Investors in Bosnia and Herzegovina (Croatia) for the concept phase consider of most important F2, F5 and F3 (F2, F3 and F4), the phase of defining and planning: F1, F2 and F5 (F1, F2, F5 and F7), performance/execution: F7, F4, F3 (F7, F4 and F5) (Fig. 2).

Contractors in Bosnia and Herzegovina (Croatia) for the concept phase consider of most important F3, F5 and F1 (F2, F5, F3 and F1), the phase of defining and planning: F1, F3 and F5 (F1, F7 and F5), performance/execution: F4, F7, F3 (F4, F3 and F7) (Fig. 3).

Project managers in Bosnia and Herzegovina (Croatia) for the concept phase consider of most important F2, F3 and F5 (F2, F5, F3), the phase of defining and planning: F1, F3 and F5 (F1, F5 and F7), performance/execution: F4, F7, F6 (F4, F3 and F7) (Fig. 4).

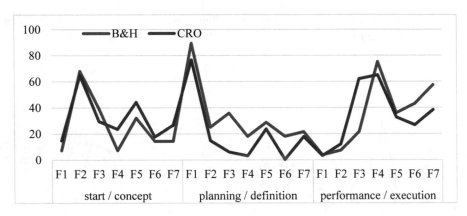

Fig. 4. Importance of key success factors in project phases from project manager in B&H and Croatia

The differences in importance are visible both between individual factors and phases of the project, as well as between countries (Figs. 2, 3 and 4).

4 Discussion and Conclusion

The top three ranked factors in Bosnia and Herzegovina are: monitoring and feedback, client acceptance and consultation, communication while in Croatia the top free factors are: communication, monitoring and feedback, and project schedule and plans (Table 1). Davis [8] in the study ranking top tree success factors on this way: communication, client consultation, client acceptance which is different compared to the ranking factors of this research (for all a respondents). In B&H two factors are the same, and in Croatia only the one is the same. This shows that there is not a big difference in understanding the importance of factors in countries the analysis was carried out.

According to their importance, the factors for phase such as conception for investors in B&H are F2, F5, in Croatia F2, F3, for contractors from Bosnia and Herzegovina they are F3 and F5, from Croatia F2, F5, and for project managers from B&H F2, F3, and from Croatia F2 and F5 (Figs. 2, 3 and 4).

This results can help participants in construction projects to focus on the key success factors which were marked as important for the phases in which they participate or are rated as important for the entire project.

References

1. Project Management Institute: 10th Global Project Management Survey, Success in Disruptive Times|Expanding the Value Delivery Landscape to Address the High Cost of Low Performance. PMI (2018)
2. Wit, A.: Measurement of project success. Int. J. Proj. Manag. **6**, 164–170 (1988)

3. Winch, G.: Managing Construction Projects. Wiley-Blackwell, Hoboken (2010)
4. Hyväri, I.: Success of projects in different organizational conditions. Proj. Manag. J. **37**, 31–41 (2006)
5. Pinto, J.K., Slevin, D.P.: Critical factors in successful project implementation. IEEE Trans. Eng. Manag. **34**(1), 22–28 (1987)
6. Jugdev, K., Müller, R.: A retrospective look at our evolving understanding of project success. Proj. Manag. J. **36**(4), 19–31 (2005)
7. Turner, J.R., Müller, R.: The project manager's leadership style as a success factor on projects: a review. Proj. Manag. J. **36**(2), 49–61 (2005)
8. Davis, K.: Different stakeholder groups and their perceptions of project success. Int. J. Proj. Manag. **32**, 189–201 (2014)
9. Turner, J.R.: Five conditions for project success. Int. J. Proj. Manag. **22**(5), 349–350 (2004)
10. Ljevo, Ž., Vukomanović, M., Rustempašić, N.: Analyzing significance of key quality factors for management of construction projects. Građevinar **69**, 359–366 (2017). https://doi.org/10.14256/JCE.1723.2016

Challenges and Perspective of Building Information Modeling in Bosnia and Herzegovina

Žanesa Ljevo[1(✉)], Suada Džebo[1], Mirza Pozder[1], and Saša Džumhur[2]

[1] Faculty of Civil Engineering, Department of Roads and Transportation,
University of Sarajevo, Sarajevo, Bosnia and Herzegovina
zanesalj@gmail.com, suada.dzebo.gf@gmail.com,
pozder.mirza@hotmail.com
[2] Dipl. Ing., IPSA Institute LLC Sarajevo, Sarajevo, Bosnia and Herzegovina
sasa.dzumhur@ipsa-institut.com

Abstract. Building Information Modeling (BIM) content has become the most advanced approach to integrating information in infrastructure projects. There is a clear trend of BIM expansion worldwide, but Bosnia and Herzegovina (B&H) is lagging behind as there are still no BIM projects done. Industry reports forecast that the wider adoption of BIM will unlock 15–25% savings to the global infrastructure market by 2025. The benefits of BIM are many and mostly refer to improved collaboration within stakeholders, reduced costs for companies as well as reduced repetitive and time-consuming procedures. This paper brings out the results of the BIM awareness survey conducted in B&H, including the ranking of expected advantages for B&H construction industry from implementation of BIM. Apparently, there are many challenges ahead Governments, construction industry and academy in the country. Therefore, this paper also draft outlines of roadmap for the BIM implementation at the national level.

Keywords: Building information modeling · Construction industry
B&H · Survey · Results

1 Introduction

What we call "Building Information Modeling" nowadays was presented although the origins of BIM concept date back to early 1970s [1]. Recently, BIM has emerged as one of the key streams in construction industry and AEC engineering. However, B&H is lagging behind as there are still no BIM projects done in the country.

The European Union (EU) has recognized potential of BIM not only to generate greater value for money (especially in public works) but also to encourage in-novation and digitalization in the construction sector. Therefore, the EU Directive 2014/24/EU on Public Procurement states that "for public works contracts and design contests, Member States may require the use of specific electronic tools, such as of building information electronic modelling tools or similar". Bosnia and Herzegovina is, as a potential EU candidate country, obliged to transpose EU Directives into its legislation.

© Springer Nature Switzerland AG 2019
S. Avdaković (Ed.): IAT 2018, LNNS 60, pp. 28–33, 2019.
https://doi.org/10.1007/978-3-030-02577-9_4

In 2017, the EU BIM task group drafted "Handbook for the introduction of Building Information Modelling by the European Public Sector" with objective to:

- build common understanding and language,
- share and promote the consistent introduction of BIM,
- encourage wider use of developed standards and common principles.

The Boston Consulting Group (BCG) estimate that by 2025 "full-scale digitalization… will lead to annual global cost savings of 13% to 21% in the design, engineering and construction phases and 10% to 17% in the operations phase" [2].

BIM is currently the most common denomination for a new way of approaching the design, construction and maintenance of buildings. It has been defined as "a set of interacting policies, processes and technologies generating a methodology to manage the essential building design and project data in digital format throughout the building's life-cycle" [3].

BIM may currently be considered the fastest developing concept in construction management. It focuses on construction market globalization, corresponding with the general trend towards globalization, and follows the also fast–developing information technology sector [4].

The current state of BIM implementation in B&H is closely related to the fact that this concept is a relatively new issue in the country.

Therefore, the ultimate aim of this paper is to recommend outlines of roadmap for the BIM adoption and implementation in the country, thus paving the way to the following specific objectives:

- assess the BIM concept,
- identify current status of BIM,
- investigate challenges and propose solutions to enhance BIM acceptance and application by the B&H engineers.

2 Overview of Literature

3D modelling began in the early 1970's based on CAD technologies developed in diverse industries. The construction industry applied 2D design initially for utilizing CAD. To enhance construction specific CAD, the concept of BIM was introduced in the early 2000's [5].

Several industry reports identify systemic issues in the construction process relating to its levels of collaboration, under-investment in technology and R&D; and poor information management. These issues result in poor value for public money and higher financial risk due to unpredictable cost overruns, late delivery of public infrastructure and avoidable project changes. The European construction sector output of €1.3tn (trillion) [6] is approximately 9% of the region's GDP and it employs over 18 million people; 95% of which are employed by small and medium sized enterprises (SME) [7]. However, it is one of the least digitalized sectors with flat or falling productivity rates [8]. The sector's annual productivity rate has increased by only 1% over the past twenty years [9]. According to the Industry Agenda, BIM is the technology-led change most likely to deliver the highest impact to the construction sector.

When engineers and contractors in France, Germany, the UK and the US were asked about their involvement with BIM for transportation infrastructure projects, over three quarters of respondents who use it (76%) report that they are creating their own models, and the remainder work with models authored by others. In this research study, both of these two categories are considered BIM users [10].

Figure 1 shows how many of the engineers and contractors who are currently working with BIM report they were (2015), are (2017), and expect to be (2019) using BIM on 50% or more of their transportation infrastructure project [10].

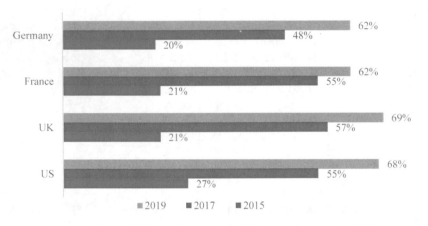

Fig. 1. Use of BIM on transportation infrastructure projects [10]

3 Research Methodology and Results

The BIM awareness survey in B&H was conducted in order to get insight into the application of BIM technology in B&H engineers practice. Results of research conducted in Croatia about awareness of Croatian civil engineering on BIM issues, application of BIM technology and understanding what BIM has served as the base research.

The ranking of importance (Relative Importance Index - RII) is the key quality factors from the perception of investors, contractors and project managers [11].

RII lies in interval between 0 and 1 - when used in ordinal grading scale and many researchers advocate this method of ranking; as RII gets higher the belonging factor is considered more important (1 being the most important and 5 being the least important – Likert scale).

The questionnaire was filled-in during a BIM seminar ("CGS Lab Connect") held in Sarajevo. Out of eighty-four questionnaires, sixty were filled-in 100% correct.

The survey sample included civil engineers (71.9% of all survey participants), architects (1.8%), surveying engineers (8.8%), traffic engineers (10.5%) and others represented in small percentages (in total 7.0%).

22% of the BIM survey participants have had less than 5 years of working experience, 37% of them from 5 to 10 years, another 29% from 11 to 20 years and 12% over 20 years of working experience.

The answers showed that over 95% participants are familiar with the BIM technology concept.

Over 60% of respondents are aware of BIM tools, most of them with Civil 3D, ArchiCAD, Plateia, Revit and Infraworks. On the other hand, for example, 46% of the survey participants are not aware that Microsoft Project is also a BIM tool (Fig. 2).

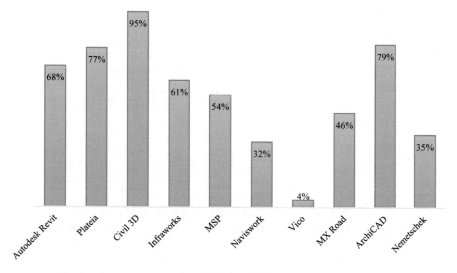

Fig. 2. Answers to question 'Which of BIM tools is familiar to You?'

We can conclude that the survey participants are aware of the BIM technology, but the majority of respondents were unaware both of the all IT tools availability on the market as well as of their BIM related purposes.

By analyzing the survey results presented in Table 1 it can be concluded that employees of B&H companies recognize the advantages of BIM application for AEC industry mostly through 'Reduction of repetitive and time-consuming procedures' (rated as the most important; RII = 0.88), 'Faster reaction on changes' (RII = 0.8500) and 'Faster reaction on changes' (RII = 0.810).

Table 1. Relative Importance Index - RII based ranking of answers to question 'Which are the advantages of BIM implementation?'

Which are the advantages of BIM implementation?	RII	Ranking
For reduction error	0.77	4
For reduction of the repetitive and time consuming procedures	0.88	1
For security measures improvement	0.76	5
For better corporation with other subjects	0.81	3
For business and project cost reduction	0.74	8
For better predictability and cost control	0.76	5
For faster participants decisions	0.76	5
For faster reaction on changes	0.85	2

When comparing the survey results with similar research conducted in Croatia there are certain similarities. So, e.g. the participants of both surveys are aware of the benefits that BIM brings and the best ranked advantage of BIM application for the industry in both surveys is the same - 'Reduction of repetitive and time-consuming procedures'.

4   Discussion and Conclusion

Results of preliminary BIM survey presented in this paper have shown that academy, AEC industry and public sector in B&H are aware of the BIM technology. However, there is a long and challenging road in front of them to implement BIM, not only when it comes to transport infrastructure, but for all AEC industry related projects.

As BIM concept has not been recognized yet in B&H Laws, the public procurement procedure for construction industry is still following Design-Bid-Build procedure, so e.g. plotted CAD drawings are still required as main design outputs. Consequently, when it comes to project implementation lack of information management, collaboration and coordination between Public Clients and designers usually appears causing different types of errors and conflicts. As a result, project implementation requires higher costs and more time than it was originally expected.

Therefore, the first steps for public sector include adoption of standards and specifications developed for work in digitalized environment as well as demonstration of BIM benefits through realization of BIM pilot projects, thus motivating whole AEC industry in B&H to implement BIM.

On the other hand, B&H AEC industry must understand that BIM benefits through creation of new values are far beyond the initial costs of BIM implementation and thus adopt digitalization as one of main strategic objectives.

Moreover, each company must comprehend all implications throughout its supply chain (e.g. training required, management of processes and systems etc.) as well as requirements for management and exchange of information.

Finally, inclusion of BIM in academic education syllabi enabling students to get familiar with BIM concept and to acquire some basic skills is of utmost importance for implementation of BIM at the national level.

Overall, it may be said that the whole process of BIM implementation at the national level should follow the worldwide best practices, where motivation, collaboration and enablement (i.e. devotion to changes in technology, work processes and behavior) have been recognized as key factors for successful BIM implementation.

References

1. Liu, Y., et al.: Understanding effects of BIM on collaborative design and construction: an empirical study in China. Int. J. Proj. Manag. **35**(4), 686–698 (2017)
2. EU BIM Task Group: Handbook for the Introduction of Building Information Modelling by the European Public Sector (2017). http://www.eubim.eu/handbook/
3. Succar, B.: Building information modelling framework. a research and delivery foundation for industry stakeholders. Autom. Constr. **18**, 357–375 (2009)
4. Galić, M., et al.: Review of BIM's implementation in some EU AEC industries. In: 13th International Conference Organization, Technology and Management in Construction, pp. 462–476. Croatian Association for Construction Management, Poreč, Croatia (2017)
5. Volk, R., et al.: Building information modeling (BIM) for existing buildings—Literature review and future needs. Autom. Constr. **38**, 109–127 (2014)
6. FIEC: Annual Report, 2017 and European Commission (2017)
7. European Construction Forum (2017)
8. Accenture: Demystifying Digitization (2016)
9. Global Institute: Reinventing Construction: A Route to Higher Productivity, February 2017
10. Dodge Data & Analytics: The Business Value of BIM for Infrastructure 2017, Smart Market Report (2017)
11. Kolarić, S., et al.: Developing a methodology for preparation and execution phase of construction project. Org. Technol. Manag. Constr. Int. J. **7**, 1197–1208 (2015)

Infrastructure for Spatial Information in European Community (INSPIRE) Through the Time from 2007. Until 2017

Nikolina Mijić$^{(\boxtimes)}$ and Gabor Bartha

Institute of Geophysics and Geoinformatics, Faculty of Earth Science and Engineering, University of Miskolc, Miskolc, Hungary
nikolinamijic7@gmail.com

Abstract. The term of infrastructure, as a mechanism of support for spatial data, has been used for the first time in the early 1990s in Canada. Today, the concept of spatial data infrastructure (SDI) has become a worldwide new paradigm for the collection, use, exchange and distribution of spatial data and information. Spatial data infrastructure has been developed through sets of spatial data, metadata, agreements for joint spatial data use and distribution, network services and related coordination activities. SDI is always present in a certain form, but the level of implementation varies according to current demand and technological readiness. Subjects can be classified at several basic levels – from personal and corporative, through local and county, to national, regional and finally, global. Today, the most important level is the national one, i.e. the National Spatial Data Infrastructure (NSDI) project (OG 16/2007) and INSPIRE Directive (Infrastructure for Spatial Information in the European Community - 2007/2/EC). Without spatial data and related services it would be impossible to manage space effectively, plan city development and infrastructure networks, monitor situation on the ground, or carry out many other activities. This paper gives an overview what has been happening throughout the time with INSPIRE Directive starting from 2007. including legislative regulations, technical requirements, assumed standards, scientific methodologies, developed data specifications and, finally, resulting software tools and services. The assessment also describes overall country-wise alignment to INSPIRE standards and services implementation throughout EU member states, thus their readiness for fully standardized data acquisition, representation and exchange on national and regional levels. Hereby represented country- specific implementation assessment includes following indicators: (a) legislative conformance with imposed INSPIRE regulations, (b) technical SDI conformance with imposed standards and data specifications, and (c) implemented INSPIRE-compliant systems, services and datasets.

Keywords: NSDI · SDI · INSPIRE · EU member states

© Springer Nature Switzerland AG 2019
S. Avdaković (Ed.): IAT 2018, LNNS 60, pp. 34–42, 2019.
https://doi.org/10.1007/978-3-030-02577-9_5

1 Introduction

Spatial data infrastructures exist for quite a long time, actually from the moment when the first spatial data were collected and presented in maps and plans [1]. With the rapid development of spatial data collecting and of communication technologies, spatial data infrastructure has become a more and more important factor in the way of spatial data usage at the level of private and public sector, or state and ultimately at global level. President Clinton's Executive Order 12906 from 1994 played an important role and a stimulus for the creation of national spatial data infrastructures. Besides national spatial data infrastructures, different initiatives at regional (EUROGI, PCGIAP…) and global level (GSDI) were also included. Development of spatial data infrastructures in separate countries is different [2]. Sets of basic spatial data also vary from country to country, and each national spatial data infrastructures is different with regard to society needs, sociological evolution, economic reality and national ambitions and priorities.

The efficient land management with sustainable development, and the planning of all land operations, demands for spatial files arrangement and modernization, and establishment of national spatial data infrastructure. Its establishment demands for full coordination and cooperation between the provider and the spatial data user, as well as between public and state institutions [3].

The INSPIRE initiative was launched in 2001, with the intention of providing harmonised sources of spatial information in support of the formulation, implementation and evaluation of community policies [4]. It relates to the base of information collected with member states in order to respond to a wide range of policy initiatives and obligations at local, regional, national and international levels.

2 INSPIRE Components

The legal framework of INSPIRE has two main levels [5]. At the first level, there is the INSPIRE Directive itself, which sets the objectives to be achieved and asks the MS to pass their own national legislation establishing their NSDIs. This mechanism of European plus national legislation allows each country to define its own means of achieving the objectives agreed upon, taking into account its own institutional char acteristics and history of development. At the second level of legislation, INSPIRE envisages technical implementing rules in the form of regulations. These are actually the main components of infrastructure:

- Metadata
- Interoperability of spatial data sets and services
- Network services (discovery, view, download, invoke) made available through the INSPIRE geoportal
- Coordination and measures for monitoring and reporting.

From 2005 onwards, and in parallel with activities to prepare the INSPIRE Directive, several drafting teams (DTs) have started to elaborate implementing rules (IRs). In addition, several thematic working groups (TWGs) have been elaborating data specification for the different themes of the three annexes of the Directive since 2008.

All the IRs take the form of a decision or regulation and must be implemented by individual member states once they are published. Each IR is accompanied by technical guidelines (TG) which, in addition to providing general support for implementation, may give directions on how to further improve interoperability.

2.1   Metadata

The INSPIRE Metadata Regulation entered into force on 24 December 2008 (European Commission 2008). December 2010, MS must provide the metadata for data sets and services listed in Annex I and II of the Directive. A revised version of the TG to implement the Regulation using EN ISO 19115 Metadata and EN ISO 19119 Services was also published on the INSPIRE web site in June 2010 [5].

2.2   Network Services

Figure 1 gives an overview of INSPIRE architecture with common network services. Network services are necessary for sharing spatial data between the various levels of public authority in the community. The INSPIRE Network Services Regulation (European Commission 2009a) was adopted by the Commission on 19 October 2009. It contains the implementing rules for discovery and view services. The TG Discovery Services (Version 3.1) and the INSPIRE View Service TG (Version 3.1), were prepared by the Network Services Drafting team and published on the INSPIRE website on 7 November 2011 [5].

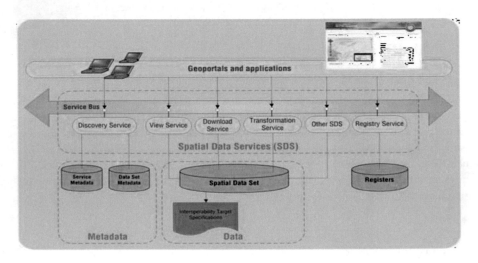

Fig. 1. INSPIRE architecture

2.3   Data Specifications

Commission Regulation No 1089/2010 (European Commission 2010b; European Commission 2011) of 23 November 2010 implementing Directive 2007/2/EC of the

European Parliament and of the Council as regards interoperability of spatial data sets and services has been published.

This Regulation concerns the interoperability of spatial data sets for Annex I spatial data themes. TGs for the spatial data themes of Annex I are available on the INSPIRE website [5].Also available are the framework documents for the INSPIRE data specifications development, updated to reflect experience with Annex I data themes. INSPIRE data specifications reached an important milestone in June 2011 with the delivery by the TWGs of the Data Specifications Version 2.0 for Annex II & III Data Themes, the launch of stakeholder consultation and start of testing the proposed specifications. Success in reaching this point while respecting tight deadlines was due to the expertise, dedication and commitment of all the experts involved and the support they received from their organizations.

Objections received during the consultation and testing period have been resolved, and the Implementing Rule Legal Act for Annex II and III has been drafted. The draft is currently under revision by the services of the Commission. When exactly the INSPIRE Committee will express an opinion on the proposed legal act depends on when the translations will be available (late 2012, or early 2013). The 3.0 versions of the draft TGs for all 25 themes covered in INSPIRE Annexes II and III were published on the INSPIRE web site on 16 July 2012.

2.4   Monitoring and Reporting

On 5 June 2009, the Commission Decision implementing Directive 2007/2/EC of the European Parliament and of the Council as regards monitoring and reporting was adopted (European Commission 2009b).

A document explaining the rationale of the selected indicators, as well as guidelines and a Microsoft Excel template for reporting, have been developed and made available. These documents can be found on the INSPIRE website in the section on monitoring and reporting.

3   Experimental Research – INSPIRE Through the Time

Monitoring and Reporting system (requirements, processes and supporting tools) is based on Article 21 of the Directive and on the 2009 Reporting Implementing Decision. Experience from the previous reporting rounds and the evaluation have shown that this system leaves room for improvement. Textual information is still quite significant in this system which may have an adverse effect on the relevance and comparability across MS of the information provided.

The aim of this country fiche template document is to enable a comparative analysis of the MS 2016 reports and action plans which will be done by JRC and EEA. Country fiches should bring reporting information from monitoring and implementation report together in a comprehensive view. The template consists of several components:

- Information extracted from the report on status of implementation and operation of the infrastructure (State of Play)

- Information extracted from the monitoring data on status of implementation (automatically generated content from INSPIRE Dashboard)
- MS action plan info: MS objectives, actions and roadmap to reach INSPIRE implementation objectives
- Summary (based on overall information including bilateral meetings)
- Specific recommendation (optional).

Country fiche template contains all legally binding information given in INSPIRE Directive, Article 21:

1. Member States shall monitor the implementation and use of their infrastructures for spatial information. They shall make the results of this monitoring accessible to the Commission and to the public on a permanent basis.
2. No later than 15 May 2010 Member States shall send to the Commission a report including summary descriptions of:
   (a) how public sector providers and users of spatial data sets and services and intermediary bodies are coordinated, and of the relationship with the third parties and of the organisation of quality assurance;
   (b) the contribution made by public authorities or third parties to the functioning and coordination of the infrastructure for spatial information;
   (c) information on the use of the infrastructure for spatial information;
   (d) data-sharing agreements between public authorities;
   (e) the costs and benefits of implementing this Directive.
3. Every three years, and starting no later than 15 May 2013, Member States shall send to the Commission a report providing updated information in relation to the items referred to in paragraph 2.
4. Detailed rules for the implementation of this Article shall be adopted in accordance with the regulatory procedure referred to in Article 22(2).

The MIF is an informal collaboration between the EU level partners (namely the European Commission, mainly Directorate-General for Environment (DG ENV) and Joint Research Center of the European Commission (JRC), and the European Environment Agency (EEA), short the EU Coordination Team "CT") and the Member State competent authorities responsible for the INSPIRE implementation. It has built on the work of the consultative process to prepare the numerous Implementing Acts [6] for the INSPIRE Directive and is now maintaining them. It also prepared useful guidance documents and exchanged good practices also with the help of EU-funded projects. Moreover, stakeholder engagement was part of the activities from the outset. In addition, the main achievements over the past years are, in particular:

- Providing guidance for Member States by developing technical guidelines.
- Corrective maintenance of the INSPIRE framework by managing and resolving issues in technical guidelines and preparing proposals for change for Implementing Acts;
- Adaptive maintenance of the INSPIRE framework;
- Development of tools supporting implementation;
- Building capacity in the Member States for INSPIRE implementation;

In addition to these discussion, the Commission services embarked in a series of bilateral meetings with many Member States (between October 2015 and April 2016) to discuss specific implementation gaps and identify ways to close them. Overall, the idea was to ask Member States to prepare specific, tailor-made action plans together with the national reports which are due in May 2016. The discussions during these bilateral meetings gave an excellent insight into the particular challenges in the different Member States and allowed for a discussion on how the Commission could assist in addressing them.

The outcome of the INSPIRE Report, together with the MIG-P orientation debate and the feedback from the dialogues, fed into the preparation of this MIWP 2017–2020. Moreover, this MIWP had to be designed in full knowledge of a number of Commission priorities, external factors and processes which can influence the further work under MIWP positively, in particular:

- The Digital Single Market initiatives with particular relevance for the INSPIRE Directive, namely the free flow of data initiative, the e-Government Action Plan and the European Interoperability Framework, where synergies can be created;
- The Better Regulation agenda driving efficiency and effectiveness whereby the INSPIRE Directive can help reducing administrative burden whilst enhancing the access to evidence for policy making and implementation;
- The Environment policy agenda based on the 7th Environment Action Programme with a strong emphasis on implementation;
- The link to EU policies and other international initiatives, in particular Copernicus, the HORIZON 2020 agenda, the United Nations Committee of Experts on Global Geospatial Information Management (UN GGIM) and GEO where INSPIRE already plays an important role.
- Agenda 2030 and the need for geospatial data in achieving and monitoring the SDGs. Also the Census 2021 will be a driver for NSIs to modernize their statistical production and use addresses, buildings or cadastral parcels to link to statistical data.
- The national eGovernment and Open Data initiatives, where convergence of efforts and alignment of implementation rules would partially address the omnipresent resource issues.

On all these and other initiatives not specifically listed here, the MIWP 2017–2020 can play an important role to contribute and can act as a platform to explore and exploit synergies to the maximum extent in a collaborative and consultative spirit that dominated in the INSPIRE implementation from the outset.

4   Vision and Objectives of INSPIRE

Vision for a European spatial data infrastructure for the purposes of EU's environmental policies and policies or activities which have an impact on the environment (Article 1 of INSPIRE Directive 2007/2/EC), is to put in place easy-to-use, transparent, interoperable spatial data services which are used in the daily work of environmental policy makers and implementers across the EU at all levels of governance as well as businesses, science and citizens to help improving the quality of the environment and leading to effectiveness gains and more simplification.

When talking about users, it is clear that public authorities dealing with the environment (e.g. from EU policy making to national implementation to local enforcement) are the initial primary beneficiary of the INSPIRE implementation. But just about any public authority that uses spatial data can benefit, such as an agriculture department or the transport authorities. In particular the collaboration between the INSPIRE implementation and the eGovernment initiatives in many countries has widened the potential user base. Eventually, academics, researchers, non-governmental organizations, businesses and citizens are also expected to benefit.

Business will most likely be encouraged to develop new electronic applications for markets interested in (quality) geospatial information - for example, providing shoppers with the locations of bank machines, insurance companies with information on flooding hazards, or cyclists with cycling shop locations, delivered through personal mobile phones.

Therefore, user demands will become more important in the strategic direction as a basis for this work programme, in addition to the continued "support for implementation" (work area 4). The main other working areas are:

- to assess the fitness for purpose of the INSPIRE framework and promote simplification (see work area 1: "Fitness for purpose": Making INSPIRE "fit for purpose" supporting solution-oriented end-user perspective);
- to deliver short term results (quick win applications) including helping to streamline reporting (which is one use case but not the only one) (see work area 2: "End-user applications" for environmental reporting and implementation);
- to ensure alignment and synergies with EU emerging policies and initiatives (see work area 3: "Alignment with EU policies/initiatives" creating a platform for cooperation).

The new strategic direction will guide the MIWP 2017–2020 and result in immediate actions so that we demonstrate that the INSPIRE Directive can be implemented in a proportionate, faster and pragmatic way. This strategy is the centrepiece of the new MIWP 2017–2020.

Given the significant scope and ambition of the INSPIRE Directive, the implementation process overall would benefit from stricter EU priority setting. This would allocate the limited resources on those issues with highest priority and where tangible benefits for environment policy can be expected. It would also strengthen the cross-border and EU dimension of the INSPIRE Directive implementation because interoperability can only be successful if all partners (EU, national, regional and local administrations) share the same priorities so that we all "pull in the same direction". Hence, when defining new actions for the MIWP the following criteria for priority setting should be considered (which would replace the prioritization template currently used):

1. Engage users!
2. Addressing emerging priorities (EC and MSs);
3. Demonstrate short term benefits of current investment;

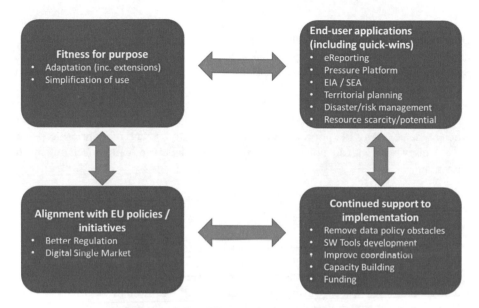

Fig. 2. The four main work areas under the INSPIRE

4. Make the INSPIRE framework more effective and better exploitable;
5. Facilitate implementation (e.g. through appropriate simplification measures);
6. Ensure sustainability of INSPIRE;
7. Adapt to changes (e.g. driven by the Digital Single Market or Better Regulation).

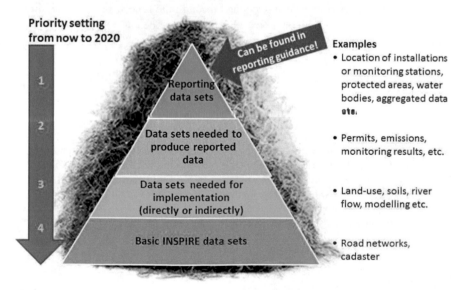

Fig. 3. Illustrative example on how EU priority setting approach as regards spatial data sets in the use case of reporting can be visualized.

As regards priority setting in relation to spatial data covered by the INSPIRE Directive, the following approach, from the EU (reporting) perspective has been introduced for discussion. The operational details and activities are discussed later (cf. Sect. 4). They do not neglect user needs for planning, running and monitoring environmental infrastructures.

Any priority setting approach has its intrinsic logic that one area is prioritised over another but that ultimately, step-by-step, all issues get addressed in a systematic and efficient manner. Any EU priorities complement any national and other priorities which are set elsewhere and do not alter in any way the legal obligations set out by the Directive.

5 Conclusion

Implementation of INSPIRE standards will begin at 2019. But in meanwhile some of the countries must develop their Geoportals and applications. Most of the EU member state submit their report every three year about monitoring and reporting. Some of EU countries didn't reach a big level of developing INSPIRE through the ISO standards.

The assessment also describes overall country-wise alignment to INSPIRE standards and services implementation throughout EU member states, thus their readyness for fully standardized data acquisition, representation and exchange on national and regional levels.

Hereby represented country-specific implementation assesment includes following indicators: (a) legislative conformance with imposed INSPIRE regulations, (b) technical SDI conformance with imposed standards and data specifications, and (c) implemented INSPIRE-compliant systems, services and datasets.

Here was shown what has been happened through the time with INSPIRE directive and standards. Also it was shown a next period of implementation and planning of the future in INSPIRE.

References

1. Groot, R., McLaughlin, J.: Geospatial Data Infrastructure: Concepts. Cases and Good Practice. Oxford University Press, Oxford (2000)
2. Phillips, J., Rajagopalan, B., Cane, M., Rosenzweig, C.: The role of ENSO in determining climate and maize yield variability in the U.S. cornbelt. Int. J. Climatol. 19, 877–888 (1999)
3. Messer, I.: INSPIRE's shift emphasis. GIM Int. 26(5), 27–29 (2012)
4. Annoni, A.: JRC and INSPIRE iteroperability. GIM Int. 20(3), 10–12 (2006)
5. http://inspire.jrc.ec.europa.eu/
6. https://ies-svn.jrc.ec.europa.eu/projects/mig-p/wiki/5th_MIG-P_meeting

Application of the Airborne LIDAR Technology on the Quarry Using AutoCAD Civil 3D Software

Nikolina Mijić[(✉)]

Institute for Geophysics and Geoinformatics,
Faculty of Earth Science and Engineering, University of Miskolc,
Miskolc 3515, Hungary
nikolinamijic7@gmail.com

Abstract. Times are quickly changing - AutoCAD Civil 3D provides rich set of geodetic tools and add-ons to dramatically speed-up surveyed data post-processing, visualization and analysis. Drone-based laser scanning speeds up the data collection stage of the workflow and, when compared to aerial photogrammetry, offers much faster turnaround of the physical quantities. AutoCAD Civil 3D allows to compute volumes and generate profile views within a matter of hours, so that stockpile quantities recorded are accurate on a set day, rather than reflecting a historic situation. Drone-based 3D laser scanning not only reduces the time spent on stockpile surveying while enhancing the safety of workers, but also offers a level of surface detail that is incomparable to that collected from total stations: 100,000+ 3D points collected in just a few minutes. AutoCAD Civil 3D enables creating TIN surface from points within RCS format point cloud scanned object created with Autodesk ReCap. Drone-mounted 3D laser scanner includes a GPS receiver and inertial measurement unit (IMU), so data can be geo-referenced to an exact location. Each operation referenced to the same co-ordinate system. Key benefit of this operation is better accuracy and traceability of these methods. Drone-mounted LiDAR represents a safe way to survey dangerous and hostile environments. Once a point cloud-based surface created within AutoCAD Civil 3D, it can be used built-in tools to perform quick volumetric calculations, and easily create alignment profile cross-sections using only polylines drawn atop of a generated TIN.

Keywords: AutoCad Civil 3D · 3D modeling · LiDAR · Point cloud
Surface

1 Introduction

While quarry and plant managers recognize the necessity of carrying out physical inventories of material stockpiles for accounting purposes, that takes up time and which has the potential to slow production. Manual surveys using total stations and traditional global positioning system (GPS) equipment are time consuming, requiring 10 to 20 shots on a typical 15,000 cubic-meter stockpile. LIDAR is today one of the most modern technology that is used in the survey and development of topographic maps for different purposes. The technology based on collection of three different sets of data. Position

© Springer Nature Switzerland AG 2019
S. Avdaković (Ed.): IAT 2018, LNNS 60, pp. 43–51, 2019.
https://doi.org/10.1007/978-3-030-02577-9_6

sensors are determined using Global Positioning System (GPS), using phase measurements in the relative kinematics, use of Inertial Measurement Unit (IMU). The last component is a laser scanner. The laser sends infrared light to the ground and was reflected it to the sensor. The time between the broadcast signal reception to the knowledge of the position and orientation sensor, allows the three-dimensional coordinates to calculate the Earth. LiDAR has a very simple principle of measurement. The scanner emits pulses with a high frequency and they are reflected from the surface back to the instrument. Mirror inside the laser transmitter is moved by rotating perpendicular to the tack allowing measurement in a wider band. Time elapsed from the emission to return every impulse and inclination angle from the vertical axis the instrument used to determine the relative position of each measured point. The absolute position sensor is determined by GPS every second. Data laser scanning combined with modern scanners and orientation to obtain three-dimensional coordinates of the laser footprint on the surface of the field [3]. Drone-based laser scanning speeds up the data collection stage of the workflow. In this paper, techniques of drone-mounted 3D laser scanning are presented. It will be used an ortho-photo input which is later processed. Along with combination of the appropriate software, a surface model based on point clouds and Delaunay triangulation will be created. Software's, which are used for creating 3D model and process of the data, are:

AutoCAD Civil 3D (https://www.autodesk.com/products/autocad-civil-3d), Pix4D (https://pix4d.com/), ReCap (https://www.autodesk.com/products/recap/overview) and Agisoft PhotoScan (http://www.agisoft.com/).

2   LIDAR Platforms for Collecting Data Sets

Airborne topographic LiDAR systems are the most common LiDAR systems used for generating digital elevation models for large areas. LiDAR was first developed as a fixed-position ground-based instrument for studies of atmospheric composition, structure, clouds, and aerosols and remains a powerful tool for climate observations around the world. Modern navigation and positioning systems enable the use of water-based and land-based mobile platforms to collect LiDAR data. Data collected from these platforms

Fig. 1. Mobile LiDAR collected from a vehicle (left) and a boat (right)

are highly accurate and are used extensively to map discrete areas, including railroads, roadways, airports, buildings, utility corridors, harbors, and shorelines. Two different techniques of collecting data, from a boat and from vehicle has been shown on Fig. 1 [8].

Airplanes and helicopters are the most common and cost-effective platforms for acquiring LiDAR data over broad, continuous areas. Airborne LiDAR data are obtained by mounting a system inside an aircraft and flying over targeted areas [1]. LIDAR positioning principle includes all procedures, which had been done before beginning of the LIDAR survey. Procedure of the LIDAR survey is to fly an aircraft or helicopter over the specific area and to operate laser scans from side to side. Inertial system keeps track of the rotations of the aircraft in the tjree axes and the GPS keeps track of the actual location on the aircraft space. Actually, results of the LIDAR survey is a set of points which, consists of easting northing elevation [5]. Figure 2 [9] is showing procedure for a LIDAR survey.

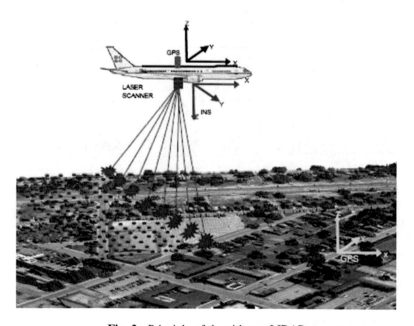

Fig. 2. Principle of the airborne LIDAR

Airborne LIDAR systems include [6] Dynamic Differential GPS receiver, Inertial measurement unit (IMU), Laser scanners, Laser rangefinders, Imaging device and Central Control Unit Management System. The original data obtained by Airborne LIDAR can be melt with all kinds of digital images after it has been processed by software, which can output all kinds of products of survey and remote sensing. The whole flow of data processing from data collection to production output can be divided into five main sections [7]: Data collection, Data pretreatment, filter and classification of point cloud, fusion and application of LIDAR and other remote sensing data, ground-object extraction and modeling based on LIDAR data. Detail flow of the LIDAR data processing shown on the Fig. 3.

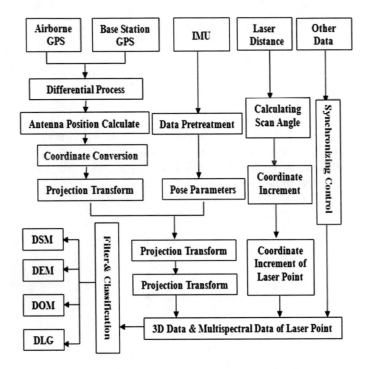

Fig. 3. Detail flow of the LIDAR data processing

3 Techniques for Creating 3D Digital Elevation Model

For Surveying and Civil Engineering, the most important applications are aerial scanning and terrestrial scanning. Terrestrial Scanning creates 3D models of complex objects, piping networks, roadways, archeological sites, buildings, bridges, etc. Aerial Scanning has many uses - measuring agricultural productivity, distinguishing faint archeological remains, measuring tree canopy heights, determining forest biomass values, advancing the science of geomorphology, measuring volcano uplift and glacier decline, measuring snow pack, and providing data for topographic maps [4]. In the Fig. 4 [8] the process of the using LiDAR techniques has been shown.

Process of creating point clouds isn't easy and takes a lot of time. Only for processing of the data set few different softwares have been used. Basis for the terrain modeling is a set of scattered points in 3D space. For subsequent modeling, various mathematical functions can be used. In any case, the representation of the terrain surface is realized by small surface elements. Terrain Modeling using the network quadrilaterals (GRID) is more suitable for organizing and storing data in the matrix form, and later to use various algorithms for data processing [2]. DEM data are commonly in raster files (Fig. 5) with formats that include GeoTiff (.tif), Esri Grid (.adf), floating point raster (.flt), or ERDAS Imagine (.img). In some cases, the data are available in a TIN format (e.g., Esri TIN). In the raster cases, they are created using point files and can be interpolated using many different techniques. The techniques used to create DEMs range from simple (e.g., nearest neighbor) to complex (e.g., kriging) gridding routines can create slightly different surface types.

Fig. 4. Point clouds used for representing and measuring roadways

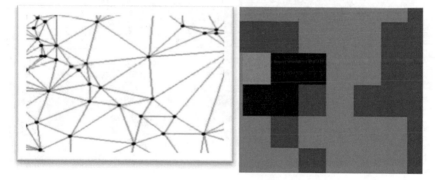

Fig. 5. Surface represented like a TIN (left) and as a raster (GRID) (right)

The most common are the surfaces created by TIN or the inverse distance weighted (IDW) routines. The appropriate interpolation method depends on the data and the desired use of the DEM [1].

4   Results and Discussions

Experimental results of this paper are based on creating of 3D models of the quarry. These 3D models are based on the point clouds which are part of airborne images created with new modern technlogy. First of all a drone was used to acquire all area of the quarry which is located in Bosnia and Herzegovina. Before of all, data sets of points which are used for this experiment are obtained from using LIDAR survey. In the section before it was explained whole process of getting LIDAR data and processing them. Detail flow of the LIDAR data process was shown on the Fig. 3. These images are processed in Pix4D software. Processing of the images is the first stage after drone shooting of the area. After processing this software gave us point clouds which can now be use for creating 3D model of the quarry. Besides of the software Pix4D, for the processing of the data Agisoft PhotoScan can also be used. During the data processing we must do the classification of the objects and vegetation and separate terrain from these objects. After processing airborne images and getting point clouds which are in . las or. laz files we must do the conversion if we want to create 3D models of the terrain. Conversion of the .las and .laz files were done in Autodesk ReCap convertor. Procedure for conversion is very easy, we just import .las or .laz files and create new project in Autodesk ReCap and export it like a .rcp files. These .rcp files can be used to import them in AutoCAD Civil 3D and analyse and create a 3D model of the surface, in this case quarry surface. Imported .rcp file in AutoCAD Civil 3D is shown on the Fig. 6.

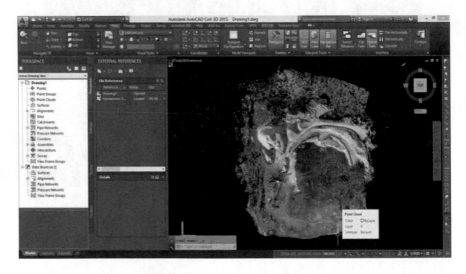

Fig. 6. Quarry processed point cloud

One can use the Create Surface from Point Cloud command to create a surface from several point clouds, selecting only the areas that one wants to include and filtering out non-ground points so they are not included in the resulting surface. When using this command, user can select entire point clouds or areas of point clouds to include in the surface. One can select areas of point clouds by using window selections, by defining polygon areas, or by selecting existing closed polylines in the drawing. On the Fig. 7 is shown a surface model of quarry.

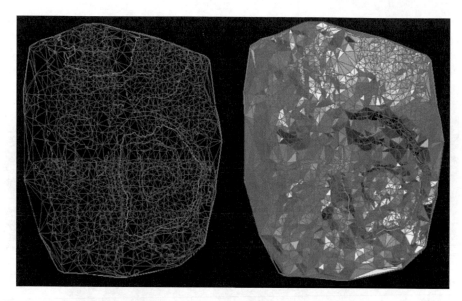

Fig. 7. Surface model of the quarry, triangulation network (left) and 3D model of surface (right)

Because drone-mounted 3D laser scanner includes a GPS receiver and inertial measurement unit (IMU), data can be geo-referenced to an exact location. This ensures that each inventory operation can be referenced to the same coordinate system and pile limits each time the inventory is performed: a key benefit when accountants require accuracy and traceability of methods. On the Fig. 8, different ways of seeing a quarry is shown.

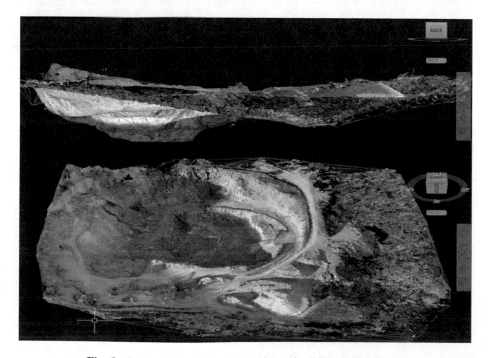

Fig. 8. Point clouds of the quarry shown in different projections

Once a point cloud-based surface is created within AutoCAD Civil 3D, it can be used built-in tools to perform quick volumetric calculations (e.g. against previously measured structure), and easily create alignment profile cross sections using nothing but polylines drawn a top of a generated TIN like it is shown on Fig. 9.

Fig. 9. Creating of the alignment using 3D surface model of the quarry

5 Conclusion

For the creating 3D models of the surface, we can use very different platforms. In this paper, it was shown how to convert and process different files. After that, use the same files and data sets for different purposes. It was used a specific case of the quarry which is located in Bosnia and Herzegovina. In this experimental research, it was used different platforms and software tools which are based on processing LiDAR data. Creating a point clouds are not easy and process of import data sets, which are recorded with drones. The main purpose of this work was to show different tools and platforms for processing LiDAR data. On the other hand, it is shown how we can create from these data cross sections after we created 3D surface model, in this case model of the quarry. The aim of this paper was to show how could be used different kind of software's to process LIDAR data. After getting row data, it was used different software's to process data and create images which are compatible with software's that are used in this research. Although, it was also, used comparison of different software has to get better accuracy and examination of the elevation but these two experiments can be the theme for discussion in the following research papers.

References

1. Carter, J., Schmid, K., Waters, K., Betzhold, L., Hadley, B., Mataosky, R., Halleran, J.: Lidar 101: An Introduction to Lidar Technology, Data, and Applications, 76 p. NOAA Coastal Services Center, Charleston (2012)
2. Janic, M., Djukanovic, G., Grujovic, D., Mijic, N.: Eartwork volume calculation from digital terrain models. J. Ind. Des. Eng. Graph. **10**, 27–30 (2015)
3. Mijic, N., Sestic, M., Koljancic, M.: CAD—GIS BIM integration—case study of Banja Luka city center. In: Advanced Technologies, Systems, and Applications, pp. 267–281. Springer (2017)
4. Rankin, F.A.: LiDAR applications in surveying and engineering. In: GIS Conference, Raleigh, NC (2013)
5. Li, S., Liu, T., You, H.: Airborne 3D imaging system. Geo-information Sci. **1**, 23–31 (2000)
6. Li, S., Xue, Y.: Positioning accuracy of airborne laser ranging and multispectral imaging mapping system. J. Wuhan Tech. Univ. Surv. Mapp. (WTUSM) **12**, 341–344 (1998)
7. Liu, J., Zhang, X.: Classification of laser scanning altimetry data using laser intensity Editor. Doard Geomat. Inf. Sci. Wuhan Univ. **30**, 189–193 (2005)
8. http://www.ncgisconference.com/2013/documents/pdfs/Rankin_Thu_130.pdf
9. http://www.tankonyvtar.hu/en/tartalom/tamop425/0027_DAI4/ch01s02.html

Seismic Assessment of Existing Masonry Building

Nadžija Osmanović[1]([✉]), Senad Medić[2], and Mustafa Hrasnica[2]

[1] Termo-beton Ltd., Breza, Bosnia and Herzegovina
nadja_turbo@hotmail.com
[2] Faculty of Civil Engineering, University of Sarajevo,
Sarajevo, Bosnia and Herzegovina
senad_medic@yahoo.com, hrasnica@bih.net.ba

Abstract. In this study nonlinear static pushover and dynamic time-history analyses of an old heavily damaged masonry building situated in Sarajevo were performed. Two numerical macro-models were created using finite element program Diana 10.1. Engineering masonry constitutive law was used to describe highly nonlinear behavior of masonry walls. The first model represents the existing damaged structure. In the second model, which represents the rehabilitated structure, R.C. floors and internal walls were added to the building. Results indicate that significant cracking occurs in the existing structure and that collapse is expected for an earthquake with PGA of 0.1 g. Seismic response of the upgraded building characterized by limited nonlinear deformations was more favorable considering that the material properties assumed for masonry were the same for the existing and for the rehabilitated building.

Keywords: Damaged building · Engineering Masonry model
Nonlinear analysis · Seismic capacity

1 Introduction

Bosnia and Herzegovina is situated in a seismically active region of South-East Europe, with maximum peak ground acceleration 0.1–0.2 g in the most of the country regions for 475 years return period. Traditional buildings were built of masonry walls and wooden floors, without reinforcement and confining RC elements. This kind of masonry buildings corresponds to seismic vulnerability classes B and C (according to EMS classification). Considering this, during earthquake significant damages are expected, with wide cracks in walls [1]. During massive reconstruction after WWII, unconfined unreinforced masonry buildings with up to 5 floors were erected. After earthquake in Skopje in 1963, the first seismic regulations were issued, and RC confining elements became usual way of construction [1]. Seismic resistance was provided by walls in mutually orthogonal directions. However, longitudinal walls were quite often avoided due to functional demands which makes these buildings rather vulnerable. Different strengthening techniques can be applied in order to enhance the load bearing and deformation capacity of existing or damaged buildings [2, 3].

© Springer Nature Switzerland AG 2019
S. Avdaković (Ed.): IAT 2018, LNNS 60, pp. 52–61, 2019.
https://doi.org/10.1007/978-3-030-02577-9_7

Masonry is anisotropic material, which has different properties in perpendicular directions (vertically to the bed joint and in the direction of the bed joint) [4].

The quality of the masonry structure has an important influence on the quality of structure. The quality of the joint between the element and the mortar affects the transfer of stress. The compressive strength of the bed joints is usually much higher than of head joints, which are often only partially filled with mortar which usually has less rigidity and greater deformability than mortar in bed joints. Also, the difference in shear stress transmission in bed joints is higher due to the higher mortar quality and better adhesion, especially due to the favorable impact of the pressure perpendicular to the horizontal joint [5]. Considering complex behavior and modelling of masonry advanced numerical models should be used. In order to gain a complete insight into structural behavior, especially in the case of cyclic loads, nonlinear analyses are necessary. In this study nonlinear analyses were performed using software Diana 10.1 [6]. Different modeling strategies can be used to analyze masonry structures [7]. Engineering Masonry model (EngMas), which is a continuum smeared failure model, was used and it was applied for walls, including cracking, crushing and shearing failure modes.

The analyzed building has three stories and an area of 14 × 14 m in plan. It is located in Sarajevo (VII seismic zone according to MCS scale which corresponds roughly to PGA of 0.1 g) and it was constructed before the Second World War. There are no horizontal and vertical confining elements in the building. The floors were wooden except above the basement which was constructed with steel profiles and concrete arches. The Institute for Materials and Constructions of the Faculty of Civil Engineering in Sarajevo conducted investigation works in 2009 and determined the condition and degree of damage of the load bearing structure [8]. During the war period (1992–1995) the building suffered significant damage [9].

Two finite element models were created. The first model was used to simulate the existing state of the load bearing structure – Model 1. In the second model inner walls and RC floors were added in order to evaluate the performance of the proposed rehabilitation – Model 2.

2 Analysis of the Damaged Building

Based on the results reported in study [8], brick and mortar strength were obtained. The brick strength was 7.5 MPa and the mortar strength was 1 MPa. Due to lack of experimental testing of other mechanical properties of full brick walls in existing buildings, recommendations from literature were used [6, 10].

The state of the damaged building and general model layout are provided in Fig. 1. Typical wall thickness is 77 cm, and the basement floor structure is 40 cm thick. The floor exists only above the basement of the damaged structure and it is modelled according to linear elastic parameters listed in Table 1. Masonry parameters used for Model 1 are given in Table 2.

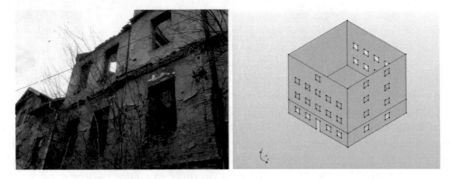

Fig. 1. Analyzed building (left) and Model 1 (right)

Table 1. The floor parameters

E – stiffness	3000 MPa
υ – poisson's ration	0.2
ρ – density	2500 kg/m^3

Table 2. Masonry parameters for Model 1

E_y – stiffness in x direction	600 MPa
E_x – stiffness in y direction	300 MPa
G_{xy} – shear modulus	190 MPa
Θ – crack angle	30°
f_{tx} – tensile strength in x direction	0.10 MPa
f_{ty} – tensile strength in y direction	0.05 MPa
G_{fty} – crack energy in y direction	5 N/m
f_{cy} – compressive strength in y direction	2 MPa
G_{fc} – compressive failure energy	15000 N/m
Φ – friction angle	32°
c – cohesion	0.1 MPa
G_{fs} – shear failure energy	20 N/m
ρ – density	1850 kg/m^3

2.1   Pushover Analysis of the Damaged Building

Nonlinear static - pushover analysis was carried out under constant gravitational load and horizontal incrementally increasing load. The important result of the analysis is the capacity curve - pushover curve, which is given as the ratio of the total transverse force and the horizontal displacement of the top of the building.

For the purpose of comparing the capacity of the structure and the earthquake demand, the capacity curve and design spectrum are shown in the same format. Based on the EC8 design spectrum (type 1, behaviour factor q = 1, damping of 5%, design ground acceleration 0.1 g, and soil type B, corresponding to the location - Sarajevo VII seismic zone), conversion to ADRS format was applied (Fig. 2).

Capacity Spectrum Method - damaged building

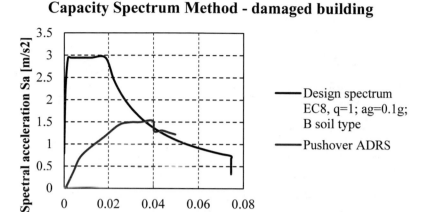

Fig. 2. Capacity spectrum method for damaged building

The capacity of the structure is equivalent to earthquake demand for the spectral displacement of 4 cm which corresponds to the roof displacement of 7.2 cm. Vertical cracks stretch from the top of the wall to the wall corners, and the maximum crack opening amounts to 2 cm at the wall corners. The failure is governed by shear. It can be concluded that the analyzed damaged structure would not withstand an earthquake with a PGA of 0.1 g without major damage (Fig. 3).

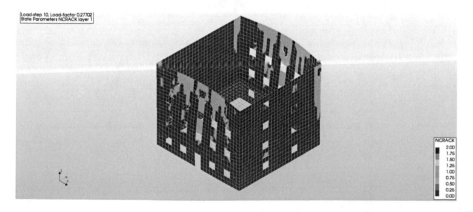

Fig. 3. Number of cracks

2.2   Time – History Analysis of the Damaged Building

The earthquake load in the y direction is given by the accelerogram shown in [6] (Fig. 4). The earthquake lasts 3 s, and the acceleration data were scaled by 0.1 g. Horizontal displacements of the top of the structure during the earthquake are shown in Fig. 5.

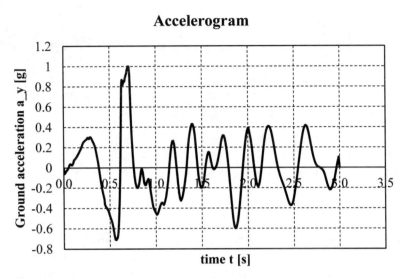

Fig. 4. Accelerogram

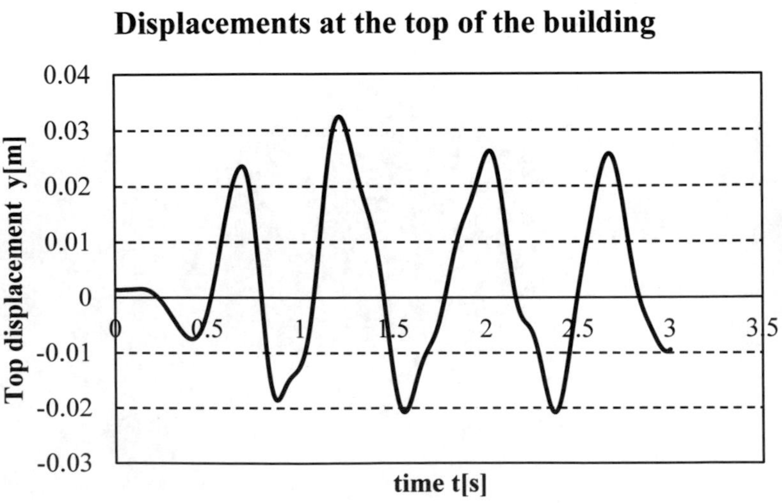

Fig. 5. Displacement at the top of the building during earthquake loading

The maximum horizontal displacement at the top of the building is less than in the case of pushover analysis. This is expected because only one accelerogram was considered. Still, maximum crack width is 1 cm, and it can be concluded that the structure would not withstand the earthquake without major damage (Fig. 6).

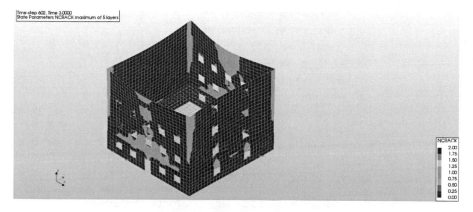

Fig. 6. Number of cracks

3   Analysis of the Rehabilitated Building

The second model was used to simulate the behavior of the rehabilitated building (Fig. 7). The load bearing structure was improved by adding the inner walls and the floors with the main properties given in Table 1. The thickness of the facade walls is 77 cm, the inner walls 46 cm. The floor structure is 25 cm thick, except above the basement where the floor thickness is 40 cm. Masonry parameters for Model 2 are listed in Table 3.

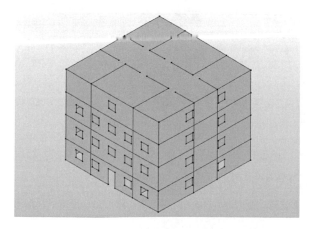

Fig. 7. Model 2

Table 3. Masonry parameters for Model 2

E_y – stiffness in x direction	600 MPa
E_x – stiffness in y direction	300 MPa
G_{xy} – shear modulus	190 MPa
Θ – crack angle	30°
f_{ty} – tensile strength in y direction	0.05 MPa
G_{fty} – crack energy in y direction	5 N/m
f_{cy} – compressive strength in y direction	2 MPa
G_{fc} – compressive failure energy	15000 N/m
Φ – friction angle	32°
c – cohesion	0.1 MPa
G_{fs} – shear failure energy	20 N/m
ρ – density	1850 kg/m³

3.1 Pushover Analysis of the Rehabilitated Building

The pushover analysis was performed applying horizontal load pattern in accordance with its first eigen vector.

Comparing the response of the damaged and the repaired structure, it is apparent that the load capacity of the strengthened one is much higher (Fig. 8). The earthquake demand is reached for quite small displacements of cca. 1 cm for the strengthened building. From the diagram it can be concluded that that the structure will withstand the earthquake before it reaches the peak strength. Diagonal cracks occur mostly around the window openings (Fig. 9).

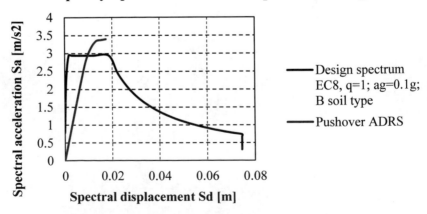

Fig. 8. Spectrum capacity method for the repaired building

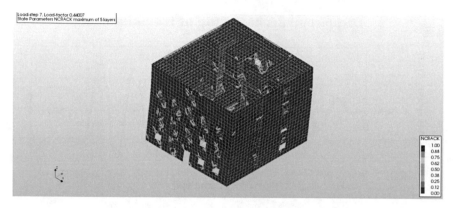

Fig. 9. Number of cracks

3.2 Time History Analysis of the Rehabilitated Building

The load in the direction x is given as accelerogram (Fig. 10) [6], and the values are additionally scaled by 0.1 g. Horizontal displacements at the top of the structure during the earthquake are shown in Fig. 11.

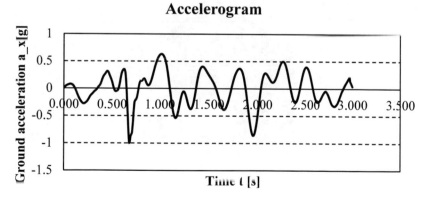

Fig. 10. Accelerogram

Displacements at the top of the building

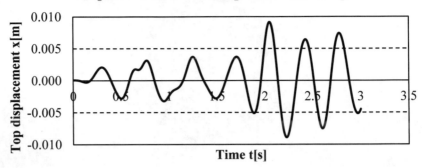

Fig. 11. Displacements at the top of the building during earthquake loading

In the case of nonlinear dynamic analysis, cracks occur around the opening in two directions due to cyclic loading (Fig. 12). Damages are considerably smaller than for the case without internal walls. The repaired building would in this case fulfill the earthquake demand without major damage.

Fig. 12. Number of cracks

4   Conclusion

Nonlinear static and dynamic analyses of an existing masonry building were carried out using a computer program based on the finite element method, Diana 10.1. The pushover analysis of the damaged structure resulted with a maximum roof displacement of 7.2 cm and crack widths of cca. 2 cm at the wall corners. It was concluded that the failure is governed by shear and that the damaged structure would not withstand an earthquake with a PGA of 0.1 g without major damage. Unlike the damaged building, the repaired building with added RC floors and interior walls has greater stiffness and load bearing capacity. It would withstand an earthquake with a PGA of 0.1 g without any major damage. The earthquake demand was reached for quite small displacements of cca. 1 cm, before the structure reached its peak strength. Material properties are the same for both models, which shows how important is the regularity of the structural system, e.g. how much the proper structural system affects the overall capacity.

Time history analyses of the damaged and the rehabilitated structures yield results similar to pushover analysis results. In the case of time history analysis, more accelerograms should be considered in order to obtain relevant results. Even though the response of the structure can be inspected at any moment during earthquake, this would consume huge amount of time and computer resources, which limits its everyday use. Pushover analysis is a practical alternative because it gives good insight into the seismic performance of the structure.

References

1. Hrasnica, M.: Damage assessment of masonry and historical buildings in Bosnia and Herzegovina. In: Ibrahimbegović, A., Zlatar, M. (eds.) Damage Assessment and Reconstruction After War or Natural Disaster. Springer, Berlin (2009)
2. Hrasnica, M., Medic, S.: Seismic strengthening of historical stone masonry structures in Bosnia Herzegovina. In: 15th World Conference on Earthquake Engineering. International Association for Earthquake Engineering (2012)
3. Hrasnica, M., Biberkić, F., Medić, S.: In-plane behavior of plain and strengthened solid brick masonry walls. Key Eng. Mater. **747**, 694–701 (2017)
4. Page, A W · The biaxial compressive strength of brick masonry. Proc. Inst. Civil Eng. **71**(2), 893–906 (1981)
5. Smilović, M.: Ponašanje i numeričko modeliranje zidanih konstrukcija pod statičkim i dinamičkim opterećenjem. Doktorska disertacija, Fakultet građevinarstva, arhitekture i geodezije, Sveučilište u Splitu (2014)
6. TNO DIANA: User's manual. DIANA FEA BV, Delft (2016)
7. Medić, S., Hrasnica, M.: Modeling strategies for masonry structures. In: Hadžikadić, M., Avdaković, S. (eds.) Advanced Technologies, Systems, and Applications II, IAT 2017. Lecture Notes in Networks and Systems, vol. 28. Springer, Cham (2018)
8. IMK: Elaborat o stanju i stepenu oštećenja nosive konstrukcije objekta u ul. Instutut za materijale i konstrukcije Građevinskog fakulteta, Univerziteta u Sarajevu, Sikirića 2, Sarajevo (2009)
9. Medic, S., Ćuric, J., Imamovic, I., Ademovic, N., Dolarevic, S.: Illustrative examples of war destruction and atmospheric impact on reinforced concrete structures in Sarajevo. In: Damage Assessment and Reconstruction after War or Natural Disaster. NATO Science for Peace and Security Series C: Environmental Security. Springer, Dordrecht (2009)
10. Sorić, Z.: Zidane konstrukcije I. Zorislav Sorić, Zagreb (2004)

Time-Dependent Behavior of Axially Compressed RC Column

Senad Medić[✉] and Muhamed Zlatar

Faculty of Civil Engineering, University of Sarajevo,
Sarajevo, Bosnia and Herzegovina
senad_medic@yahoo.com, muhamed.zlatar@gmail.com

Abstract. Short-term and long-term behavior of axially compressed RC column is presented in this paper. Rheology of concrete is studied on a typical symmetrically reinforced cross-section. Reliable assessment of creep and shrinkage deformations is essential in verification of serviceability limit state. Initial stress distribution characterized by higher concrete stress evolves over time and results with stress relief in concrete and significant increment of compression in reinforcement. Different analysis procedures which pertain to effective modulus method, rate-of-creep method, age adjusted effective modulus method and step-by-step method were critically compared. Age adjusted effective modulus method was found to be the optimal choice regarding practical implementation and precision.

Keywords: RC column · Rheology · Compressive stress distribution
Age adjusted effective modulus

1 Introduction

When concrete is exposed to long-term load, deformation gradually increases over time, and can ultimately exceed its initial (instantaneous) value. Therefore, a reliable estimate of the instantaneous and time-dependent deformation is of key importance for satisfying the serviceability limit state. If the temperature and stress remain constant, the deformations increase due to creep and shrinkage. Creep strain depends on sustained load, while shrinkage deformation is stress-independent. To accurately predict these effects, reliable data for creep and shrinkage properties of the particular concrete mix and analytical/numerical procedures for inclusion of rheology are necessary.

Data on shrinkage and creep can be found in the literature, however, the comparison of data indicates significant differences (coefficient of variation up to 20%) [1]. On the other hand, experimental testing is not practical for designers because it is long-lasting and it is not guaranteed that the tested concrete will be identical to that which is used in the structure.

At some point in time t, the total deformation of concrete consists of several components that include instantaneous (elastic) deformation ε_e, creep ε_{cr}, shrinkage ε_{sh}, and temperature deformation ε_t. Although not strictly true, we consider these deformations independent and we calculate them separately and finally combine. In case the temperature is constant, we have (1):

© Springer Nature Switzerland AG 2019
S. Avdaković (Ed.): IAT 2018, LNNS 60, pp. 62–72, 2019.
https://doi.org/10.1007/978-3-030-02577-9_8

$$\varepsilon(t) = \varepsilon_e(t) + \varepsilon_{cr}(t) + \varepsilon_{sh}(t) \tag{1}$$

The strain components of a concrete sample exposed to sustained compressive stress σ_{c0} applied at time τ_0 are illustrated in Fig. 1. Shrinkage strains develops immediately after concrete casting or at the end of curing period ($t = \tau_d$). A sudden jump in the strain diagram (instantaneous or elastic strain) is caused by application of stress and it is followed by a gradual increase in strain due to creep. From Fig. 1, it can be concluded that the magnitude of the final strain is several times (typically 5 times) larger than the magnitude of the elastic strain.

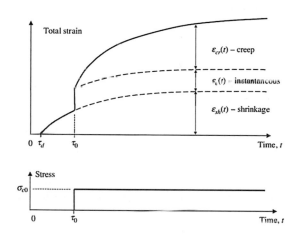

Fig. 1. Concrete strain components under sustained load [1]

The internal forces due to the imposed (or restrained) deformation (support settlement) in a statically undetermined structure are proportional to stiffness. However, due to the creep, internal actions caused by deformations decrease with time [2]. On the other hand, the creep will not cause the redistribution of load induced internal forces provided that creep characteristics are uniform throughout the structure. The creep effect in this case is similar to a gradual and uniform reduction in the modulus of elasticity. If the structure contains concrete parts of different ages, the cross-sectional forces are redistributed from areas with stronger creep into the area with a weaker creep.

Creep can drastically affect the distribution of the stress in the RC cross-section. For example, creep reduces the effects of shrinkage and temperature (eigenstresses) and stresses due to imposed deformation. Also, due to compatibility requirement between concrete and the bonded reinforcement, the creeping part of the section is relieved, and the reinforcement is additionally compressed. For low-strength steel, it is possible that the stresses in longitudinal reinforcement reach yielding point at service load level. To prevent buckling of longitudinal bars at the service load condition, it is necessary to closely space the lateral reinforcement, usually formed as closed ties or helices. Adverse effects of creep pertain to larger deflections and shortening of the prestressing cables. It is important to note that the creep deformation does not affect the strength at the ultimate limit state, that is, the primary effects of time dependent deformations relate to the service limit state.

2   Analysis Procedures

Creep capacity of concrete is usually measured using the creep coefficient, φ (t, τ). In a concrete specimen subjected to a constant sustained compressive stress, $\sigma_c(\tau)$, first applied at age τ, the creep coefficient at time t is the ratio of the creep strain to the instantaneous strain and is given by (2):

$$\varphi(t, \tau) = \frac{\varepsilon_{cr}(t, \tau)}{\varepsilon_e(\tau)} \tag{2}$$

Therefore, the creep strain at time t caused by a constant sustained stress $\sigma_c(\tau)$ first applied at age τ is (3):

$$\varepsilon_{cr}(t, \tau) = \varphi(t, \tau)\varepsilon_e(\tau) = \varphi(t, \tau)\frac{\sigma_c(\tau)}{E_c(\tau)} \tag{3}$$

where E_c (τ) is the elastic modulus at time τ. The creep function J (t, τ) is defined as the sum of the instantaneous and creep strains at time t produced by a sustained unit stress applied at τ (4):

$$J(t, \tau) = \frac{1}{E_c(\tau)}[1 + \varphi(t, \tau)] \tag{4}$$

Then, the strains caused by sustained stress can be determined as (5):

$$\varepsilon_e(\tau) + \varepsilon_{cr}(t, \tau) = J(t, \tau)\sigma_c(\tau) = \frac{\sigma_c(\tau)}{E_c(\tau)}[1 + \varphi(t, \tau)] = \frac{\sigma_c(\tau)}{E_e(t, \tau)} \tag{5}$$

where E_c (t, τ) is known as the *effective elastic modulus*.

Creep strains of concrete at service loads ($\sigma_c < 0.45$ f_{ck}) are proportional to stress, so the principle of superposition is frequently used to calculate the deformation caused by a time-varying stress history. The principle of superposition states that that the strain produced by a stress increment applied at any time τ i is not affected by any stress applied either earlier or later. According to the principle of superposition, the total stress-dependent strain in concrete at time t (elastic and creep strains) can be written as (6):

$$\varepsilon_e(t) + \varepsilon_{cr}(t) = \sum_{i=0}^{1} \frac{\Delta\sigma_c(\tau_i)}{E_c(\tau_i)}[1 + \varphi(t, \tau_i)] = \sum_{i=0}^{1} J(t, \tau_i)\Delta\sigma_c(\tau_i) \tag{6}$$

If stress increments are infinitesimal, then the sum turns to integral and we obtain the integral-type creep law [3] given in (7):

$$\varepsilon(t) = \int_{\tau_0}^{t} J(t,\tau)d\sigma_c(\tau) + \varepsilon_{sh}(t) = \int_{\tau_0}^{t} \frac{1 + \varphi(t,\tau)}{E_c(\tau)} d\sigma_c(\tau) + \varepsilon_{sh}(t) \qquad (7)$$

Next, we investigate the effects of creep and shrinkage on an axially loaded 30/30 cm massive column made of C30/37 and reinforced with $A_s = 18$ cm^2 (B500S bars). The compressive force F is equal to 1000 kN, and the load is applied at the age $\tau_0 = 14$ days. Time variations of creep and shrinkage coefficients as well as change of Young's modulus and concrete strength were calculated using MATLAB [4].

2.1   Effective Modulus Method

The simplest and oldest method for including creep in structural analysis is the effective modulus method (EMM). In EMM, the integral-type creep law (7) is approximated by assuming that the stress-dependent deformations are produced only by a sustained stress equal to the final value of the stress history, that is:

$$\varepsilon(t) = \int_{\tau_0}^{t} \frac{1 + \varphi(t,\tau)}{E_c(\tau)} d\sigma_c(\tau) + \varepsilon_{sh}(t) \approx \frac{1 + \varphi(t,\tau_0)}{E_c(\tau_0)} \sigma_c(t) + \varepsilon_{sh}(t) \qquad (8)$$

Creep is treated as a delayed elastic strain and is taken into account simply by reducing the elastic modulus of concrete with time. A time analysis using the effective modulus method is nothing more than an elastic analysis in which E_e (t, τ_0) is used instead of E_c (τ_0). Shrinkage may be included in this elastic time analysis in a similar way as a sudden temperature change in the concrete would be included in a short-term elastic analysis [1]. According to the EMM, the creep strain at time t (8) depends only on the current stress in the concrete $\sigma_c(t)$ and is therefore independent of the previous stress history. The ageing of the concrete has been ignored. For an increasing stress history, the EMM overestimates creep, while for a decreasing stress history, creep is underestimated. Equation (8) is valid only when the concrete stress is constant in time. In such cases, the EMM gives excellent results.

Governing equations that need to be respected are equilibrium (9), compatibility of strains (10) and constitutive law (11).

$$F = F_c(t) + F_s(t) = \sigma_c(t) \cdot A_c + \sigma_s(t) \cdot A_s \qquad (9)$$

$$\varepsilon_c(t) = \varepsilon_s(t) \qquad (10)$$

$$\varepsilon_c(t) = \frac{\sigma_c(t)}{E_e(t, \tau_0)} + \varepsilon_{sh}(t)$$

$$\varepsilon_s(t) = \frac{\sigma_s(t)}{E_s}$$

$$(11)$$

If we denote effective modular ratio E_s/E_e (t, τ_0) as α_e and geometric reinforcement ratio as ρ, stresses in concrete and reinforcement are equal to (12):

$$\sigma_c(t) = \frac{F}{A_c \cdot (1 + \alpha_e\rho)} - \frac{E_s \cdot \rho \cdot \varepsilon_{sh}(t)}{(1 + \alpha_e\rho)}$$

$$\sigma_s(t) = \frac{\alpha_e \cdot F}{A_c \cdot (1 + \alpha_e\rho)} + \frac{E_s \cdot \varepsilon_{sh}(t)}{(1 + \alpha_e\rho)}$$

$$(12)$$

Creep coefficients, shrinkage strains and effective modulus variation are shown in Table 1. Variation of stress and strain in concrete and steel bars obtained by implementing EMM is given in Table 2.

Table 1. Creep coefficients, shrinkage strains and effective modulus variation in EMM

$(t - \tau_0)$ [day]	$\varphi(t, \tau_0)$	$\varepsilon_{sh}(t - \tau_0) \times 10^{-6}$	$E_e(t, \tau_0)$ [MPa]
0	0	−30	30643
10	0.67	−99	18308
30	0.93	−150	15914
70	1.17	−209	14145
200	1.50	−309	12263
500	1.77	−409	11075
10 000	2.13	−603	9794

Table 2. Results of axially compressed column analysis using EMM

$(t - \tau_0)$ [day]	$\sigma_c(t)$ [MPa]	$\sigma_s(t)$ [MPa]	$\varepsilon_c(t)$ [‰]	$\varepsilon_e(t)$ [‰]	$\varepsilon_{cr}(t)$ [‰]	$\varepsilon_{sh}(t)$ [‰]
0	−9.72	−69.4	−0.347	−0.317	0	−0.029
10	−8.79	−115	−0.579	−0.287	−0.193	−0.098
30	−8.39	−135	−0.677	−0.274	−0.254	−0.150
70	−8.00	−155	−0.776	−0.261	−0.305	−0.209
200	−7.44	−183	−0.916	−0.243	−0.364	−0.309
500	−6.95	−207	−1.038	−0.227	−0.401	−0.409
10 000	−6.17	−246	−1.233	−0.201	−0.429	−0.603

2.2   Rate of Creep Method

The rate of creep method (RCM) is on the assumption that the rate of change of creep with time, $d\varphi(t, \tau_0)/dt$, is independent of the age at loading, τ. This means that creep curves for concrete loaded at different times are assumed to be parallel. Although this assumption is not true, however, the advantage of the method lies in the fact that only a single creep curve is required to calculate creep strain due to any stress history. The rate of change of creep depends only on the current stress and the rate of change of the creep coefficient and is given by (13):

$$\dot{\varepsilon}_{cr}(t, \tau_i) = \frac{\sigma_c(t)}{E_c(\tau_0)} \dot{\varphi}(t, \tau_0) \tag{13}$$

The rate of change of the instantaneous strain at any time depends on the rate of change of stress (14):

$$\dot{\varepsilon}_e(t) = \frac{\dot{\sigma}_c(t)}{E_c(\tau_0)} \tag{14}$$

It is further assumed that shrinkage develops at the same rate as creep (i.e. the creep and shrinkage curves are affine) (15):

$$\frac{d\varepsilon_{sh}(t)}{dt} = \frac{\varepsilon_{sh}(\infty)}{\varphi(\infty, \tau_0)} \dot{\varphi}(t, \tau_0) \tag{15}$$

Since behavior of the material is described in the rate format, equilibrium and compatibility can be written as (16, 17):

$$\dot{F} = \dot{F}_c + \dot{F}_s = \dot{\sigma}_c A_c + \dot{\sigma}_s A_s = 0 \tag{16}$$

$$\dot{\varepsilon}_s = \dot{\varepsilon}_c \tag{17}$$

By integrating, the stress in concrete is equal to (18):

$$\sigma_c(t) = (\sigma_c(\tau_0) + S)e^{-\frac{\alpha\rho}{1+\alpha\rho}\varphi(t,\tau_0)} - S$$
$$S = \frac{\varepsilon_{sh}(\infty) \cdot E_c}{\varphi(\infty, \tau_0)} \tag{18}$$

Creep coefficients and shrinkage strains variation are shown in Table 3. Variation of stress and strain in concrete and steel bars obtained by implementing RCM is given in Table 4.

Table 3. Creep coefficients and shrinkage strains variation in RCM

$(t - \tau_0)$ [day]	$\varphi(t, \tau_0)$	$\varepsilon_{sh}(t - \tau_0) \times 10^{-6}$
0	0	0
10	0.67	−191
30	0.93	−262
70	1.17	−330
200	1.50	−424
500	1.77	−500
10 000	2.13	−603

Table 4. Results of axially compressed column analysis using RCM

$(t - \tau_0)$ [day]	$\sigma_c(t)$ [MPa]	$\sigma_s(t)$ [MPa]	$\varepsilon_c(t)$ [‰]	$\varepsilon_e(t)$ [‰]	$\varepsilon_{cr}(t)$ [‰]	$\varepsilon_{sh}(t)$ [‰]
0	−9.82	−69.2	−0.321	−0.321	0	0
10	−8.44	−133	−0.667	−0.275	−0.201	−0.191
30	−7.95	−157	−0.789	−0.259	−0.268	−0.262
70	−7.49	−181	−0.904	−0.245	−0.329	−0.330
200	−6.88	−211	−1.056	−0.225	−0.407	−0.424
500	−6.41	−235	−1.175	−0.209	−0.465	−0.500
10 000	−5.78	−266	−1.329	−0.189	−0.537	−0.603

2.3 Age-Adjusted Effective Modulus Method

In order to take into account ageing of concrete, a simple adjustment was proposed by Trost and later developed by Bažant [1]. Due to ageing of concrete, creep deformations of gradually loaded specimen are significantly smaller than that resulting from abruptly applied stress. The earlier the specimen is loaded, the greater the final creep strain. A reduced creep coefficient $\chi (t, \tau_0) \varphi (t, \tau_0)$ can therefore be used to calculate creep strain if stress is gradually applied, where $\chi (t, \tau_0)$ is called the ageing coefficient $(0.4 < \chi (t, \tau_0) < 1)$. Modulus of elasticity is now equal to (19):

$$\tilde{E}_e(t, \tau_0) = \frac{E_c(\tau_0)}{1 + \chi(t, \tau_0)\varphi(t, \tau_0)} \tag{19}$$

Ageing coefficient is usually assumed to be equal to 0.65 for creep problem with constant loading or 0.8o in case of constant deformation (relaxation problem).

Creep coefficients, shrinkage strains and age-adjusted effective modulus (AEMM) variation are shown in Table 5. Variation of stress and strain in concrete and steel bars obtained by implementing AEMM is given in Table 6.

Table 5. Creep coefficients, shrinkage strains and effective modulus variation in AEMM

$(t - \tau_0)$ [day]	$\varphi(t, \tau_0)$	$\varepsilon_{sh}(t - \tau_0) \times 10^{-6}$	$E_e(t, \tau_0)$ [MPa]	$\tilde{E}_e(t, \tau_0)$ [MPa]
0	0	−30	30643	30643
10	0.67	−99	18308	21310
30	0.93	−150	15914	19133
70	1.17	−209	14145	17429
200	1.50	−309	12263	15522
500	1.77	−409	11075	14263
10 000	2.13	−603	9794	12856

Table 6. Results of axially compressed column analysis using RCM

$(t - \tau_0)$ [day]	$\sigma_c(t)$ [MPa]	$\sigma_s(t)$ [MPa]	$\varepsilon_c(t)$ [‰]	$\varepsilon_e(t)$ [‰]	$\varepsilon_{cr}(t)$ [‰]	$\varepsilon_{sh}(t)$ [‰]
0	−9.72	−69.4	−0.347	−0.317	0	−0.029
10	−8.77	−117	−0.585	−0.286	−0.200	−0.098
30	−8.35	−138	−0.689	−0.272	−0.267	−0.150
70	−7.93	−159	−0.794	−0.259	−0.326	−0.209
200	−7.32	−189	−0.947	−0.239	−0.399	−0.309
500	−6.78	−216	−1.082	−0.221	−0.450	−0.409
10 000	−5.91	−260	−1.299	−0.193	−0.503	−0.603

2.4 Step-by-Step Method

Step-by-step method (SSM) is based on an incremental form of the superposition principle where continuous stress history is described by a step-wise function. As previously shown, concrete stresses reduce in time, which means that the initial load increment is compressive, and all others are tensile. The SSM method is general and can be used to predict behavior due to any stress history using any desired creep and shrinkage curves. Detailed development of governing equations is given in [5] and will be omitted here. Finally, stress and strain distributions are shown in Table 7.

Table 7. Results of axially compressed column analysis using SSM

$(t - \tau_0)$ [day]	$\sigma_c(t)$ [MPa]	$\sigma_s(t)$ [MPa]	$\varepsilon_c(t)$ [‰]	$\varepsilon_e(t)$ [‰]	$\varepsilon_{cr}(t)$ [‰]	$\varepsilon_{sh}(t)$ [‰]
0	−9.72	−69.4	−0.347	−0.317	0	−0.029
10	−8.74	−118	−0.590	−0.286	−0.214	−0.098
30	−8.33	−139	−0.69	−0.273	−0.269	−0.150
70	−7.91	−160	−0.798	−0.261	−0.327	−0.209
200	−7.28	−191	−0.956	−0.243	−0.405	−0.309
500	−6.74	−218	−1.093	−0.227	−0.457	−0.409
10 000	−5.78	−266	−1.332	−0.201	−0.530	−0.603

3   Comparison of Results and Conclusion

The comparison of results obtained using different methods is given in Figs. 2, 3 and 4.

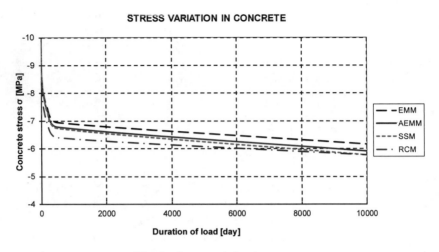

Fig. 2. Stress variation in concrete

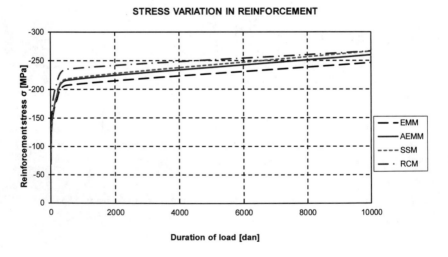

Fig. 3. Stress variation in reinforcement

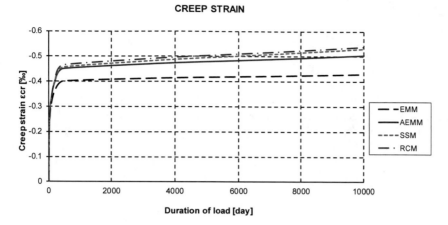

Fig. 4. Creep strain

It is obvious that all methods predict a significant redistribution of stress between the concrete section and the reinforcement. In the analyzed example of a reinforced concrete axially compressed column, the stresses in the concrete are significantly reduced with time, while the stresses in the reinforcement are increased. For example, if we take the results of the AEMM method, at the time of the loading (initially) the concrete section of the cross section takes over 87% of the external force. After 10.000 days under sustained load, this value is reduced to 53%. In the same period, the stress in the reinforcement increased from 69 MPa to 260 MPa, which is a characteristic result of the redistribution of stress that occurs in such RC element.

From the above diagrams it can be seen that the effective modulus method (EMM) underestimates the creep deformation for the stress history that decreases in time, and results with a complete creep recovery after removal of the load. Also, EMM predicts the smallest creep deformation at any given moment of time, as well as the smallest eventual stress redistribution. In contrast, the RCM overestimates creep as it does not allow any reversible deformation. This method creates the largest creep deformations as well as the largest redistribution.

The approximations of AEMM method and SSM lie between these extreme results. SSM is the most accurate, but also the most time consuming, because it depends very much on time discretization (the number of steps) and the number of creep coefficients. Therefore, this method is only suitable for computer implementation. AEMM method is a combination of the simplicity of an effective modulus method and the accuracy of SSM method that is achieved by additional ageing coefficients. It would be correct to determine the ageing coefficient using SSM method at each time step. However, this does not actually happen in practice because it represents a huge time loss, and by adopting certain constant values, the results are sufficiently acceptable for practical problems. Finally, based on the calculated results, AEMM is the best choice in terms of efficiency and accuracy.

References

1. Gilbert, R.I., Ranzi, G.: Time Dependent Behaviour of Concrete Structures. Spon Press, London (2011)
2. Zlatar, M.: Betonske konstrukcije I. Građevinski fakultet, Sarajevo (2012)
3. Rüsch, H., Jungwirth, D.: Stahlbeton-Spannbeton Band 2, Berücksichtigung der Einflüsse von Kriechen und Schwinden auf das Verhalten von Tragwerken. Werner (1976)
4. MATLAB R2010a. www.mathworks.com
5. Medic, S.: Time-dependent behavior of axially compressed RC column (2016)

Experimental Testing and Numerical Modeling of Semi-prefabricated RC Girder of Grbavica Stadium Eastern Grandstand

Senad Medić[1]([⊠]), Muhamed Madžarević[1], and Rasim Šehagić[2]

[1] Faculty of Civil Engineering, University of Sarajevo,
Sarajevo, Bosnia and Herzegovina
senad_medic@yahoo.com, madzar@bih.net.ba
[2] Calypso doo, Sarajevo, Bosnia and Herzegovina
rasimsehagic@gmail.com

Abstract. Experimental in-situ and laboratory testing of semi-prefabricated RC girder used for erection of the new football stadium Grbavica eastern grandstand is presented in the study. The girder is a continuous L-shaped beam assembled in two phases. First, a thin precast simple span Omnia slab was laid on main frames and used as formwork. Next, concrete was cast in place in order to form the final geometry of the cross section and the continuous beam. Static test load required by design was applied and deflections were measured on site. 3PB test was conducted on a characteristic beam at the structural laboratory of the Faculty of Civil Engineering in Sarajevo. Nonlinear finite element model was created in Diana 10.1 employing total strain-based crack model for concrete and von Mises plasticity for reinforcing steel and verified against the experimentally obtained pushover curve. The model was capable of tracing the complete load path from linear phase up until failure. Numerically obtained crack pattern was realistic. Failure mechanism characterized by yielding of reinforcement and crushing of concrete was confirmed in the computational model.

Keywords: RC semi-prefabricated beam · In-situ and laboratory testing Numerical modeling

1 Introduction

Stadium Grbavica of the football club Željezničar in Sarajevo was heavily damaged during the war. It was eventually rehabilitated, however, the spectator capacity remained small. The old eastern grandstand was demolished and the new one was erected with significant support of loyal fans who voluntarily donated material assets into this project. The structure is made of reinforced concrete and it's dimensions in plan are 103.50×22.35 m (Fig. 1). The structure consists of walls, beams and folded slabs and it is laid on foundation strips. The materials used are concrete C25/30 and reinforcement B500.

The focus of this paper is on L-shaped stand girders which are connected and form a folded plate. The girder is a continuous beam assembled in two phases. First, a thin precast simple span Omnia slab 5 cm thick was laid on main frames and used as

© Springer Nature Switzerland AG 2019
S. Avdaković (Ed.): IAT 2018, LNNS 60, pp. 73–79, 2019.
https://doi.org/10.1007/978-3-030-02577-9_9

Fig. 1. Eastern grandstand of Grbavica stadium in Sarajevo

formwork. Next, concrete was cast in place in order to form the final geometry of the cross section and the continuous beam (Fig. 2). A typical girder was tested at the Institute for materials and structures of the Faculty for Civil Engineering, University of Sarajevo [1].

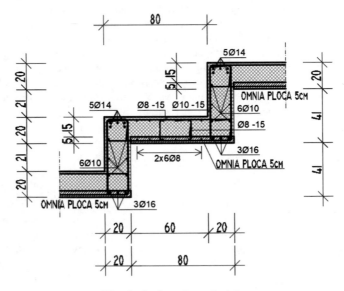

Fig. 2. L-shaped stand girder

2 Experimental Model

The stand girder was examined until failure at the Institute for Materials and Structures. Although the actual girder span was 8.2 m, the supports were set at 6.0 m due to the boundaries of the base plate for which the test is performed (Fig. 3). The force was applied over steel profile and plates in order to evenly distribute the loading across the entire cross section (Figs. 3 and 4).

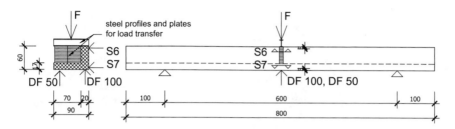

Fig. 3. Test set-up

The load-displacement diagram in Fig. 5 shows that the change in stiffness or the appearance of the first crack occurs at a vertical load equal to 55 kN i.e. a bending moment of 82.5 kNm. The girder was designed for a load of 5 kN/m² in the service state. This load on the continuous stand girder produces the actual bending moment of about $ql^2/10 = 33.6$ kNm. Thus, for the service condition the girder remains elastic and in state I (no cracks) with the safety coefficient of $82.5/33.6 = 2.45$. This is a very good fact because the maximum dynamic factor of a realistic impact is approximately 1.7. The girder has lost its load bearing capacity at a vertical load of 260 kN (M = 390 kNm). The safety coefficient for the ultimate limit state is $390/33.6 = 11.6$.

Fig. 4. View of the girder and test frame

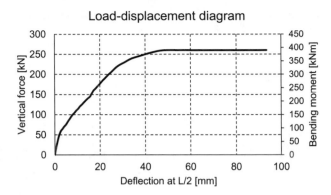

Fig. 5. Load-displacement diagram

The cracking pattern is shown in Fig. 6. Cracks occurred due to bending, and there were no shear cracks. Concrete in the compressed zone was crushed and the reinforcement buckled (Fig. 6).

Fig. 6. Cracking pattern (left) and buckling of compressed reinforcement (right)

3   Numerical Model

3D model of the tested girder was made in program Diana 10.1 [2] using solid finite elements for concrete and truss elements for reinforcement (Fig. 7).

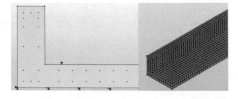

Fig. 7. Side view with reinforcement bars (left) and detail of finite element mesh (right)

Concrete was modeled using total strain-based crack model with parameters listed in Table 1, while the steel reinforcement parameters are given in Table 2. Concrete strength was determined experimentally by loading Ø10 concrete cores until failure and the mean value of tests was 32.2 N/mm². Concrete in the tested beam was confined by shear reinforcement and the obtained compressive strength was increased by 10% due to confinement. Stresses in concrete and reinforcement at the ultimate limit state are shown in Figs. 8 and 9, respectively.

Table 1. Parameters of concrete

Name	Value
Material model	Total strain-based crack model
Young's modulus	31900 N/mm²
Poisson's ratio	0.167
Mass density	2.4e−09 T/mm³
Crack orientation	Rotating
Crack bandwidth specification	Rots
Tensile curve	Linear-ultimate crack strain
Tensile strength	3 N/mm²
Ultimate strain	0.0035
Compressive strength	35 N/mm²

Table 2. Parameters of reinforcement

Name	Value
Material class	Reinforcements and pile foundations
Material model	Von Mises plasticity
Young's modulus	200000 N/mm²
Yield	VMISES
Plastic hardening	Plastic strain-yield stress
Strain-stress diagram	0 571 0.06 667 N/mm²
Hardening hypothesis	Strain hardening
Hardening type	Isotropic hardening

Stresses in concrete and reinforcement at the ultimate limit state are shown in Figs. 8 and 9, respectively. The crushing strain in concrete and yielding deformations in steel were attained. Crack widths are given in Fig. 10.

Fig. 8. Stresses in concrete [N/mm²]

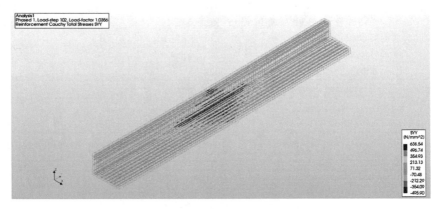

Fig. 9. Stresses in reinforcement [N/mm²]

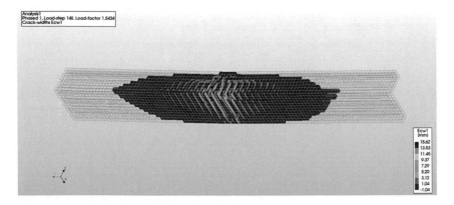

Fig. 10. Crack widths [mm]

4   Comparison of Results and Conclusion

The comparison of load-displacement curves obtained experimentally and numerically is given in Fig. 11.

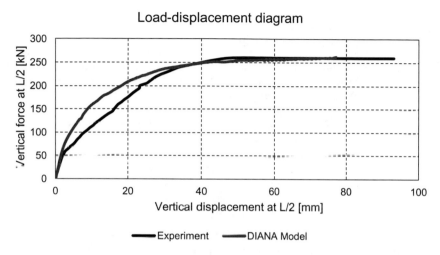

Fig. 11. Load-displacement relationship – comparison of experimental and numerical result

The model was capable of tracing the complete load path from linear phase up until failure. Numerical model predicts initial stiffness and the ultimate load bearing capacity very well. After occurrence of the first crack, the stiffness abruptly changes in the experimental curve, which is not described by the model. Numerically obtained crack pattern was realistic. Cracks occurred due to bending, and there were no shear cracks. The yielding of reinforcement and crushing of concrete observed in the experiment were also obtained by the numerical model. The assumed constitutive models proved very useful and could be used for predicting ultimate load bearing capacity of similar RC structure.

From a practical point of view, the installed girders fulfill all necessary criteria for a certificate of occupancy. The structure has the required load bearing and deformation capacity to resist the load conditions assumed in the design. Moreover, the girder remains uncracked for the service load level.

References

1. IMK: Elaborat o ispitivanju probnim opterećenjem konstrukcije istočne tribine stadiona "Grbavica" u Sarajevu. Instutut za materijale i konstrukcije Građevinskog fakulteta, Univerziteta u Sarajevu (2017)
2. TNO DIANA: User's Manual. DIANA FEA BV, Delft, The Netherlands (2016)

Analysis and Visualization of the 3D Model – Case Study Municipality of Aleksandrovac (Serbia)

Mirko Borisov[1]([⊠]), Nikolina Mijic[2]([⊠]), Zoran Ilic[1], and Vladimir M. Petrovic[3]

[1] Faculty of Technical Sciences, University of Novi Sad,
Trg Dositej Obradovic 6, 21000 Novi Sad, Republic of Serbia
mirkoborisov@gmail.com, mborisov@uns.ac.rs,
geoingbc@gmail.com
[2] Faculty of Earth Science and Engineering, University of Miskolc,
Egyetemvaros, Miskolc H-3515, Hungary
nikolinamijic7@gmail.com
[3] Department for Ecology and Technoeconomics, Institute of Chemistry,
Technology and Metallurgy, University of Belgrade,
Njegoševa 12, 11000 Belgrade, Republic of Serbia
vladimirpetrovic.gis@gmail.com

Abstract. This paper describes analysis and visualization of the 3D model of municipality Aleksandrovac that made using new technologies. First of all, the geospatial features of the municipality Aleksandrovac were analyzed. There are paying special attention on the geomorphological and hydrological characteristics of the given area. For the creation and display of 3D terrain models, from topographic maps of certain dimensions there are used original data. The quality and loyalty of the elevation model of the terrain depends of the data which are collected, i.e. on the scale of the original maps, but also on the way of interpretation and visualization of the 3D model. On the other hand, the organization and structure of data are influencing on the creation of the 3D model. In this paper were applied different techniques of the data structure. There are used different methods for visualization and 3D modeling of the municipality Aleksandrovac, and creation of the GRID and TIN model.

Keywords: The city of Aleksandrovac · Geospatial features · 3D model Analysis and visualization

1 Introduction

The paper analyzes the geospatial features of the municipality Aleksandrovac and the procedure for visualizing 3D terrain models based on available and collected data. Considering the long tradition of altitude display of the terrain in the form of the contour lines and angles, the extraction of the contour line is still an available option in the application of new technologies. This is especially interesting for geodetic and cartographic experts, but also for users from other professions.

© Springer Nature Switzerland AG 2019
S. Avdaković (Ed.): IAT 2018, LNNS 60, pp. 80–92, 2019.
https://doi.org/10.1007/978-3-030-02577-9_10

However, with the advantage of modern technologies, the technology of 3D model design and its geo-visualization are increasingly used [1].

This paper consists of the three parts. In the first part, the geospatial features of the studied area were analyzed, i.e. geographical characteristics of municipality Aleksandrovac. The second subheading of this paper has applicative character. It provides concepts of digital terrain models, i.e. methods of data collection and the process of the 3D terrain modeling. It also describes the software environment, which they are using for modeling of the terrain.

This paper give an overview of the original data (topographic and cartographic backgrounds) and the methodology of 3D modeling of the terrain. The third part of the paper deals with geo-visualization and application of the 3D modeling of the terrain. Analyzes and interpretations of 3D models have been realized. Many applications realized in practice through appropriate examples. At the end of the paper, the most important conclusions were drawn.

2   Geospatial Features of the Municipality Aleksandrovac

The municipality of Aleksandrovac belongs to the Rasinski district, which is located in the central part of the Republic of Serbia. It is located between the mountains Kopaonik, Zeljin, Goc and Jastrebac, bordering with the municipalities of Brus, Raska, Vrnjacka Banja, Trstenik and Krusevac.

The area of the municipality Aleksandrovac is 387 km², with 26 522 inhabitants according to the census from 2011. On the territory of the municipality Aleksandrovac, there are no railway paths, and through the general route, the roads of the second line are:

II-119 Kruševac-Josanicka Banja, according to the municipalities of Raska and Novi Pazar, Exit P-118 Stopanja-Brus to Kopaonik and DP P-222 Krusevac-Brus-Aleksandrovac-Goc towards Zeljin and Vrnjacka Banja [2].

Many hills and valleys surround area of the municipality Aleksandrovac. Terrain degradation is distinct and ranges from 180 to 1785 m above sea level. The lowest corner of the municipality Aleksandrovac is 186 m in the cave of Pepeljuga, in the eastern part, and the highest one on Zeljin on the Rogava cave is 1784 m above sea

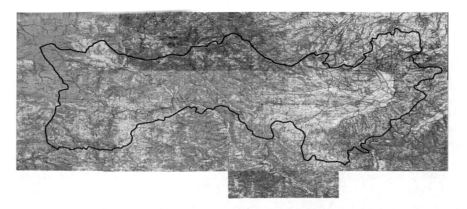

Fig. 1. Geographical area of the municipality Aleksandrovac

level [2]. Mostly hills are mountainous and low-mountain areas characteristic of the eastern part of the municipality, while the regions with a distinctly mountainous character are characterized by the western part of the municipality Aleksandrovac. This area was shaped by river flows after the Pannonian sea withdrew (Fig. 1).

The territory has two morphological units called Donja and Gornja Zupa. Donja Zupa is a puppet in the basin of 160 km^2 and is divided into three smaller elongated valleys along the streams of Pepeljuga, Kozetinska and Drenacka Reka. Gornja Zupa is located above Donja Zupa and covers the most part of the territory of 227 km^2. Unlike Donje, Gornja Zupa has an accentuated hilly-mountainous character that is characteristic of settlements of broken morphological structures with spacious forests and watercourses [3]. In the mountainous part, in the valleys of the major river flows, there are certain composite valleys, which is especially characteristic of the upper flow of the Rasina river, where from Mitrovo Polje under Zeljin and Goc, a narrow quince enters Plesko field, according to Budilovina and Milentija.

The municipality of Aleksandrovac has a moderate continental climate with some elements of the Mediterranean climate. Varieties of this kind are generally called "parish climates" characterized by mild, moderately cold winters with relatively low snow and without extremely low temperatures with warm, dry and long flights. Mean annual temperature in the general Aleksandrovac is 11.6 °C, the maximum average monthly temperature in July is 22.1 °C, while the average monthly average temperature in January is 0.4 °C. The average annual insolation amount in the county is 1774 h, the longest average inoculation is 753 h, while the lowest average insolation has winter (169 h). From April to June, the most severe period lasts, while the winds are rare throughout the year, and are mostly represented by the northwestern wind [2].

When it comes to hydrologic characteristics, the more significant watercourse is represented by the Rasina river, the right tributary of the Western Morava, where it flows to Krusevac. Other significant river flows are Josanicka river, Vratarica, Pepeljuga, Kozetinska and Drenacka river.

However, during the period of May until October, it often happens that the other rivers, except Josanica, are dried by the karst-limestone composition of the soil. Kozetanska river is formed in the lower part of the village Kozetin near of the municipality Aleksandrovac, less than two points are: Popovacko-Nursing (right Kozetin) and Latkovacko-Puhovackog stream (left Kozetan). The valleys of the Kozetinska, Drenacka and Pepeljuga rivers, as well as their tributaries, represent areas of accumulation. In the alluvial plain Pepeljuse, there are groundwater at a shallow depth.

Mainly groundwater is expressed in the form of normal aquifers or water reserves depend on the geological structure. For example, mountain of Zeljin is very rich with water. It has huge number of water sources during the year and the streams and rivers are building significant water network. Hydrological, the area is divided into two gravitational areas: Rasinsko, which mostly includes the northern and northeastern part, and Ibarsko, which occupy the southern and southeastern part. Rasina is 92 km long and has a basin of 981 km^2. It springs on the slopes of Goc and Zeljin, at an altitude of 1340 m from the springs of Velika and Burmanska rivers, and it flows into the western Morava 5 km downstream of Krusevac. The basin of this river has a distinctly asymmetrical shape, because apart from the river Zagriza, which rises from the mountain Goc on the left side of the spring, all other tributaries touch it on the right.

The Burmese river that also flows east and the river Vranju meets outside the territory of this municipality. The river Burmese is formed from the Great river that flows from it to the right, after the source below Zeljin, in its upper course is called Smrecka river. As for the Ibarian gravitational area, the main river is Josanicka river (Josanica), whose total length is 17 km. It springs on the slopes of Zeljin and Kopaonik and comes from Plocka (Konjska) river and Brajkovina, which meet in the village of Jelakci. Through the territory of the municipality of Aleksandrovac, it flows in a length of about 6 km, forming at the very end of the municipality of Drenska, while the remaining 11 km runs through the territory of the municipality of Raska. Hydrographic features of the given space have been elaborated in more detail for the development of a hydrological model, or calculated water flows based on 3D terrain model.

The soil characteristics of the municipality Aleksandrovac are particularly interesting. Namely, the western part of the municipality of Aleksandrovac is characterized by stony land with the most represented type of soil underground, covered with numerous forests and pastures located on a slightly hilly terrain. The eastern and northern part of the municipality is characterized by a loose land, a type of enriched meadows that is suitable for growing vegetable crops, in contrast to the western part that is suitable for fruit growing and viticulture. Already mentioned, alluvial soil is present in the valley and its Pepeljusa in the eye, characterized by a different composition and a considerable humidity, are also suitable for the cultivation of vegetables.

In the central and eastern parts of the municipality of Aleksandrovac there are also manure, land characteristic for mildly stratified terrains up to 600 m above sea level, which is the ideal basis for cultivating vines. As the height increases, the gardens turn into brown subsoil, forest lands that characterize the mountainous part of the municipality. It is also important to mention the erosion processes that are mainly expressed on steep terrain, but with intensive biological processes, primarily afforestation, with time significantly reduced areas affected by erosion [3].

Biogeographic features are also interesting. As already mentioned, agricultural crops are mainly characteristic of the lower parts of the municipality of Aleksandrovac, primarily fruit growing and viticulture, but also cereals, vegetables, fodder and industrial plants. Mountain areas are characterized by natural vegetation - one third of the municipality of Aleksandrovac is under forested areas, where 62.2% of the forests are social and 37.8% are private. In the forests of the forest, forests account for around 98%, and coniferous with about 2%. Within the decade, the share of beech is about 93%, oak about 6% and other species about 1%. In conifers, the most common are pine (about 85%) and dishes (about 9%), while spruce (about 3%) and other conifers (about 3%) are represented to a lesser extent. Other forest natural resources (medicinal plants, forest fruits, mushrooms, hunting wild and etc.), are not used sufficiently and rationally. Most commonly, mushrooms are collected, while other fruits of nature are very poorly represented due to large migrations of rural population from the mountainous region [2].

3   Design of the 3D Model of the Terrain in the Municipality Aleksandrovac

With the advantage of the new technologies, there are developed just few techniques for spatial data collection data for the purpose of a digital terrain model (remote sensing, laser scanning radar recording, and GPS).

However, scanning geodetic and cartographic layers, is still a rational and very economical method for collecting data [4]. First of all, topographic maps and maps provide numerous geomorphologic features, characteristic points and structural lines [5]. The practical part of this paper refers to the 3D representation of the terrain, i.e. the procedures for digitalization of cartographic layers of certain dimensions that cover the area of the municipality of Aleksandrovac. In geographical terms, three types of terrain can be distinguished: flatland, mountain and mountainous land. In addition, the terrain characteristics are complex and geomorphologic diversity, with a fairly pronounced network of watercourses and landforms (Fig. 2).

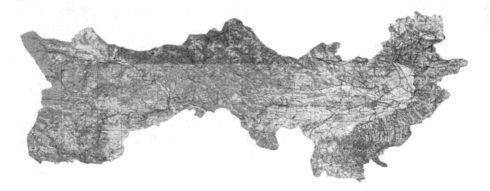

Fig. 2.   Test area used for creating of the 3D model of the terrain

The software and software solutions ArcGIS and QGIS were used to collect, process and create 3D terrain models. The QGIS software environment (formerly known as "Quantum GIS") is a GIS open source computer application that allows visualization, management, editing and analysis of a wide range of different data formats. It is also important to note that QGIS supports various operating systems, including Mac OS X, Linux, BSD and Windows.

QGIS software was used primarily in the procedures of digitizing content and vectorization, and ArcGIS applied to create a digital terrain model. The analysis of 3D terrain model was also made with ArcGIS, ArcMap 10.1 and ArcScene 10.1. The possible in the middle are many analyzes and using software environments QGIS. However, far better results are achieved using the ArcGIS software environment [6]. After scanning (resolution of 300 dpi), i.e. translating an analog paper card into digital form, the next step is georeferencing.

Georeferencing is the process of translating a scanned raster substrate into a target cartographic projection - in a specific case reproducing in 7. Gauss-Krüger zone based on known parameters of Affine transformation. All sheets of topographic maps and maps have been merged and cut off with the border of the territorial municipality of Aleksandrovac [4].

After that, the longest part of 3D modeling is followed, which is the vectorization of contour lines and angles (Fig. 3).

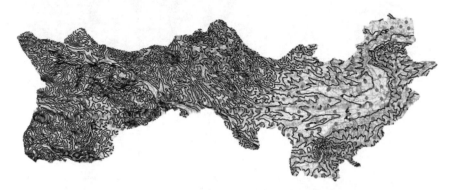

Fig. 3. Vectorization process of the contour lines and heights

Prior to that, in the QGIS software environment, has been done is digitization of the angle and characteristic details. Incorporating two new layers into shape format. Layers are a logical partition of the detail of the drawing in such a way that the user can display and hide them at the same time as the user environment. Also, a new column Z is added in the attribute table, which refers to the node height of the data. Since the contents of vectorization has been done, approaches to the interpolation and the process of obtaining a 3D terrain model. In the ArcGIS software environment, you can access Arc Toolbox and select 3D Analyst Tools, or Raster Interpolation and Topo to Raster [7].

Identical option in the software environment QGIS, is the Raster Interpolation DEM. DEM *(Digital Elevation Model)* conversion is a method of translating original geometry into a digital surface topography model field area in a raster format, in which the color of the pixels is in direct correlation with the elevation of the terrain.

Digitized contour lines and angles are inserted for initial data. Depending on the method of interpolation and interpretation, different results are obtained [8].

The digital model shown is obtained by the Topo to Raster method, and later a 3D model for displaying in ArcScene is generated using the Raster to TIN option (3D Analyst Tools → Conversion → From raster → Raster to TIN). Resulting TIN (Triangular Irregular Network) the model is an interpolated representation of terrain morphology in a three-dimensional form, and as a vector record used in a large number of GIS applications, and also a basis for visualizing the target model (Fig. 4). Also, many terrain analyzes were realized through the raster data model (Fig. 5).

Fig. 4. 3D model of the municipality Aleksandrovac (TIN)

Fig. 5. 3D model of the municipality Aleksandrovac (DEM – Raster)

4   Application and Geovisualization of the 3D Model of the Terrain

New technologies provide the possibility of creating 3D models based on the projected geo database. In addition, they enable a variety of applications and 3D visualization, which is implemented interactively and according to the needs of users. Visualization of data in the form of 3D is one of the highest quality and most efficient the way to model and display the geospatial reality [8]. In Fig. 6, illustrated is a shaded relief model, and the terrain models are shown in Fig. 7.

With 3D models it is possible to perform numerous and varied analyzes.

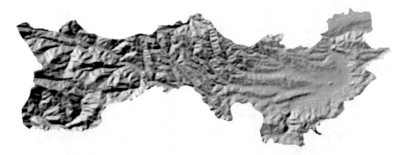

Fig. 6. Shaded 3D model of the municipality Aleksandrovac

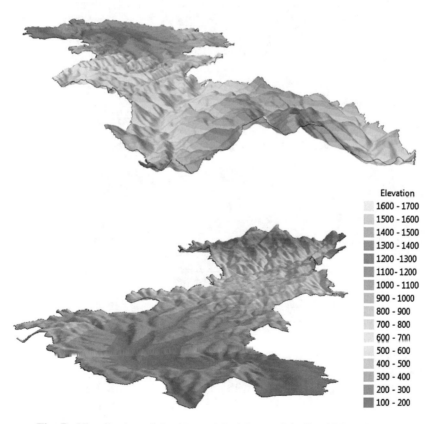

Elevation
1600 - 1700
1500 - 1600
1400 - 1500
1300 - 1400
1200 -1300
1100- 1200
1000 - 1100
900 - 1000
800 - 900
700 - 800
600 - 700
500 - 600
400 - 500
300 - 400
200 - 300
100 - 200

Fig. 7. Visualization of the 3D model of the municipality Aleksandrovac

The analysis include:

- interpolation of altitudes in the given points,
- drawing of longitudinal and cross sections,
- calculating volumes,
- calculation of certain morphological parameters of the terrain surface (inclination, aspect, curvature),

- calculation of hydrological parameters (flow direction, flow accumulation, drainage network calculations and delineation of the basins), the calculation of the track record and the development of visibility maps [9].

Here are some of the possible applications and analyzes 3D terrain model. One of the advantages of 3D models is that it is easy to generate longitudinal or transverse profiles. Just drag the desired line for the intersection of height and can be created automatically longitudinal or transversal profile (Fig. 8) and P place using the options *Line of Sight* is possible to generate a line of sight from the initial observation point to the desired target (Fig. 9).

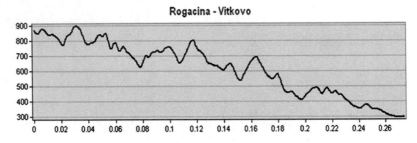

Fig. 8. Cross-section profile Rogacina-Vitkovo

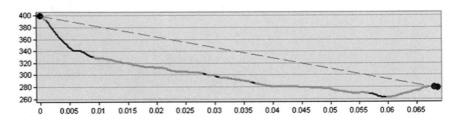

Fig. 9. Line of visibility on arbitrary selected profile

Also, an option *Spatial Analyzer Tools Surface Viewshed*, an analysis of the visibility from an arbitrarily chosen point was performed.

For example, visibility from the peak Siljaja located in the western part of the municipality of Aleksandrovac and with an altitude of 1282 m (Fig. 10).

Fig. 10. Analysis and visibility display from the point Siljaja

One of the most important geomorphological parameters is the slope. Slope of the terrain is a vertical angle that covers the surface with a horizontal plane [3].

As a result of many processes and, above all, the impact of endogenous and exogenous processes. Also, there are different algorithms for calculating the slope of the terrain. For example, in DEM models with like this in the output raster cells has its own value of slope. The smaller value and the higher value of the slope, showing the steeper terrain.

The slope can be from break up the percentage or the degree there. In Fig. 11, an example is given of the slope in degrees calculated using the Options that are located in the *Spatial Analyst Tools → Surface → Slope*.

Fig. 11. Map slope of the municipality Aleksandrovac

Exposition represents the position (orientation) of the surface relief in comparison to the world. Determination of exposure is significant, and can be determined in relation to the four main directions of the world (north, east, south and west) and four additional directions (northeast, southeast, southwest, and northwest). Namely, is calculated for each triangle of the TIN, or for each cell of the grid of the grid [9]. Exposure field can have values of 0° (north direction)–360° (again north direction). Value of each cell of the grid exposure indicates an orientation surface of the field depending on the angle of inclination (Fig. 12).

The basic steps in obtaining information on hydrological phenomena on the ground, consisting of filling depressions in DEM, determining the direction of swelling, determining the accumulation of water, based on the accumulation of runoff are calculated riverbeds (Fig. 13).

The boundaries of river basins are meant to include all rivers in the river basin. This process involves several steps. As we said the first step is to fill depressions in DEM in *Spatial Analyst Tools → Hydrology → Fill*. Then the obtained raster determine the direction of runoff: *Spatial Analyst Tools → Hydrology → Flow Direction*. After that, the Lady determining the accumulation of runoff: *Spatial Analyst Tools → Hydrology*

Legend

Straight (-1)
North (0-22.5)
Northeast (22.5-67.5)
East (67.5-112.5)
Southeast (112.5-157.5)
South (157.5-202.5)
Southwest (202.5-247.5)
West (247.5-292.5)
Northwest (292.5-337.5)
North (337.5-360)

0 2 4   8   12   16
Kilometers

Fig. 12. Map of the terrain exposure of the municipality Aleksandrovac

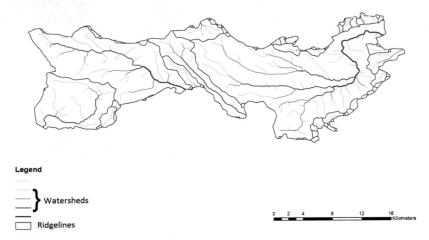

Legend

} Watersheds

▭ Ridgelines

0 2 4   8   12   16
Kilometers

Fig. 13. Map of streams and watersheds based on DEM calculation of the municipality Aleksandrovac

→ *Flow Accumulation* and determine the watershed through the options: *Spatial Analyst Tools → Hydrology → Basin*.

Finally received raster images via the *Conversion Tools → From Raster → Raster to Polygon* and performed data conversion into the vector format i.e. *shapefile*.

Also possible is the combination of the 3D terrain models with various layers such as watercourses, orthophotos, etc. In this paper is displayed switch the 3D model with topographic map (Fig. 14).

Fig. 14. 3D model of the municipality Aleksandrovac overlapped with topographic maps [10, 11]

5   Conclusion

The analysis of geospatial features of the municipality Aleksandrovac and the application of new technologies, were developed and presented as a specific 3D model of the area. Unlike traditional display of geo-modern way of visualization is more effective and more acceptable for many users. The creation of a 3D model is a demanding job and involves the use of modern computer technology that 2D model complements the third dimension, and thus provides a better perception of space. The major focus of the work is dedicated to the visualization and analysis of data in order to display a variety of practical features.

First of all, these are the methods of obtaining data on the height and slope of the terrain at any point, drawing the profile and to obtain new information on the ground.

In the other hand, progress of computer technology and internet connections, it is expected that a greater number of ways of obtaining 3D models and visualization with multimedia content.

Finally, these techniques should be better used in the near future. These new methods and technologies should be represented in the surveying profession, and even in science and economy. The combination of GIS and web mapping geovisualization today is one of the most promising information technology.

References

1. Li, Z., Zhu, Q., Gold, C.: Digital Terrain Modeling – Principles and Methodology. CRC Press, Florida (2005)
2. Lutovac, M.V.: Župa aleksandrovačka-antropogeografska ispitivanja, SANU, Srpski etnografski zbornik, Naselja i poreklo stanovništva, knjiga 43, Beograd (1980)
3. Manojlović, P., Dragićević, S.: Praktikum iz geomorfologije. Geografski fakultet Univerziteta u Beogradu, Beograd (2002)
4. Arrighi, P., Soille, P.: From scanned topographic maps to digital elevation models. In: Proceedings of Geovision 1999, International Symposium on Imaging Applications in Geology (1999)
5. Republički geodetski zavod/RGZ: Monografija, Geodetska delatnost u Srbiji 1837–2012, Beograd, Srbija (2012)
6. Environmental Systems Research Institute/ESRI: Using Arc GIS 3D Analyst, User Guide, Redlands, USA (2010)
7. Environmental Systems Research Institute/ESRI: ArcGIS for Desktop10.x, Korisničko uputstvo. GDI Press, Beograd (2015)
8. Šiljeg, A.: Digitalni model reljefa u analizi geomorfometrijskih parametara- primer PP Vransko jezero. Doktorska disertacija, Prirodnoslovni fakultet Univerziteta u Zagrebu, Hrvatska (2013)
9. Borisov, M., Petrovi, V.M., Vuli, M.: Vizuelizacija 3D modela geopodataka i njihova primjena. Geodetski glasnik **48**(45), 29–45. Sarajevo, BiH
10. www.rgz.gov.rs. Accessed 18 Dec 2017
11. http://www.vgi.mod.gov.rs. Accessed 22 Dec 2017

Data Quality Assessment of the Basic Topographic Database 1: 10000 of the Federation of Bosnia and Herzegovina for Land Cover

Slobodanka Ključanin[1]([✉]), Zlatko Modrinić[2], and Jasmin Taletović[3]

[1] Faculty of Civil Engineering, Department of Geodesy,
University of Sarajevo, Sarajevo, Bosnia and Herzegovina
slobodanka63@yahoo.com
[2] Geoinformatika d.o.o. Split, Split, Croatia
info@geoinformatika.hr
[3] Institute for Development Planning of Canton Sarajevo,
Sarajevo, Bosnia and Herzegovina
Jasmin.Taletovic@zpr.ks.gov.ba

Abstract. Based on the new model of the Topographic Information System (TIS) (2015), the creation of the basic topographic database of the Federation of Bosnia and Herzegovina at 1: 10000 (BTD) has been initiated (2016). For this purpose, the selected pilot area was Goražde located in the Bosnian-Podrinje canton, which was used for the preparation of Methodology and procedures for establishing and maintaining a basic topographic database. While creating the BTD for the pilot area, it became clear that it is almost impossible to get information about the subject of Land Cover. The data available for this subject is of very poor quality. On the other hand, it was noticed that no national institution has categorized this subject. As TIS recommends the CORINE classification for the Land Cover subject, even that few data that could have been collected quickly became meaningless. The goal of this article is to evaluate the quality of BTD data for Land Cover for the Sarajevo Canton area. Data quality assessment implies an estimate of: the origin of data, position and height accuracy, accuracy of attributes, completeness of data, logical consistency, semantic and time accuracy. For the purposes of this article, we used the data collected for the preparation of the Study: "Inventory of the condition, development of a database for the coverage and land usage method for the Sarajevo Canton in GIS technology" and other available data sets for Land Cover for the Sarajevo canton area.

Keywords: Basic topographic database (BTD) · Land Cover
CORINE classification

1 Introduction

Land Cover data by definition are data on the physical or biological coverage of the Earth's surface, taking into account artificial surfaces, agricultural areas, forests, wetlands and water bodies. In this way, they differ from land use data (INSPIRE Directive,

© Springer Nature Switzerland AG 2019
S. Avdaković (Ed.): IAT 2018, LNNS 60, pp. 93–103, 2019.
https://doi.org/10.1007/978-3-030-02577-9_11

Annex III, topic number 4) [1]. The land cover data set consists of a collection of land cover units. These units can be geometrically represented by points, polygons or raster cells (which results with two basic models, one for vector data, and one for gaps). The land cover data set is also linked to the code list (e.g. CORINE Land Cover code list). CORINE Land Cover, as well as most regional and national land cover data, can be displayed using the model [1].

Land Cover is one of the main themes of any topographic map, regardless of its scale. However, the collection of Land Cover data in our country is problematic. Namely, the competent institutions do not consider the land cover as indicated above, but it is divided into a number of regulations on the classification of agricultural land, forest land and forests, urban land, etc. There are no Land Cover-Vegetation Maps that could provide the necessary information to understand the current situation. Monitoring changes in nature based on data collected in this manner requires data on vegetation from a few years. There are no Land Cover charts available that can serve as a measure of urban growth, water quality modeling, forecasting and assessment of flood effects and storm storms, wetland monitoring and potential damages from sea level growth, monitoring of changes in land cover with environmental impacts or creating links with socioeconomic changes such as increasing the number of residents [2].

In view of the above, when creating the Topographic Information System (TIS), it was not possible to apply any rulebook that could cover the necessary Land Cover data. The problem is solved by overcoming the CORINE Land Cover classification. For the territory of Bosnia and Herzegovina (B&H), the CORINE project was implemented in two installments (2000 and 2006), so there are certain data sets that can be used for making topographic maps [3]. The basic objective of the European Environment Agency (CORINE) is to collect data on Land Cover and Land Use for the territory of Europe, applying a unique methodology for data collection [4].

Another problem is present, and occurs when assessing spatial data quality for making topographic maps. There is a general consensus that quality is a subjective term, and each user decides if a set of data possesses good quality when it meets expectations, that is, that it uses the purpose. Unfortunately, data producers can not foresee the expectations of all future users [5]. Therefore, it is recommended to use ISO 8402: 1994 [6] which "fully characterized by a product that can meet the required or implied end user needs." In the domain of geospatial data, it includes numerous measurements including spatial accuracy, time accuracy, consistency, completeness, scope, and attribute accuracy.

2   Assessment of Spatial Data

Unlike the situation of tens of years ago when digital space data did not exist or were unavailable, today's problem is becoming more and more to find the data that will meet our needs. Information on the quality of these data from which it is possible to assess the extent to which this data corresponds to the needs of the user is very difficult and sometimes impossible to find. Due to the lack of information, expertise and the cost of implementing quality control, few years ago the quality control was not given the necessary attention [7].

Assessment of quality of spatial data (i.e. appropriate topographic maps) stored in the database in B&H has not yet gained in importance. Principles for the accuracy of analogue topographic maps are generally applied. So the assessment of the accuracy of topographic maps is still divided into two quality elements:

1. accuracy of general information and
2. geometric accuracy.

The accuracy of general information is, in practice, very difficult to define because it is "impossible to find a mathematical expression for the accuracy of general information. Only the wrong number can be determined with respect to the total number of data of the same type. The mistake of generalization is also very difficult to formulate with a mathematical formula that would quite objectively evaluate the quality of work" [8]. Geometric accuracy was determined by an appropriately selected topographic map scale.

Given the current technology, and the expertise of staff who collect, process and visualize spatial data, it is apparent that we lack the rules for assessing the accuracy of digital spatial data. Therefore, it is necessary to define the elements of the assessment of the quality of digital spatial (topographic) data to enable efficient exchange and integration of data from different sources in order to create new values (e.g. new topographic products) [9]. Similarly, the accuracy of analogue topographic maps can be estimated here and it can generally be divided into - accuracy of general information and geometric accuracy. However, the elements of the accuracy of the general information are more numerous than it was for analogue topographic maps.

In the world, the problem of defining spatial data quality elements has been noticed much earlier than in B&H, so the CEN (Center Europeen des Normalisation, 1992) accepted the five elements that are in the NIST norm. In addition, the United Nations Environment Program (UNEP) has developed a classification of spatial data quality [10].

According to Guptill and Morrison [10] there are seven elements of spatial data quality:

1. origin of data
2. positioning accuracy
3. attribute accuracy
4. completeness of the data
5. logical consistency
6. semantic accuracy
7. time accuracy/information.

2.1   Origin of Data

In the analogous time of creating cartographic products with field data, additional documents such as drafts and technical reports were submitted. With the arrival of digital technology information about the origin of data falls into the shadows. So with the digital record we do not get more information about the actual source of data. Therefore, it is necessary that a digital metric record creates a set of metadata containing:

1. source (person, institution, investor, date of data collection, quality of data downloaded)
2. reference surface (area to which the data are mathematically supported)
3. spatial data attributes (scale, resolution, accuracy and precision)
4. coordinate systems (used to determine the position of objects in space)
5. cartographic projection (for the source map - it is necessary to know in which projection it was made)
6. spatial data corrections (number and type of correction performed over the collected data, e.g. atmospheric, radiometric correction, digitization scales, etc.)
7. used transformations
8. format in which data is kept etc. [9].

2.2   Position Accuracy

Location, attributes, and other data about a real object are nowadays commonly stored in databases. This way they are available to current and future users. What matters is that these facilities must meet the following requirements:

1. they are uniquely defined in the defined coordinate system
2. that they are classified (object classes within a particular theme)
3. define geometry (point, line, polygon)
4. define metadata on:
   a. start/end time of the life cycle of the object (date of entry into the database/update, deletion),
   b. methods of data collection,
   c. data storage methods,
   d. completeness i.e. the amount of data of a given set of objects represented in the database in relation to the actual number,
   e. topological correctness,
   f. the quality of the data collected and
   g. output data quality [9].

2.3   Attribute Accuracy

The attribute is a fact about a place, a collection of places or an object on the Earth's surface. Spatial resolution is unconditionally included in the definition of some point attributes such as population density or Land Cover of an area within which observation and a certain environment are derived. While each attribute in the spatial data base is by definition associated with a point, line, or surface, the attribute resolution process and its relationship to the geometry of the dot, line, or surface may be complex and in some cases unknown. Geometric rendering of an object is not sufficient to determine procedures that lead to object property uncertainties.

The attribute quality directly relates to the concept of uncertainty of attribute data. Undesirableness is defined as the measure of the range of the attribute value that can be derived as a result of repeated measurements, measurements by alternative instruments or methods, or repeated interpretations of various observers, alternative interpretations of processing by alternative algorithms, etc. [10].

2.4   Completeness

Completeness is one of the five quality data elements defined by the National Committee for Digital Cartographic Data Standards (NCDCDS) as a standard for digital map data [11]. The NCDCDS norm defines completeness as an attribute describing the relationship between objects stored in the data set and the abstract universe of all objects [10].

Data set (e.g. map) for some application may be complete or incomplete. Therefore, we need to distinguish data quality and suitability for use, i.e. two types of completeness: completeness of data (caused by omission error and measurable data quality element) and completeness of model (which is an aspect of convenience for use).

2.5   Logical Consistency

Logical consistency deals with the logical rules of the structure and spatial attributes and describes the compliance of some data with other data in the set (database). Depending on the structures used, different methods for testing logical consistency in spatial data sets can be applied. In addition, there are tests of the consistency of database attributes, metric and affiliation tests, topological tests and relationship design tests and consistency checks [9].

2.6   Semantic Accuracy

Semantic accuracy is one of the elements of the quality of digital spatial data sets. Semantics is a branch of linguistics that studies the meanings, changes in meaning, and rules that determine relationships between sentences or words and their meanings. More generally, semantics is a study of the relationship between signs and symbols and what they mean. Semantic accuracy refers to the quality of the description of geographic objects according to the chosen model. Semantic accuracy includes concepts commonly known as completeness (omitted data or emission, completeness), consistency (valid semantic constraints), actuality (changes over time) and attribute accuracy (accuracy of qualitative or quantitative attributes) [12].

2.7   Time Information

Time information when data elements are collected or revised are an important factor for data quality judgment. Although users often want the latest information, historical information is needed (information about changes over time) needed to study any processes. The spatial data collection method affects the time information. For example, it differentiates the date of aero-recording from the date of registration of the right of ownership [11].

2.8   The Matrix Quality Assessment Data

The need for different visualization of data also requires different modeling of source data. Any matrix can be accepted in model quality estimates [13].

In order to visualize the real world, in a certain scale and style, we have to carry out conceptualization and then visualization (visualization can be different, depending on the needs of the courthouse, e.g. topographic view, thematic view of real world data). In this case, we select the objects from the real world and the way we will visualize it:

1. Space (objects in space),
2. Time (type of time object) or
3. Attribute (e.g. use of color for vegetation type, road rank, etc.).

This is followed by a data matrix. It is also possible to create a matrix data model:

1. Quality model (clarity, precision, completeness, consistency, resolution) and
2. Data quality (accuracy, completeness, consistency, resolution).

The data matrix model encompasses uncertainty of relational connections. The quality model refers to the quality of representation of the complex reality from which the data is collected. The quality model includes the assessment of the appropriate spatial representation of objects and the level of detail presented [9].

3   Land Cover Spatial Data Quality Estimation

After accepting the Strategy by the Federal Administration for Geodetic and Real Property Affairs (hereinafter FGA) a new Topographic Information System (TIS) of Federation Bosnia and Herzegovina (FB&H) was created. TIS was created in compliance with INSPIRE specifications, with certain deviations in order to comply the legal regulations of FB&H. The connection to Real Estate Cadaster Database (RECDB) is defined using the methodology for the establishment of Basic Topographic Database (BTD) based on the pilot project "BTD Bosnia-Podrinje Canton" [14].

One of the conclusions that came out of the pilot project is that the Land-Cover theme is overwhelmingly overlapping with the Land Use theme, and the data are timeless and incomplete. They are unexpected because they use data that owners themselves report for land use tax payments (which after 1995 are not binding) and are incomplete because they compare cadastral data and data from other sources (which is why there is a dilemma that is current and reliable). The problem arises when attempts are made to obtain data from institutions that collectively collect, process and store data on Land Cover because there are no uniform criteria for data collection. The aforementioned shortcomings were attempted to overcome by comparing data from CORINA Land Cover for B&H, but there is a problem as the dissolution of the data is different.

Since TIS recommends the CORINE classification for the Land Cover theme, a further pilot project (for the purpose of this article) has been implemented, the basic task of which is to evaluate the Land Cover theme data quality for the Canton Sarajevo area. For the purpose of this pilot project, data was collected for the purposes of the study: "List of conditions, development of database for coverage and use of land for Sarajevo Canton in GIS technology" and other available data for Land Cover for the Sarajevo area.

3.1 Estimate the Accuracy of the Data from RECDB and CLC

The accuracy of the data captured by RECDB for the BTD Land Cover needs was made by spatial and attributive comparison of two sets of data: cadastral data on land use and land plot data from the Study.

The data from the Land Cover Study are downloaded as "as such" and are not further processed because they are all necessary processing done during the Study. The database contains landline attributes by Corine Land Cover (CLC) standard, surface and geometric component polygon (region). The CLC code list for the EU Inspire standard contains three levels, while the code list formed from the Land Cover Study report also contains the 4th level of division.

Following the recommended spatial accuracy assessment matrix, it was necessary to determine whether data from these two sets of data could be compared and analyzed at all. For this purpose, basic data source settings have been compared (see Table 1a and b).

From Table 1a and b it can be concluded that data is taken from two institutions, using the same mathematical basis and same geometry, but collected by different geodetic methods and different accuracy.

Table 2 shows placement accuracy of the data through the unambiguity of the data, their classification and the geometry definition. One of the most important tasks in this pilot project was the testing and analysis of data classification by Land Use attributes and Land Cover attributes, i.e. the rule of linking the attribute values of two different code lists. The classification model is made by using logical attribute comparison methods at a number of available CLC markers, while for certain doubly coupled items the reversibly used and original results obtained by geometric cross-section of two spatial databases but not by breaking the rule of logical selection. Although at first glance the same code list looks similar, it has been shown that the fourth level of the CLC code list represents pre-detailed elaboration of the attribute for linking the cadastral code of land use. Still, the general conclusion is that over 60% of object classes match.

Table 3 shows the evaluation of the accuracy of the general information i.e. metadata selected two sets of data. It can be seen from the data sets that data on the life cycle of data (at least the beginning of the life cycle - in the Study). Data on data collection methods and data storage are also provided.

About the accuracy of the attribute, semantic accuracy has no data[1]. There is also no estimate of fully collected data. Topological examinations of the FGA are carried out in a regular review of cadastral parcels, while a database for the Study was required to comply with a small number of topological rules. It was also found that both sets of data satisfy the required accuracy of BTD 1: 10000.

---

[1] Although it can be assumed that there were errors detected and their correction.

Table 1. Data from two data sources

(a)

		DATA SOURCE					
Institution		Federal Administration for Geodetic and Real Property Affairs (FGA)					
DATA SOURCE — Department for Planning the Development of Canton Sarajevo (DPDCS)			Continuous data collection	One-time data collection	Reference ellipsoid (BESEL 1841)	Coordinate System (EPSG 31276)	Cartographic Projection (Transfers Mercator)
		Continuous data collection	FGA				
		One-time data collection		DPDCS			
		Reference ellipsoid (BESEL 1841)			FGA/ DPDCS		
		Coordinate System (EPSG 31276)				FGA/ DPDCS	
		Cartographic Projection (Transfers Mercator)					FGA/ DPDCS

(b)

		DATA SOURCE					
Institution		Federal Administration for Geodetic and Real Property Affairs (FGA)					
DATA SOURCE — Department for Planning the Development of Canton Sarajevo (DPDCS)			Land Use	Land Cover	Primary data collection	Secondary data collection	Digital data features (scale, resolution, precision, etc.)
		Land Use	FGA				
		Land Cover		DPDCS			
		Primary data collection			FGA		
		Secondary data collection				DPDCS	
		Digital data features (scale, resolution, precision, etc.)					FGA/ DPDCS

Table 2. Positional accuracy

POSITIONAL ACCURACY OF DATA		POSITIONAL ACCURACY OF DATA			
	Institution	Federal Administration for Geodetic and Real Property Affairs (FGA)			
	Department for Planning the Development of Canton Sarajevo (DPDCS)		Data are un-ambiguous	Data are classified	Data geometry is defined
		Data are un-ambiguous	Yes		
		Data are classified		Yes	
		Data geometry is defined			Yes

Table 3. Metadata

METADATA		METADATA								
	Institution	Federal Administration for Geodetic and Real Property Affairs (FGA)								
	Department for Planning the Development of Canton Sarajevo (DPDCS)		Data Lifecycle	Defined collection methods	Defined methods of keeping	Attribute Accuracy	Semantic accuracy	Completeness	Topological Correctness	Output data quality
		Data Lifecycle	Yes							
		Defined collection methods		Yes						
		Defined methods of keeping			Yes					
		Attribute Accuracy				No				
		Semantic accuracy					No			
		Completeness						No		
		Topological Correctness							Yes	
		Output data quality								Yes

4   Conclusion

The pilot project tried to give the Land Cover (Use) accuracy estimates for two sets of data taken from two different institutions - the Federation Geodetic Agency of Bosnia and Herzegovina and the Department for Planning of the Development of Canton Sarajevo. Given that the data was collected for different purposes: for the attribution of Land Use cadastral parcels (FGA) and for Land Cover inventory in Sarajevo Canton (PDCS), they were collected by various geodetic methods, and data of different accuracy and different classifications was collected. The question arises: is it possible to compare these two sets of data at all and whether they can be used for the purposes of the Basic Topographic Database (BTD)? After a detailed examination, it was found that the classification of the object classes corresponds to over 60% of the cases. The same Reference ellipsoid (BESEL 1841), Coordinate System (EPSG 31276) and cartographic projection (Transfers Mercator) were used. Thus, both sets of data have a common mathematical basis and the data are uniquely determined. It has also been found that Land Use and Land Cover themes have the same defined geometry (polygon). Land Use and Land Cover placement accuracy (regardless of the different geodetic data collection methods) is greater than the required accuracy for BTD.

When referring to the accuracy of the general attributes, we must say that the information on attribute accuracy, semantic accuracy, and completeness is missing. Metadata on collection, processing, and data retention methods are available for both data sets. Data Lifecycle Information also exists for both data sets, with Land Cover data only having the beginning of a lifecycle since the data is only collected once, while for Land Use data there are information on the start/change/end of the life cycle. Logical Consistency Data (only available for Land Use) and topological data testing were not available for both sets of data in the same way (e.g. a higher number of topological rules is prescribed for Land Use than for Land Cover).

From all of the above, it can be concluded that in Bosnia and Herzegovina there is a tradition of estimating topographic (spatial) data quality, with certain additional metadata estimates lacking in certain additional metadata estimates - attribute accuracy, semantic accuracy and completeness. Since there are no clear rules for testing the topological accuracy of data sets (each institution is governed by its own rules) it would be necessary to make certain specifications on which to avoid it.

References

1. INSPIRE theme register-Land cover. http://inspire.ec.europa.eu/theme/lc. Accessed 2 Mar 2018
2. NOAA. What is the difference between Land cover and Land use? https://oceanservice.noaa.gov/facts/lclu.html. Accessed 2 Mar 2018
3. Taletovic, J., Đuzo, F., Vojnikovic, S., Ljuša, M., Custovic, H.: Basic principles, a methodological approach CORINE Land Cover in B&H and Analysis of results CLC2000 and CLC2006. Union of Associations of Geodetic Professionals in Bosnia and Herzegovina. Geodetski glasnik, vol. 45, no. 42, Sarajevo, Bosnia and Herzegovina (2012)

4. European Environment Agency. CORINE Land cover. Publication. https://www.eea.europa.eu/publications/COR0-landcover. Accessed 5 Mar 2018
5. Blower, J., Masó, J., Díaz, D., Robert, C., Griffiths, G., Lewis, J., Yang, X., Pons, X.: Communicating thematic data quality with web map services. ISPRS Int. J. Geo-Inf. **4**, 1965–1981 (2015). https://doi.org/10.3390/ijgi4041965, ISSN 2220-9964, www.mdpi.com/journal/ijgi/. Accessed 25 Mar 2018
6. International Organization for Standardization. Quality management and quality assurance. https://www.iso.org/standard/20115.html. Accessed 22 Feb 2018
7. Divjak, D., Baricevic, V.: The Role of Quality Control in Creating Spatial Data Infrastructure (2011). http://www.kartografija.hr/3nipp_sazetci_pregled.hr/items/15.html. Accessed 2 Apr 2018
8. Peterca, M., Radoševic, N., Milisavljevic, S., Racetin, F.: Cartography. Military Geographic Institute, Belgrade. The Socialist Federal Republic of Yugoslavia (1974)
9. Kljucanin, S., Posloncec-Petric, V., Bacic, Ž.: Basic of Spatial Data Infrastructure. Dobra knjiga, Sarajevo, Bosnia and Herzegovina (2018)
10. Guptill, S.C., Morison, J.L.: Spatial data quality elements In: Lapaine, M. (ed.) Spatial Data Quality Elements. The State Geodetic Administration of the Republic of Croatia, Zagreb (2001). Translation of the book Guptill, S.C., Morrison, J.L. (eds.): Elements of Spatial Data Quality (1995)
11. Mollering, H.: A Draft Proposed Standard for Digital Cartographic Data. National Committee for Digital Cartographic Standards, American Congres on Surveying and Mapping Report #8 (1987). https://pubs.usgs.gov/of/1987/0308/report.pdf. Accessed 2 Apr 2018
12. Salge, F.: Sematic accuracy In: Lapaine, M. (ed.) Spatial Data Quality Elements. State Geodetic Administration of the Republic of Croatia, Zagreb (2001). Translation of the book Guptill, S.C., Morrison, J.L. (eds.): Elements of Spatial Data Quality (1995)
13. Haining, R.: Spatial Data Analysis: Theory and Practice. Cambridge University Press (2003, 2004)
14. Kljucanin, S.: The new topographic information system and establishing the basic topographic database of the Federation of Bosnia and Herzegovina. In: Advanced Technologies, Systems, and Applications II: Proceedings of the International Symposium on Innovative and Interdisciplinary Applications of Advanced Technology (IAT). Springer (2018). ISBN-13: 978-3319713205, ISBN-10: 3319713205

Determining Effective Stresses in Partly Saturated Embankments

Haris Kalajdžisalihović[✉], Hata Milišić, Željko Lozančić,
and Emina Hadžić

Faculty of Civil Engineering,
Department of Water Resources and Environmental Engineering,
University of Sarajevo, Sarajevo, Bosnia and Herzegovina
kahariss@gmail.com, hata.milisic@gmail.com,
zeljko.lozancic@gmail.com, eminahd@gmail.com

Abstract. Different layers of heterogeneous materials inside embankments lead the mentioned materials into different state of stresses. On the hypothetical example of the embankments made of clay and drainage layer, the calculation of the effective stresses in the body of the embankment with the drainage layer will be shown. Within the framework of the presented research, a model was created in the Geo Studio environment that performs a water filtration budget in an unsaturated environment, based on the Finite Element Method for non-linear 2D Laplace equation. The results of the model show the distribution of effective stresses and displacements for embankments loaded only by their own weight.

Keywords: Flow in unsaturated zone · Finite element method
Effective stresses · Laplace equation · Ideally elastic soil behavior

1   Introduction

Gravity acceleration is one of the most important factor for the movement of water through the porous material and the deformation of rock and soil mass. Aforesaid applies also to embankment dams that will be the subject of study in this paper. One of the aims is to analyze impact of changes in hydraulic gradient on the sensitivity of solutions of the potential failure of heterogeneous embankment slope.

For saturated/unsaturated groundwater flow, the stand alone "FLUID" model in the Geo Studio soft-ware was used, based on finite element method (FEM), which calculates hydraulic gradients, pressures and saturations. In addition, "SOLID" module is used, based on FEM, on the same computational network scheme and in the same software package as "FLUID" model. "SOLID" model results are displacements, strains and stresses distribution. Solutions of the "FLUID" model are used as input in the "SOLID" model. Solving the model to the previously mentioned method is called "UNCOUPLED" model.

In this paper impacts of changes in hydraulic gradients and effective stresses depending on the location of the drainage layer are shown.

S. Avdaković (Ed.): IAT 2018, LNNS 60, pp. 104–112, 2019.
https://doi.org/10.1007/978-3-030-02577-9_12

2 Mathematical Model

Some required assumptions for fluid and soil behavior. Gaseous phases will not be processed, as well as processes between the phases. The isothermal processes for liquid and solid phase are assumed. Density and viscosity of water is constant value. Solid phase is isotropic heterogeneous media, elastic model of constitutive law, as well as in terms of hydraulic conductivity.

2.1 Decoupling Fluid-Solid Equation

This section will present decoupled equations of fluid flow, for motion of homogenous, incompressible fluid in a heterogeneous, ideally elastic soil. Equation are given for 2D problem, but problem could be spatially generalized for 3D:

$$m - m_0 = \frac{3\rho_0(v_u - v)}{2GB(1 + v_u)}\left[(\sigma_{11} + \sigma_{yy}) + \frac{3p^*}{B(1 + v_u)}\right] \tag{1}$$

$$p^* = \rho_w gh = p + \rho_w gy \tag{2}$$

Conservation of mass of fluid for transient motion as follows:

$$\frac{\partial q_i}{\partial x_i} + \frac{\partial m}{\partial t} = 0 \tag{3}$$

Substituting (1) and (2) to (3):

$$\frac{1}{\rho_0 g}\frac{\partial}{\partial x_i}\left(K\frac{\partial p^*}{\partial x_i}\right) = \frac{3(v_u - v)}{2GB(1 + v_u)}\frac{\partial}{\partial t}\left[(\sigma_{xx} + \sigma_{yy}) + \frac{3p^*}{B(1 + v_u)}\right] \tag{4}$$

where:

p - hydrostatic pressure,
h - head,
ρ_0, ρ_w - mass density of water,
g - gravitational acceleration,
K - coefficient of permeability,
B - Skempton pore pressure coefficient,
G - shear modulus,
σ_{xx}, σ_{yy} - total stresses in x and y,
v_u - undrained Poisson ratio,
m - mass of pore fluid per unit volume of medium,
m_0 - constant value of m measured at some referent state.

In the case of porous medium steady state fluid flow, the right hand side of the Eq. (4) is zero. Hence, for steady state flow, fluid pressures are not related to stresses and Laplace's Eq. (5) could be solved, unlike in transient conditions.

2.2   Mathematical Model of Fluid Motion

First assumption is a very slow movement of fluid through a porous medium. For this approach Darcy's law can be applied. Substituting Darcy's law in conservation of mass of fluid we obtain Laplace's equation:

$$\frac{\partial}{\partial x}\left(K\frac{\partial h}{\partial x}\right) + \frac{\partial}{\partial y}\left(K\frac{\partial h}{\partial y}\right) = 0 \tag{5}$$

In unsaturated soils we can make decoupling of coefficient of permeability on relative (which is related to capillary ability) and saturated one:

$$\frac{\partial}{\partial x}\left(K_sK_r(h)\frac{\partial h}{\partial x}\right) + \frac{\partial}{\partial y}\left(K_sK_r(h)\frac{\partial h}{\partial y}\right) = 0 \tag{6}$$

where:

K_s - saturated coefficient of permeability (fixed)
$K_r(h)$ - relative coefficient of permeability.

Material conductivity-pressure function is used. Various authors have contributed to the development of these functions, Maulem, Green, Brook-Corey, Fredlund-Xing, Van Genuchten. The last mentioned author derived a closed form equations for saturation and hydraulic conductivity. In this paper the Van Genuchten model is shown.

Model Van Genuchten (1980)

Van Genuchten [1] provided the relative conductivity, Kr, as a function related to negative pore pressure, with corresponding adjustable parameters, a, n, m, which are the result of experiments:

$$K_r = \frac{\left[1 - \left(a\psi^{(n-1)}\right)(1 + a\psi^n)^{-m}\right]^2}{\left[((1 + a\psi)^n)^{\frac{m}{2}}\right]} \tag{7}$$

where m is calculated as:

$$m = \frac{n-1}{n} \tag{8}$$

where:

a, n - coefficients due to material,
ψ - negative pore water pressure.

2.3   Mathematical Model of Solid

Strains and Compatibility

Relative displacements caused by external and internal loads are given to:

$$\varepsilon_{xx} = \frac{\partial u_x}{\partial x} \tag{9a}$$

$$\varepsilon_{yy} = \frac{\partial u_y}{\partial y} \tag{9b}$$

$$\varepsilon_{yx} = \frac{1}{2}\left(\frac{\partial u_y}{\partial x} + \frac{\partial u_x}{\partial y}\right) \tag{9c}$$

where:

u_x and u_y: displacements of solid,
ε_{xx} and ε_{yy}: axial strains of solid,
ε_{xy}: shear strain.

By differencing these Eqs. (9a), (9b) and (9c) gives kinematical continuity equation:

$$\frac{\partial^2 \varepsilon_{xx}}{\partial y^2} + \frac{\partial^2 \varepsilon_{yy}}{\partial x^2} = 2\frac{\partial^2 \varepsilon_{yx}}{\partial x \partial y} \tag{10}$$

Stress Equilibrium

In steady-state flow motion in porous media stresses are also steady state, so the equations take the form of the traditional Cauchy continuum equations of static.

Effective Stresses

The definition of effective stress is given as the following mathematical expression:

$$\sigma'_{ij} = \sigma_{ij} + \alpha p \delta_{ij} \tag{11}$$

where:

σ'_{ij} - effective stress,
σ_{ij} - total stress,
δ_{ij} - Kroneker symbol,
α - coefficient.

Various authors (Terzaghi [2], Passman and McTigue [3] and many others) gave coefficient α that depends of the module of elasticity of the entire porous medium compared to module of elasticity of solid particles, and some additional equations that could be applied for transient analysis. For steady state analysis, coefficient α take the value $\alpha = 1$.

Effective Stress Equilibrium

Substitution of static equilibrium stress (11) in Cauchy equations, and add (12):

$$h = \frac{p}{\rho_w g} + y \tag{12}$$

results:

$$\frac{\partial \sigma'_{xx}}{\partial x} + \frac{\partial \sigma'_{xy}}{\partial y} = \alpha \rho_w g \frac{\partial h}{\partial x} \tag{13a}$$

$$\frac{\partial \sigma'_{yy}}{\partial y} + \frac{\partial \sigma'_{xy}}{\partial x} = (\rho_s - \rho_w)g + \alpha \rho_w g \frac{\partial h}{\partial y} \tag{13b}$$

where:

ρ_s - bulk density of soil-fluid mixture.

As in last mentioned Eqs. (13a) and (13b) there are two equations and three variables. The exact solvable system of equations is determined using stress-strain relation. In this research elastic constitutive law is used.

Terzaghi defined that the strains depend only on effective rather than total stresses. For Biot's equations [4], for the selected type of Hooke's law, relationships between deformation and the effective stresses are:

$$\varepsilon_{xx} = \frac{1}{E} \left[(1 - v^2)\sigma'_{xx} - v(1 + v)\sigma'_{yy} \right] \tag{14a}$$

$$\varepsilon_{yy} = \frac{1}{E} \left[(1 - v^2)\sigma'_{yy} - v(1 + v)\sigma'_{xx} \right] \tag{14b}$$

$$\varepsilon_{xy} = \frac{1 + v}{E} \sigma'_{xy} \tag{14c}$$

where:

v - Poisson ratio
E - elasticity.

The first equation of compatibility of effective stresses we get by combining (14a)–(14c) and Eq. (10), and second one by differentiating Eq. (13a) with respect to x and (13b) with respect to y. Combining these two equations we obtain the compatibility equation of effective stresses:

$$\left(\frac{\partial^2}{\partial x^2} + \frac{\partial^2}{\partial y^2} \right) (\sigma'_{xx} + \sigma'_{yy}) = \frac{1}{1 - v} \left[\rho_w g \left(\frac{\partial^2 h}{\partial x^2} + \frac{\partial^2 h}{\partial y^2} \right) + g \frac{\partial \rho_s}{\partial y} \right] \tag{15}$$

Displacements

Equations (13a), (13b) and (15) are complete and solvable, but it is simpler to model the problem of displacement. The equations given by (9a)–(9c) and substituted in (14a)–(14c):

$$\sigma'_{xx} = \frac{E}{1 + v} \left(\frac{\partial u_x}{\partial x} \right) + \frac{vE}{(1 - 2v)(1 + v)} \left(\frac{\partial u_x}{\partial x} + \frac{\partial u_y}{\partial y} \right) \tag{16a}$$

$$\sigma'_{yy} = \frac{E}{1+v}\left(\frac{\partial u_y}{\partial y}\right) + \frac{vE}{(1-2v)(1+v)}\left(\frac{\partial u_x}{\partial x} + \frac{\partial u_y}{\partial y}\right) \tag{16b}$$

$$\sigma'_{xy} = \frac{E}{1+2v}\left(\frac{\partial u_x}{\partial y} + \frac{\partial u_y}{\partial x}\right) \tag{16c}$$

If this Eqs. (16a), (16b) and (16c) are substituted in the Eqs. (13a) and (13b):

$$\frac{E}{2+2v}\nabla^2 u_x + \left[\frac{vE}{(1-2v)(1+v)} + \frac{E}{2+2v}\right]\left(\frac{\partial^2 u_x}{\partial x^2} + \frac{\partial^2 u_y}{\partial x \partial y}\right)$$
$$= \rho_w g \frac{\partial h}{\partial x} \tag{17a}$$

$$\frac{E}{2+2v}\nabla^2 u_y + \left[\frac{vE}{(1-2v)(1+v)} + \frac{E}{2+2v}\right]\left(\frac{\partial^2 u_y}{\partial y^2} + \frac{\partial^2 u_y}{\partial x \partial y}\right)$$
$$= (\rho_s - \rho_w)g + \rho_w g \frac{\partial h}{\partial y} \tag{17b}$$

3 Numerical Model

3.1 Numerical Model for Fluid Motion

First approximation is linearization of Nonlinear Laplace's equation and solve it by the finite element method (FEM). At the required tolerance of solutions (in this paper, 0.01 m) is performed by repeating the procedure linearization, and in every step model solves FEM for a given network of elements. When tolerance is satisfied, procedure stops. In post-processing part, model calculates hydraulic gradients.

3.2 Numerical Model for Solid

Displacements are defined by Eqs. (17a) and (17b). Equation is linear, and, due to selected constitutive law, the model solves them in a single step.

3.3 Example

Network is triangular with linear finite elements. There are 997 elements connected to 559 nodes. Dark presented (brown) elements represent clay and the bright (yellow) sandy material. Boundary conditions are: constant level of water on the left, upstream, side is 6 m, and the right, downstream 2.5 m.

It is important to note, when calculating "UNCOUPLED" model, network of elements in both models, "FLUID" and "SOLID", must be identical, or we need to interpolate values of gradients in "solid" model nodes.

4   Numerical Results

In the hypothetical example of the embankment, i.e. the cross sections, dimensions are: in bottom 21 m, 5 m wide on top with a 1:1 slope, composed of two materials: clay and the drainage layer (Fig. 1). Results of calculations are shown in (Figs. 2, 3, 4, 5 and 6). The selected material parameters are as follows (Table 1):

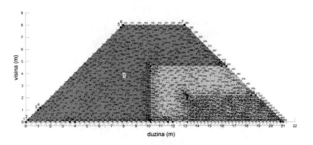

Fig. 1. Cross section of hypothetical sample of embankment. Network of elements is shown.

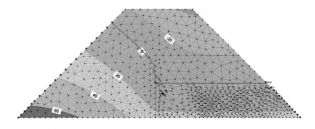

Fig. 2. Iso pressure diagram based on solved «FLUID» model by FEM.

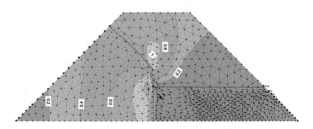

Fig. 3. Iso saturation diagram based on solved «FLUID» model by FEM

Results of effective stresses are shown directly, as average of sum of two mean effective stresses (Fig. 6).

Values of pressures (Fig. 2), i.e. the level of water table, while going down the pressures increase, and upwards decrease, saturations (Fig. 3).

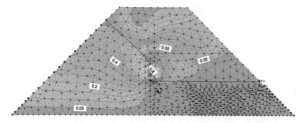

Fig. 4. Iso failure potential diagram solved in «SOLID» model by FEM. Load that include is only own weight of soil.

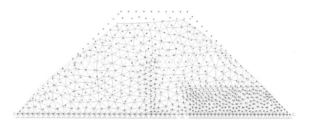

Fig. 5. Deformation diagram solved in «UNCOUPLED» model by FEM.

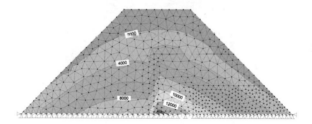

Fig. 6. Mean effective stresses in Uncoupled model.

Table 1. Values of parameters of materials

Material	E (kPa)	v	α	n	θr	θs
Hygiene sandstone	5000	0.3	0.152	1.17	0.02	0.46
Beit Netofa Clay	10000	0.2	0.79	10.40	0.14	0.37

5 Conclusion

Locating the drainage layer downstream facilitates impact on: smaller value of buoyance and smaller vertical displacements on downstream side of embankment. It implies smaller value of failure potential in that area.

References

1. van Genuchten, M.T.: A closed form equation for predicting the hydraulic conductivity of unsaturated soils. Soil Soc. Am. J. **44**, 892–898 (1980)
2. Terzaghi, K.: Die berechnung der Durchassigkeitsziffer des Tones aus dem Verlauf der hydrodynamischen Spannungsercheinungen. Sitzungsber. Akad. Wiss. Wien Math. Naturwiss. Kl. Abt. 2A **132**, 105–124 (1923)
3. Passman, S.L., McTigue, D.F.: A new approach to the effective stress principle in Compressibility Phenomena in Subsidence. In: Saxena, S.K. (ed.) Engineering Foundation, New York, pp. 79–91 (1986)
4. Biot, M.A.: General theory of three dimensional consolidation. J. Appl. Phys. **12**, 155–164 (1941)

Different Possibilities for Modelling Cracked Masonry Structures

Naida Ademovic[1(✉)] and Marijana Hadzima-Nyarko[2]

[1] Faculty of Civil Engineering, Department of Materials and Structure,
University of Sarajevo, Sarajevo, Bosnia and Herzegovina
naidadem@yahoo.com
[2] Faculty of Civil Engineering, Department of Materials and Structure,
University J.J. Strossmayer in Osijek, Osijek, Croatia
mhadzima@gfos.hr

Abstract. The finite element method (FEM) is a numerical method used for solving different problems in engineering field. The starting point being a continuum media of the structure, meaning that the structure is undamaged. However, in order to be able to represent the actual stated of the structure, with all of its defects and cracks, it is necessary to incorporate these anomalies. This all with the aim that the model is a good representative of the real structure. This paper gives and overview of several numerical approached in the modelling of discontinuities on the masonry structures.

Keywords: Masonry · Modelling · Finite element model · Cracks
Discontinuities

1 Introduction

Material which has been used for ages and still is being used is masonry. This material is used today in many counties and frequently it has a loadbearing function, besides being used as a nonstructural element, or as infill material in buildings with reinforced concrete frames. This is different from country to country, depending on the history of masonry structures and construction techniques. Looking at the historical buildings made of masonry defects have to exist. Defects on masonry structures are of different degrees and levels, however the concept of defects is rather subjective. Different effects can cause cracking to occur (unit properties, climatic effects, thermal or moisture movement, foundation settlement, poor construction load transfer etc.). The most frequent source of masonry failure is really cracking. Connection between the causes and crack formation and its development of masonry structures has been investigated for the last 180 years and still is.

Today with the vast expansion of cities and construction of different underground structures has risen the awareness of the engineering community about the existing masonry structures. These underground activities pose risks to unreinforced masonry structures (URM). In order to determine the actual response of the building it is of the utmost importance to incorporate in the model the existing damages to the structure and then perform adequate analyses. If this is not done the output scenario would not reflect

© Springer Nature Switzerland AG 2019
S. Avdaković (Ed.): IAT 2018, LNNS 60, pp. 113–120, 2019.
https://doi.org/10.1007/978-3-030-02577-9_13

the actual state of the structure, possibly causing additional defects and possible collapse of the structure. Burlant et al. [1] after the damages which was observed on the structures founded on clay soils caused by severe droughts, classified damage of the masonry buildings into three categories in dependency of ease of repair: aesthetic, serviceability, and stability.

Several elements make the modelling procedure of masonry structures a rather challenging and difficult task. First of all, the constitutive model for masonry, as a complex heterogeneous anisotropic material, composed of masonry elements which can be of different shapes and materials (clay, stone, etc.), bonded or not bonded with mortar joints (lime, cement, etc.), usually is rather intricate. The properties of each material in any structure will vary. These variations may have an effect on the mechanical response to applied load or environmental changes (e.g. humidity and temperature) [2]. Additionally, for the existing structures the correct mechanical and physical characteristics have to be determined. If historical structures of cultural heritage are to be evaluated and assessed nondestructive and minor destructive tests have to be applied. One has to be careful with this as well, as the obtained data may not be enough to be representative. Unavoidable difficulty is determination of the actual loading level and distribution [3] as the building due to its long history has experienced different defects, cracks, causing redistribution of loads, degradation of material due to weathering and other actions. Load transfer and distribution may activate specific resisting phenomena (contact problems, friction, eccentric loading) [4].

2   Selection of a Modeling Approach

Due to the existence of numerous numerical methods it is important to know-how to select the appropriate method for a specific problem. It will depend upon the structure that is being analyzed; the balance between accuracy and simplicity; available input data; available financial resources and time, as well as the experience of the modeler [5].

The requirement of assessing exiting structure and predicting their in-service behavior led to development of various models of different levels of complexity. Here only some discrete methods trying to take into account the existing damages in the masonry structures will be briefly discussed indicating their advantages and disadvantages. The discrete element models that will be briefly elaborated are: Distinct element method, Discrete deformation analysis (DDA) and Non-Smooth Contact Dynamics (NSCD).

3   Discrete Element Methods (DEM)

The discrete element method was developed by Cundall in 1971 for jointed rock, which was modelled as an assemblage of rigid blocks. As stated by Cundall [6], and Cundall and Hart [7] a numerical technique is said to be a discrete element model if: it consists of separate, finite-sized bodies, so-called discrete elements, each of them being able to displace independently from each other, so the elements have independent degrees of freedom; the displacements of the elements can be large; and the elements can come into contact with each other and loose contact, and these changes of topology are automatically detected during the calculations. So, basically there are two elementary components: the elements, and the contacts between them.

In the discrete element method, a rock mass is represented as discrete blocks, and discontinuities are treated as interfaces between bodies. The contact displacements and forces at the interfaces are calculated by tracing the movements of the blocks. Applied loads or forces to a block system can cause disturbances that propagate and result in movements.

This approach found its successful application in the analysis of masonry arch bridges [8–10]. Most of these analyses took into account static loads or statically equivalent seismic loads. Additional research has to be done with the application of dynamic loads [11] and [12]. In all of the analysis it has been confirmed that the collapse modes are governed mechanism, where deformability of the elements is negligible.

Several methods have been developed, each one having its own peculiarities, benefits and disadvantages, which have shown to be suitable for solving problems including discontinuities.

3.1   Distinct Element Method

This method is developed form the original work done by Cundall and it represents an explicit method based on finite difference principles that can model complex, non-linear behaviors. The concept of this method is incorporated into of the commercial software UDEC and 3DEC, developed by Itasca Ltd. [13]. This software can perform 2D and 3D analysis, exposed to either static or dynamic loading. The discontinuous medium is represented as an assemblage of discrete blocks (deformable or rigid) while the discontinuities are treated as boundary conditions between blocks, interacting through unilateral elasto-plastic contact elements which follow a Coulomb slip criterion for simulating contact forces. The method is based on a formulation in large displacement (for the joints) and small deformations (for the blocks), and can correctly simulate collapse mechanisms due to sliding, rotations and impact. Large displacements along discontinuities and rotations of blocks can occur, the existing contact can be lost and the new one can occur.

Models may contain a mix of rigid or deformable blocks. Deformable blocks are defined by a continuum mesh of finite-difference triangular zones. These zones are actually continuum elements behaving according to a prescribed linear or nonlinear stress-strain law. The relative motion of the discontinuities is also governed by linear or nonlinear force-displacement relations for movement in both the normal and shear directions. Mortar joints are represented by zero thickness interface between the blocks. Contact is obtained by a set of point contacts models and a "soft approach" is assumed, meaning blocks can overlap when in compression. In this way the mesh between the blocks and joints is not required as well as no need for remeshing methodology to update the size of contacts when large relative displacements come upon [6].

The main advantage of this approach is the possibility of following the displacements and determining the collapse mechanism of structures made up of virtually any number of blocks [14]. This method has been used for various masonry applications, from masonry wall panels, masonry-infilled steel frames with openings, stone masonry arches and aqueducts and column-architrave structures under seismic actions and reliability for non-linear materials under large displacements. This method does not need a lot of computational memory and it is suitable for parallel processing [15]. It has been determined that the calibrated parameters depend on the scale of the structures, so they do not always give a reliable solution if the calibrated parameters are applied to models of different scale [16].

Application of recent research on masonry arch bridges utilizing this method is presented in [17], and in masonry wall panels [18], which showed good agreement with the experimental results (Fig. 1).

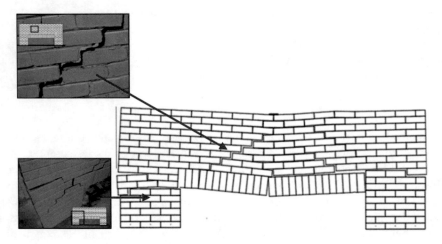

Fig. 1. Failure mode for panel predicted using UDEC [18].

3.2   Discontinuous Deformation Analysis (DDA)

Discontinuous Deformation Analysis was as well developed for solving problems in rock mechanics and geotechnical engineering. This method was proposed by Shi and Goodman [19] in 1984 and further developed by Shi [20]. On contrast to other discrete element techniques, DDA uses an implicit algorithm based on global stiffness. It is usually formulated as a work-energy method, using the principle of minimum potential energy. This permits equilibrium of blocks to be governed by the contact equations and considers friction. Contact are considered as rigid, so called "hard contact approach" meaning no interpenetration of blocks is allowed.

Here, as the methods takes into account the internal forces of the blocks' mass, large displacements and deformations are considered under dynamic loadings as well, besides static ones. Several extensions and updates were done to the original method proposed by Shi [20]. Three extensions were done by Lin et al. [21] (those being: improvement of block contact, calculation of stress distributions within blocks and block fracturing.

Application of this method in the last years has been seen in the analysis of masonry arch bridges. The first analyzed bridge by this method was the Mosca's bridge built in 1823 [22] (Fig. 2).

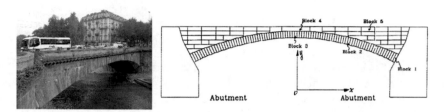

Fig. 2. (a) Mosca's bridge [23] (b) DDA model of Mosca's bridge [24]

At the beginning of the 21st century Thavalingam et al. [25], Bićanić et al. [26] indicated that this method is a good alternative for assessment structural integrity of masonry arch bridges and the prediction of the ultimate load and failure mode analysis. It was seen that backfill and the changes in the lateral stiffness have an important influence on the failure load [26].

3.3   Non-smooth Contact Dynamics (NSCD)

Jean et al. [27] introduced the Non-Smooth Contact Dynamics method or shortly contact dynamics (NSCD). It as well uses an implicit algorithm for dynamic equations and the Signorini relation (a complementary relation) as a non-smooth modelling of unilateral contact, and Coulomb law as a dry friction law. As this method uses few large time steps, it requires a substantial number of iteration at each time step.

The main four features of the standard NSCD method are firstly, that, for each candidate to contact, the relative velocity and the reaction force are related through a unilateral Signorini-like relation and a frictional Coulomb like relation. Secondly, linear relation is formed between the relative velocity and the reaction force through a linearized form of the dynamical equation. Thirdly, for each candidate to contact solution is obtained by the 'Signorini, Coulomb, standard'. Finally, updating from a candidate to the next one is done by the 'standard' solution [28]. Chetouane et al. [29]

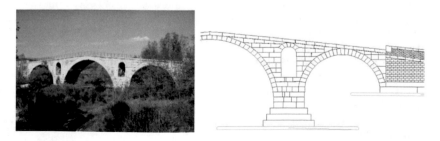

Fig. 3. (a) Pont Julien (France—1st century BC) [29] (b) NSCD model Pont Julien [29]

showed the benefits of the rigid approach as being faster in respect to the deformable one, but at the same time being less realistic. In the future, it is planned to produce a mixed rigid/deformable blocks approach. Application to masonry arch bridges was done be Chetouane et al. [29] and Acary et al. [30] (Fig. 3).

4   Comparison and Conclusions

Modelling of masonry has been attracted a great amount of research works in the few last decades. When modelling damaged masonry DEM (static and pseudo-static loading) has found its superior application in relation to finite element methods (FEM). One of the important advantages of DEM is its mesh independency and does not have convergence problems when large displacements are addressed. On the other hand generation of a model in DDA is rather complex, requires large computational time, high input data making it less favorable in the modelling procedure. As well, the disadvantage of DEM is as well computational effort and input material data. Non-Smooth Contact Dynamics method (NSCD) is not widely used in civil engineering problems. It application if we may say stated some 15 years ago. There is a need for further development in this field, validation and valorization of results.

References

1. Burland, J.B., Broms, B.B., de Mello, V.F.B.: Behavior of foundations and structures, state-of-the-art report. In: 9th International Conference on Soil Mechanics and Foundation Engineering II, Tokyo, Japan (1977)
2. Sarhosis, V., Oliveira, D.V., Lourenco, P.B.: On the mechanical behavior of masonry. In: Sarhosis, V., Bagi, K., Lemos, J.V., Milani, G. (eds.) Computational Modeling of Masonry Structures Using the Discrete Element Method, pp. 1–27. IGI Global (2016)
3. Lemos, J.V.: Modeling stone masonry dynamics with 3DEC. In: Konietzky, H. (ed.) Modeling Stone Masonry Dynamics with 3DEC, pp. 7–12. Taylor & Francis Group, London (2004)
4. Roca, P., Cervera, M., Gariup, G., Pela', L.: Structural analysis of masonry historical constructions. Classical and advanced approaches. Arch. Comput. Methods Eng. **17**, 299–325 (2010)
5. Lourenco, P.B.: Computations on historic masonry structures. Prog. Struct. Eng. Mater. **4**(3), 301–319 (2002)
6. Cundall, P.A.: A computer model for simulating progressive, large-scale movements in blocky rock systems. In: Proceedings of the International Symposium on Rock Fracture, Nancy, October 1971, vol. 1, paper no. II–8, pp. 129–136. International Society for Rock Mechanics (ISRM) (1971)
7. Cundall, P., Hart, D.: Numerical modelling of discontinua. J. Eng. Comput. **9**, 101–113 (1992)
8. Lemos, J.V.: Assessment of the ultimate load of a masonry arch using discrete elements. In: Middleton, J., Pande, G.N. (eds.) 1995 3rd International Symposium on Computer Methods in Structural Masonry, Lisbon, Portugal, pp. 294–302. Books and Journals International, Swansea (1996)

9. Tóth, A.R., Orbán, Z., Bagi, K.: Discrete element analysis of a stone masonry arch. Mech. Res. Commun. **36**, 469–480 (2009)
10. Kassotakis, N., Sarhosis, V., Forgács, T., Bagi, K.: Discrete element modelling of multi-ring brickwork masonry arches. In: 13th Canadian Masonry Symposium, pp. 1–11. Canada Masonry Design Centre, Halifax (2017)
11. Lemos, J.V.: Discrete element modelling of the seismic behaviour of stone masonry arches. In: Pande, G.N., Middleton, J., Kralj, B. (eds.) Computer Methods in Structural Masonry – 4, pp. 220–227. E&FN Spon, London (1998). Proceedings of the 4th International Symposium Numerical Methods Structural Masonry, STRUMAS IV, Florence, September 1997
12. DeLorenzis, L., DeJong, M.J., Ochsendorf, J.: Failure of masonry arches under impulse base motion. Earthq. Eng. Struct. Dynam. **36**(14), 2119–2136 (2007)
13. ITASCA: 3DEC - Universal Distinct Element Code Manual. Theory and Background. Itasca Consulting Group, Minneapolis (2004)
14. Cundall, P.A.: Formulation of a three-dimensional distinct element model. Part I: a scheme to detect and represent contacts in a system composed of many polyhedral blocks. Int. J. Rock Mech. **25**, 107–116 (1988)
15. Giordano, A., Mele, E., De Luca, A.: Modelling of historical masonry structures: comparison of different approaches through a case study. Eng. Struct. **24**, 1057–1069 (2002)
16. Alexandris, A., Protopapa, E., Psycharis, I.: Collapse mechanisms of masonry buildings derived by the distinct element method. In: Proceedings of the 13th World Conference on Earthquake Engineering, paper no. 548 (2004)
17. Sarhosis, V., Oliveira, D.V., Lemos, J.V., Lourenco, P.B.: The effect of skew angle on the mechanical behaviour of masonry arches. Mech. Res. Commun. **61**, 53–59 (2014)
18. Sarhosis, V., Sheng, Y.: Identification of material parameters for low bond strength masonry. Eng. Struct. **60**, 100–110 (2014)
19. Shi, G.-H., Goodman, R.E.: Discontinuous deformation analysis. In: Proceedings of 25th U. S. Symposium Rock Mechanics, pp. 269–271. SME/AIME, Evanston (1984)
20. Shi, G.H.: Discontinuous deformation analysis: a new numerical model for the statics and dynamics of block system. Ph.D. thesis. Department of Civil Engineering, University of California, Berkeley, CA (1988)
21. Lin, C.T., Amadei, B., Jung, J., Dwyer, J.: Extensions of discontinuous deformation analysis for jointed rock masses. Int. J. Rock Mech. Min. Sci. Geomech. Abstr. **33**(7), 671–694 (1996)
22. Ma, M.Y., Pan, A.D., Luan, M., Gebara, J.M.: Stone arch bridge analysis by the DDA method. In: Arch Bridges, pp. 247–256. Thomas Telford, London (1995)
23. https://commons.wikimedia.org/wiki/File:Ponte_Mosca_Torino.JPG
24. Ma, M.Y., Pan, A.D., Luan, M., Gebara, J.M.: Seismic analysis of stone arch bridges using discontinuous deformation analysis. In: 11th World Conference on Earthquake Engineering, paper no. 1551, pp. 1–8 (1995)
25. Thavalingam, A., Bićanić, N., Robinson, J.I., Ponniah, D.A.: Computational framework for discontinuous modelling of masonry arch bridges. Comput. Struct. **79**(19), 1821–1830 (2001)
26. Bićanić, N., Stirling, C., Pearce, C.J.: Discontinuous modelling of masonry bridges. Comput. Mech. **31**, 60–68 (2003)
27. Jean, M., Moreau, J.J.: Dynamics of elastic or rigid bodies with frictional contact and numerical methods. In: Blanc, R., Suquet, P., Raous, M. (eds.) Proceedings of the Mecanique, Modelalisation Numerique et Dynamique des Materiaux, pp. 9–29. Publications du LMA, Marseille (1991)

28. Allix, O., Daudeville, L., Ladevèze, P.: Proceedings of MECAMAT: Mechanics and Mechanisms of Damage in Composites and Multi-Materials, Saint-Etienne, 15–17 November 1989, p. 143 (1989). Baptiste, D. (ed.)
29. Chetouane, B., Dubois, F., Vinches, M., Bohatier, C.: NSCD discrete element method for modelling masonry structures. Int. J. Numer. Methods Eng. **64**, 65–94 (2005)
30. Acary, V., Jean, M.: Numerical simulation of monuments by the contact dynamics method. In: Workshop on Seismic Performance of Monuments, DGEMN-LNEC-JRC, Monument 1998, Lisbon, Portugal, November 1998, pp. 69–78 (1998)

Importance and Practice of Operation and Maintenance of Wastewater Treatment Plants

Amra Serdarevic[(⊠)] and Alma Dzubur

Department of Water Resources and Environmental Engineering,
Faculty of Civil Engineering, University of Sarajevo,
Sarajevo, Bosnia and Herzegovina
`amra.serdarevic@gf.unsa.ba`

Abstract. The main goal of the design, construction and operation of wastewater treatment plant (WWTP) is to ensure effluent quality by meeting the parameter values determined by legislation. Wastewater treatment plant is typically related to high cost of construction and equipment as well as for monitoring process, maintenance and operation in regular working mode. It causes economic and social pressure, even in developed countries. Therefore, engineers, water utility company and plant operation staff are looking for optimization, creative, cost-effective and environmental corresponding solution for appropriate technology and equipment. Operation activities are there to provide that a WWTP produces the desired quality and quantity of treated water and meets the standards, while maintenance are the activities to ensure regular and efficient work of equipment to achieve the sustainable operational objectives. For example, small, simple treatment plant with low capital costs may have high operational expenses and therefore will have higher total cost compare with an alternative technology. WWTPs in underdeveloped countries and developing countries are usually confronted with problems in operation and maintenance after testing period, when water utility company and local community must secure sufficient funds to cover high operation costs. That is the starting point for cutting down operation and maintenance protocol and costs. The consequences are often serious and significantly affect the increase in costs for the purpose of repairing the damage and demand a lot of efforts to bring back a WWTP into the normal operation. This paper is an overview of importance and practice for monitoring process, basic requirements of operation control and maintenance with review of situation with WWTPs in BiH and example of the Sarajevo WWTP "Butila".

Keywords: Monitoring · Operation · Maintenance · Effluent · Standards
WWTP

1 Basic of Wastewater Treatment Plants

The progressive deterioration of water resources and water scarcity gave the Wastewater Treatment Plant (WWTP) fundamental role and importance in the large number of activities undertaken in domain of water quality protection.

© Springer Nature Switzerland AG 2019
S. Avdaković (Ed.): IAT 2018, LNNS 60, pp. 121–137, 2019.
https://doi.org/10.1007/978-3-030-02577-9_14

First step in wastewater treatment process is collection. All wastewater should be collected and directed to a central point by collection systems and after that to treatment plant. The basic function of wastewater treatment is to speed up the natural processes by which wastewater is purified. Treatment processes are generally divided into the physical, chemical and biological. Treatment for domestic wastewater and industrial wastewater after pre-treatment usually combines primary, secondary and tertiary stage/treatment technologies, including appropriate facilities and equipment shown on the Fig. 1.

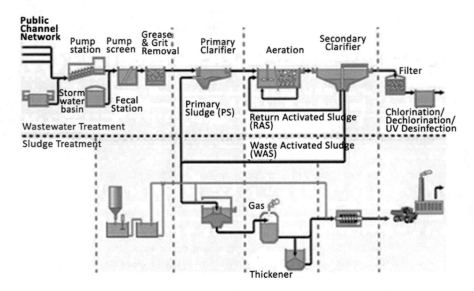

Fig. 1. Flow scheme of a conventional large-scale activated sludge system [1]

At the primary stage, large floating objects and solids are removed and settled from wastewater. After wastewater enters a treatment facility, coarse and fine screens remove floating objects that could clog pumps, small pipes, and downstream processes. Generally, screens are placed in a channel and inclined towards the flow of the wastewater. After screens, wastewater flows further into grit chamber where sand, grit and small stones settle to the bottom. Grit chamber could be combined with grease removal on the top of the aerated grit chamber.

This unit is very important, especially in cities with combined sewer systems. Large amounts of grit and sand can cause serious operating problems, such as clogging of devices, excessive wear of equipment, damage of pumps or decrease tank volume. After grit chamber, wastewater flows into the sedimentation tank and suspended solids gradually sink to the bottom. This mass of settled solids is called primary sludge and should be removed and particularly treated. Primary treatment can reduce BOD by 20 to 30% and suspended solids by up to 60%.

The secondary treatment uses biological processes to remove dissolve organic matter. Microbes consume the organic matter as food converting it to carbon dioxide, water, and energy [2]. While secondary treatment technologies vary from suspended growth (activated sludge) to attached growth systems, the final phase of each involves an additional process to remove suspended solids.

The activated sludge process is a proved and widespread technology. There are many technology options in developed activated sludge methods and many of them related to energy consumption. An adequate supply of oxygen is necessary for the activated sludge process to be effective. The oxygen is generally supplied by mixing air with the sewage and biologically active solids in the aeration tanks by one or more of several different methods. From the aeration tank, the treated wastewater flows to the secondary tank and the excess biomass is removed. Some of the biomass is recycled to the head of the aeration tank to provide designed concentration of the suspended solids or to support process of denitrification, while the remind biomass (exceed sludge) is treated before disposal or reuse.

Secondary treatment can remove up to 85% of BOD and total suspended solids.

The highest level of the wastewater treatment is tertiary treatment to remove contaminants or specific pollutants. Tertiary treatment is typically used to remove phosphorus or nitrogen, which can cause eutrophication. Tertiary treatment can remove 99% of all impurities from sewage, but it is expensive process.

Currently, nearly all WWTP's provide a minimum of secondary treatment, but for some recipients, the discharge effluent of secondary treatment still degrades water quality. Advanced treatment technologies should be applied with combination of secondary biological treatment to remove nitrogen and phosphorus.

Nowadays, new pollution problems have placed additional problems and stronger limits on wastewater treatment systems. Pollutants, such as heavy metals, toxic substances and chemical compounds are more difficult to remove from water. To meet effluent standards and return more usable water to lakes and streams, new technologies and methods for removing pollutants are being developed. Advanced wastewater treatment techniques comprise all biological treatment capabilities for removing organic pollution as well as nitrogen and phosphorus in combination with physical-chemical separation techniques such filtration, carbon adsorption and reverse osmosis. These wastewater treatment processes, in combination or alone, can achieve almost any degree of desired pollution elimination [3]. Effluents purified by appropriately chosen and carefully designed wastewater treatment can be discharged into the water body or it can be reused for industrial, agricultural, recreation or even for water supplies.

2 Importance of WWTP's Maintenance and Operation Control

Wastewater treatment plants are no longer just structure with equipment for chosen technology in the aim to meet the standards, eliminate odour, flies, etc. Wastewater treatment facilities are more complex, raw water is more difficult to treat and there is an increasing expectation in service, operation and control in exploitation. Therefore the costs for normal operation of a WWTP are usually rather high. Thus the very important roles in WWTP working life are an economic operation, control and maintenance, all together in the purpose for reliable operation of WWTP to meet regulatory requirements.

All of this leads to the importance of the proper protocol of the maintenance and operation control, analyses of the system requirements and improvement of the operation.

According to literature, studies and many reports from the existing WWTP, maintenance is usually the weak point in operation of the WWTP. Poor maintenance of the equipment and insufficient spare parts, absence of responsibility and the lack of funds for the procurement of chemicals and the maintenance of the plant can cause serious problems, failures and significant costs [3].

Thus, it is very important to provide detailed manual, an adequate staff training, work controls, audits, safety procedures, incident management and other specific requirements of the client (operator) related to the operation of the plant.

Brief overview of the WWTP's operation and maintenance with the example of operation of the WWTP Butila in Sarajevo, Bosnia and Herzegovina are presented in this paper.

2.1   Operation of the WWTP

Operations are the activities to make sure the plant produces the desired quality and quantity of treated water and meets the current legislation. Wastewater collection and treatment system must be operated as designed to adequately meet standard of the effluent and protect water quality. Operators of the WTTPs manage a complex system of machines, often using control boards, to transfer or treat wastewater.

WWTP is in operation every day, without any brake in performance. Company and licensed operators are responsible for the proper operation of the plant, including analytical testing, engineering, and budget and administration issues.

The operation manual contains a list of items, with their positions, that must be monitored on a regular basis, as well as the frequency to monitor.

The list of daily and monthly tasks should include sample collection and on-site readings for monitoring purposes. Influent and effluent flow meter readings must be performed daily since the pumps and equipment hour meter readings should be taken weekly.

The development of equipment for automatic measure allows the continuous measurement of the following parameters, important for the proper operation of the wastewater treatment plant [10]:

- Water level;
- Flow,
- Temperature,
- Electrical conductivity;
- Concentration of dissolved oxygen;
- Turbidity.

The way and position where it should be installed must be described in the manual, in accordance with plant design and technology.

The raw sewage inflow as well as final effluent should be measured as a minimum flow measuring requirements. There could be installed more flow meters between tanks, for recirculation, side streams, etc.

Applied metering instruments should be described and specified in the mechanical manuals for installation and maintenance requirements. Tables and graphs should be included in the manual (various flowrates vs water depths/level) to provide manual determination of the flow rate in case the corresponding meter is out of order.

Also, it is highly important to mark where samples for water quality should be collected as well as the frequency of such collections. The various tests that need to be performed must be determined in accordance with standards and procedures for the wastewater laboratory analyses.

Operation methods and requirements for each part of the WWTP such as preliminary treatment, primary settlement, biological treatment and sludge treatment, should be described in detail with emphases on shape, upward flow rate, inlet and outlet of the structure, equipment and method of operation. All drawings and mechanical manuals should be available on the site, particularly pumps, valves, weirs, etc.

For example, activated sludge reactors normally consist of different compartments for anaerobic/anoxic, anaerobic, and clarification processes or all processes can occur in the same reactors (SBR). These reactors are usually fitted with an assembly of mechanical equipment. Every piece of the equipment should be explaining in detail, with reference to the mechanical manual (for regular maintenance). Methods of adjusting certain concentration in the reactor effluent, like pH, oxygen content, NH4-N, etc. in line with required limits, should be also defined for normal operation of the WWTP.

The specific description usually refers to the methods how to determine and achieve value of the mixed liquor suspended solids (MLSS) at which plant must be operated (g/L).

2.2 Maintenance of the WWTP

Maintenance are the activities that should be undertaken to ensure the plant equipment operates continuously and efficiently to achieve operational objectives. Wastewater treatment plant must provide reliable and stabile service and avoid equipment breakdowns. Breakdowns of the equipment usually can be avoided if operators regularly inspect the equipment, pipelines, inlet and outlet of the system.

Systematic and preventive maintenance uses data obtained through the inspections of the equipment before equipment failures occur.

Based on the proper instruction and educated staff, good maintenance program will reduce breakdowns and contribute to the cost-effective operation of the WWTP [11].

It is necessary to describe in detail all action required from operator regarding the position and method of final effluent discharge. The position and method of taking samples or manual control should be clearly stated and referred to the main drawings. That includes all equipment with their technical specifications, for example pumping systems.

Each pump in the plant needs to be carefully maintained, with specification and explanation of their function and how they are controlled in accordance to the mechanical manual (like influence of change in head due to the water levels or valves alterations, etc.)

Also, it is very important to state means of preventing some possible or specific difficulties and provide operators with suggestions how to solve or react on it.

3   Wastewater Treatment Plant "Butile" – Sarajevo (BiH)

Sarajevo is the capital of Bosnia and Herzegovina with a population of 400.000.

Responsibility for the water supply system, sewerage system and wastewater treatment at the central Wastewater Treatment Plant of Sarajevo is assigned to the Public enterprise "Vodovod i Kanalizacija" (ViK) (Water Management and Sanitation) by the Federal Ministry for Spatial Planning and Environment through the Government of Canton Sarajevo.

Wastewater treatment plant (WWTP) of Sarajevo was built and started with testing operation in 1984. The plant has been used until 1992. with good operational results. Due to the war in Bosnia and Herzegovina from 1992–1995, the wastewater treatment plant completely stopped with operational activities. The wastewater treatment plant was completely devastated and ruined. In order to improve the water quality of Miljacka river and Bosna river as watercourses which receive effluent from the wastewater treatment plant of Sarajevo it was very important to rehabilitate and reconstruct the Sarajevo wastewater treatment plant.

3.1   Existing Sewerage System

The Central Sarajevo Sewerage System managed by ViK, serves about 90% of the population of Canton Sarajevo. The trunking system is approximately 46 km length, with diameters of the main pipelines in the range of 500 mm to the maximum of 2000 mm. Some parts of the collection system (particularly in Old town and Centre) were constructed 100 years ago. The aged sewerage systems mainly collect combined sanitary and storm water drainage. This system causes problems such as, overloading of sewers and accumulation of large amount of grit and inert materials especially during heavy rains that could cause blockages of the screw pumps and screens. They could also cause problems in the aerated grit chamber and manholes.

Those problems were tackled by solution provided with the pre-treatment and pre-screening facilities upstream of the raw water before the pumping station (inlet). These facilities consists rectangular, horizontal flow grit channels to remove heavier grit particles and sets of coarse and medium screens.

3.2   Technical Details of the Reconstructed WWTP Butila

The Waste Water Treatment Plant (WWTP) Butila in Sarajevo was reconstructed and officially opened on 22 May 2017, marking the final phase of the EU-funded project "Reconstruction of the Waste Water Treatment Plant 'Butila' in Sarajevo".

The main objective of this project, supported by the EU funds, is to reduce the pollution of surface waters from urban wastewater which was discharged without treatment directly to the Rivers Miljacka and Bosna.

The overall budget of the reconstruction amounted to EUR 25.6 million with the EU contribution of EUR 13 million. The project supported the repair and replacement of primary and secondary sewers in Sarajevo, the efficiency improvement of the waste water collection network in the Sarajevo Canton, the pollution reduction of surfaces waters from urban wastewater which is discharged without treatment to the rivers Miljacka and Bosna, as well contribution to the improvement of the living conditions for the population in the Sarajevo Canton and at the downstream settlements of the river Bosna [7].

The existing WWTP, which was constructed in 1984, has been designed, rehabilitated and upgraded in 2 phases:

Phase 1: Rehabilitation of the existing WWTP to re-establish the original capacity and treatment level (secondary treatment), 600.000 PE.
Phase 2: Upgrade of the WWTP for nutrient removal to achieve compliance with EU standard, 650.000 PE.

Expecting amount of the raw water is approx. 169.500,00 m^3/day. The average dry weather flow is 2, 00 m^3/s and peak wet weather flow is between 3, 9 up to 5, 2 m^3/s [5, 8].

There are 25 facilities of wastewater treatment plant for pre-treatment, primary, secondary, sludge treatment processes, laboratory and associated administrative premises [4, 8].

Reconstruction of the wastewater treatment plant in Sarajevo based on the pre-war facility includes following technology and technical phases for the wastewater and sludge treatment (Fig. 2):

1. Preliminary and Primary treatment (mechanical treatment)
2. Secondary treatment (biological treatment)
3. Sludge treatment:

   – Primary thickener
   – Anaerobic digestion
   – Disposal of the digested sludge
   – Mechanical dehydration
   – Energy utilisation from sludge.

The design parameter for organic, suspended solids loading in the influent of WWTP is based on 600.000 PE (Phase I) and assessed industrial load. The organic load in the influent of WWTP is as following [4, 7, 8]:

- Total load BOD5 is 36.000 kg/day
- Total load COD is 72.000 kg/day
- Total load TSS is approx. 42.000 kgTSS/day
- Total Kjeldal Nitrogen is approx. 6600 kg TKN/day.

Fig. 2. Panorama of the reconstructed Wastewater Treatment Plant Butila [7]

The layout of the WWTP Sarajevo is shown on the Fig. 3. The testing and operational period of the plant was started in spring 2016.

The wastewater treatment plant was designed to meet the standards of current legislation and allowed concentration and load in the effluent of the WWTP according to the class of recipient and number of PE.

For the river Bosnia, according to the Regulation for discharging effluent into the watercourses in FBiH [6] the concentration of the BOD5 and TSS should not be larger than limited concentration for:

- Total suspended solids (TSS) 35 mg/L
- Organic load (BOD$_5$) 25 mg/L

3.3 Operation and Maintenance of the WWTP Butila

Operational control WWTP Butila is carried out according to the protocol compliant with requirements of relevant legislation [6].

Monitoring, control and adjustment operation of the units of the WWTP are carried out with support by hydro-mechanical equipment and software for automatization of the installed equipment. Hydro-mechanical equipment is divided into measurement and regulation equipment (in hydraulic sense) and measuring equipment of technological parameters. Following are monitored at the plant: flow, level, pressure, temperature, oxygen and pH values (Table 1). Table 1 lists the measuring devices installed at the main control points while at the plant there are still a number of control/measuring points between structures and in facilities for wastewater treatment and sludge.

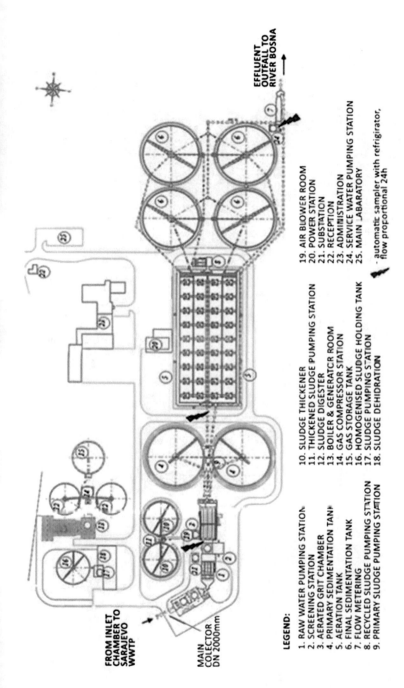

Fig. 3. Scheme of Wastewater Treatment Plant Butila [4]

LEGEND:

1. RAW WATER PUMPING STATION
2. SCREENING STATION
3. AERATED GRIT CHAMBER
4. PRIMARY SEDIMENTATION TANK
5. AERATION TANK
6. FINAL SEDIMENTATION TANK
7. FLOW METERING
8. RECYCLED SLUDGE PUMPING STATION
9. PRIMARY SLUDGE PUMPING STATION

10. SLUDGE THICKENER
11. THICKENED SLUDGE PUMPING STATION
12. SLUDGE DIGESTER
13. BOILER & GENERATOR ROOM
14. GAS COMPRESSOR STATION
15. GAS STORAGE TANK
16. HOMOGENISED SLUDGE HOLDING TANK
17. SLUDGE PUMPING STATION
18. SLUDGE DEHIDRATION

19. AIR BLOWER ROOM
20. POWER STATION
21. SUBSTATION
22. RECEPTION
23. ADMINISTRATION
24. SERVICE WATER PUMPING STATION
25. MAIN LABARATORY

- automatic sampler with refrigirator, flow proportional 24h

FROM INLET CHAMBER TO SARAJEVO WWTP

MAIN COLECTOR DN 2000mm

EFFLUENT OUTFALL TO RIVER BOSNA

Table 1. List of the equipment of the main points and flow measuring, WWTP Butila

Nr.	Component/location	Type of measurement	Performance/capacity/measuring range	Medium characteristics
Level measurement				
1	Raw water PS	LS	0.25 m	Raw sewage (200–1000 mg/l TSS)
2		C - ULS	0–6.2 m	
3	Coarse screen station	CD - ULS	0–2.5 m	
4	Fine screen station	CD - ULS	0–1 m	
5	Aerated grit and grease chamber	C - HT	0–6 m	Grease/water mixture (1–2% DS)
6	Activated sludge PS	C - UT	0–4.5 m	Return activated sludge (0.7–1% DS)
7		LS	0.25 m	
8	Mechanical sludge thickening	C - HT	0–1 m	Th.excess & primary sludge (6% DS)
9	Thickened sludge tank	C - ULS	0–3 m	
10	Primary sludge PS	C - ULS	0–4.3 m	Primary sludge (2–3% DS)
11	Mixed sludge tank	C - ULS	0–5 m	Mixed primary, excess sludge, scum (1.5% DS)
Flow measurement				
1	Overflow chamber	C - ULS	0–4.5 m	Raw sewage (200–1000 mg/l TSS)
2	Aerated grit and grease chamber	C - ULS	0–10000 m^3/h	
3	Distribution chamber of primary sedimentation tanks	C - ULS	0–1 m	
4	Secondary sedimentation tanks	C - ULS	0–3000 m^3/h	
5	Effluent measurement	CF for venturi channel	0–5000 m^3/h	Final effluent (10–35 mg/l TSS)
6	Primary sludge PS	EMF, DN 150	0–300 m^3/h	Primary sludge (2–3% DS)
7	Activated sludge PS, WAS pumps' outlet	CF, EMF, DN200	0–300 m^3/h	Return activated sludge (0.7–1% DS)
8	Mechanical sludge thickening	EMF, DN 150	0–100 m^3/h	Mixed primary, excess sludge and scum (1.5% DS)

(continued)

Table 1. (*continued*)

Nr.	Component/location	Type of measurement	Performance/capacity/measuring range	Medium characteristics
9	Thickened sludge pumping station	EMF, DN 100	0–30 m³/hr	Thickened excess & primary sludge (6% DS)
10	Mechanical sludge dewatering	EMF, DN 125	0–60 m³/h	Digested sludge (ca. 4% DS)

ULS - ultrasonic level sensor; PS - pumping station; HT - hydrostatic type; LS - level switch; C – continuous; EMF - electromagnetic flow meter; CF - continuos flow; CD - continuos differential

Fig. 4. Flow measurement for venture canal, with ultrasonic sensor (20.04.2018) [9]

As one of the most important parameter continuously measured is the inflow to the plant and the amount of water at the outlet of the plant.

With the development of technology, nowadays most often applied are three modern types of water flow meters in open canals:

- Radar meter that measures water speed without water contact and ultrasound that measures water level without water contact
- Ultrasonic meter, which by ultrasound measures the speed of water in contact with water and ultrasound measures the water level without water contact
- Electromagnetic flow meter, which electromagnetically measures the water speed in a siphon tube full of water without water contact.

All three types of meters have advantages and disadvantages so that each has its field of application in drainage systems. All meters have a large flow measurement range of up to several thousand litres per second. The measuring device consists of a sensor and a control unit with telemetry outputs of 4–20 mA, LCD - a flow meter and a total flow indicator (Fig. 4) and logger for measuring memorization and power supply 230 V AC or 12 V DC respectively 24 V DC.

Ultrasonic meter is recommended at the output of wastewater treatment plant and in larger canals with not too polluted water (especially regarding floating matters).

Measurement of flow at the entrance to WWTP Butile is carried out by an ultrasonic flow meter mounted on the overflow chamber of entrance building. Measurement is continuous, with application of ultrasonic level sensor. The amount of water smaller or equal to 5.2 m^3/s is taken further to mechanical treatment. In the case of inflows larger than 5.2 m^3/s, excess water is drained (bypass) and released directly into the recipient – river Miljacka.

The flow measurement is further controlled at the exit (outlet weir) from the grease and grit chambers.

Water level measurement is also continuous, with ultrasonic level sensor. Measuring the flow rate at this position controls the amount of water entering biological treatment. The biological process is significantly more sensitive to the oscillation of the

Fig. 5. Position of the probe for measuring the concentration of oxygen and temperature in the aeration tank (20.04.2018.) [9, 10]

wastewater inflow than the mechanical treatment. This is primarily reflected in the parameters of organic pollution, nutrients, dissolved oxygen and other parameters essential for a stable purification process.

At the exit from the plant, behind the subsequent sediment, a venture water meter is installed. The water meter measures the amount of purified water – effluent (Fig. 4), with ultrasonic level sensor (ultrasonic flow meter).

Table 2. Equipment installed on the main point of pressure, temperature, oxygen and pH measurement and control, WWTP Butila

Nr.	Component/location	Type of measurement	Performance/capacity/measuring range	Medium characteristics
Pressure measurement				
1	Gravel trap blowers' outlet	PIRCA	0–2 bar	Ambient air
2	Air blower room, blowers' outlet			-
3	Blower station for aeration tank, blowers' outlet pipe	CP	0–1 bar	Compressed air (430 mbar, 80 °C)
4	Service water pumping station	PRS	0–7 bar	Service water
5	Gas room	CP	100 mbar	Biogas
6	Sludge digesters		50 mbar	
Temperature measurement				
1	Gravel trap blowers' room	TTCV	0–50 °C	Ambient air
2	Air blower, blowers' room			
3	Fine screening station room			
4	Blower station for aeration tank, blowers room			
5	Sludge circulation	C		Digested sludge (3–4% DS)
6	Sludge digesters			
Oxygen measurement				
1	Activated sludge tanks	continuous measurement of dissolved oxygen	0–10 mg O_2/l	Activated sludge (2000–4000 mg/l TSS)
pH measurement				
1	Sludge circulation	online pH measurement on DN 300 pipe	0–14 pH	Digested sludge (3–4% DS)

C – continuous; CP - continuos pressure; PRS - pressure switch; TTCV - temperature thermostat for control of ventilator

For all types of measuring devices, their regular maintenance is very important. On WWTP Butila, Endress Hauser ultrasonic flow meters are used, which have a high degree of accuracy and precision of the measured size.

Fig. 6. Level and flow measuring of return activated sludge (20.04.2018) [9]

In addition to the above mentioned locations for measuring the flow of wastewater, there are a number of other control points for flow measurement as well as checkpoints for sludge recirculation, dosing of the polymer, water consumption for maintenance services and others (Table 1).

In addition to flow measurement at the plant, regular measurement of other parameters are required for proper operation and maintenance. The parameters that are continuously measured during the operation of the device are temperature, dissolved oxygen (Fig. 5.), pH value, conductivity and pressure. Equipment installed at WWTP Butila for flow measurement, tank water levels, and other parameters is displayed in Table 2. The table shows the main measuring points and the range of operation for the installed equipment.

Specific equipment and the measurement-control technique are used on the sludge treatment line. The sludge that is deposited in the secondary sedimentation tank is transferred to the pumping station for recirculation and excess sludge, whereby with the help of the built-in equipment and the automated system the separation is carried out according to the aeration tank (recirculated sludge) and to the tank of mixed sludge (excess sludge). Level measurement is performed for this process as well as measurement of the flow of excess sludge. Control is applied to regulate the quantity of sludge that are returned to the process and quantities that represent excess sludge and require further processing, before final disposal. The equipment for measuring the sludge level and recirculation of the sludge to the aeration tank, installed on the wall of the active sludge pump station, is shown on Fig. 6.

Monitoring and control of the devices in Butile is automatic via Supervisory Control and Data Acquisition (SCADA system). SCADA system is connected on main program WinCC Siemens. In this way, it is possible to monitor, control and adjust certain process parameters, as well as all the basic elements of the system (such as ON/OFF of individual elements of the system). All facilities of hydraulic equipment and measuring equipment are connected to the SCADA system, via PLC cards, which sends information with the signal, e.g. about the state of the level or flow, expressed in meters for the levels and m^3/h for flows. Importance to maintain the equipment regularly is vital because of their accuracy and precision, which is manifested through the annual calibration and verification of the devices.

Fig. 7. Laboratory for water and sludge analyses at the WWTP Butila – Sarajevo [5]

Fig. 8. Outlet of the WWTP Butile Sarajevo - clear water (12/10/2017) [9]

Measuring devices maintains authorized service (in the case of a plant in Sarajevo it is Endress Hauser). Considering that WWTP Butila –Sarajevo, still works within the warranty period (until November 2018), expert team PPOV Butila has already started the process of obtaining verification and calibration of measuring devices.

Each measuring device is associated with alarms, indicating the levels in individual units of the plant.

Table 3. List of the automatic samplers on the WWTP Butila

Nr.	Component/location	Type of measurement	Medium characteristics
Automatic sampler			
1	Aerated grit and grease chamber - downstream of fine screen	automatic sampler with refrigerator, flow proportional 24 h	Raw sewage (200–1000 mg/l TSS)
2	Inlet to biology - distribution chamber of aeration tanks		
3	Outlet - effluent measurement		Final effluent (10–35 mg/l TSS)

In addition to parameters that are measured automatically and continuously (water levels, water flow, temperatures, dissolved oxygen, electrical conductivity and others), during the treatment process, there is also manual control of water quality parameters in tanks with laboratory equipment for tests on the site.

Water quality control at the entrance (influent) and at the exit from the plant (effluent), and the control of process parameters in the biological unit is carried out in accordance with the laboratory testing program.

Wastewater and sludge analyses are performed in laboratories same as the case of WWTP Butila, Sarajevo. Some quality parameters of wastewater are binding (requirement under the legislation), and some are monitored for better insight into the processes and are used for the stabilization and improvement of the device operation.

Laboratory analyses of water and sludge, which are done on WWTP Butile, relate to the following quality parameters: temperature ($^\circ$C), suspended solids according to Imhoff, pH value, electrical conductivity, BOD_5, COD, SS, TS, TSS, ISV, TOC, NH_4-N, PO_4-P/TP, NO_2-N, NO_3-N and TN (Figs. 7 and 8).

Wastewater samples are taken as composite samples, by automatic samplers (automatic sampler with refrigerator, proportionally by 24-h flow). Automatic samplers are installed in positions according to Table 3.

4   Conclusions

Presented overview of the wastewater treatment operation and maintenance emphasizes complexity and importance of these highly important issues. Nowadays many treatment plants are under operation throughout the world, so there are lot of experiences regarding the technology and protocol of the operation control and maintenance. However, the maintenance of the WWTP is usually the weakness point in operation of WWTP. Decision and selected treatment process should not only be based on effluent values and maintenance but also on the system simplicity and cost effectiveness in normal operation of WWTP. Considering current environmental problems, it is not unrealistic to believe that the trend of development of new or improvement of existing WWTPs technology will be continued all over the world. At the same time, loads on existing plants are expected to increase due to the growth of urban areas. This situation demands more efficient treatment procedures for wastewater including sophisticate and automatized measurement and operation system control. However, it should be taken into account that apart from capital investments it is also very important to secure sufficient funds for normal operation and maintenance of the both existing and new WWTP for each year. Important step is continuous education and training of the staff and exchange their knowledge between operators on the site, conferences, seminars, etc.

All decisions related to the financing, selection of technology, construction as well as operation and maintenance of a wastewater treatment are within responsibility of the relevant authorities and key stakeholders (public and utility companies, private sector, industry, local community, etc.). Water pollution control and actions to preserve clean water should be regarded as an imperative issue for whole society.

References

1. http://archive.sswm.info/category/step-rrr-business-development/module-2-sector-inputs/technological-options/technological-19. Uploaded 20 Feb 2018
2. Serdarevic, A., Dzubur, A.: Wastewater process modeling. Coupled Syst. Mech.-Int. J. Interact. Coupled Syst. 5(1), 21–39 (2016). http://dx.doi.org/10.12989/csm.2016.5.1.021. ISSN: 2234-2184 (Print) 2234 2192 (Online)
3. Serdarevic, A., Dzubur, A.: Wastewater process modeling. In: ECCOMAS- Thematic Conference, Proceedings of 2nd International Conference on Multi-scale Computational Methods for Solids and Fluids, Sarajevo, 10–12 June 2015
4. Serdarevic, A., Sulejmanagic, I.: Reconstruction of the wastewater treatment plant of Sarajevo, damage assessment and reconstruction after natural disasters and previous military activities. In: Proceedings NATO-ARW 983112 International Conference, Građevinski fakultet u Sarajevu, Sarajevo, BiH., Springer, Nedherlands, pp. 476–485 (2008)
5. Dizdarević, A.: Način rada na Postrojenju za prečišćavanje otpadnih voda u Butilama, Svjetski dan voda, Zenica, mart (2017)
6. Uredba o uslovima ispuštanja otpadnih voda u okoliš i sisteme javne kanalizacije, ("Službene novine FBiH", broj 101/15 i 01/16)
7. https://www.youtube.com/watch?v=UdZT3mi-nFE. Uploaded 20 Mar 2018
8. Main project: Sarajevo WWTP Butila. IPSA Institut d.o.o. Sarajevo, July 2014
9. Serdarevic, A., Dzubur, A.: Private collection of the photos of the WWTP Butila (2017/2018)
10. Dzubur, A., Serdarević, A.: Kontrola i održavanje PPOV – Primjer PPOV Butile, Sarajevo, BiH – 5. Konferencija "Održavanje 2018", Zenica, BiH, 10–12 maj 2018
11. Tchobanoglous, G., Franklin, L.B., Stensel, H.D.: Wastewater Engineering Treatment and Reuse, 1819 p., 4th edn. Metcalf & Eddy, Inc., McGraw-Hill Education (2003)

Mathematical Modeling of Surface Water Quality

Hata Milišić$^{(\boxtimes)}$, Emina Hadžić$^{(\boxtimes)}$, Ajla Mulaomerović-Šeta$^{(\boxtimes)}$,
Haris Kalajdžisalihović$^{(\boxtimes)}$, and Nerma Lazović$^{(\boxtimes)}$

Faculty of Civil Engineering, Department of Water Resources and
Environmental Engineering, University of Sarajevo, Sarajevo, Bosnia and
Herzegovina
hata.milisic@gmail.com, eminahd@gmail.com,
kahariss@gmail.com, nerma.ligata@gmail.com,
ajla.mulaomerovic@gf.unsa.ba

Abstract. Water is one of the main elements of the environment which determine the existence of life on the Earth, affect the climate and limit the development of civilization. Water resources management requires constant monitoring in terms of its qualitative-quantitative values. Water quality models are important tools to test the effectiveness of alternative management plans on the water quality of water bodies. One of the tools that are used to solve problems of surface water pollution is modeling of changes which take place in rivers waters and associated water quality changes. In the last thirty years a rapid development of mathematical modeling of water resources quality has been observed. A number of computer models have been designed which are successfully applied in practice in many countries, including B&H. The main aim of this study was to develop and demonstrate use of a water quality model as a tool for evaluation of alternative water management scenarios for the river basin of Neretva, B&H. MIKE 11 model has demonstrated its applicability to simulation of pollution in streams, and therefore is an appropriate tool for decision making related to the quality of water resources.

Keywords: Water pollution · Mathematical models · MIKE 11
Quality water

1 Introduction

Water is an important element for life on the earth. It is an essential natural resource for environmental sustenance. Water quality modeling plays a vital role in water quality studies. The problem of reducing pollution and improving water quality can be solved by using the appropriate mathematical models and their implementation in software [1]. For decades, water-quality models have been used as tools to assess the combined effects of advection, dispersion, reaeration, and selected chemical and biological reactions on stream water quality. These models tended to use steady-state representations of stream hydraulics and included only a small number of reactions [2]. Recent advancements in computer technology, however, have allowed more complex and

© Springer Nature Switzerland AG 2019
S. Avdaković (Ed.): IAT 2018, LNNS 60, pp. 138–156, 2019.
https://doi.org/10.1007/978-3-030-02577-9_15

dynamic water-quality models to be built. As these tools have become more capable, their utility in helping scientists, regulators, and river managers understand stream processes also has increased. Models are now commonly used to assess pollutant transport, quantify source and sink processes, determine assimilative capacities, and design regulatory compliance schemes. In these endeavors, today's models are often sufficiently accurate to be useful, particularly when simulating advection/dispersion, water temperature, conservative transport, and simple eutrophication processes [2]. There has been developed a lot of mathematical and numerical models to simulate the water quality. These models are able to simulate the real situation at streams, however the range of the reliability and accuracy of the results is very wide [3].

The aim of this paper is to point out the importance of surface water quality conservation and the possibility of using numerical models for estimating the current and forecast of future water quality statuses of the Neretva River in Bosnia and Herzegovina. The MIKE 11 software package was used for all simulations. MIKE 11 is an industry standard for simulating flow and water level, water quality and sediment transport in rivers, irrigation canals, reservoirs and other inland water bodies. It is a comprehensive engineering tool with a wealth of capabilities provided in a modular framework [3]. Based on the aforementioned, the individual aims of this research are as follows:

- Modeling and simulation of spatial and temporal changes of conservative (electrical conductivity) and non-conservative water quality parameters (oxygen regimen) of the Neretva River on the river reach section between the HPP (Hydro Power Plant) Mostar and MS (Measuring Station) Žitomislić (MIKE 11 Model calibration and verification);
- Prediction of different management scenarios for reducing pollution in the Neretva River (Simulation of future water quality status without and with Wastewater Treatment Plant using MIKE 11 model).

Analysis of observation and simulation results is presented in this paper. The results of modeling and simulation can be further exploited for the long-term and complete solution of the problem of water quality management, and thus the environment as a whole.

2   Surface Water Quality Models

Predicting the spread of contaminants is important for managing and protecting rivers and streams. Surface water quality models can be useful tools to simulate and predict the levels, distributions, and risks of chemical pollutants in a given water body. The modeling results from these models under different pollution scenarios are very important components of environmental impact assessment and can provide a basis and technique support for environmental management agencies to make right decisions [4, 5].

The field of the water quality models is very large and includes different kind of models in dependence of the level of the input and output information, the complexity of the modelled events, the modelled water body, the used mathematical methods, the type of the basic equations, the aim of the modelling, the structure of the modelling system, the scale of interest in steady-state or non-steady state conditions and others. Many different types of water quality models are available so it is not possible to give a simple classification (Fig. 1) [1].

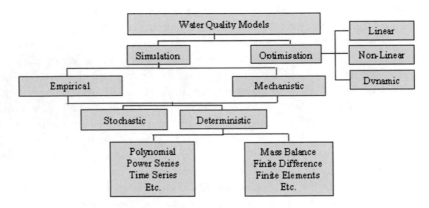

Fig. 1. Classifications of water quality models [1]

With the development of model theory and the fast-updating computer technique, more and more water quality models have been developed with various model algorithms. Up to date, tens of types of water quality models including hundreds of model software have been developed for different topography, water bodies, and pollutants at different space and time scales. Surface water quality models including the Streeter-Phelps model, QUASAR model, QUAL model, WASP model, CE-QUAL-W 2 model, BASINS model, MIKE model, and EFDC model (Table 1), were widely applied worldwide [5]. When using numerical modeling quite different preliminary assumptions and modeling techniques can used for the solution of individual water quality problems. Most of the computer programmes are compiled as the combination of hydrodynamic river analysis system and transport-dispersion modules. The selection of the mathematical model for the individual problem solution must issue from the catchment, stream and pollution data available and from the anticipated accuracy of results [6, 7].

One of the most well known European 1-D modelling system of the pollutant transport in the river network is MIKE 11, developed by the Danish Hydraulic Institute [8]. MIKE 11 is a professional Engineering software package for the simulation of flows, sediment transport and water quality in estuaries, rivers, irrigation systems and other water bodies. MIKE11 has been designed for n integrated modular structure with basic computational modules for hydrology, hydrodynamics, advection - dispersion, water quality and cohesive and non-cohesive sediment transport. It also includes modules for the surface runoff. MIKE11 has a well developed graphical user interface integrated with the pre - and postprocessors that support the system interaction with the GIS [8].

Table 1. Main surface water quality models and their versions and characteristics [5].

Models	Model version	Characteristics
Streeter-Phelps models	S-P model Thomas BOD-DO model; O'Connor BOD-DO model; Dobbins-Camp BOD-DO model	Streeter and Phelps established the first S-P model in 1925. S-P models focus on oxygen balance and one-order decay of BOD and they are one-dimensional steady-state models
QUAL models	QUAL I; QUAL II; QUAL2E; QUAL2E UNCAS; QUAL 2 K	The USEPA developed QUAL I in 1970. QUAL models are suitable for dendritic river and non-point source pollution, including one-dimensional steady-state or dynamic models
WASP models	WASP1-7 models	The USEPA developed WASP model in 1983 WASP models are suitable for water quality simulation in rivers, lakes, estuaries, coastal wetlands, and reservoirs, including one-, two-, or three-dimensional models
QUASAR model	QUASAR model	Whitehead established this model in 1997. QUASAR model is suitable for dissolved oxygen simulation in larger rivers, and it is a one-dimensional dynamic model including PC_QUA SAR, HERMES, and QUESTOR modes
MIKE models	MIKE11 MIKE 21 MIKE 31	Denmark Hydrology Institute developed these MIKE models, which are suitable for water quality simulation in rivers, estuaries, and tidal wetlands, including one-, two-, or three dimensional models
BASINS models	BASINS 1 BASINS 2 BASINS 3 BASINS 4	The USEPA developed these models in 1996. BASINS models are multipurpose environmental analysis systems, and they integrate point and nonpoint source pollution. BASINS models are suitable for water quality analysis at watershed scale
EFDC model	EFDC model	Virginia Institute of Marine Science developed this model. The USEPA has listed the EFDC model as a tool for water quality management in 1997. EFDC model is suitable for water quality simulation in rivers, lakes, reservoirs, estuaries, and wetlands, including one-, two-, or three-dimensional models

3   Materials and Methods

3.1   Study Area

The Neretva River is the largest karst river in the catchment of the Adriatic Sea. The total length of the Neretva River is 225 km, the majority of which, a total of 203 km, is in B&H. Only the last 22 km of the river pass through Croatia on its way to the Adriatic Sea. The Neretva River has an average annual discharge of 194.4 m³/s, while the total area of the hydrogeological basin catchment in the B&H Federation (one of the two B&H entities) is 5745 km². The Neretva River is an important source of water supply, hydroelectricity generation, tourism and recreation and irrigation. Water regime in the lower course of the Neretva River is very complex because of the downstream side impact and built hydropower system with the upstream side. A characteristic of the Neretva River is its great disparity in annual (or seasonal) water [9].

Wastewater Pollution of the Neretva River from Municipal Sources. Urban growth and industrial, agricultural and power development have had a negative impact on the ecology of the Neretva basin, in particular the Delta wetlands. Untreated municipal wastewater from communities along the Neretva (population of the Neretva catchment area is 350, 000 out of which 115,000 is in urban areas along the River) is responsible for about a third of the Neretva and its tributaries being classified as polluted. Industrial wastewater from metal processing industries in the Mostar area, harbor operations in the Delta and intensive farming along the river banks have also contributed to the pollution [9].

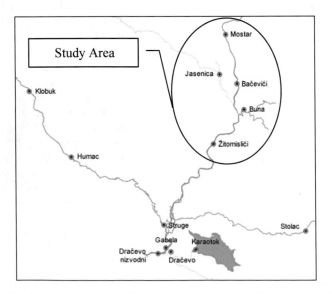

Fig. 2. Study area - map of the respective catchments of the Neretva river

The largest city on the Neretva River is Mostar. The population of the Mostar region has dropped from 127,000 to around 100,000 in the years following the 1992–95 war. Currently in Mostar, there is no waste water treatment plant. And only some quarters of the city have functioning sewerage systems. A total of 50 km of new sewerage piping have been built in recent times, but this entire sewerage system still discharges its untreated waste water into the Neretva River at some 35 locations [9].

The only collector for collecting wastewater and precipitation is the Neretva River. This includes the lateral flow of water into the Neretva river from all small and large rivers. The main pollutant of the river's water is extremely large and uncontrolled use of pesticides in agricultural land near or near the river. The second major pollutant is untreated wastewater from settlements, cities and surrounding industries. In spite of the aforementioned non-existent pollution control efforts, the water quality index of the Neretva River is still assessed in the II Class category [9].

The condition of the Neretva River and its tributaries and other constructed ponds have degraded over the years in term of water pollution, river environment and ecosystem. There is, therefore, an urgent need to improve the rivers water quality and its environment in order to maintain Mostar City and the Neretva River as an important tourist destination, and hydroelectric power generation.

3.2   Numerical Modelling

Numerical models are to be successfully calibrated and properly applied and it is to be improved our understanding of the complex interactions among different parameters such as temperature, biological oxygen demand, dissolved oxygen, salinity, and eutrophication in the fresh water and sea water environment.

DHI's MIKE 11 model is a professional engineering software package for the simulation of flows, water quality and sediment transport in estuaries, rivers, irrigation systems and other water bodies. MIKE 11 is a one-dimensional, fully dynamic modelling tool. The hydrodynamic (HD) module forms the basis for most add-on modules including the Advection-Dispersion module and Water Quality module [8].

3.2.1   Hydrodynamic Module

The Hydrodynamic module is the core of the MIKE 11 system that provides complete solution for the Saint Venant equations or either of the two simplified versions called the kinematic and the diffusive wave approximations (DHI MIKE11, 2003). Full and complete dynamic equations are given below in the following expressions [8–10]:

$$\frac{\partial A}{\partial t} + \frac{\partial Q}{\partial x} = q \tag{1}$$

$$\frac{\partial Q}{\partial t} + \frac{\partial}{\partial x}\left(\frac{\alpha Q^2}{A}\right) + g \cdot A \frac{\partial h}{\partial x} - g \cdot A(S_0 - S_f) = 0 \tag{2}$$

Where the given variables are as follows: A-cross sectional area, t-time, Q-discharge, x-distance downstream, q-lateral inflow, g-gravitational acceleration, h-depth, α-velocity coefficient, S0-bottom branch slope i Sf-energy level slope.

The MIKE 11 solution of the continuity and momentum equations is based on an implicit finite difference scheme developed by Abbott and Ionescu (1967). The scheme is setup to solve any form of the Saint Venant equations – i.e. kinematic, diffusive, or dynamic. The water level and flow are calculated at each time step, by solving the continuity equation and the momentum equation using a 6-point Abbot scheme with the mass equation centered on h-points and the momentum equation centered on Q-points. By default, the equations are solved with 2 iterations. The first iteration starts from the results of the previous time step and the second uses the centered values from the first iteration. The number of iterations is user specified [8] (Fig. 3).

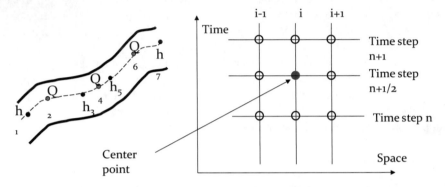

Fig. 3. Numerical scheme - 6 point Abbott-Ionescu scheme [11]

Cross sections are easily specified in both area and longitudinal location through the user interface. The water level (h points) is calculated at each cross section and at model interpolated interior points located evenly and specified by the user-entered maximum distance. The flow (Q) is then calculated at points midway between neighboring h-points and at structures [8].

The hydraulic resistance is based on the friction slope from the empirical equation, Manning's or Chezy, with several ways of modifying the roughness to account for variations throughout the cross sectional area [8, 12].

Boundary types include Q-h relation, water level, discharge, wind field, dam break, and resistance factor. The water level boundary must be applied to either the upstream or downstream boundary condition in the model. The discharge boundary can be applied to either the upstream or downstream boundary condition, and can also be applied to the side tributary flow (lateral inflow). The lateral inflow is used to depict runoff. The Q-h relation boundary can only be applied to the downstream boundary [8, 12].

3.2.2   Advection-Dispersion Module

The mathematical model consists of one - dimensional advection - dispersion mass balance equation of the given pollution parameter, corresponding initial and boundary conditions. The assumptions in such models are [8, 9]:

- the density of polluted water is constant and similar to "clean" water density;
- the substance is well mixed over the cross section;
- only longitudinal hydrodynamic dispersion occurs.

The Advection-Dispersion module is based on a one-dimensional equation for the conservation of the mass of dissolved or suspended particles or material (for example, salt or cohesive sediment). The behaviour of non-conservative substances, which decay, can also be simulated in the AD module. The one-dimensional advection-dispersion equation, which takes into account the first order of decay effect, is given in the following expression below [8, 9, 13]:

$$\frac{\partial AC}{\partial t} + \frac{\partial QC}{\partial x} - \frac{\partial}{\partial x}\left(AD\frac{\partial C}{\partial x}\right) = -AKC + C_2 q \tag{3}$$

Where the given variables are as follows: Q- discharge $[L^3 T^{-1}]$, C- concentration $[ML^{-3}]$, D- dispersion coefficient $[L^2 T^{-1}]$, A- cross sectional area $[L^2]$, K- first order decay factor or the linear coefficient of decay $[T^{-1}]$, C_2- source/sink of pollution $[ML^{-3}]$, q- lateral inflow $[L^2 T^{-1}]$, x – distance along watercourse [L], and t- time [T].

Equation (3) describes two transport processes: the advective (or convective) and the dispersive. The equation is resolved through the numerical application of the scheme on finite difference, giving a very insignificant numerical dispersion as its result. The application of this calculation scheme in the AD model is stable even for using a large Peclet number [8, 9, 13]:

$$Pe = v\frac{\Delta x}{D} > 2 \tag{4}$$

The time and space discretization should be chosen in a way so that the convective Courant number (Cr) is less than the number 1.0:

$$Cr = v\frac{\Delta t}{\Delta x} < 1 \tag{5}$$

The AD module uses output data from the Hydrodynamic module as its input data, while the dispersion coefficient (D) is described as a function of the mean average velocity, as it depicted in equation six below [8, 9, 13]:

$$D = av^b \tag{6}$$

Where the given variables are as follows: a – dispersion factor, b – dispersion exponent.

The paper deals with 1-dimensional numerical model MIKE 11 and its response on various values of dispersion coefficient. This parameter is one of the most important input data for simulation of pollution spreading in streams. Getting fair value, however, is in practice very difficult [14].

Longitudinal dispersion is difficult to determine as it depends upon too many variables and their nonlinear inter-relationships. A large disparity exists between the values of dispersion coefficients obtained for idealized and simplified systems (such as irrigation channels) and for rivers [10]. Such a disparity suggests that the processes contributing to dispersion in rivers are not well understood. The knowledge of accurate value of longitudinal dispersion coefficient D is important for determining self-purifying characteristics of streams, devising water diversion strategies, designing treatment plants, intakes and outfalls, and studying environmental impact due to injection of polluting effluents into the stream [14].

3.2.3   Water Quality Module

Water quality (WQ) module describes the basic processes of river water quality in areas influenced by human activities, e.g. oxygen depletion and BOD levels as a result of organic matter loads [15].

The WQ-module is coupled to the AD module, which means that the WQ module deals with the chemical/biological transforming processes of compounds in the river and the AD module is used to simulate the simultaneous transport process. The WQ module solves a system of coupled differential equations describing the physical, chemical and biological interactions in the river. The relevant water quality components must be defined in the AD editor [8, 15].

The water quality processes include modeling of DO and BOD with nutrients, COD with nutrients, eutrophication, heavy metals, iron-oxidation, extended eutrophication and nutrient transport. The component which involves the modeling of DO and BOD corresponds to different levels (six levels) of increasing complexity as shown in Fig. 2. Phosphorus and coliform components can also be added to any level of complexity. Concentrations of DO and BOD were calculated in MIKE 11 by taking into consideration advection, dispersion and the most important biological, chemical and physical processes [8, 15].

BOD – DO Model

Modeling Dissolved Oxygen

The dissolved oxygen model (DO model) includes the processes reaeration and bacterial decomposition of the BOD. The dissolved oxygen model at the level one can be written as follows [8, 15]:

$$\frac{dC}{dt} = K_a(C_s - C) - K_{BOD}L \tag{7}$$

Where:

C - concentrations of dissolved oxygen (mg O2/l), Cs - concentration of saturated oxygen (mg O2/l), Ka - coefficient of reaeration (1/day), KBOD - degradation coefficient BOD5 (1/day), L – concentrations of BOD5.

Modeling Biochemical Oxygen Demand

Modeling the biochemical oxygen demand of BOD5 at the level one within the (WQ) water quality module includes biological oxidation or degradation, or

$$\frac{dL}{dt} = -K_{BOD}L \tag{8}$$

Where:

L - concentration of BOD5, KBOD - organic matter decomposition coefficient expressed by BOD and calculated according to the following expression [8, 15]:

$$K_{BOD} = K_1 L \frac{C^2}{K_2 + C^2} \theta_{BOD}^{(T-20)}$$ (9)

Where:
K1 is the BOD oxidation coefficient, C is the concentration of dissolved oxygen, K2 is the coefficient of the influence of dissolved oxygen concentration on the BOD distribution coefficient and θ_{BOD} is the temperature coefficient of impact.

4   Model Application of the Neretva River

The model streaming area or zone where the model will be implemented represents the section of the Neretva River immediately downstream from the reservoir of the Mostar Power Plant to the Žitomislići Water Level Station (See Fig. 4).

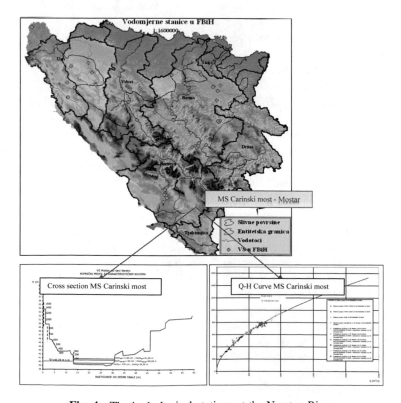

Fig. 4. The hydrological stations at the Neretva River

The input data have been used to development of the numerical model of the Neretva River are [16, 17]:

- Digitalized maps of the Neretva River scaled 1:1000, and information on cross sections along the axis of the Neretva River's model streaming area – from Mostar HydroPower plant to MS Žitomislići.
- Registered hydrological data (water discharge and water level), input of wastewater discharges and tributary flows as well as data on water quality from the following measuring and water quality stations:
  - Sutina/Raštani – immediately downstream from HE Mostar Power Plant
  - Carinski most
  - Bačevići
  - Buna on the river Buna
  - Žitomislići

The considered river reach has a length of 26 km. The numerical model will also take into account the main tributaries of the Neretva River in the section under observation: the Radobolja, Jasenica and Buna rivers. This research paper also takes into account the discharge of city waste water (communal and industrial, in part) from the urban core of Mostar. These discharges of waste water are dumped directly into the Neretva River without any kind of prior waste water treatment. The location selected for the discharge of Mostar's communal waste water is located immediately downstream from the Mostar Water Level Station [16, 17].

MIKE-11, a one-dimensional hydrodynamic simulation program developed by the Danish Hydraulic Institute (DHI) was utilized to model stream flow and water quality processing in the Neretva River. To initiate the process of modeling all the identified streams of the study area, details of the main catchment of Neretva river, physical characteristics of the river such as cross sectional levels and dimensions, longitudinal bed profiles and slopes; meteorological and hydrological data such as rainfall; parameters and coefficients of the affected water quality components; and pollution inputs in term of pollutant concentration was compiled and analyzed. The model was applied to simulate electrical conductivity, DO and BOD, at two monitoring stations, namely Bačević and Žitomislić [16, 17].

Hydrodynamic flow analysis is performed within the hydrodynamic module (HD), which is the core of the system. For the analysis of transport, the advective-dispersive module (AD) was applied, while the analysis of the change in the concentration of oxygen parameters was realized by the WQ model, which relies on the solution of the hydrodynamics of flow and penetration, ie solutions obtained from the HD and AD modules. The considered area, to which the MIKE 11 model is applied, is located in the middle course of the Neretva river downstream of HPP Mostar to MP Žitomislići. The total length of the considered section is about 26 km. The length of the section from HPP Mostar to the railway bridge, where MP Bačevići is located is about 14 km, and from the MP Bačevići to MP Žitomislići 12 km. The most important tributaries in the considered area are Radobolje, Jasenica and Buna [16, 17].

For the application of the MIKE 11 model, the geometric, hydraulic, hydrological and qualitative characteristics of the analyzed watercourse segment, as well as the quantities and dynamics of pollutant release, are defined. For the modeling of hydrodynamic processes, that is, simulation of flow and water level along the Neretva river, it was necessary to define the values of the coefficient of friction (roughness), while the values of dispersion coefficients should have been predicted for the modeling of transport processes. For water quality modeling, unstable (non-conservative) parameters, subject to decomposition processes, BOD5 and dissolved oxygen O2, were selected as key quality parameters, and the organic decomposition factors and reaction coefficients were to be defined.

4.1   Model Calibration and Verification

In this study, the pollution characteristics in the Neretva River are compared through analysis of observation results and modeling. Numerical simulations were performed with the MIKE 11 software.

The focus of this paper is the calibration and verification of solute transport models using measured data and prediction of different management scenarios for reducing pollution in the Neretva River (Simulation of future water quality status without and with Wastewater Treatment Plant). During this research work, the MIKE 11 numerical model has been used to simulate different states of water quality in certain selected segment of the Neretva River (Power plant Mostar – MS Bačevići – MS Žitomislići). The key parameters on water quality, BOD and dissolved oxygen (DO), were modelled, as was electrical conductivity, which is an additional conservative parameter used to calibrate the dispersion coefficient (D). This is a research paper that describes the BOD-DO model, which takes into account advection, dispersion and the most important biological, chemical and physical processes [9, 16, 17].

To calibrate the advection-dispersion (AD) model it was necessary to, beforehand, calibrate the hydrodynamic model of the Neretva River. Therefore, the model's calibration encompassed by this work has the primary task of defining the following values or variables: the coefficient of roughness, the dispersion coefficient, as well as the coefficients for organic material decay and reaeration.

The calibration of the hydrodynamic (HD) model (See Fig. 5) was implemented using data on water level values collected from the MS Bačevići. The entry data into the model (the upstream boundary condition) was the discharge from the Hydroelectric Power Plant Mostar–Water Wave 28 – on 31 March 2005, while the downstream boundary condition was the curve water level graph registered at the Žitomislići hydrological station (Measurement station). Verification of the model (See Fig. 5) using water discharge data from the aforesaid MS Bačevići on 28–30 June 1979 (which is during the season of low water) [9, 16, 17].

Calibration of the AD/WQ model (See Figs. 6 and 7) was conducted on the basis of available experimental data (the simultaneous measurement of BOD, DO and conductivity parameters in June of 1979) [9, 16, 17]. Data on water quality, collected in 2005 and statistical analyzed (mean average values), was used for the verification of the AD part of the model [9, 16, 17] (Fig. 8).

It is clear that calibration and verification represent the bulk of the procedure for model development and testing, once an experimental data set has been obtained. There is, however, no guarantee that the validity of the model extends beyond the sample data set against which it has been calibrated. Validation is, then, the testing of the adequacy

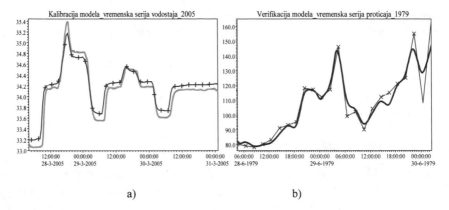

a)   b)

Fig. 5. Calibration and verification of HD model at MS Bačevići - comparison between measured and calculated (a) water levels and (b) water discharges (blue line represents a measurement and the solid line (green and red) the model result [16, 17]

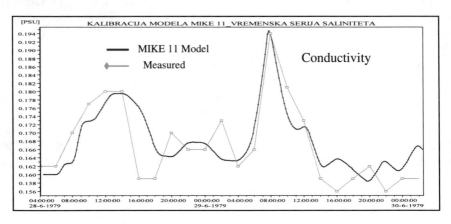

Fig. 6. Calibration of AD model - comparison between simulated and observed conductivity at station Bačevići (28–30 June 1979) [16, 17]

of the model against a second, independent set of field data. Because validation thus entails the design and implementation of new experiments, it is unfortunately a step in the analysis that is all too rarely attempted.

5   Results and Discussions

We have developed a numerical model used software MIKE 11 that solves the one-dimensional Saint-Venant equations and the Advection-Dispersion-Reaction equation to study the pollutants transport in the Neretva River.

The results of the numerical simulations in this paper are presented and are shown below. During the implemented simulations, the calibration of the coefficient of

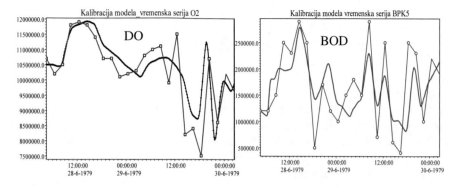

Fig. 7. Calibration of WQ model - comparison between simulated and observed DO and BOD at station Bačevići (28–30 June 1979) [16, 17]

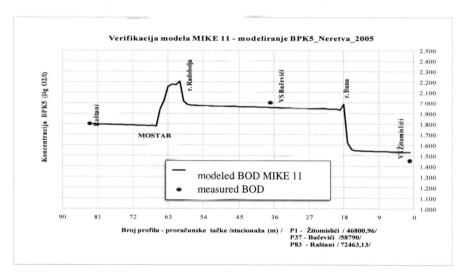

Fig. 8. Verification of WQ model/longitudinal profile/comparison between simulated and observed BOD at study reach of Neretva River (2005 Year) [11]

roughness was carried out by making a comparison of calculated and registered (measured) water levels at the Bačevići Measurement Station. The various coefficients of roughness (both lengths and depths of the cross sections) are result of this calibration. The coefficient of roughness also takes into account the resistance of local structures protruding into the river (bridges and other structures) so that it was not necessary to undertake separate calculations for each individual structure.

The model results show agreement with measurements of water level and discharge, as well with water quality parameters. Therefore, it is considered that the developed model can be implemented and applied to different situations for this study area and others rivers with similar characteristics.

The coefficients of resistance, calculated through the calibration of the hydrodynamic model, are in the following range $n \approx 0,030–0,085$ m-$^{1/3}$s. The results of calibration and verification of the hydrodynamic model (HD) are shown in Fig. 5. The model results show agreement between measured and simulated values of both water discharges (b) and water levels (a) at the measuring station Bačevići.

As shown in Fig. 5, the hydrodynamics numerical results correspond fairly well with field measurements, which demonstrate that the model results are consistent and reliable to the real river behaviour. Therefore, it is considered that the developed model can be implemented and applied to different situations for the studied area [16, 17].

After calibration and verification of hydrodynamic parameters of model we made calibration of advective-dispersive parameters of model. Calibration of the AD Model gave values for the coefficients of dispersion, which range from 130–280 m^2/s (Fig. 6).

Likewise, the water quality modules were validated by comparison with field measurements, observing that the model results are consistent with these measurements and they are in the same order of magnitude. After this step it is possible to carry out the calibration of coefficients of decay for BOD and DO parameters at MS Bačevići (Table 2).

Table 2. Water quality monitoring at river neretva during 1979 year [11].

Water quality parameter	Measuring profile HPP Mostar			Measuring profile Bačević		
	Min.	Aver.	Max	Min.	Aver.	Max
COD (mg/l)	1.9	3.30	5.06	0.95	4.10	6.7
DO (mg/l)	7.7	10.6	12.2	7.5	10.6	13.8
BOD (mg/l)	0.1	1.2	2.2	0.4	1.8	4.0
Ammonia (mg/l)	0.02	0.04	0.16	0.02	0.07	0.29
Nitriti (mg/l)	0.001	0.0014	0.002	0.001	0.0022	0.006
Nitrate (mg/l)	0.10	0.48	0.95	0.10	0.40	0.90
Conductivity (µS)	334	343	386	324	352	392

The decay factor received through the model's calibration has the following value 0, 5–1, 5 1/day, which is a normal value for natural rivers such as the Neretva River (a river with expressed turbulence). The coefficient of reaeration has a mean average value of about 2,0 1/day. Comparative view of measured and simulated data for DO and BOD can clearly be seen in Fig. 7.

Statistical analysis of modelled values and observations for the water level and water quality parameters have given us the following coefficients of correlation (See Table 3).

Data on water quality, collected in 2005 was used for the verification of the WQ of the model.

Table 3. Statistical analysis of MIKE 11 model calibration results

Model parameters	Coefficient of correlation R^2
Water level (H)	0.946
Conductivity [µS]	0.623
DO [mg/l]	0.658
BOD [mg/l]	0.512

For the assessment of the future status of watercourses and the effects of waste water, after the implemented calibration and verification of the numerical model, a simulation of the future status of the water quality of the Neretva River was made for the planning period from 2012 to 2032. (planning threshold in 2022). Two simulation scenarios were performed representing dry season:

– Scenario 1 (without the planned WWTP).
– Scenario 2 (with planned WWTP).

Figure 7 shows some of the obtained results (line with point represents a measurement value and the solid line the model result).

Figures 9 and 10 show some of the obtained results.

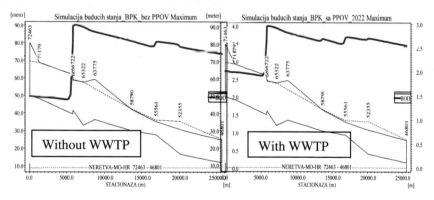

Fig. 9. Longitudinal profile - simulation future water quality status for BOD at station Bačevići (2022 Years) without and with planned WWTP

Figure 9 shows a comparative view of the results for management options according to scenarios 1 and 2. In the figure it can be clearly seen what effects are achieved for the scenario with the built I phase WWTP. Maximum concentrations do not exceed MPC (Maximum Permissible Concentration) for either scenario 1 or scenario 2, but scenario 2, downstream along the river, improves water quality on a larger segment reach.

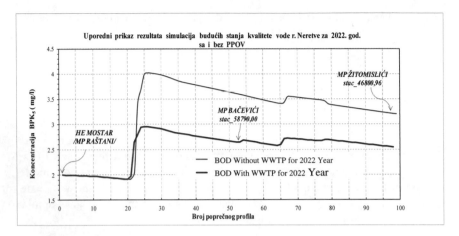

Fig. 10. Comparative results for modelled BOD without and with WWTP at station Bačevići (red line represents the model result without WWTP and blue line represents the model result with planned WWTP)

6 Conclusions

MIKE 11 model has demonstrated its applicability to simulation of pollution in streams, and therefore is an appropriate tool for decision making related to the quality of water resources.

The following can be concluded on the basis of the results of the implemented simulations:

- The accuracy of the calibration results of the model is mostly influenced by input data (pollutant load, quantity and composition of wastewater, location and method of indentation, characteristics of the receiver, etc.), then the reliability of the estimated model parameters, the structure (complexity) of the model and, finally, the calibration quality.
- As a basis for a precise model of transport of pollution, it is important to have a well-established hydrodynamic model. Although this has been established in the model presented in this paper, the modeling of transport-qualitative processes to unstable water quality parameters, such as BPK5 and O2, cannot be precisely performed.
- Modeling of these unstable parameters is not useless, as it provides a good insight into the changes over time and along the stream, and approximate estimates of future water quality conditions can be given depending on changing input data (boundary conditions).
- In the city of Mostar (approx. population of 100,000) there is no wastewater treatment plant, yet (it's under construction). However, this model can also be used to predict the water quality of the Neretva River even after the construction of a wastewater treatment plant. In other words, the future capacity level for water treatment can be planned using this model.

- Although the quality of the Neretva river water is already in the existing state, according to all modeled indicators, it can be assessed as satisfactory, by constructing the entire drainage system and the first and second phase of the waste water treatment plant of the City of Mostar (100 000 and 175 000 ES) improvement of the water quality of the Neretva river at an effective flow of 50 m^3/s.
- If certain decisions are made on the basis of the results of modelling the advection and dispersion of pollutants into the Neretva River, then we need to take into account propagation of uncertainty of estimated parameters of the model in a final complete solutions. These decisions, for example, could be on the management of water resources in the watershed area or other activities undertaken with the goal of ecological protection.

This paper addresses the important problem of river pollution and pollutant propagation, which has to be attended with the help of predictive tools (pollutant transport models) in order to develop pollution assessment and counteracting systems and to take correct management decisions.

References

1. Ziemińska-Stolarska, A., Skrzypski, J.: Review of mathematical models of water quality. Ecol. Chem. Eng. S. **19**(2), 197–211 (2012). https://doi.org/10.2478/v10216-011-0015-x
2. Riahi-Madvar, H., Ayyoubzadeh, S.A.: Developing an expert system for predicting pollutant dispersion in natural streams. In: Vizureanu, P. (ed.) Expert Systems (2010). ISBN: 978-953-307-032-2
3. Marusic, G.: A study on the mathematical modeling of water quality in "River-Type" aquatic systems. WSEAS Trans. Fluid Mech. **2**(8), 80–89 (2013)
4. James, A.: An Introduction to Water Quality Modelling, 2nd edn. Wiley, New York (1993)
5. Orlob, I., Gerald, T.: Mathematical modeling of water quality: streams, lakes, and reservoirs, Copyright © 1983 International Institute for Applied Systems Analysis (1982)
6. Martin, J., McCutcheon: Hydrodynamics and Transport for Water Quality modelling. CRC Press, New York (1999)
7. Zheng, C., Gordon, B.: Applied Contaminant Transport Modelling. Van Nostrand Reinhold, New York (1995)
8. Danish Hydraulic Institute: Mike 11. A Modelling System for Rivers and Channels. User Guide. Danish Hydraulic Institute, Hørsholm (2007a). Danish Hydraulic Institute: ECOlab, WQ Templates, Scientific Description. Danish Hydraulic Institute, Hørsholm (2007b). Danish Hydraulic Institute: ECOlab, User Guide. Danish Hydraulic Institute, Hørsholm (2007c)
9. Milišić, H., Kalajdžisalihović, H.: Pollutant dispersion modelling in natural rivers. In: Conference Proceedings - 5th IWA Eastern European Water Professionals Conference for Young and Senior Water Professionals, The International Water Association (IWA) - Kiev, Ukraina, 26–28 June 2013., str. 127–134 (2013)
10. Ruzgiene, I., Ruzgas, T.: Mathematical modeling of water quality change in eastern european river. Int. J. Sci. Environ. Technol. **3**(3), 861–866 (2014). ISSN 2278-3687 (O)
11. Bandić, H.: Analiza primjene numeričkih modela za simulaciju transporta zagađenja u vodotocima - Magistarski rad, Fakultet građevinarstva, arhitekture i geodezije Sveučilišta u Splitu (2012)

12. Andrei, A., et al.: Numerical limitations of 1D hydraulic models using MIKE11 or HEC-RAS software – case study of Baraolt River, Romania. IOP Conf. Ser.: Mater. Sci. Eng. **245**, 072010 (2017)
13. Liang, J., et al.: MIKE 11 model-based water quality model as a tool for the evaluation of water quality management plans. J. Water Supply: Res. Technol.—AQUA (2015). https://doi.org/10.2166/aqua.2015.048
14. Ayyoubzadeh, S., et al.: Estimating longitudinal dispersion coefficient in rivers. Expert Syst. Appl.: Int. J. Arch. 36(4) (2009)
15. Radwan, M., et al.: Modelling of dissolved oxygen and biochemical oxygen demand in river water using a detailed and a simplified model. Proc. Intl. J. River Basin Manag. **1**(2), 97–103 (2003)
16. Milišić, H., Kalajdžisalihović, H.: Numeričko modeliranje i simulacija transportra zagađenja Neretvom – Časopis "VODOPRIVREDA", Beograd, Srpsko društvo za odvodnjavanje i navodnjavanje, Broj 258–260, str. 199–206 (2012)
17. Milišic, H., Hadžić, E., Lazović, N.: Application modeling and assessment of water quality in a natural rivers. In: Conference Proceedings - 2nd International Conference on Multiscale Computations for Solids and Fluids, 10–12 June 2015 - Sarajevo, Bosnia and Herzegovina (2015)

Method of Annual Extreme and Peaks Over Threshold in Analysis of Maximum Discharge

Ajla Mulaomorević-Šeta$^{(\boxtimes)}$, Nerma Lazović, Emina Hadžić,
Hata Milišić, and Željko Lozančić

Department of Water Resources and Environmental Engineering,
University of Sarajevo, Sarajevo, Bosnia and Herzegovina
ajla.mulaomerovic@gf.unsa.ba, nerma.ligata@gmail.com,
eminahd@gmail.com, hata.milisic@gmail.com,
zeljko.lozancic@gmail.com

Abstract. The comparative results of defining high waters with a probabilistic approach are presented in the paper. High waters are defined using two most commonly used methods that are of interest for the rational dimensioning of the corresponding types of hydrotechnical objects and systems: the method of annual extremes and the method of peaks/thresholds. The method of annual extreme treats the theoretical distribution functions commonly used in hydrological practice: Normal (Gaussian), Log-Normal (Galton), Pearson 3, Log-Pearson 3, and Gumbel's distribution, and the final selection of the function is based on the results of the Kolmogorov test, i.e. agreement of the empirical and theoretical probability distribution functions. For the threshold method, a Poisson-Weibull model with a Poisson distribution for the peak occurrence frequencies and a two-parameter Weibull's distribution for peaks height was used, which for the maximum discharge gives a three-parametric distribution function. Comparative results of high waters according to these methods are given to 11 gauge stations in Vrbas river basin. Basin areas are from 200 up to almost 5300 km^2, and observation duration from 16 to 47 years.

Keywords: High waters · Probabilistic methods · Method of annual extreme
Peak/threshold method · Return period · Vrbas river

1 Introduction

Flood protection in basins of smaller watercourses in Bosnia and Herzegovina has always been in the second plan. Protection measures were mostly local in nature [1], limited to larger settlements or more important industrial facilities. Existing flood protection facilities could not always provide protection. The problem of estimating flood water is even more relevant after the floods of May 2014. At present, there are no regulations or recommendation concerning high water of return period T ($Q_{max,T}$) determination in Bosnia and Herzegovina.

In this paper, statistical methods are used to determine high waters Qmax,T, which are based on the use of historical data on the occurrence of large waters only. Such data are subjected to statistical analyzes, with the ultimate goal of constructing a line of

© Springer Nature Switzerland AG 2019
S. Avdaković (Ed.): IAT 2018, LNNS 60, pp. 157–174, 2019.
https://doi.org/10.1007/978-3-030-02577-9_16

probability of occurrence of large waters, i.e. to determine the exceedance probability of certain quantities of flows.

In hydrological practice, the available data sets are far short of the required [2], so the probability of relatively rare occurrences is obtained using extrapolations of the probability function. In order to avoid unwanted surprises, which could result from unreliable estimates of high waters, practical solutions are reduced either to the selection of a large water of a relatively long return period, or to the adoption of a value from the upper limit of 95% confidence interval for a certain return period.

In this paper, two statistical methods are employed to determine high waters in Vrbas river basin: method of annual extrema taking into account only one, extreme events, in a year, and peak over threshold method considering data above a threshold without losing the valuable information that could be given and other observed flow.

2   Materials and Methods

The aim of this study was executed based on observation series of annual maximum events ($Q_{max,god}$) and peak over threshold (POT) ξ for the analyzed basins The observation series covered the multi-year period (Table 1, Fig. 2). For annual maxima, the values of $Q_{max,T}$ were determined among five distribution laws (Gauss, Galton, Log-Pearson, Pearson and Gummbel), and final selection is based on the results of Kolmogorov test for which is proved best matching with empirical function (Weibull). Statistic parameters are estimated using moment method.

Beside annual maxima approach, $Q_{max,T}$ are determined using another statistical method [3, 4], peak over trashold, employing Poisson distribution law describing peak occurrence frequency and Waibull 2 parameters function for peaks value.

On Fig. 2 there are input data for calculation. Red bars represent annual maxima, with position on begin of a year, and blue bars represent peaks (Fig. 1).

3   Statistical Analysis

Statistical analysis was conducted using method of annual maxima values $Q_{max,god}$. The term annual extreme refers to the largest instantaneous flow rate in a particular river profile, registered during the calendar (more frequent) or hydrological year. By taking all such extremes out of the available number of years, a time series of maximum annual flows is formed Q_1^m, Q_2^m, $\dots$, Q_{i-1}^m, Q_i^m, Q_{i+1}^m, $\dots$, Q_N^m which forms the basis for any further statistical analysis of maxima annual flows by the method of annual extremes, whereby it is understood that the time series members (maximum annual flows) are random and mutually independent.

The aim of the analysis is to determine the likelihood of the phenomenon, that is, the probability distribution function of maxima annual flows, which is achieved by finding the probability distribution function [5, 6]:

Table 1. Review of available data (annual maxima and peaks) on average daily discharge on gauge station in Vrbas basin

Basin	Gauge station	Period	Interruption	Observation (years)
Vrbas	Gornji Vakuf	1946–1988	1965	42
Vrbas	Daljan	1972–1990	1974, 1977, 1981	16
Vrbas	Han Skela	1971–1990		20
Janj	Otoka	1986–1990	1976, 1983, 1984	20
Janj	Sarići	1963–1990	1980–1984	23
Pliva	Majevac	1967–1989		22
Pliva	Volari	1971–1990		20
Vrbas	Kozluk Jajce	1971–1989		19
Vrbas	Banja Luka	1962–2015	1968, 1972, 1974–1977, 1981, 1984–1996	34
Vrbanja	Vrbanja	1961–2015	1971 1973, 1975, 1992–1996	47
Vrbas	Delibašino Selo	1962–2015	1970, 1972, 1976–1977, 1979, 1981, 1987	47

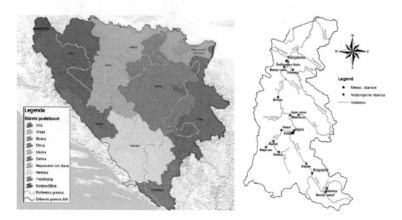

Fig. 1. Main river basin in Bosnia and Herzegovina (left); Gauge station position on Vrbas river basin (right) [7]

$$\Phi(Q) = P[Q \geq q] \tag{1}$$

For the empirical distribution function, a Weibull's probability distribution function is defined as

$$\Phi_e(Q_m) = \frac{m}{N+1} \tag{2}$$

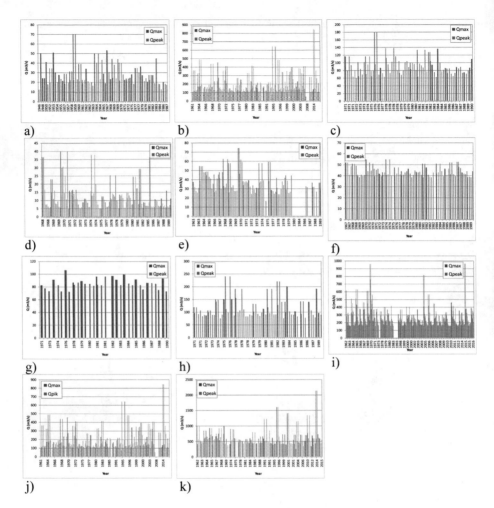

Fig. 2. Annual maxima and peak recorded on relevant gauge stationa: (a) Gornji Vakuf; (b) Daljan; (c) Han Skela; (d) Otoka; (e) Sarići; (f) Majevac; (g) Volari; (h) Kozluk Jajce; (i) Banja Luka; (j) Vrbanja; (k) Delibasino Selo

where m represents the regular number of flows in the ordered sample (ascending order), and N total length (equal to the number of years of observation) N.

Below are the expressions used to calculate the basic numerical characteristics [8] for the series of maxima annual flows on treated watercourses, whose values can be found in Table 2.

Average value:

$$Q_{avr} = E(Q) = \frac{1}{N} \sum_{i=1}^{N} Q_i \qquad (3)$$

Table 2. Numerical characteristics of annual maximum discharge on relevant gauge stations for available data

Original data				Gauge station	Log data			
Y_{avr}	S_y	C_S	C_V		Y_{avr}	S_y	C_S	C_V
29.65	12.58	1.02	0.42	Gornji Vakuf	1.44	0.18	0.10	0.12
86.33	35.70	0.96	0.41	Daljan	1.90	0.17	0.44	0.09
104.75	27.06	0.84	0.26	Han Skela	2.01	0.11	0.15	0.05
19.90	10.57	0.63	0.53	Otoka	1.24	0.24	−0.11	0.19
43.13	12.94	0.38	0.30	Sarići	1.61	0.14	−0.63	0.09
46.30	5.13	−0.11	0.11	Majevac	1.66	0.05	−0.31	0.03
85.49	9.88	0.17	0.12	Volari	1.93	0.05	0.01	0.03
150.23	42.72	0.45	0.28	Kozluk Jajce	2.16	0.12	0.01	0.06
430.65	184.01	1.54	0.43	Banja Luka	2.60	0.16	0.50	0.06
268.15	157.02	1.29	0.59	Vrbanja	2.36	0.26	−0.21	0.11
709.04	358.31	1.84	0.51	Delibašino Selo	2.81	0.19	0.52	0.07

Variance:

$$Var(Q) = \frac{1}{N}\sum_{i=1}^{N}(Q_i - Q_{avr})^2 = \frac{1}{N}\sum_{i=1}^{N}Q_i^2 - Q_{avr}^2 \qquad (4)$$

Standard deviation:

$$S = \sqrt{Var(Q)} \qquad (5)$$

Coefficient of variation:

$$c_v(Q) = \frac{S}{Q_{avr}} \qquad (6)$$

Skewness:

$$c_s = \frac{\frac{1}{N}\sum_{i=1}^{N}(Q_i - Q_{avr})^3}{S^3} \qquad (7)$$

Based on the calculated statistical parameters of maxima annual flows, five distribution functions, commonly used in hydrology, are applied: Normal (Gaussian), Log-Normal (Galton), Pearson III, Log-Pearson III and Gumbel's distribution functions. Which of the above theoretical distribution functions are likely to describe the sample depends on the goodness of adjusting the theoretical to empirical function of probability distribution. To test the goodness of fit, Kolmogorov test was used, which is based on the verification of the maximum difference, i.e. the maximum deviation between the theoretical and empirical functions given as:

$$D_N = \max|\Phi_e(x_i) - \Phi_t(x_i)| \quad for\ i = 1,\ 2,\ \ldots,\ n \quad -\infty < x < \infty \tag{8}$$

Where $\Phi_e(x_i)$ and $\Phi_t(x_i)$ are empirical (Weibull) and theoretical probability of exceedance, respectively.

When the relationship is met:

$$P[D_N \leq D_0] = P[D_N \leq D_{1-\alpha}] = 1 - \alpha \tag{9}$$

it means that the largest difference D_N is relatively small because it is smaller than some critical value of $D_{1-\alpha}$ for the significant level α (α usually takes 5% in hydrological practice) (Fig. 3).

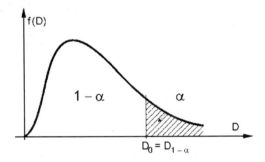

Fig. 3. Region of rejection of H_0 hypothesis for the adopted significance level α [5]

Using the term (8), the maximum difference D_N of the empirical and theoretical distribution functions are compared with the critical value D_0 (which is the function of degrees of freedom and risk coefficient α) in order to assess which region belongs to, i.e. whether the zero hypothesis is accepted (that the differences are not significant), or rejects it (Table 3).

If D_N is small enough (with an acceptable level of risk α), it implies that the theoretical distribution function $F_t(x)$ is well adapted, so that the H_0 hypothesis is accepted, i.e. the sample volume N with the empirical distribution function $F_e(x)$ belongs to a population whose continuous distribution function $F_t(x)$.

Comparing D_N to D_0 it is concluded that every theoretical distribution function satisfies Kolmogorov test at significante level of 5%. Ultimately, the function that has the least relative deviation from the empirical function has been adopted.

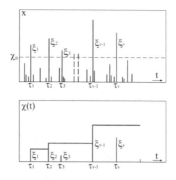

Fig. 4. Illustration of the random process χ (t) [6]

Calculation of high waters in function of return period (T) is done employing frequency factor K_T, i.e.: dodat uz $y(T)$

$$y_T = y_{avr} + K_T S_y \tag{10}$$

Where

y_T is annual maxima for return period T (or its logarithm, in case of employing Galton or log-Pearson III function),
K_T frequency factor depending on distribution law and return period
S_y standard deviation of annual maxima (or standard deviation od logarithm of annual maxima).

In case of Log function, $Q_{max,T} = 10^y$, otherwise $y_T = Q_{max,T}$.
Results of annual maxima for selected gauge station is presented in Fig. 7.

4 Peak Method

In the analysis of high waters by the annual maxima method, a sample of maximum flows is formed by taking only one extreme data from a year, while all other data are rejected, which is the main objection to this method. Clearly, the selected extreme in one year can be overcome several times in next year, and according to the theory of extremes, this data is rejected so that a smaller sample of maximal flows for statistical analysis is obtained and thus loses the valuable information that could be given and other observed flows. This defect of the annual extreme method eliminates the peak method that takes into account all the extreme values of hydrologic random variables that exceed a limit (base value or threshold) no matter how many times they appear in a year. The advantage of this method is that it can provide some additional information in analysing the occurrence of maxima annual flows such as the distribution of the number of peaks (peak occurrence frequency), the distribution of peak height, etc.

The occurrence of the values of maximum flows X which are greater than some values in the time interval (0, t) - for example a year, is analysed. This phenomenon is a

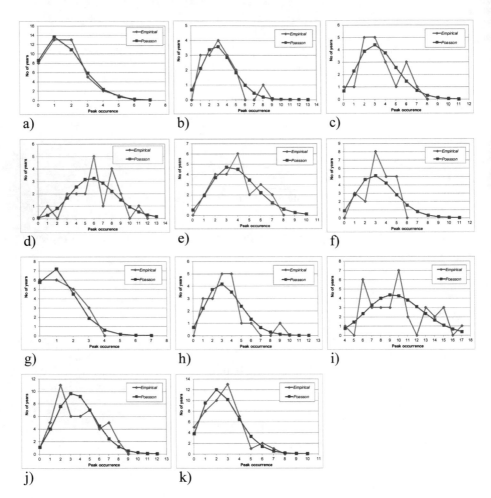

Fig. 5. Poisson distribution of the number of peaks (peak occurrence frequency) for relevant stations (a) Gornji Vakuf; (b) Daljan; (c) Han Skela; (d) Otoka; (e) Sarići; (f) Majevac; (g) Volari; (h) Kozluk Jajce; (i) Banja Luka; (j) Vrbanja; (k) Delibasino Selo

typical random process because obviously it cannot be predicted with certainty in which year a certain value will occur.

This random process is defined by the term [6]:

$$\chi(t) = \max\{\xi_v\} \quad \text{where} \quad \xi = X - x_B \quad \tau_v \le t. \quad v = 1, \ldots \eta_t \tag{11}$$

One realization of this random process is shown in Fig. 4.

In the upper part of Fig. 4 a chronological diagram of randomly variable X, i.e. flows above threshold (χ_B) registered for example in one year, is shown. It is noted from the figure that there are v registered values of random variable X are higher than the base value χ_B. The occurrence of the first value greater than χ_B with peak height

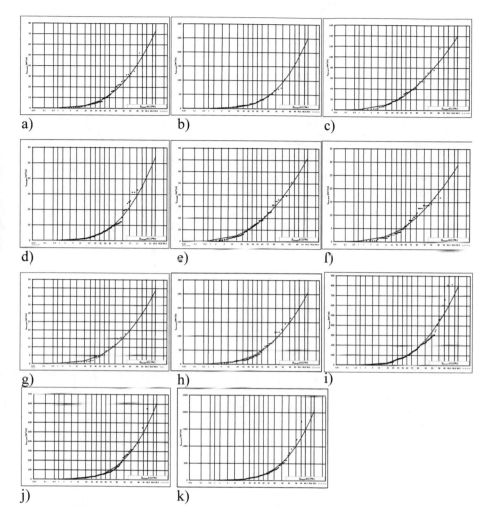

Fig. 6. Peaks height distribution (dots for empirical, blue line for theoretical 2-parameters Weibull's distribution) for relevant stations (a) Gornji Vakuf, (b) Daljan; (c) Han Skela; (d) Otoka; (e) Sarići; (f) Majevac; (g) Volari; (h) Kozluk Jajce; (i) Banja Luka; (j) Vrbanja; (k) Delibasino Selo

$\xi_1 = \chi_1 - \chi_B$ recorded at the moment τ_1, other values χ_2 at a time τ_2, and χ_v with peak height ξ_v at the moment τ_v.

The bottom part of Fig. 4 shows the dependence of the random process $\chi(t)$ in the realization of the diagram on the upper part of the Fig. 4. Since $\chi(t)$ represents the maximum value of the height $\chi(t)$ in the interval $(0, \tau_i)$, the size $\chi(t)$ in the interval $(0, \tau_1)$ is ξ_1. In the interval $(0, \tau_2)$ it is ξ_2 since it is $\xi_2 > \xi_1$. In the interval $(0, \tau_3)$, it is also ξ_2 since it is maximum value of three registered peak values. The value $\chi(t)$ for the total observed interval $(0, t)$, for example a year, according to Fig. 4, is ξ_{v-1} because it is the maximum value of all peak height.

Table 3. Absolute differences D_N between empirical and theoretical probability distribution functions, critical value D_0 for 5% significant level and adopted distribution

Gauge station	Gaus	Galton	P III	L- P III	Gumbel	D_0	Adopted
Gornji Vakuf	0.128	0.048	0.061	**0.042**	0.059	0.21	L-P III
Daljan	0.175	0.159	0.129	**0.132**	0.144	0.38	L-P III
Han Skela	0.108	0.090	**0.081**	0.084	0.085	0.33	P III
Otoka	0.198	**0.103**	0.162	0.109	0.143	0.32	GALTON
Sarići	0.089	**0.064**	0.076	0.073	0.084	0.29	GALTON
Majevac	0.138	0.139	0.137	**0.136**	0.163	0.31	L-P III
Volari	0.130	0.126	0.128	**0.126**	0.151	0.33	L-P III
Kozluk Jajce	0.149	0.134	**0.129**	0.134	0.144	0.34	P III
Banja Luka	0.186	0.116	0.106	**0.083**	0.120	0.24	L-P III
Vrbanja	0.123	0.085	0.067	0.078	**0.062**	0.21	GUMBEL
Delibašino Selo	0.174	0.084	0.054	**0.050**	0.104	0.20	L-P III

The occurrence of the maximum values of the extreme flows (χ) in the interval (0, t) is described by the distribution function:

$$F_t(x) = P[\chi(t) \leq x] \tag{12}$$

In order to calculate this distribution function, it is necessary to analyse two random variables:

- the number of extremes greater than the time interval (0.t) (i.e. peak occurrence frequency)
- peak height (extremes above selected base value).

4.1 Peak Occurrence Frequency

The number of peaks in the time interval (0, t) is a random size whose values can be 0. 1. 2. ... i.e. will have a distribution of probability [6]:

$$\eta_t : \begin{pmatrix} 0 & 1 & 2 & \cdots \\ p_0 & p_1 & p_2 & \cdots \end{pmatrix} \tag{13}$$

Characteristics of set $\{\eta_t = v\}$ are:

$$t > 0 \ (\eta_{t=i}) \cap (\eta_{t=j}) = 0, \quad i \neq j \ \bigcup_{v=0}^{\infty} (\eta_t = v) = \Omega_\eta = M \tag{14}$$

where Ω_η is the space of elementary events.

The density law of the probability distribution of peaks occurrence frequency in the time interval (0, t) is defined by the expression:

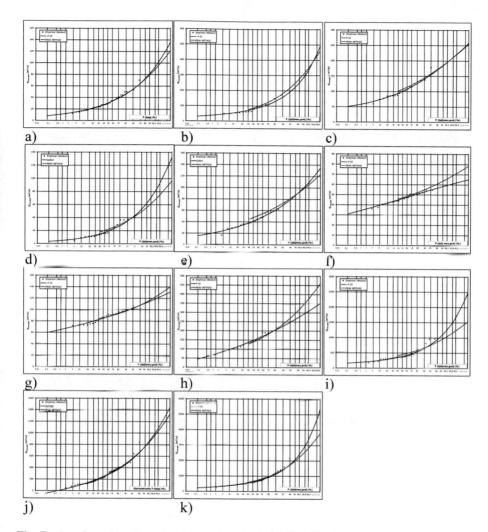

Fig. 7. Annul maxima determined applied methods (a) Gornji Vakuf; (b) Daljan; (c) Han Skela; (d) Otoka, (e) Sailel; (f) Majevac; (g) Volari; (h) Kozluk Jajce; (i) Banja Luka; (j) Vrbanja; (k) Delibasino Selo

$$p_v(t) = P[\eta_t = v] \tag{15}$$

Taking into account that the phenomenon of the number of peak occurence is a Markov process of a discrete type, and introducing the notion of the time-delay time function of the peaks $\lambda(t, v)$, the law of the distribution function (15) can be expressed depending on the shape of the function $\lambda(t, v)$. Thus, for the function of peaks occurrence frequency in time with the shape

$$\lambda(t, v) = \lambda(t) \tag{16}$$

and the probability of the number of peaks (i.e. occurence frequency) is expressed by:

$$p_v(t) = e^{-\lambda(t)} \frac{[\lambda(t)]^v}{v!} \tag{17}$$

which represents Poisson's law of probability distribution with a variable parameter $\lambda(t)$ which represents the average (expected) number of peaks in the time interval $(0,t)$. Expected value of the number of peaks $E(\eta_t)$ and the corresponding variance $Var(\eta_t)$ are:

$$E(\eta_t) = \lambda(t) \quad Var(\eta_t) = \lambda(t) \tag{18}$$

In engineering practice, this feature of the Poisson distribution is used to select the threshold value. Namely, by varying the threshold, one can choose the threshold at which the number of peak is distributed according to Poisson's law i.e. the condition $Var(\eta_t)/E(\eta_t) = 1$ must be satisfied (which follows from Eq. 18). In practical calculations, Poisson's distribution can be adopted if $0.8 < Var(\eta_t)/E(\eta_t) < 1.2$ [6].

The values that make up a series of peaks must be independent, which means that no flows can be taken from two, even more consecutive days, because they belong to the same hydrological event. A series of peaks consists of a different number of data for each year, due to which the distribution of a series of peaks is not directly comparable to the distribution of the corresponding series of annual maxima

The number of peaks in a year (peaks occurrence frequency) is discrete random variable that can take the values $\eta = 0, 1, 2, \ldots$ and consequently has a distribution of probability:

$$p = \{\eta = k\} = p_k . \quad k = 0, 1, 2, \ldots . \tag{19}$$

If a series of peaks is observed for N years, then the number of peaks in each year are $\eta_1, \eta_2, \ldots, \eta_N$, and the total number of peaks during N years is $M = \eta_1 + \eta_2 + \ldots + \eta_N$.

The average (expected) value and varians of the number of peaks (peaks occurrence) are:

$$\lambda = \bar{\eta} = E(\eta) = \frac{1}{N} \sum_{i=1}^{N} \eta_i = \frac{M}{N}; \tag{20}$$

$$Var(\eta) = \frac{1}{N} \sum_{i=1}^{N} (\eta_i - \bar{\eta})^2 \tag{21}$$

Peaks occurence frequency (empirical) (Eq. 19) and Poisson distribution function (Eq. 17) is shown on Fig. 5, and characteristic values of peaks occurence are represented in Table 4.

Table 4. Statistical characteristics of the number of peak (peak occurrence) and height on selected watercourses

Gauge station	Q_B (m³/s)	M	Occurrence frequency			Peaks height			Weibull's parameters	
			$E(\eta)$	$Var(\eta)$	I_d	$E(\xi)$	$Cv(\xi)$	$Cs(\xi)$	a	b
Gornji Vakuf	18	67	1.60	1.21	0.92	10.5	1.0	1.6	0.99	10.46
Daljan	43	51	3.19	1.83	1.05	27.1	1.0	1.9	0.83	24.46
Han Skela	62	68	3.40	1.85	1.00	28.9	0.8	1.3	1.29	31.27
Otoka	5	123	6.15	2.43	0.96	6.7	1.1	2.0	0.90	6.38
Sarići	23	88	3.83	1.77	0.82	14.7	0.8	1.0	1.28	15.82
Majevac	38	76	3.30	1.29	0.51	5.9	0.8	0.8	1.27	6.31
Volari	81	25	1.25	1.07	0.92	6.8	1.0	1.3	1.04	6.88
Kozluk Jajce	75	64	3.37	1.92	1.10	39.9	1.0	1.5	1.04	40.62
Banja Luka	150	333	9.79	3.29	1.11	106.6	1.1	2.7	0.95	103.91
Vrbanja	9?	179	3.81	2.13	1.19	86.1	1.2	2.8	0.83	77.68
Delibašino selo	400	119	2.53	1.63	1.04	206.6	1.3	3.0	0.79	181.50

4.2   Peaks Height Distribution Function

From the expression for the mean value of the number of peaks given by the equation [9, 10]:

$$E(\eta) = \lambda = \frac{M}{N} \tag{22}$$

it follows that during the (N) years of observation, $M = \lambda N$ values of random variable Q with values above threshold Q_B are recorded. These values are called peaks and their height is defined by the term $\xi = Q - Q_B$.

Editing peak height data by size, a statistical sequence of randomly variable realizations is formed

$$\Xi: \quad \xi_1, \ \xi_2, \ \dots, \ \xi_i, \ \dots, \ \xi_M \tag{23}$$

where

$$\xi_1 \leq \xi_2 \leq \ \dots \leq \ \xi_i \leq \dots \leq \ \xi_M \tag{24}$$

which allows calculating the empirical probability distribution function using the expression (the same one as for empirical distribution of annual maxima):

$$H_e(\xi) = P[\Xi \geq \xi] = \frac{m}{M+1} \tag{25}$$

where m is positions of random variable in the ordered sample, and M the total number of peaks.

The theoretical function of the peaks height distribution is defined by the expression [11, 12]:

$$H(\xi) = P[\Xi \leq \xi] \tag{26}$$

i.e.

$$\Phi(\xi) = P[\Xi \geq \xi] = 1 - H(\xi) \tag{27}$$

In order to define this probability distribution function, the number of peaks (peak occurrence frequency) in the interval $(0, \xi)$ is observed, using a random variable μ_ξ whose values can be 0, 1, 2. ... i.e. μ_ξ will have a distribution:

$$\mu_\xi: \begin{pmatrix} 0 & 1 & 2 & \dots \\ p_0 & p_1 & p_2 & \dots \end{pmatrix} \tag{28}$$

The distribution of the number of peaks in the time interval $(0, \xi)$ is defined by the expression:

$$p_n(\xi) = P[\mu_\xi = n] \tag{29}$$

Taking into account that the number of peaks is a Markov process of a discrete type, and introducing the notion of the time-delayed-peak response function $k(\xi, n)$, the law of the cumulative distribution function of the probability $H(\xi)$ and probability of the number of peaks $h(\xi)$ (Eq. 29), if $k(\xi, n) = k(\xi)$ is selected, can be expressed as [12]:

$$\Phi(\xi) = 1 - e^{-K(\xi)} \tag{30}$$

$$h(\xi) = k(\xi) \cdot e^{-K(\xi)} \tag{31}$$

where

$$K(\xi) = \int_0^\xi k(s)ds \tag{32}$$

It is obvious that the distribution of the peaks height directly depends on the shape of peaks accurence frequency distribution $k(\xi)$ in the interval $(0, \xi)$. Experience with the use of the peak method in analyzing the maximum values of the random variable has shown that in the analysis of the statistical series peaks, we can successfully apply Weibull's distribution, the Gudric distribution, and the two-parameter log-Normal distribution. In this paper, the peaks height distribution is described by the Weibull theoretical function, which is expressed as [12]:

$$H(\xi) = 1 - e^{-\left(\frac{\xi}{b}\right)^a}$$

(33)

i.e.

$$k(\xi) = \frac{a}{b}\left(\frac{\xi}{b}\right)^{a-1}$$

(34)

The values of the unknown parameters (a) and (b) are determined from the following two equations giving the expressions for calculating the mean value of the peak height $\bar{\xi}$ and the corresponding variation coefficient $c_{v\xi}$:

$$\bar{\xi} = b \cdot \Gamma_1; \quad c_{v\xi} = \frac{\sqrt{\Gamma_2 - \Gamma_1^2}}{\Gamma_1}$$

(35)

Based on the calculated parameters a and b, it was possible to calculate the values of the of peaks height cumulative distribution function $\Phi(\xi) = 1 - H(\xi)$ assuming Weibull two-parameter distribution.

Empirical (Eq. 25) and theoretical peaks height distribution (Eq. 34) for relevant stations are shown in Fig. 6, while values parameters a and b of Weibull's distribution are presented in Table 4.

4.3   Distribution Function of the Annual Maxima

After defining the distribution of peaks occurrence and peaks height distribution, it is possible to access the third step in applying the peak method, i.e. defining the distribution function of the annual extrema, which is a combination of the two distributions.

The distribution function of the extreme values in the interval (0,t) is defined by the expression:

$$F_t(x) = P[\chi(t) \le x]$$

(36)

Considering the characteristic of the set $[\eta_t - v]$ and assumptions:

- series of peaks $\xi_1, \xi_2, \ldots, \xi_v, \ldots$ consist of independent random variables with identical distribution $H(\xi) = P[\Xi \le \xi]$, that is $H(x) = P[X \le x]$, since it is $\Xi = X - x_B$
- for each $v = 1, 2, \ldots$ series $\{\xi_v\}_1^\infty$ is independent of the number of previous reports. ie. from $\tau(v)$ and $\tau(v+1)$

For the peaks occurence defined by Poisson's law (Eq. 17), the distribution function of the extremes will be [12]:

$$F_t(x) = e^{-\lambda(t)\cdot[1-H(x)]}$$

(37)

Using the above function it is possible, based on a series of extreme values (greater than the number of years), to obtain a relationship that allows defining the return period T (x) over the years:

$$T(x) = \frac{1}{1 - F_t(x)} \tag{38}$$

When using the peak method, peaks occurence frequency and peaks height distribution are determined. The parameters of those distributions define the theoretical distribution function of the annual maxima flows that is compared with the empirical distribution of the annual maxima values. Since the theoretical distribution function should not have more than three parameters, it is desirable that the distribution of peaks occurence frequency and height is one-parameter or two-parameter [12]. In this paper, a Poisson-Weibull model with a Poisson (one parameter) distribution for the peaks occurence frequency and a two-parameter Weibull's distribution for the peaks height are selected, which for the maxima flows gives a troparametric distribution function as:

$$F_t(Q) = P[Q \le Q_{\max}] = e^{-\lambda(t)[1-H(Q)]} \tag{39}$$

i.e.

$$\Phi_t(Q) = P[Q \ge Q_{\max}] = 1 - e^{-\lambda(t)[1-H(Q)]} \tag{40}$$

Based on the Eq. 39, a calculation of the theoretical function of the distribution of maxima flows for all investigated basins has been carried out. Graphic representation of the cumulative probability distribution functions $F_t(Q)$ is shown on Fig. 7, together with adopted theoretical function in annual maxima method and empirical distribution of annual maxima discharges.

5   Comparation of Maximum Annual Discharge Determined Annual Maxima and Peak Over Threshold Method

In this paper, probabilistic (statistical) methods were used to define maximum annual discharge probability function. These methods are based on the use of only historical data (historical samples) on the occurrence of large waters subjected to statistical analysis with the ultimate goal of constructing the functions of distribution of probability of occurrence of large waters.

In the opinion of a large number of hydrologists, the extrapolation of probability distribution functions can only be allowed up to (3–5)·N, where N is the length of the available sequence. For Bosnia and Herzegovina, this would mean that for an average length of 30 years, the maximum flow could be defined to the maximum of the return period of 150 years. In the opinion of some hydrologists, the defect of short hydrological sequences can be avoided using peak over threshold (POT) method that analyze all extreme values that exceed a limit (threshold), regardless of how many times they reported in a year.

In this paper, comparative results of high waters calculations are given by both probabilistic methods (annual extreme method and peak method) on 11 water gauges with basin areas from 200 to 5300 km^2 and lengths from 16 to 47 years in order to see the possible advantage of peaks methods over annula maxima method, for the observation lengths that are commonly encountered in Bosnia and Herzegovina.

Comparative results of the application of the threshold/peak method and the method of annual maxima can be made with the following preliminary conclusions:

- On all gauge stations with a satisfactory level of data for the application of the method of annual extrema it is possible to apply the method of peaks;
- Kolmogorov's test has confirmed that all five probability distribution functions (Gaus, Galton, Pearson, Log-Pearson and Gumbel) well describe the empirical (Weibull) function adopting 5% significate level
- In most cases, best matching with empirical function is achieved for Log Pearson III function (6 out of 11), and for the rest 5, Log Pearson III is second best,
- For peak method, a Poisson-Weibull model with a Poisson distribution for the peak occurrence frequency and a two-parameter Weibull's distribution for peak height was used, which for the maximum gives a three-parameter distribution function
- Applied methods give the same results in cases where the mean number of peak per year is less than 2, as in the case of VolarI (E(η) = 1.25) and Gornji Vakuf (E (η) = 1.6),
- Biggest difference in results are for gauge station Banja Luka, with mean number of peak per year E(η) = 9.79, for which annual maxima method gives higher values, and difference between two methods increase as return period increase.

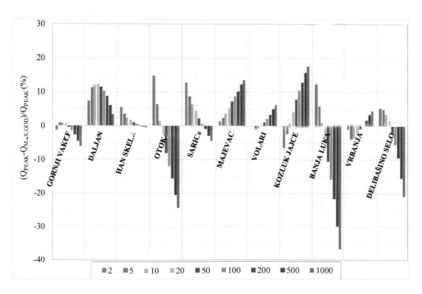

Fig. 8. Percentage difference in maximum annual flows according to applied methods in a function of return period

– On relevant stations, up to return period of 50 years applied methods differences between applied method are up to cca 10%, and peak method gives a slightly higher value
– The differences in the results by the methods increase with the increase in the return period
– At Daljan station with the shortest observation, peak method gives higher values in relation to the method of annual maxima (Fig. 8).

Bearing in mind previous conclusions, research and testing of the peak method on other basins should be continued, by varying the threshold value (by increasing the frequency of peak occurrence) and applying other common theoretical functions of peaks occurrence frequency and peaks height distribution, and by assessing the impact of the length of the historical set of observations on the estimated values of the maximum annual flows.

References

1. Anđelić, M., Bonacci, O., Đorđević, N., Hrelja, H., et al.: Maksimalno vjerovatne velike vode. Jugoslovensko društvo za hidrologiju i Zavod za hidrotehniku Građevinskog fakulteta u Sarajevu, Sarajevo (1986)
2. Bonacci, O.: Predavanja iz hidrologije na postdiplomskom studiju Građevinskog fakulteta Sveučilišta u Sarajevu (2000)
3. Fejzić, Đ.: Mogućnosti i primjeri primjene metode pragova u hidrotehničkoj praksi, Seminarski rad na Građevinskom fakultetu u Sarajevu (2008)
4. Hrelja, H.: Analiza kiša kratkog trajanja za potrebe definiranja oticajna sa urbanih površina, Zavod za hidrotehniku Građevinskog fakulteta u Sarajevu, Sarajevo (1984)
5. Hrelja, H.: Vjerovatnoća i statistika u hidrologiji, Građevinski fakultet Univerziteta u Sarajevu (2000)
6. Hrelja, H.: Inženjerska hidrologija, Građevinski fakultet u Sarajevu (2007)
7. Prohaska, S., Topalović, Ž., Mulaomerović-Šeta, A., Lončarević, D.Ž.: Izrada mapa opasnosti i mapa rizika od poplava u slivu rijek Vrbas u BiH, Aneks 2, Pregled i analiza hidroloških podataka i razvoj hidrolođkog modela (2016)
8. Mulaomerović, A.: Uporedna analiza rezultata primjene metode godišnjih ekstrema i metode pragova u definiranju velikih voda. Diplomski rad na Građevinskom fakultetu u Sarajevu (2008)
9. Radić, Z., Mihailović, V., Plavšić, J.: Uporedna analiza statističkih metoda za proračun velikih voda, 16. Savetovanje SDHI i SDH, Donji Milanovac, Srbija (2012)
10. Vukmirović, V.: Analiza vjerovatnoće pojave hidroloških veličina, Građevinski fakultet Beograd i Naučna knjiga, Beograd (1990)
11. Zelenhasić, E.: Theoretical probability distributions for flood peaks, Colorado State University Fort Collins, Colorado (1970)
12. Zelenhasić, E.: Inženjerska hidrologija, Naučna knjiga, Beograd (1991)

Numerical Investigation of Possible Strengthening of Masonry Walls

Venera Simonović[1(✉)] and Goran Simonović[2]

[1] Polytechnic Faculty University of Zenica, Zenica, Bosnia and Herzegovina
venera.simonovic@ptf.unze.ba
[2] Faculty of Civil Engineering, Institute for Materials and Structures,
University of Sarajevo, Sarajevo, Bosnia and Herzegovina
goransimonovic@yahoo.com

Abstract. The aim of this paper is to present comparative analysis of possible strengthening of masonry wall with two different approaches: with steel grid and damper and with RC panels with FRC dowels. For both proposed approaches, advantages and well as disadvantages are presented. The model compounded from n-link elements and fined elements is used for the model of the masonry wall. Such model is possible to apply in most commercial software for general purpose use. It is possible to simulate the wall rocking, sliding and toe crushing with developed model. Both types of strengthening are applied afterwards, as staged construction. Strengthenings are modelled with nonlinear n-link elements. The results of the investigation show that the crucial influence on quality of the strengthening systems has the choice of the stiffness of elements which need to provide adequate strengthening of the wall.

Keywords: Masonry · Finite element modeling · Dampers
Fibre reinforced concrete

1 Introduction

For modeling of masonry walls, a model was originally developed by Simonović [5] and inproved by Simović [4]. With the model, it is possible to simulate the opening of the coupling due to overturning moment, crushing and sliding. Combined fracture can not be simulated directly, but if bearing capacity of spring is adopted as V_{rd}, for known normal force such kind of sliding is simulated directly as sliding. Different modeling strategies are shown in Fig. 1.

The model is originally adjusted to use in SAP2000 software [1], but with slight modification it can be used in all universal engineering software that has finite element walls for the base, and nonlinear springs for connection between the finite elements. With the model, it is possible to achieve the interaction of all constructive elements, plates, panels, beams, walls, and to form complex spatial models and use different materials, which is a basic advantage over the use of highly specialized software for the calculation of masonry structures.

The wall model [2] itself is quite simple and is based on the postulates of material resistance. Wall behavior is simulated by finite elements through software application.

© Springer Nature Switzerland AG 2019
S. Avdaković (Ed.): IAT 2018, LNNS 60, pp. 175–181, 2019.
https://doi.org/10.1007/978-3-030-02577-9_17

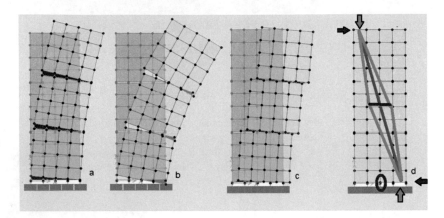

Fig. 1. The behavior of the basic model of the wall: (a) opening of the cracks due to tension, (b) overturning of the wall as a ridged body, (c) toe crashing of the wall, (d) sliding the wall by the coupling.

In a coupling where it is assumed that non linarites will occur, the above mentioned connections are introduced. The wall is divided into the network of finite elements, and the nodes of the mesh are bonded to the n-link connecting elements.

The normal force in the wall is equal to the resultant of normal stresses in the cross-section of the wall at the length of the wall (Fig. 2a). The in-plane, bending moment is the same as the moment that normal stresses over the axes 3. By dividing the wall length on the n segments (Fig. 2b), the integral dependency can be substitute with discrete values which can be written for enough number of wall segments as shown in Eq. 1:

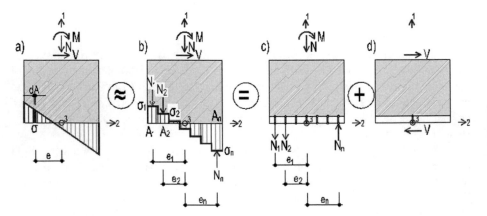

Fig. 2. Definition of forces for: (a) homogeneous wall, (b) wall coupling divided into n segments, (c) wall coupling with introduced connections for the transfer of normal force, (d) wall coupling with the introduced connections for the transfer of shear forces.

$$N = \int_A \sigma \cdot dA \approx \sum_{i=1}^{n} \sigma_i \cdot A_i = \sum_{i=1}^{n} N_i$$

$$M = \int_A \sigma \cdot e \cdot dA \approx \sum_{i=1}^{n} \sigma_i \cdot A_i \cdot e_i = \sum_{i=1}^{n} N_i \cdot e_i$$

(1)

Each n-link element is assigned appropriate bearing capacity. Wall stresses are thus approximated with the forces in n-link elements (Fig. 2c). Axial n-link element which is subjected to tension, will not transfer such force into the rest of the model. If the n-link element is subjected to pressure which is less than the wall segment strength, the n-link element is still in linear range and is elastic. Increasing the compression and reaching the compression strength, the model starts to yields. The model can also slide what is achieved applying n-link connection (Fig. 2d). The forces in n-link elements are obtained by model calculation. The wall model yields when one of the conditions is met in Eq. 2:

$$M_{Rd} \geq M = \sum_{i=1}^{n} N_i \cdot e_i$$

$$N_{i,Rd} = f_d \cdot A_i \geq N_i$$

$$V_{Rd} \geq V$$

(2)

2 Strengthening of the Wall with Dampers

The principal idea is to build a flat steel truss next to the wall that needs to be reinforced. Mechanic damper is also placed in the diagonal of such a grid. Grid elements are linear elements and they can be modeled into the software as n-link elements. Each such element can be assigned its characteristic thru dependence: normal force-displacement.

The total behavior of the model should be closely related to the behavior of its components (Fig. 3), what can be used for control of the obtained results. Namely, the analysis is done for the wall without reinforcement first and then for reinforcement only. Summation of the forces in certain parts of the model for the same displacement, the total force should be obtained by the model's calculation as a whole. All available dampers can be used as mechanical dampers, and their

behavior in the model can easily be simulated by the connecting elements in the way it was previously presented in the text. Some of the DC90 [3] producers are shown in the following illustration (Fig. 4):

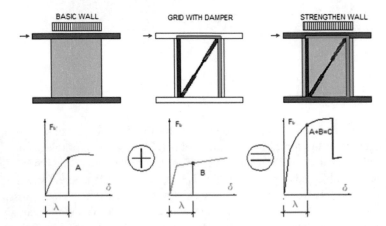

Fig. 3. The behavior of the model shown as the sum of the base shear of the single wall and base shear of the truss for the same displacement at the top of the wall.

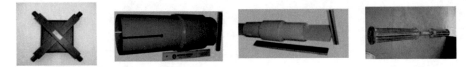

Fig. 4. Different types of mechanical absorbers.

3   Strengthening of the Wall with RC Panels and FRC Dowels

For this case reinforcement of the wall, the panel of reinforced concrete is added on existing structure. The behavior of such a panel is simply numerically simulated if it is assumed that the bending moment is transmitted to steel anchor and the transverse force is overtaken by dowels of fiber reinforced concrete.

The total model behavior should be close to the behavior of its entirety (Fig. 5) as described in the case of damper strengthening. The behavior of the fiber reinforced concrete dowel can be obtained and defined using experimental results performed at Civil Engineering Faculty University of DžemaL Bijedić in Mostar [8].

This connection is further described as "S connection". Another type of connections in the model is assumed to be so called "CT connection". As axial elements "CT connections" transmits pressure or tensile stresses only. The analyses of anchors bodies was not the topic of the work, the aim was to research the behavior of such reinforced wall (Fig. 6).

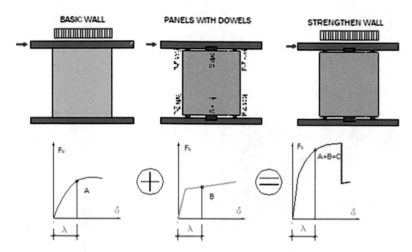

Fig. 5. The behavior of the model shown as the sum of the base shear of the single wall and base shear of the panels for the same displacement at the top of the wall.

4   The Results of the Numerical Investigation

The research was carried out on simple wall models, family buildings, and residential buildings. Details can be found in the literature [5–7] while one of the example is shown in the Fig. 7. Wall strengthen with panels and was strengthen with dampers are shown to the left, and the capacity curve is shown to the right. Obviously, both strengthening systems result in a significant increase in the capacity curve of strengthen masonry wall. The advanced model is stiffer compared to un- strengthen wall. The behavior of the strengthen model is close to the elastoplastic behavior and yielding

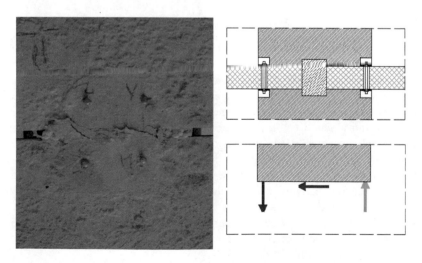

Fig. 6. Testing of FRC elements and schematic explanation of the proposed system.

occurs when transverse force reaches its bearing capacity. It is important to emphasize that the non-reinforcing wall behavior was characteristic of slim walls, i.e. fracture was due to effects of rocking.

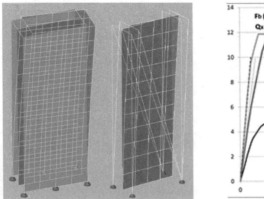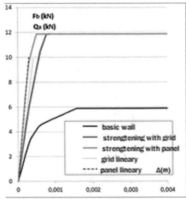

Fig. 7. Comparison of obtained results.

By introducing strengthening, the walls become somehow reinforced, the eccentricity of the normal force in the wall is reduced, the surface of the stress block is greater, the existence of the n-link that transmit the tension control the openings of the couplings, but many other benefits of reinforcement are also gained. These models can be calculated using dynamic analysis as well. There was no significant difference between the results obtained by pushover analysis and time history for different earthquake accelerograms.

5   Conclusions

The analysis of strengthening elements of masonry buildings should be systematically accessed. In the first step, it is necessary to perform detailed linear analysis, then nonlinear, then assume the strengthening and with approximate procedure to determine how effective these systems are, and then consider the positive effects of its application. If the reinforcements do not provide the aimed effect, it is necessary to seriously consider the introduction of other rigid constructive elements. The construction of a new, rigid wall,, eventually a beam over the wall, where feasible, can represent more effective protection against the negative impacts of the accelerating earthquake force.

Strengthening need to be designed so that they are effective in the planned overtaking of forces for relatively small displacement. In order for the strengthening elements to be satisfy the required demands, it must be quite rigid that it is not favorable from the economic point of view. Of course, all numerically-researched strengthening options should, first and foremost, be tested on experimental models, and further research should be lead in that direction.

References

1. CSI Computers and Structures Inc, SAP2000 – Integrated software for structural analysis and design, Berkley, USA
2. Hrasnica, M., Medic, S.: Finite element modeling of experimentally tested solid brick masonry walls. In: 16 ECEE, Thessaloniki (2018)
3. Petraskovic, Z., Gocevski, V.: Seismic Analysis of existing masonry structures reinforced with "SYSTEM DC90" Dampers. In: SE-EEE 1963-2013, Skopje, Macedonia (2013)
4. Simonović, G.: Proračunski model za trodimenzionalnu analizu seizmičke otpornosti zidanih zgrada, Faculty of Civil Engineering, University of Sarajevo (2014)
5. Simonović, V.: Numerička analiza seizmičke otpornosti zidanih zgrada primjenom spojeva od mikroarmiranog betona i mehaničkih dampera, Faculty of Civil Engineering, Mostar University, BiH (2017)
6. Simonovic, V., Šahinagić-Isović, M., Selimotić, M., Simonović, G.: Numerical analisys of seismic resistance of masonry buildings using passive dampers. In: 16 ECEE, Thessaloniki (2018)
7. Simonovic, V., Šahinagić-Isović, M., Selimotić, M., Simonović, G.: Numerical analisys of seismic resistance of masonry buildings using connections of fiber reinforced concrete. In: 16 ECEE, Thessaloniki (2018)
8. Šahinagić-Isović, M.: Posebne vrste betona-mikroarmirani betoni, Univerzitet "Džemal Bijedić", Građevinski fakultet, Mostar (2015)

River Restoration – Floods and Ecosystems Protection

Emina Hadžić[1]([✉]), Hata Milišić[1], Ajla Mulaomerović-Šeta[1],
Haris Kalajdžisalihović[1], Dženana Bijedić[2], Suvada Jusić[1],
and Nerma Lazović[1]

[1] Department of Water Resources and Environmental Engineering,
University of Sarajevo, Sarajevo, Bosnia and Herzegovina
eminahd@gmail.com, hata.milisic@gmail.com,
nerma.ligata@gmail.com, {ajla.mulaomerovic,
haris.kajajdzisalihovic,suvada.jusic}@gf.unsa.ba
[2] Architecture Faculty, University of Sarajevo,
Sarajevo, Bosnia and Herzegovina
dzenanabijedic@gmail.com

Abstract. Rivers have always been the most important source of water for man. And not only were that, often in its natural state, through natural retention and bayou, the rivers have the best flood defense. It is also important to note that coastal vegetation has had a significant impact on the purification of water that is infiltrated from the waterways into the groundwater. Human development, population growth, urbanization, climate change have led to a significant drop in river health at the global level. In the zone of settlements, the rivers have undergone major morphological and hydraulic changes, which ultimately led to disturbances of the river's ecological status and the loss of sociological role of the river in the urban environment. In this connection, this paper will give an overview of the most common mistakes made in river regulation over the past period. It will also highlight the ways and possibilities of reconstructing the river from the position of ecologically sustainable development and flood protection. It is reflected in the passive and active ways of restoring the river, and the importance of applying the principle of integral river management in the context of water recovery.

Keywords: River restoration · River ecosystem · Climate change
Floods

1 Introduction

People have always settled river valleys and river coasts. The first civilizations were born in a fertile land near Tigris and Euphrates in Mesopotamia, the Nile in Egypt and the Yellow River in China. Possibility to use watercourses as a waterway, more favorable climatic conditions in river valleys, favorable conditions for agricultural production, usees of water potential for the production of electricity in the industry, usees of river water for cooling thermal power plants, fishing, recreation and water sports are some of the reasons that attracted people to the river valleys. Rivers have

© Springer Nature Switzerland AG 2019
S. Avdaković (Ed.): IAT 2018, LNNS 60, pp. 182–191, 2019.
https://doi.org/10.1007/978-3-030-02577-9_18

always been the most important source of fresh water for humans. And not only that, the rivers in their natural state, through natural retention and sleeves, were also and still are the best defense against floods.

Social, economic and political development both in the past and today, largely associated with the availability and distribution of fresh water of the river. However, very often, and especially in the last decades, rivers are mentioned as risks to human, either in the case of flooding of large waters, or because of their increasing pollution. Human development, population growth, urbanization, climate change, are just some of the causes that have led to a significant decline in the health of rivers on a global scale.

As a result of human activities, hydrological cycles are disrupted both globally and locally, and this is particularly characteristic of river basins in urban areas. Generally, the urbanization of river basins increases the impermeable surfaces, reduces green and other surfaces that have retention action, resulting in an increase in the coefficient of swelling, shortening the time of the concentration of large waters and modifying the surface and groundwater regimes [1]

In addition to changes in surface runoff and worsening flood hydrograph whom to safely pass through the city, others dramatic impact of urbanization on river ecosystems as well as the deterioration of water quality.

Rivers have become recipients of waste waters of settlements and industries, household waste is most often directly dumped into them, and in the case of high intensity precipitation, the very dirty streets and roads in the settlements are rinsed, and then through the sewage system for rain water, they are poured into the river. Various chemical fertilizers and pesticides, which are mainly used without planning in agriculture, infiltrated into the groundwater and surface water.

The hydraulic structures in the riverbed are dividing the rivers, which significantly changes the natural water regime and sedimentation regime, which, on the other hand, has a negative effect on aquatic ecosystems.

2   Objectives of Water Regulation Work

Very often natural characteristics of the watercourse are not suitable from the aspect of water use, as well as the use of the land on which they are located. Therefore, there is a need for their regulation. Protection from floods, torrents and erosion effects on the riverbed and surrounding objects can be commonly referred to as hydrotechnical regulation. These processes can be consequential activities due to certain human failures but also often occur as a need protection from the natural characteristics of the watercourse.

River regulation (or watercourse regulation, river engineering) is the process of applying planned activities to the modify of the position of the watercourse, the hydrological characteristics of the river or flow regime, in order to achieve the set goals. Watercourse regulation can have multiple objectives as well as a multifunctional character. In order to achieve the objectives of regulation work, it is sometimes necessary to make only minimal corrections of the natural characteristics of the watercourse. There is often a need for major changes, and in some cases it is necessary to form completely new riverbeds on the larger lengths of the river valleys. The goals of regulation work have the greatest importance for defining criteria for the selection of basic concepts and elements of regulation works. Due to the fulfillment of the regulatory objectives, a significant change of the natural elements of the watercourses is often necessary. The degree of satisfaction of the regulatory objectives does not only depend on the technical conditions, but it is necessary to have economic and social justification. In most cases, watercourse regulation requires large investment funds, and the analysis of economic and social justification requires special attention.

It is not enough that preliminary studies of the river basin and watercourses only identify problems and suggest technical solutions, but it is necessary to determine and to what extent the individual watercourse parameters need to be changed. It is particularly important to evaluate all possible significant environmental impacts, which in some cases constitute limiting elements in selecting the most favorable solution.

With the works on the regulation of the river, the natural conditions are disturbed, as well as the mutual relations of individual watercourse elements, and the attitude of the watercourse towards the environment. The works on the river regulation can cause significant changes in the condition of the already built water management facilities in the watercourse, and to influence the conditions for the construction and use of the planned future facilities. On the basis of the above, it can be concluded that the work on watercourse regulation is carried out within the wider water management system, that these works cannot be viewed as isolated from other water management actions, which is why they should be harmonized with a general water management solution, if it is defined by the development strategy of the water sector, or otherwise appropriately defined.

Watercourses are an important element of the environment that needs to be protected and preserved, and during the regulations activities, it is necessary to maintain the tendency of the slightest change of the natural conditions. On the contrary, each of the objectives of the regulation of watercourses imposes the need for one or more water flow parameters to be changed. It is very important that in advance, before approaching the regulation works of each river, they are complexly examined, and impartially and correctly evaluate all the effects of planned works on the environment.

When we have clearly defined objectives of regulatory works, known natural conditions (watercourse characteristics and water environment) and parameters of the water management system, as a whole, it is possible to make good decisions about the character of regulation activities (Fig. 1), as well as the tendencies in the selection of elements technical solutions.

The specificity of the regulation of the river is that after the work is carried out, the nature of the response is followed, which is not easy and often impossible to predict. Disruption of a natural watercourse regime with works in the watercourse or by changing the water and sediment regime results in a series of short or long-term morphological processes leading to the establishment of a new equilibrium state (Fig. 2).

Solving problems in river regulation requires knowledge, experience, synthesis skills, engineering intuition, and compulsory use of previously acquired knowledge from numerous scientific disciplines such as: fluid mechanics, statistics, hydraulics, hydrology. Therefore, the approach to planning river regulation must be integral, "sustainable" and multidisciplinary.

a)   b)

Fig. 1. Regulatory works on: (a) river Nišava in Niš, (b) river Miljacka in Sarajevo

a)   h)

Fig. 2. Problems with river sediment (a) river Sava, [4], (b) river Nišava

3 Floods and Urban Regulation

Although the floods, since the very first civilizations, were a major threat to the human community, this problem was actualized after several major floods in various parts of Europe and the world in the last decade of the twentieth century, followed by high damages and losers of human life, [2]. Regardless of the causes of the flood, the probability of flooding always exists, so this phenomenon as a natural phenomenon cannot always be prevented, no matter how secure the defense system and prevention measures were, [3].

The causes of floods are numerous, and it can generally be said that the floods are caused by natural occurrences and artificial influences, [4]. Th Strong floods cause climatological natural phenomena such as rainfall - rain, melting snow and ice, or their combined action. In addition to climatic causes, other natural causes of flooding can be phenomena such as earthquakes, landslides, water drain at the river mouth due to waves, etc. [5].

The amount of water (precipitation), their spatial distribution, intensity and duration of precipitation are the main climatic causes of floods. In addition to these causes, the occurrence of flooding is influenced by the receiving capacity of the watercourse or water supply network to receive and continue to drain water, the situation in the whole catchment area, especially in the area right next to the water course, the weather conditions before the start of precipitation, ground cover and topography, [6].

However, the anthropogenic impact on floods must not be forgotten because it is large. According to Popovska and Đorđević [1], narrowing the inundation areas and even the main river bed, due to the construction various urban contents in the coastal zone, the man seriously deteriorated the hydraulic flow conditions in the zone of the settlement, thus increasing the levels of large waters of the same return periods.

On the other hand, the construction of increasingly expensive and security-sensitive facilities in settlements has been subject to ever more stringent criteria in terms of the required reliability of flood protection: from the probability of high water 2% (fifty years old large water) in the conditions of smaller settlements, to the probability of about 0.2% (five-year-old large water), in the conditions of the largest urban centers, in which the floods would cause great economic, ecological, and sociological consequences in a much wider area than the direct flood zones, [7].

The worrying tendency, which is increasingly prevalent in transition countries, is that, under the pressure of the owners of capital, which are in an interesting relationship with decision-makers at all levels of government, urban planning and regulatory plans are changing, in order to reduce or completely eliminate and destroy the green spaces due to the construction of profitable facilities. This leads to a radical deterioration of climate and living conditions in cities. When the former wide river corridors, often with well-groomed vegetation, significantly narrow down and barriers with high buildings (this is always done under the 'development' slogan), the conditions of air flow through the city and its ventilation are worsening radically, with ever more severe consequences for human health [1].

Climate change is particularly noticeable in the water domain, as more and more obvious changes are made, primarily in terms of worsening the distribution of precipitation and runoff, both in space and in time. Particularly unfavorable is the deterioration of extreme hydrological phenomena. Due to excessive heating of surface water masses in certain areas, especially in the 'hot seas', increased evaporation, and disturbances of a lot of persistent ocean currents circulation, there appear significant extreme concentrations of air humidity and the formation and movement of turbulent air masses, which are manifested by enhanced excretion precipitation of very high intensities.

In a number of areas of the world there are precipitations and over 200 mm/day, which cause much bigger and ruinous watercourses, but also create lunches that destroy entire urban areas. In such circumstances, river corridors and riverbed for large waters, that are dimensioned by to former precipitation and water regimes, become insufficiently capacity in the new circumstances [1] (Fig. 3).

4   Access to River Regulation

By the mid-20th century river regulation are mainly carried out in such a way that, in the foreground, were placed human interests. Due to the increase of material possibilities, the company, the improvement of machinery for carrying out the works, and

Fig. 3. Damage from floods in Europe in 2002, [4]

the growing need to regulate watercourses, there is an increasing the tendency is expressed that the regulatory work is carried out in the shortest possible time and that the full effects of regulatory work are achieved in a very short time. Therefore, riverbeds are generally rigorously changed (Fig. 4).

Fig. 4. The regulation of river Zujevina

Since the end of the 20th century, the regulation of the river also takes care of environmental protection as well as the benefits for people, and many river regulation projects have the goal, only, the river restoration or protection of nature. Namely, in the selection of control elements, the importance of the criteria for the preservation of aquatic ecosystems is of increasing importance (Fig. 5).

The regulation of open watercourses carried out in classic ways did not meet expectations, both in terms of engineering and environmental protection. Innovative approaches must strive integral and comprehensive settlement of this issue [5].

a)   b)

Fig. 5. River Cheonggyecheon in Seoul, in the culmination of destruction (a), after restoration works (b), [1]

5 River Restoration

Historically, activities on watercourse recovery are relatively recent. The process of restoration that began in the 1970s and 1980s was first started in developed countries such as the United States, the EU countries and Australia. However, significant improvements in the implementation of various watercourse restoration measures can also be found in developing countries. Later the start of watercourse recovery in these countries, according to Speed et al. [8], does not necessarily mean that watercourses are less polluted, but often means that industrialization and agricultural growth were priorities in relation to environmental conservation.

In most of the highly developed countries, which had the greatest impact on river ecosystems with their activities, the restoration process was practically necessarily, and the most frequent answers to the problem of watercourse degradation were measures aimed at improving the water quality of the watercourses, preserving the existing ecosystem function and limiting or reducing human influences on the rivers. Such measures deal with resolution of the effects of spot and diffuse pollution, excessive water pumping, unplanned development of the catchment area and especially coastal zones, with the aim of improving water quality and reducing the risk of floods.

In cases of severe water degradation, when the ecosystem function was not able to return to the desired level with such measures, more direct intervention was undertaken, to make changes in the physical structure of the watercourses (e.g. improvement of the habitat), removal or reduction of the impact of obstacles within the watercourse, increase of river flow and afforestation parts of the basin and the coastal zone, [8].

More recently, the watercourse recovery process is characterized by the integral resolution of the resulting problems with other human activities within the catchment area, giving the importance of maintaining the ecosystem function even in the highly developed catchment areas. Such integration towards Gilvear et al. [9], despite all efforts, acknowledges the limited possibilities of watercourse recovery in which likely human impact will remain dominant. In highly developed catchment areas, with several competing users, the process of watercourse recovery often needs to achieve multiple, and sometimes, contradictory, goals, [9]. For example, recovery targets may at the same time include improving water quality, improving the benefits of fostering urban development or recreational activities, flood protection, promoting biodiversity and improved navigation. These concurrent targets, according to Speed [8], require the balancing of the natural functions of a river with special human needs and may require compromises in the planning process. They also require an agreement with a few stakeholders on compromises in setting priorities and objectives. Recovery goals should be defined by an interdisciplinary approach by decision makers with the consensus of interdisciplinary technical teams and other participants in social and political life. They should be the integration of two important groups of factors: (i) relating to the future conditions to be achieved (ecological reference status), (ii) and which resulted from the knowledge of social, political and economic values in the considered basin or part of the basin.

According to Speed et al. [8] setting goals and tasks of recovery should often be an iterative process where objectives and tasks need to be re-evaluated in order to achieve the best results of the applied response measures (Fig. 6).

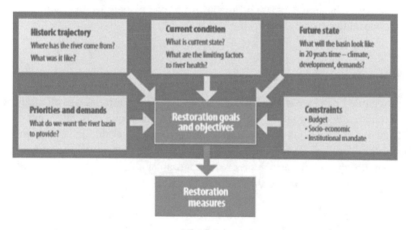

Fig. 6. Considerations in setting goals and tasks of recovery, [8]

In line with the goals set and the real possibilities for their implementation, the recovery of the watercourse should be planned. In doing so, it is necessary to distinguish several possibilities for watercourse recovery: restoration, rehabilitation and remediation. The first and most demanding activity is restoration, according to Popovska and Đorđevic [1], means the return of the river to the original ecological state, according to all relevant ecological parameters (flow regime, substrate bottom, aquatic and aquatic ecosystems, ambient conditions). According to Wade et al. [10], restoration is focused on the reconstruction and return of the intact physical, chemical and biological status of the watercourse. In its purest sense means a complete structural and functional return to the state before the disorder, [10]. Because in the most cases, this task is not realistic, it is resorted to - rehabilitation. Rehabilitation is most often a realistic and achievable activity and includes works and measures that significantly improve the ecological conditions in the river and approach the former balance ecological conditions. It is a very complex, long-lasting and expensive activity, but it is increasingly being treated as inevitable, to avoid environmental, social, and political collapse. However, already carried out rehabilitation works in a series of metropolises of the world show that such works have an economic development significance, while sociological and political significance is undoubted [1].

According to Wade [10], rehabilitation points to a process that can bc defined as a partial functional and/or structural return of a former or pre-degrading condition, especially in terms of environmental conditions. In conclusion, rehabilitation measures only relate to changes in some elements within the degraded aquatic system, but they still aim to return the ecosystem closer to the original state.

If the level of degradation of the river ecosystem is so high that even the rehabilitation of the system is not feasible, remediation should be approached. Remediation implies such an improvement of ecological conditions, so that the river system is translated into a new ecosystem, but with a significantly better status than it was anthropogenic degraded river system. The remediation process should be done very often on rivers in urban conditions, radically channeled and ecologically destroyed - dead rivers, which should again be made attractive to people, but with some other ecological conditions compared to the original [1].

6   Conclusions

The importance of preserving the water quality of watercourses, as well as preserving the amount of water, is gaining more and more weight, especially after understanding the lack of it for numerous human needs. Unfortunately, pollution of watercourses is increasing day by day. Raising human awareness about the ways, measures and possibilities of preserving water resources, in general, with all the technical and technological measures that are being implemented in society, must inevitably be raised to a higher level.

Therefore, neither engineering tasks were on the protection of river water, protection against water, water use, nor in water management, are not simple Watercourses are unsteady flows with very frequent changes of water and sediment in time, but also can be with significant changes in water quality in time and space. Therefore, activities on the river regulations cannot be considered isolated from other water management actions, and therefore they should be harmonized with a general water management solution, if it is defined by the development strategy of the water sector, or otherwise appropriately defined.

Watercourses are an important element of the environment that needs to be protected and preserved, so it is necessary to keep the tendency of as little change in natural conditions as possible in the regulations of works In contrast, each of the objectives of regulating works in a watercourse imposes the need for one or more water flow parameters to be changed. It is very important to advance in advance, before approaching the regulation of each river, all the effects of the planned works on the environment are complexly considered, impartially and correctly, valued.

The specificity of the regulation of the river is that after carrying out the work, the following is a response of nature that is not easy and often impossible to predict. The disturbance of the natural water regime (by changing the water and sediment regime) results in a series of short-term or long-term morphological processes leading to the establishment of a new state of the watercourse. Activities on watercourse recovery are becoming more and more frequent, as the destructive attitude towards aquatic ecosystems has become unsustainable and has harmful effects on humans. However, it must be noted that such activity is very expensive, demanding, often with uncertain outcomes, and that it must be strategically planned in order to achieve the required effects.

It is necessary to carry out work on the regulation taking into account the ecosystems and the health of the river. Since we cannot protect ourselves from floods, it is necessary to use natural retention and return them to rivers wherever possible (example Lonjsko polje, Croatia). Urban planning must be in the service of environmental protection and conservation of aquatic and terrestrial ecosystems. In any activities in the watercourse, or in the catchment area, it is necessary to apply the concept of environmentally acceptable, integrated water management. Such an approach would reduce the negative effects of human interventions in the environment.

References

1. Popovska, C., Đorđević, B.: Rehabilitacija reka - nužan odgovor na pogoršanje ekoloških i klimatskih uslova. VODOPRIVREDA 0350-0519, vol. 45, pp. 261–263 (2013)
2. Ivetić, M., Petković, S.: Forum voda 2014, Naučno-stručni skup Poplave u Srbiji, maj 2014, 4–5 Novembar 2014, Beograd (2014)
3. Kuspilić, N., Oskoruš, D., Vujnović, T.: Jednostavna istina – rijedak hidrološki događaj. Građevinar **66**(7), 653–661 (2014)
4. Kuspilić, N.: Regulacije vodotoka, Skripta za studente, Zagreb (2009)
5. Bonacci, O.: Ekohidrologija vodnih resursa i otvorenih vodotoka. Graevinski fakultet u Splitu, ISBN 953-6116-27-8 (2003)
6. Imamović, A.: Uzroci poplava u slivu rijeke Bosne s osvrtom na poplave u maju 2014.godine, ANUBiH, Sarajevo (2015)
7. Đorđević, B.: Realizacija razvoja vodoprivredne infrastrukture u skladu sa strategijom iz Prostornog plana Srbije. Vodoprivreda, N0 234–236, s.215–226 (2008)
8. Speed, R., Li, Y., Tickner, D., Huang H., Naiman, R., Cao, J., Lei G., Yu, T, Sayers, P., Zhao, Z., Yu, W.. River restoration: a strategic approach to planning and management. UNESCO, Paris (2016)
9. Gilvear, D.J., Casas-Mulet, R., Spray, C.J.: Trends and issues in delivery of integrated catchment scale river restoration: lessons learned from a national river restoration survey within Scotland. River Res. Appl. **28**(2), 234–246 (2012)
10. Wade, P.M.: Management of macrophytic vegetation. In: Calow, P., Petts, G.E. (eds.) The rivers handbook, vol. 1, pp. 363–385. Blackwel Science, Oxford (1994)

Seismic Analysis of a Reinforced Concrete Frame Building Using N2 Method

Emina Hajdo[✉] and Mustafa Hrasnica

Faculty of Civil Engineering, Institute for Materials and Structures,
University of Sarajevo, Sarajevo, Bosnia and Herzegovina
emina.hajdo@gmail.com, hrasnica@bih.net.ba

Abstract. Earthquake as a dynamical loading on the structures, which occurs accidentally has always been an interesting topic to engineers in everyday engineering practice, as well as to researchers and scientists. In order to obtain the best possible seismic response of a structure, it is necessary for a structural engineer to choose and design the structural system correctly. Further, the seismic response of the structure can be obtained using one of the well-known seismic analysis methods. Engineers in practice commonly use less complicated analysis methods to provide the results that are accurate enough for the design requirements. Lateral force method and response spectrum analysis are frequently used seismic analysis methods in practice, whereas researchers prefer time history analysis method or pushover analysis [1, 2]. In this paper we present a practical application of a nonlinear method called N2. This method makes a connection between two seismic analysis procedures: pushover analysis of a multi degree of freedom (MDOF) model and the response spectrum analysis of an equivalent single degree of freedom (SDOF) model. N2 method gives us a possibility to use practical analysis procedures in order to obtain seismic response of a structure. We analyse seismic response of an eight-story reinforced concrete frame building. The design of the analysed structure is carried out fully in accordance with the seismic request and in compliance with the capacity design provisions.

Keywords: Earthquake · Seismic engineering · Response spectrum analysis Pushover analysis · N2 method

1 Introduction

Capacity spectrum method was developed by Freeman [3]. Using the graphical procedure, the capacity of the structure is compared to earthquake demand. The graphical representation provides a visual prediction how the structure will respond in case of an earthquake. The capacity of structure is represented by a force-displacement curve, which is obtained using nonlinear pushover analysis. The development of the N2 method was proposed by Fajfar and Fischinger [4, 5], and later the procedure was updated [6]. The N2 method is a variant of the capacity spectrum method based on the non-elastic spectrum (N indicates that it is a nonlinear calculation, and the number 2 that two mathematical models are applied). This method is used to calculate the target displacement of the structure exposed to the earthquake. The N2 method combines the

© Springer Nature Switzerland AG 2019
S. Avdaković (Ed.): IAT 2018, LNNS 60, pp. 192–202, 2019.
https://doi.org/10.1007/978-3-030-02577-9_19

pushover analysis of the multi-degrees of freedom model, with the spectral analysis of an equivalent system with one degree of freedom [7–10].

2 N2 Method Overview

Further we give an overview with computation steps of N2 method.

I. Input data

- Properties of a structure
- Elastic acceleration spectrum

A multi degree of freedom model is applied. The seismic demand, which is the effect of the earthquake to the structure, is determined by the elastic acceleration spectrum S_{ae}.

II. Demand spectrum in ADRS form

- Elastic spectrum
- Non-elastic spectrum for constant ductility

It is necessary to determine the non-elastic spectrum in the form of acceleration-displacement (AD). For an elastic SDOF system, (pseudo) acceleration spectrum S_{ae} can easily be coupled with S_{de} spectrum, for corresponding period T, and a fixed value of viscose damping.

$$S_{de} = \omega^2 S_{ae} = \frac{T^2}{4\pi^2} S_{ae} \tag{1}$$

In the case of non-elastic SDOF system with bilinear force-displacement curve, spectral acceleration S_a and spectral displacement S_d can be determined as follows:

$$S_a = \frac{S_{ae}}{R_\mu} \tag{2}$$

$$S_d = \frac{\mu}{R_\mu} S_{de} = \frac{\mu}{R_\mu} \frac{T^2}{4\pi^2} S_{ae} = \mu \frac{T^2}{4\pi^2} S_a \tag{3}$$

where μ is ductility factor, R_μ is a reduction factor due to ductility. For a simple N2 method, a bilinear spectrum is used to determine the reduction factor R_μ.

$$R_\mu = (\mu - 1)\frac{T}{T_C} + 1, \quad T \leq T_C \tag{4}$$

$$R_\mu = \mu, \qquad T \geq T_C \tag{5}$$

III. Pushover analysis

- Assume a displacement form Φ
- Determine the distribution of horizontal forces along the height
- Determine the relation between the transverse force V and the displacement at the top of the building D_t

The pushover analysis is carried out so that the structure is subjected to monotonically increasing horizontal forces, which represent the inertial forces that would occur in structure during the ground motions. By gradual increase of lateral forces, some structural elements reach their capacity, and the stiffness of the structure decreases [11, 12]. Using the pushover analysis, we obtain a characteristic nonlinear relation between the total transverse force and the displacement of the MDOF system. Commonly the transverse force is base shear V and the displacement is the top displacement D_t.

In the N2 method, it is assumed that the horizontal force at the i-th floor is proportional to a component of the assumed displacement mode Φ_i which is multiplied by the mass of the floor m_i.

$$\mathbf{P} = p\mathbf{M}\Phi \quad P_i = pm_i\Phi_i \tag{6}$$

where $\mathbf{M}$ is the mass matrix, and Φ is the assumed displacement shape.

IV. Equivalent single degree of freedom model

- Convert MDOF parameters Q to SDOF parameters $Q*$
- Assume an approximate relation between elastoplastic force and displacement
- Determine the equivalent mass $m*$, load capacity F_y^*, displacement D_y^* and period $T*$
- Determine the capacity curve (acceleration-displacement curve)

Equation of motion of the equivalent SDOF system is:

$$m^*\ddot{D}^* + F^* = -m^*a \tag{7}$$

Where $m*$ is equivalent mass of SDOF system:

$$m^* = \Phi^T\mathbf{mI} = \sum m_i\Phi_i \tag{8}$$

$D*$ and $F*$ are displacements and forces of the equivalent SDOF system:

$$D^* = \frac{D_t}{\Gamma} \tag{9}$$

$$F^* = \frac{V}{\Gamma} \tag{10}$$

V is the base shear of the MDOF system:

$$V = \sum P_i = \mathbf{\Phi}^T \mathbf{m} \mathbf{I} p = p \sum m_i \Phi_i = pm^*$$
(11)

The constant Γ represents the factor of participation of a particular vibration mode.

$$\Gamma = \frac{\mathbf{\Phi}^T \mathbf{m} \mathbf{I}}{\mathbf{\Phi}^T \mathbf{m} \mathbf{\Phi}} = \frac{\sum m_i \Phi_i}{\sum m_i \Phi_i^2} = \frac{m^*}{\sum m_i \Phi_i^2}$$
(12)

In order to determine the idealized relation between force and displacement for an equivalent SDOF system, an engineering assessment is required. Eurocode 8, Appendix B, gives instructions for determining this curve [13]. The initial stiffness of the idealized system is determined in such a way that the surfaces below the actual and idealized force-displacement curves are equals (Fig. 1).

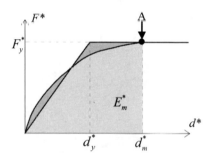

Fig. 1. Force – displacement curve of idealized system

The elastic period T* of an idealized SDOF system, with a bilinear force-displacement relation is defined as:

$$T^* = \frac{2\pi}{\omega^*} = 2\pi \sqrt{\frac{m^* D_y^*}{F_y^*}}$$
(13)

F_y^* and D_y^* are load capacity and corresponding displacement.

Finally, the capacity diagram in the ADRS format is obtained by dividing force values F^* in F^*-D^* diagram.

$$S_a = \frac{F^*}{m^*}$$
(14)

V. Seismic demand for SDOF model

- Determine the reduction factor R_μ
- Determine the displacement demand $S_d = D^*$

The reduction factor can be obtained with:

$$R_\mu = \frac{S_{ae}(T^*)}{S_{ay}} \tag{15}$$

And the required ductility is defined as follows:

$$\mu = \frac{S_d}{D_y^*} \tag{16}$$

If the elastic period T* is greater than or equal to T_C, then the required non-elastic displacement is equal to required elastic displacement, and required ductility is equal to the reduction factor (Fig. 2).

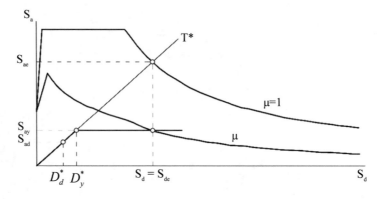

Fig. 2. Elastic and non-elastic demand spectra in relation to capacity diagram

$$S_d = S_{de}(T^*) \qquad T^* \geq T_C$$
$$\mu = R_\mu \tag{17}$$

If the elastic period is less than T_C, which is typical for lower and stiffer buildings, the required ductility can be obtained with:

$$\mu = (R_\mu - 1)\frac{T_C}{T^*} + 1 \tag{18}$$

Required displacement is:

$$S_d = \mu D_y^* = \frac{S_{de}}{R_\mu}\left(1 + (R_\mu - 1)\frac{T_C}{T^*}\right) \tag{19}$$

VI. Global seismic demand for MDOF model

- Convert the displacement of the SDOF model to the maximum displacement of the MDOF model $D_t = \Gamma D^*$

VII. Local seismic demand

- Perform pushover analysis of the MDOF until the top displacement D_t
- Determine local values (e.g. floor displacements, rotations) corresponding to value D_t

The required displacement of the SDOF system is transformed into the global required maximum displacement of the top D_t of MDOF system using the Eq. (9). The maximum displacement of the top D_t represents a target displacement.

VIII. Response rating (damage estimation)

- Comparison of local and global seismic demands with the capacity.

3   Numerical Example

In this example we analyse regular eight-story R.C. fame building. There are six frames in the transversal, and four frames in the longitudinal direction. The frame spans are 6 m in both directions. The height of the building is 29.6 m. Beams have a rectangular cross-section 40/45 cm (b/h), and columns have 60/60 cm square cross-section. Thickness of slabs is 18 cm. First natural period of the building is $T_1 = 1.29$ s [14]. All structural elements are designed respecting the regulations and demands of Eurocode 2 and Eurocode 8 (Fig. 3).

The building is located in the VIII seismic zone according to EMS-97, with the peak ground acceleration PGA of 0.2 g, with soil type B according to EC8. The behaviour factor q = 4 for the medium ductility class (DCM) is selected. The elastic and design spectra for VIII seismic zone and for soil type B are given below [15]. The analysis is performed using software SAP2000 [16] (Figs. 4 and 5).

The mass of floors, starting from the lowest level upwards, are $m_1 = 554$ t, m_2–$m_8 = 546$ t and $m_9 = 510$ t. It is assumed that the displacement form corresponds to the first vibration mode:

$$\Phi^T = [0.12 \quad 0.28 \quad 0.43 \quad 0.57 \quad 0.70 \quad 0.81 \quad 0.90 \quad 0.96 \quad 1.00]$$

The distribution of lateral forces along the height of the building is obtained using Eq. (6).

$$P^T = [0.130 \quad 0.300 \quad 0.460 \quad 0.610 \quad 0.749 \quad 0.867 \quad 0.964 \quad 1.028 \quad 1.000]$$

For the adopted distribution of lateral forces, a pushover analysis is performed. This analysis gives the relation between the total force at the foundation level V – base shear force, and the displacement of the top of the building D_t (Fig. 6).

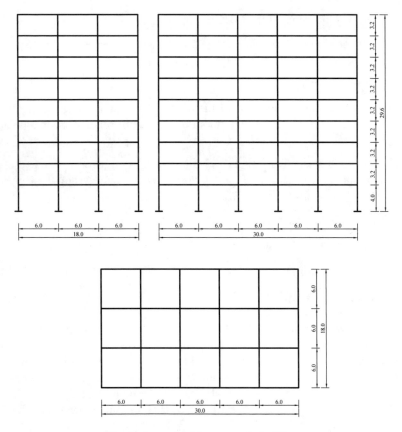

Fig. 3. Layout of the example building

Next step is the transformation of the MDOF system into a SDOF system. Equivalent mass m^* and transformation factor Γ have the following values:

$$m^* = 3115\,[t] \qquad\qquad \Gamma = 1.29$$

Then we obtain idealized curve F*-D* which is bilinear (Fig. 7).

From the obtained curve we can read the values of force and displacement at the yield limit: $F_y^* = 3850\,kN \qquad D_y^* = 8.5\,cm$

Elastic vibration period can be determined as:

$$T^* = 2\pi\sqrt{\frac{m^* D_y^*}{F_y^*}} = 1.65\,s$$

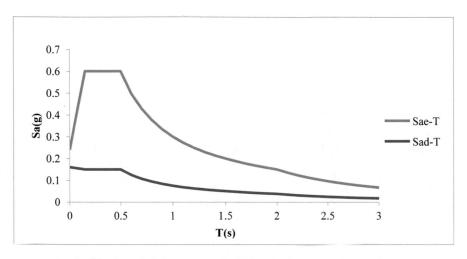

Fig. 4. Elastic and design spectra for VIII seismic zone and ground type B

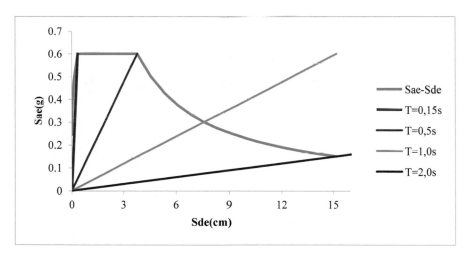

Fig. 5. Elastic spectra in ADRS format

The capacity curve is obtained by dividing the force values of idealized bilinear pushover curve with the value of equivalent mass (Fig. 8):

$$S_{ay} = 0.13g$$

In the case of unlimited elastic behaviour of the structure, the earthquake requirement is represented by intersection point of the elastic spectrum and the straight line issued from origin, corresponding to the elastic period $T^* = 1.65$ s of the equivalent single degree of freedom system. The obtained values are $S_{ae} = 0.182$ g and $S_{de} = 12.55$ cm. The reduction factor is:

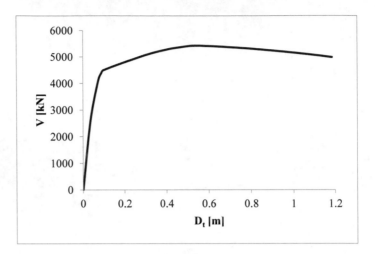

Fig. 6. Pushover curve

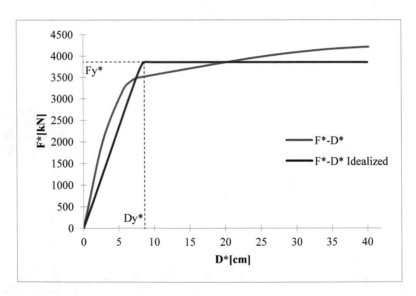

Fig. 7. Idealized capacity curve

$$R_\mu = \frac{S_{ae}(T^*)}{S_{ay}} = \frac{0.182g}{0.13g} = 1.4$$

The period of the equivalent SDOF system is $T^* = 1.65$ s, which is greater than $T_C = 0.5$ s, and the rule of equal displacements between elastic and nonlinear structure is applied:

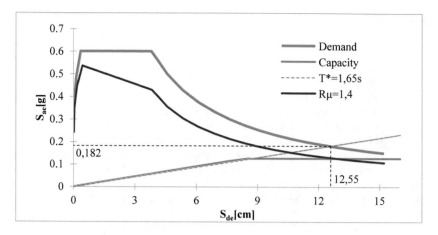

Fig. 8. Earthquake demand and capacity of the structure

$$\mu = R_\mu = 1.4$$

$$S_d = S_{de}(T^*) = 12.55 \, \text{cm}$$

The earthquake demand for the equivalent system with one degree of freedom is graphically represented by the intersection point of the capacity curve and the earthquake demand for $\mu = 1.4$.

The displacement at the top of the building - the target displacement, will be obtained from the shift of the equivalent SDOF system:

$$D_t = \Gamma D^* = 1.29 \cdot 12.55 = 16.2 \, \text{cm}$$

Therefore, the value of the target displacement of the top of the building is 16.2 cm.

4 Conclusion

The N2 method can be considered as a framework that correlates the computation using a pushover method with a response spectrum analysis. It represents a practical procedure for the assessment of the behaviour of a structure. Formulation of the method in the acceleration-shift form allows a clear interpretation of the procedure. It gives insight into the seismic performance of the structure. The results obtained using this method are sufficiently accurate if the structure predominantly oscillates in its first form.

The application of the N2 method in analysed problem is limited to the analysis of symmetrical structures. It could be adopted to consider higher vibration modes, and to analyse unsymmetrical buildings as well.

References

1. Hrasnica, M.: Aseizmičko građenje. Građevinski fakultet Sarajevo (2012)
2. Chopra, A.K.: Dynamics of Structures, Theory and Applications to Earthquake Engineering. Prentice Hall, Upper Saddle River (1995)
3. Freeman, S.A.: Development and use of capacity spectrum method. In: Proceedings of the 6th U.S. National Conference on Earthquake Engineering, Seattle (1988)
4. Fajfar, P., Fischinger, M.: Nonlinear seismic analysis of RC buildings: implications of a case study. Eur. Earthq. Eng. **1**, 31–43 (1987)
5. Fajfar, P., Fischinger, M.: N2 – a method for nonlinear seismic analysis of regular bulidings. In: Proceedings of the 9th World Conference on Earthquake Engineering, Tokyo, Kyoto, vol. 5, pp. 111–116. Maruzen, Tokyo (1988)
6. Fajfar, P., Gašperšič, P.: The N2 method for the seismic damage analysis of RC buildings. Earthq. Eng. Struct. Dyn. **25**, 31–46 (1996)
7. Fajfar, P., Fischinger, M.: N2-a method for non-linear seismic analysis of regular buildings. In: Proceedings of Ninth World Conference on Earthquake Engineering, vol. 5, Tokyo-Kyoto, Japan (1998)
8. Fajfar, P.: Capacity spectrum method based on inelastic demand spectra. Earthq. Eng. Struct. Dyn. **28**, 979–993 (1999)
9. Fajfar, P.: A nonlinear analysis method for performance based seismic design. Earthq. Spectra **16**(3), 573–592 (2000)
10. Fajfar, P., Fischinger, M., Isaković, T.: Metoda procjene seizmičkog ponašanja zgrada i mostova. Građevinar **52**, 663–671 (2000)
11. Chopra, A.K., Goel, R.K.: A modal pushover analysis procedure for esimating seismic demands for buildings. Earthq. Eng. Structural Dyn. **31**, 561–582 (2000)
12. Čaušević, M., Zehentner, E.: Nelinearni seizmički proračun konstrukcija prema normi EN 1998-1:2004. Građevinar **59**, 767–777 (2007)
13. Eurocode 8 (EC8), EN 1998-1: Design of structures for earthquake resistence – Part 1: General rules, seismic actions and rules for buildings. CEN European Comittee for Standardization, December 2004
14. Drkić, A.: Nelinearna seizmička analiza nesimetričnih višekatnih zgrada sa armiranobetonskim okvirima. Master rad, Građevinski fakultet Sarajevo (2014)
15. Alendar, V.: Projektovanje seizmički otpornih armiranobetonskih konstrukcija kroz primere. Građevinski fakultet Univerziteta u Beogradu, Beograd (2004)
16. SAP2000 CSI Analysis Reference Manual, Computers and Structures, Inc. University Avenue Berkeley, California 94704, USA (1995)

Selection, Effectiveness and Analysis of the Utilization of Cement Stabilization

Edis Softić[1(⊠)], Elvir Jusić[1], Naser Morina[1,2],
and Muamer Dubravac[3]

[1] Department of Construction, Technic University,
University of Bihać, Bihać, Bosnia and Herzegovina
edis.softic@bih.net.ba, elvir-jusic@hotmail.com
[2] Gjilan, Republic of Kosovo
[3] Department of Construction, Polytechnic University, University of Zenica,
Zenica, Bosnia and Herzegovina
muamer.db@hotmail.com

Abstract. The analysis of the utilization of the cement stabilization in the lower bearing layers of the roadway and its impact on the recess size in the lower link layers has been conducted in these papers and has been measured through a certain timeline during exploitation. The analysis has been conducted on the cement stabilization compound formula along with modus of its execution in the field. Thereafter, the analysis and measurement of the rut of certain roadway sections, that have been stabilized with cement in the lower link layers and other sections that haven't been managed with similar traffic load, have been approached in these papers as well. The results are presented in the papers that exhibit the significance of the cement stabilization utilization in the lower bearing layers in addition to its influence on safety and possibility of increasing the design period of the aforementioned, as well as the roadway's depth value.

Keywords: Cement stabilization · The lower bearing layers
The lower link layers · Ruttings

1 Introduction

The state of the roadway's surface primarily shows us the visual projection of the roadway construction, in most cases of the finishing lower layer of construction that was given based on facts on key parameters of load capacity, consumption, sustainability, safety of the roadway and its conduct in exploitation.

Taking in consideration the aspect of economical standpoint of a certain country and its development, the visual state of the road shows us the general growth of society and is an indicator of national economy's development in high degree.

For the past twenty years or so, in Bosnia and Herzegovina, a certain tendency is present that showcases investments in reconstruction, maintenance and modernization of already existing roads rather than investing in and building new ones. Building smaller subsections of highway, that are prospering for couple of years now, is the only exception. Roadway constructions are multilayered systems with an installed mechanized modus of work and a purpose of transporting static and dynamic traffic loads to

© Springer Nature Switzerland AG 2019
S. Avdaković (Ed.): IAT 2018, LNNS 60, pp. 203–212, 2019.
https://doi.org/10.1007/978-3-030-02577-9_20

the bottom structure without damaging deformities of the sub-grade. The bearing layers stabilized with cement are in advantage to the rest of the stabilized layers, because they reduce the impact of the sub-grade capacity on the roadway construction capacity, also they enable the construction on weaker bearing soil and the utilization of local materials for installation, all of which should imperatively have a determining role in decreasing recess within roadways or ruts.

Cement stabilization is immensely profitable solution for rural roads. A problem is often current in actual practice because of the deficiency of certain project-technical documents, which should be made based on given preliminary list of prioritized sections. Similar was case where a measurement has been conducted on the size and length of a rut in section of the road M4 connecting Donja Orahovica – Šićka Petlja (loop) at the entrance of Tuzla, and at the length of 26 km. Ruts are dents/recess created by wheel trails and they often appear because of the inadequate base and frequent movement of cargo vehicles. When filled with water, it easily leads to water wedging (Aquaplaning).

2   Cement Stabilization as a Rut's Base

Cement stabilization is a solid, uniform load-bearing layer for existing and future load-bearings. Stabilizes the lower bearing layers using only one stabilizer-cement. Characteristics of cement stabilization are:

- a verified solution
- Low cement content
- The ability to recycle used asphalt pavements (fulldepth reclamation FDR)

The main advantages of cemnet stabilization are reflected through:

- reduced thickness of upper and lower bearing layers
- Affordable Recycling of used asphalth pavements (FDR)

Cement-stabilized foundation provides significant savings compared to conventional alternatives (Fig. 1).

Layers of cemented stabilized grain materials are at first glance similar to the concrete, but in fact they differ from it. Unlike concrete, cement-stabilized layers contain a much smaller amount of cement (3–5%), depending on the characteristics of the grain material. The aftermarth of such a small amount of cement is the incomplete grain hardening of the cement mortar and the large cavities. Tension of this material is not great, nor does it work with temperature changes. Therefore, these layers can work without splitting. Cement as a binder is used to make cement-stabilized load-bearing layers, which, in the presence of water, bonds different types of natural stone materials used as aggregates. Cement-stabilized blends are made of:

- The aggregate
- Cement
- Water.

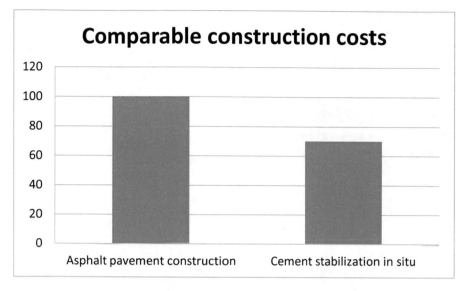

Fig. 1. Comparison of construction costs for asphalt pavement construction

Stabilization of soil or building material during road construction has a significant impact on the features of the road itself. The treated route is more stable, more resistant and longer lasting. It's possible to stabilize soil and/or aggregate (pebble/sand/dung or existing pavement in reconstruction of roads). Stabilization is performed by mixing the soil/aggregate with the stabilizer binders. The mixture is then compacted with rolls to achieve optimum compactness. In practice, two stabilization processes are used: process with the production of stabilized material on the stationary plant and installation of finisher on-site assembly by recycling machine Considering the fact that the existing material or cold recycling process (to the extent possible) treats the material found on the construction site itself, manipulation and transportation are far fewer. Here, additional, important advantages of stabilization/cold recycling are generated:

- reducing dust on site
- increasing the versatility/use of secondary materials
- preservation of natural resources (reduction of the use of river aggregates and stone)
- Reduction of costs and energy.

2.1   Cement

Cement is proclaimed the oldest binder since technological stabilization of soil – invention in 1960s. It can be regarded as a primary substance for stabilization or as a hydraulic binder, considering it can be used singularly to get the required stabilizing effect. Cement reactions do not depend on soil's minerals but rather on its response to water that is at disposal in every soil. It is one of the reasons why cement is used for stabilization of wide spectrum of the soil/plot, lot. Many kinds of cement are available on the market, however usually the choice to which kind of cement one wants to use depends on the type of soil/land being treated and the desired final solidity. Process of hydration begins when cement is mixed with water and the rest of the components, which results in solidification phenomena. With solidification the cement will overlay the ground acting as glue, but it won't change the initial structure of the soil. The hydration reaction dwells slowly on the cement grains' surface while its central parts may remain non-hydrogenated. Cement hydration is a complex process with a chain of complex chemical reactions. This process can be affected by:

- presence of strange materials and impurities
- water/cement ratio
- temperature of nourishment
- presence of mineral additives
- specific surface of measurement.

Depending on the additives in the mixture, the final outcome on binding and solidity gain on cement stabilization can vary. For that reason, this should be taken in consideration while making the mixture, all in the goal of achieving the desired solidification. Calcium silicates, C3S and C2S are the two major minerals of a regular Portland cement, that are responsible for solidification development. Calcium hydroxide is another one of the products of hydration of the Portland cement, which still reacts with "pucolic" materials available in stabilized soil/land. Cement stabilized soils/lands have the next advanced properties:

- reduced cohesiveness
- reduced volume of expansion (compression)
- increased solidity.

3   Ruts on Roads

The very ride through ruts is uncomfortable and eases the driver to lose control over the vehicle hence the driver needs to hold onto the steering wheel tightly. If possible, the driver needs to elude the ruts without falling out of one's traffic lane. Driving over the rut makes the driver briefly lose base which only increases the period of braking and its distance. Shifting from one traffic lane to the other in case of the ruts present, should be achieved in reduced speed and at a sharp angle. What are the possible dangers of driving on a road because of the rut on a roadway?

- Occurrence of the water wedge (aquaplaning) and water skiing
- Losing control over the vehicle
- Longer braking distance.

Thereafter, it is concluded that ruts are plastic deformation on roadway's surface that appear in wheel trails under the power of the road's load. Their occurrence affects reduction of traffic safety, the comfort of the ride, and the roadway construction sustainability. They appear in a relatively early stage of utilization on every type of flexible roadway constructions. The creation of the rut can be an outcome of the later compaction (consolidation) of roadway's layer under the road's load power, mechanical deformation of the base under the roadway's construction and shimmering deformation in the asphalt mixture (Table 1).

Table 1. An index of the rut has been presented in the following chart [5]

Depth of trap (mm)	To 10	10–20	Over it 20
RUTI	*1*	*2*	*3*

Grade number 1 – reflects a solid state of roadway's surface on which it isn't necessary to make any adjustments or it reflects on the smallest part of the surface for which the adjustments may be put off for some time without any damaging consequences.

Grade number 2 – presents mediocre state of roadway's surface that is – it is advised to make some adjustments of maintenance in the lack of some or another possibility…

Grade number 3 – presents a bad state of roadway's surface which is on demand for significant adjustments to be done, even the reconstruction of the roadway.

In these papers the research on influence of the asphalt's mixture composition, of the types AB 11, AB 16 and AB 16s with or without previous interference in the makings of stabilized cement mixtures, has been conducted. After defining the model of asphalt's mixture composition, the testing and measurement of the creation of ruts and it's resistance on the samples of different composition have been carried out along with the analyzed results with the goal of discovering the dependence of the composition on the property, that is the depth of the rut.

Picture number 3 – shows the layers of the roadway constructions stabilized or not stabilized with one of the different types of binders (Fig. 2).

Fig. 2. Overview of stabilized and non-stabilized layered structures of layer bonds [1]

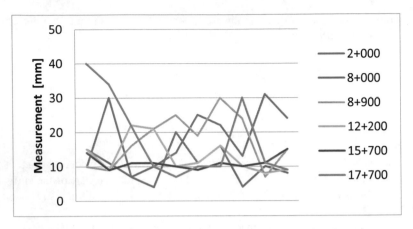

Diagram 1. The measurement of the rut on the open part of the lane

4   Composition of Asphalt Mixtures and Cement Stabilization on the Inspected Sections of the Roadway

For rut's sanitation, the cement stabilization of the roadway has been envisioned, which had been done on Šićki Brod loop and all that has proven the fact that ruts are almost all evanished from the roads. In this chart the granulometric property of stone aggregate is shown (Table 2):

Table 2. Shows the granulometric property of the stone aggregate for cement stabilization

Sieve	Residue [gr]	Residue [%]	Passage [%]
45	0	0	100
31.5	110.1	2.29	97.71
22.4	473	9.84	87.87
16	654.3	13.61	74.27
11.2	640.5	13.32	60.95
8	638.7	13.28	47.67
4	1125.9	23.41	24.26
2	444.2	9.24	15.02
1	225.4	4.69	10.33
0.71	78.7	1.64	8.69
0.5	66.5	1.38	7.31
0.25	81.7	1.7	5.61
0.125	64.2	1.34	4.28
0.09	45.8	0.95	3.33
0.063	46.5	0.97	2.36
bottom	113.4	2.36	

In this next chart the granulometric property of the stone aggregate for cement stabilization is shown (Table 3):

Table 3. Shows mass of cement in lab's casserole

Rotor blend depth (cm)	The mass of cement in the laboratory mat size 60 x 60 cm [kg]
15	2.07
16	2.21
17	2.34
18	2.48
19	2.62
20	2.76
21	2.89
22	3.03
23	3.17
24	3.31
25	3.45
26	3.58
27	3.72
28	2.86
29	4
30	4.13

5   The Results of Inspection on the Ruts on Highway Section M4 (Donja Orahovica – Šićka Petalj (Loop))

The obtained results were evaluated on the spot at the section of the highway M4 Donja Orahovica – loop Šićki Brod. The results were compared to the roadway construction on Šićki Brod loop where the cement stabilization of roadway construction was implemented.

In this chart the results of the inspected and measured ruts of the highway's section Donja Orahovica – loop Šićki Brod are shown (Table 4).

In this next diagram the results of the measurement of the rut on an open lane of the highway is shown (Diagram 1).

In the next diagram the results of the measurement of the rut at the crossroads in Lukavac are shown (Diagrams 2 and 3).

In the next chart the mean values of measurement ruts by chainage, crossroads and loops are shown.

Chart number 5: mean values of inspected ruts by chainage, crossroads and loops are shown.

Mean values of measured ruts in characteristic changes (Diagram 4).

In the next diagram mean values of the rut are shown (Fig. 3).

Table 4. The measurement of the rut at the highway section Donja Orahovica – loop Šićki Brod are shown

The height of the cave [mm]

Stationing

	2 +000	8 +000	8 +900	12 +200	15 +700	17 +700	The crossroads in Lukavac	Lace Šićki Brod
1	10	15	10	15	14	40	30	1
2	30	11	9	9	9	34	60	0
3	7	7	16	22	11	22	70	1
4	10	4	21	21	11	10	65	0
5	14	20	25	10	10	7	50	0
6	25	11	19	11	9	10	70	0
7	22	16	30	16	11	10	90	1
8	13	4	24	10	10	30	80	1
9	31	10	7	8	11	11	75	0
10	24	8	15	9	15	9	60	0

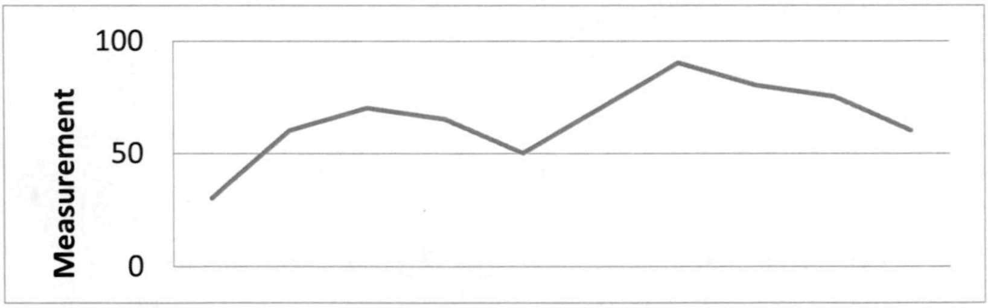

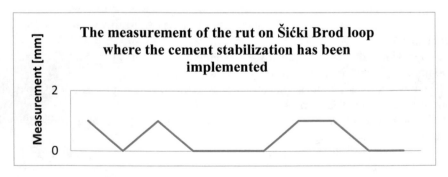

Diagram 2. The measurement of the rut at the crossroads in Lukavac

Diagram 3. The measurement of the rut on Šićki Brod loop where the cement stabilization has been implemented

Stationing

2+00	8+00	8+90	12+20	15+70	17+70	Intersecti on in Lukavac	Loop Šički brod
0	0	0	0	0	0		
18.6	10.6	17.6	13.1	11.1	18.3	65	0.4

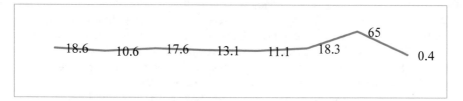

Diagram 4. Mean values of the measured rut

Fig. 3. Shows the measurement of the ruts on highway section M4 (D.O – Š.B. loop)

6   Conclusion

When designing roadway constructions, the utilization of cement stabilization has immensely important role with guaranteed multiple functionalities. It is especially vital to emphasize, which has been in this case confirmed as well, that while the utilization of stabilization is present – local materials "in situ" are getting strengthened which isn't possible without special advancements, even in the basic layers of roadway construction. Ground with weaker bearing capacity is more worthy in this case because the cement with mixture boosts the load capacity and in that way diminishes the role of stability and the module of compressibility of the sub-grade at the roadway construction's capacity. The specificity of the source material that is to be used to conduct the cement stabilization is of crucial importance when the reliability and the quality of derived stabilized layer of the roadway, is in question.

In this paper the measurement of the rut has been conducted at the typical places of the highway M4 (photo attached, Sect. 5) from the places D.O. – Š.B. loop, whose results have been shown in a chart and in a diagram prolifically.

It is evident that the values of ruts are highly greater than it should be allowed, even with the fact that the same section has been overlaid with asphalt/concrete approx. seven years ago. On the other hand, the size of the rut on the road outline on which cement stabilization has been conducted (road section Š. Loop of Kreka estate) were measured which produced minimal values of the rut and serves to prove the thesis of profitability, reliability, and confirmed technical and technological solution which has never proven wrong, not even under the hefty load capacity and of which inspected sizes of rut witness.

Compound formula is presented in a way that it entirely satisfies upholding criteria for determination of certain types of cement stabilization that are nowadays functioning as design criteria assuring the correctly made designs next to modern construction functioning. Surely technology advancements have to be taken in consideration especially because they impose the utilization of the new materials such as metal fibers, hovering ash or high pressure furnace slag that offers the same part in improvements of operation and the durability of roadway construction, while at same time it's properties and impact on roadway construction haven't been inspected.

References

1. Investing technical documentation of CEMEX BiH doo
2. Strineka, A., Brkić, J., Sekulić, D.: Influence of composition on deformability of asphalt
3. Barišić, I., Rukavina, T., Dimter, S.: Cement stabilization - characterization of materials and project criteria
4. Investing Techniques Documentation d.o.o. Roading Gračanica i d.o.o. Arapovac puts Čelić
5. Jokanović, I.,Zeljić, D., Mihajlović, D.: Evaluation of the state of the road from the technical and user aspect

Inventarization of the Benchmarks NVT II Network in the Field of the Republic of Srpska and Application of DGNSS Technology

Kornelija Ristić[1(✉)], Sanja Tucikešić[1], and Ankica Milinković[2]

[1] Faculty of Architecture, Civil Engineering and Geodesy,
University of Banja Luka, Vojvode Stepe Stepanovica 77/3, 78 000 Banja Luka,
Bosnia and Herzegovina
{kornelija.ristic,sanja.tucikesic}@aggf.unibl.org
[2] Vekom Geo Ltd., Trebinjska 24, 11 000 Belgrade, Serbia
ankica.milinkovic@vekom.com

Abstract. Implementation of new geodetic datum represents a very complex and long-term process, which implies a systemic approach, good organization and coordination of multiple tasks. Republic Administration for Geodetic and Property Affairs of the Republic of Srpska following contemporary theoretical and practical achievements in European countries and countries in the region has joined the implementation of new geodetic reference systems in the whole territory. The implementation of the chain of business divided into the entire legislative, technological and organizational units will provide a unique mathematical and physical basis for horizontal and vertical positioning, gravimetric and astronomical works and determination of geoids for the needs of the state survey and cadastre of real estate as well as for engineering and technical works for scientific purposes. This paper gives an overview of the results and experiences that have been achieved so far through the realization of the inventory of the benchmark of the second high accurate leveling network (NVT II) and for the needs of designing and performing works on the establishment of the new third high accurate leveling network (NVT III) in the territory of the Republic of Srpska.

Keywords: Benchmark · Inventarization · NVT · Leveling network

1 Introduction

High accurate leveling belongs to the most precise and most demanding geodetic measurements [4]. Reconstruction of altitude systems represents a periodical process, in which the renovation of the reference altitude system is made, due to the obsolete height data. Works on the inventory of the benchmarks of the existing high accurate leveling network NVT II comprised a series of elements from the field revision (determining the physical condition of the benchmark in the field) to collecting new information on the benchmarks and their geographic environment using the Global Navigation Satellite System technology in the ETRS89 reference system.

© Springer Nature Switzerland AG 2019
S. Avdaković (Ed.): IAT 2018, LNNS 60, pp. 213–223, 2019.
https://doi.org/10.1007/978-3-030-02577-9_21

The high accurate leveling network established between 1970 and 1973 forms the basis of the vertical reference system in the Republic of Srpska. Given the past 40 years, and the fact that stabilized leveling points (benchmarks) are subject to different geodynamic processes, it is expected that a certain percentage of the benchmark will be damaged and destroyed, and some of the benchmarks in the meantime become dysfunctional.

Bearing in mind that the NVT II network was established for the territory of the former SFRY, the inventory of the benchmark was performed along the leveling lines of the NVT II network that spread in the territory of Republic of Srpska: Leveling line (number 5) *Bosanski Petrovac - Bosanska Krupa - Kostajnica* in the territory of Republic of Srpska (Entity line Blatna - Dobrljin - Kostajnica), Leveling line *Jajce – Banja Luka – Okučani,* leveling line (number 13) *Bosanski Petrovac – Ključ – Jajce,* leveling line (number 12) *Strizivojna – Doboj – Maglaj- Kaonik,* leveling line *Doboj – Tuzla – Zvornik,* leveling line *Kuzmin – Bijeljina – Janja – Lešnica – Loznica,* leveling line *Podromanija – Vlasanica – Zvornik,* leveling line *Blažuj – Sarajevo – Podromanija – Rogatica – Ustripača,* leveling line *Ustiprača – Višegrad – Dobrun,* leveling line *Ustiprača – Foča – Brod,* leveling line *Brod – Avtovac,* leveling line *Avtovac – Bileća – Trebinje – Dubrovnik.*

Leveling lines include mareographs (mareographic benchmarks), fundamental and node benchmarks, and as a whole they comprise a series of bounded leveling figures in the leveling network (Klak et al. [2], Klak et al. [3]).

2   History of Levelling Work in the Territory of the Republic of Srpska

The first leveling measurements on the territory of the Republic of Srpska and in the countries that were part of the Austro-Hungarian Monarchy was a precise leveling network (Austrian Precision Nivelman - APN). This network was carried out in the period 1899–1907, the Vienna Military Geographical Institute (VMGI) produced four closed leveling polygons.

Other leveling measurements were carried out during 1929, the Military Geographical Institute of the Kingdom of Yugoslavia, in the territory of RS, has leveled the polygons of *Sarajevo-Sokolac-Ustipraca-Kifino Selo-Trebinje-Dubrovnik* and *Kifino Selo-Mostar.*

The third and fourth leveling measurements were made during the former Yugoslavia after the Second World War in the period from 1948 to 1952, the Federal Republic of Yugoslavia Military Geographic Institute, the FRY General Directorate of Geodesy and the Republic Geodetic Authority (RGU) of BiH, in the territory of the Republic of Srpska, renewed the VMGI trains of the first leveling measurements and carried out the work of NVT, precision leveling (PL) and the technical leveling of increased accuracy (TLIA). These are the first high accurate leveling network, which was carried out in the period from 1946 to 1963, and the second high accurate leveling network made in the period from 1967 to 1973.

After 1973, systematic works in the field and sense of updating and restoration of benchmarks and their re-leveling were not performed in the territory of the Republic of Srpska.

3   High Accurate Levelling II

At the Advisory on the high accurate leveling network held in Belgrade in 1967, the NVT I revision was being prepared and the project of the new leveling network of high accurate NVT II was developed. The area of Bosnia and Herzegovina was included in the project for the former Yugoslavia.

The project envisages a number of fundamental benchmarks, i.e. that all nodal benchmarks should be stabilized as fundamental benchmarks. The total length of the levelling polygon is 9 824 km with 27 node benchmarks. The length of the levelling polygons in Bosnia and Herzegovina is 1 966,9 km with 2182 benchmarks, of which there are 10 fundamental benchmarks and one normal benchmark in Maglaj. The normal benchmark is located in Maglaj as the center of the former Yugoslavia in a seismically and geologically stable area.

Adjustment of the NVT II network was performed by the Geodetic Faculty of the University of Zagreb. Along the leveling polygons connecting the mareographs and the leveling polygon that connects Metković with Maglaj, we measured the acceleration of gravity. And for the other leveling polygons, except *the Bosanska Krupa - Kostajnica, Bosanska Krupa - Bosanski Petrovac and Bosanski Petrovac – Šibenik* polygons, the acceleration of gravity was measured. Calculation of acceleration of gravity was carried out on the basis of Faye's anomaly maps using a map in the polyhedron projection in the scale of 1: 200,000.

Due to the lack of knowledge of the exact values of acceleration of gravity, the absence of digital models of density and relief, it is proposed for practical use to use normal heights (Molodensky), and the geopotential heights to fit our network into the European Leveling Network (UELN). The normal benchmark in Maglaj is connected by a precise leveling with mareographs in Split and Dubrovnik. Later on, leveling connected the other mareographs in Koper, Rovinj, Bakar, Split in the port, Split at Marjan, Dubrovnik and Bar. Vertical datum II NVT, i.e. the mean level of the Adriatic Sea on individual mareographs, is 3.7.1971, and from the sea swinging data measured from 12.2.1962 until 21.9.1980. At the end of the 20th century, it was confirmed that the average sea level in the mareograph in Trieste is about 12 cm lower than a properly defined mean sea level, which is the result of only one year measurements. For the appropriate determination of the medium sea level it is necessary to perform measurements over a period of 18.61 years.

4   New Leveling Network – NVT III

One of the more important components of determining geoids is the establishment of leveling network in the whole of territory of Bosnia and Herzegovina. Apart from leveling measurements, this network demands the measurements by the GNSS technology. These two processes need not run at the same time, given that the leveling points are going to be permanently stabilized. Also, there are going to be the values of gravitational acceleration determined on all points of the leveling network by means of relative gravimeters. The new Order III network comprises the following scope of activities: recognition and stabilization of leveling points, leveling measurements, determining benchmark coordinates by applying the GNSS technology, and gravimetric measurements.

5 The Results of the Analysis of the Current State of the Benchmarks of the NVT II Network

The area of the Republic of Srpska is covered by the second high accurate leveling network (NVT II), which was made up of 955 benchmarks. The analysis of the existing state of affair shows that there are only 425 benchmarks in use today (44.5%). This is due to the fact that many of these have been plastered, or the objects on which they were positioned have been ruined or destroyed (churches, mosques, bridges, edifices etc.). As for those in the vicinity of frontiers, they have been either taken out or the rocks carrying them have been damaged. Determining the physical state of the benchmarks have been performed on site. To track them down, the existing location descriptions and benchmark coordinates located by means of a manual GNSS receiver have been used. The overview of the analysis of the existing state of affairs upon leveling lines is provided in Table 1.

For every preserved NVT II network benchmark, there has been positioning within the permanent GNSS stations network of the Republic of Srpska performed by means of the GNSS technology, in relation to the ETRS89 referent system, and in one of the following ways: by means of the network RTK method, the benchmark coordinates have been determined throughout three measured 30-s sessions (if it was possible to position a GNSS antenna directly on the benchmark, and by means of the differential GNSS method (if the benchmark is in the location where GNSS measurements are not possible due to physical obstacles and signal blocking).

By comparing ellipsoid heights (h), gained through either of the aforementioned methods, with sea levels (H) of the existing leveling points, the undulations (N) in the territory of the Republic of Srpska have been calculated, with the average value of 45 m (Fig. 2). As far as the maximum and minimum values are concerned, they are 57 and 40 m respectively (Fig. 1) (Table 2).

6 Differential GNSS and the Network of Permanent Stations

Generally speaking, determining the absolute position of GNSS is a much less precise method than the method of relative positioning between two stations. This is due to most active errors that are spatially and temporally correlated and can be assessed by the receiver whose location is already known (base or referent station).

There are three categories of error sources: errors related to distance, errors related to time, and uncorrelated errors. Time-related errors are covered by synchronized or

Table 1. Presentation of an analysis of the current situation in leveling lines in the territory of the RS

Leveling lines	Total number of benchmarks	Number of destroyed benchmarks	Percentage of destroyed benchmarks
Bosanski Petrovac – Bosanska Krupa - Kostajnica	64	36	3.8%
Jajce – Banja Luka – Okučani	118	70	7.3%
Bosanski Petrovac – Ključ – Jajce	56	33	3.5%
Strizivojna – Doboj – Maglaj-Kaonik	81	51	5.3%
Doboj – Tuzla – Zvornik	78	29	3.0%
Kuzmin – Bijeljina – Janja – Lešnica – Loznica	43	21	2.2%
Podromanija – Vlasanica – Zvornik	128	79	8.3%
Blažuj – Sarajevo – Podromanija – Rogatica – Ustiprača	116	87	9.1%
Ustiprača – Višegrad – Dobrun	52	31	3.2%
Ustripača – Foča – Brod	32	18	1.9%
Brod – Avtovac	83	41	4.3%
Avtovac – Bileća – Trebinje – Dubrovnik	104	34	3.6%
Total:	955	530	55.5%

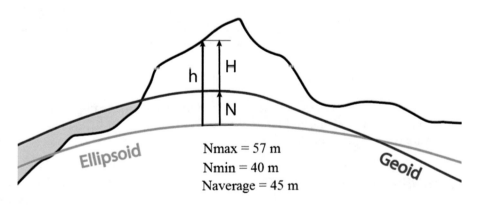

Nmax = 57 m
Nmin = 40 m
Naverage = 45 m

Fig. 1. Ellipsoid height (h), orthometric height (H) and geoid height (N)

Fig. 2. Graphic presentation of the benchmarks of the leveling network NVT II in the territory of the Republic of Srpska in the inventory process.

almost synchronized observations, whereas uncorrelated errors have effect on both indirect receivers, and it is necessary to calibrate them. As for distance-related errors, these are mostly errors of ephemerides and propagation, and they are nearly identical for receivers that are close enough. The latter can be eliminated by utilizing differential measuring techniques. Instead of absolute coordinates, coordinate differences are determined in relation to a known referent station. There are four concepts to be singled out: 1. Usage of data of one or more referent stations for subsequent processing (relative GPS), 2. Usage of location or latitude correction from code measurements on referent stations in real time (common differential GPS), 3. Usage of code latitudes and bearing phases data from a referent station in real time (precise differential GPS or Real-Time Kinematic GPS – RTK), and 4. Usage of referent data from referent stations network in real time (differential GPS, known as Multiple Referent Station or Network RTK).

One of major drawbacks of the DGPS is a fact that the influence of a certain error source, such as orbit, ionosphere, and troposphere, is getting bigger by the increase of distance from a referent station.

7 DGNSS Concept

Differential positioning of GNSS (DGNSS) is a technique of positioning in real time in which two or more receivers are used, and it is applied in GNSS processing in order to increase motivating factor behind the development of DGNSS was the presence of Selective Accessibility (SA) on a GPS signal. The SA was implemented so that it

Table 2. Results of field audit of the benchmark in leveling lines in the territory of the RS

Leveling Lines in the territory of the RS	Results of field audit of the benchmark
Leveling line Bosanski Petrovac – Bosanska Krupa - Kostajnica	This leveling line was set up on the railway Dobrljin - Bihać. In Novi Grad, a fundamental benchmark was destroyed, where a new facility was built
Leveling line Jajce – Banja Luka – Okučani	From the entity line, along the Vrbas River on the left, a leveling line was set up that was flooded to Bočac by the construction of the Bočac hydroelectric power plant
Leveling line Bosanski Petrovac – Ključ – Jajce	The inventory was carried out to the entity line in Velečevo. Fundamental benchmark F.B.-1071 in Zableće was destroyed by the construction of a new road ("Avnoj road"). This leveling line went through the old road under the new bridge, so most NVT II benchmarks were destroyed and inaccessible
Leveling line Strizivojna – Doboj – Maglaj- Kaonik	The subject of inventory were benchmarks in the territory of Republika of Srpska (Šamac - Doboj entity line). In the part of this level line from Modriča to Doboj, the benchmarks were set before the construction of a new road, so most of the NVT II benchmarks were destroyed and inaccessible. The fundamental benchmark in Modriča (Dobor tower) is stable, and one reper of the micro-network is destroyed
Leveling line Doboj – Tuzla – Zvornik	This leveling line from Doboj to Tuzla went through the railway line to the entity line (Petrovo - Novo selo). From the entity line (Mahala) to Zvornik benchmarks of NVT II are stable and placed on a new road
Leveling line Kuzmin – Bijeljina – Janja – Lešnica – Loznica	On the bridge over the Sava River in Rača, on the left and right, the NVT II benchmarks were destroyed by the construction of a new roadway. Some of the benchmarks were destroyed by the renovation of the facades at the facilities to the Janje River
Nivelmanska linija Podromanija – Vlasanica – Zvornik	In the part of this leveling line (Podromanija - Sokolac - Han Pijesak - Vlasanica), NVT II benchmarks were set up before the construction of a new roadway and most were destroyed. There are only those benchmarks that are on the objects and away from the road. In the second part of this leveling line (Vlasanica - Milići - Konjević Polje - Drinjača), the majority of benchmarks are stable. In the part of the levelling line from Drinjača to Zvornik the benchmarks are mostly in culvert. Culvert were largely renovated and expanded so most of them were destroyed
Leveling line Blažuj – Sarajevo – Podromanija – Rogatica – Ustiprača	In this leveling line, NVT II benchmarks were set up before the construction of a new roadway. On the section of the old road from the entity border, through Bulog to Ljubogošće, the benchmarks of NVT II are largely stable. On the section of Ljubogošća - Wet, the benchmarks

(*continued*)

Table 2. (*continued*)

Leveling Lines in the territory of the RS	Results of field audit of the benchmark
	NVT II have been stable only on buildings that did not perform reconstruction of the facade. The benchmarks NVT II on the section Mokro – Crvene Stijene - Podromanija are stable along an old roadway, on which access is possible. Fundamental benchmark F.B.-1085 in Podromanija is located above the intersection of regional roads, and the micro network benchmarks were destroyed by the construction of it. The benchmarks NVT II, which were at a greater distance from the old Podromanija - Rogatica road, are stable. Fundamental benchmark F.B.-1086 in Rogatica is stable, and one benchmark of the micro network is destroyed. On the section of the levelling line Rogatica - Ustiprača, the benchmarks NVT II were placed on the left side of the river Rakitnica to Mesić, and from Mesić to Ustiprača they were placed along the former narrow-gauge railway. It was not possible to perform any surveying in this part of the levelling line
Leveling line Ustiprača – Višegrad – Dobrun	Big changes took place on this leveling line. Railway facilities were destroyed in the lace of Ustiprača. Downstream the river Drina, the line of the NVT II benchmark was on the right side of the Drina (old road) to Višegrad. The new road is on the left side of the river Drina. From Ustiprača to Ajdanović on the right side of the Drina River there is a number of NVT II benchmarks, the other benchmarks are submerged by the construction of the Višegrad hydroelectric power plant. On the leveling line from Višegrad to Dobrun, the NVT II benchmarks were placed along the new road and they are stable. Fundamental benchmark F.B.-1087 in Dobrun is persistent as well as benchmarks of the micro network
Leveling line Ustripača – Foča – Brod	On this leveling line, NVT II benchmarks on railway installations (train stations) were destroyed. Inventarisation was done only for benchmarks in the territory of Republika Srpska
Leveling line Brod – Avtovac	On this leveling line, NVT II benchmarks were set up before the construction of a new section of the road. The old road went through Čemerno, where one part of the old road (from the river Sutjeska to the passage Čemerno), the road was impassable. From the passage Čemerno to the new section of the road there are constant benchmarks
Leveling line Avtovac – Bileća – Trebinje – Dubrovnik	On this leveling line, NVT II benchmarks are largely permanent. Only those benchmarks were destroyed, where the supporting wall was enlarged

would degrade GPS performances on purpose, with the improved precision as of the date of its termination, 2 May 2000.

We differentiate between DGNSS methods based on code (getting position at the level of metre) and those based on a bearing phase, which were dominant and which are further explicated.

It has been already mentioned, that the DGPS is a technique used for improving position determining, by applying corrections provided by a GPS tracking station (referent station). There are different procedures used for generating differential corrections: 1. corrections in the area of position, 2. comparison of a referent station GPS position with its a priori known position, 3. corrections in the measured area of observed pseudolatitudes on all visible satellites are compared with derived spans from known satellites and receivers, and 4. corrections in the area of spatial condition of measurement from several referent stations are used for the assessment of the condition of vector and influence within operating area.

The first procedure is rather simple and not so flexible, because it can be applied only if the same satellites are used both on a referent and on rover receiver, and it is applicable for short distances only/for that reason, this procedure is rarely used. As opposed to it, the second one is very flexible and operative within the radius of several hundred kilometers from a referent receiver, whereas the third is the most flexible one, allowing for the usage of the WADGPS (Wide Area Differential GPS) for precise applications in geodetic systems (referent stations network) (Fig. 3).

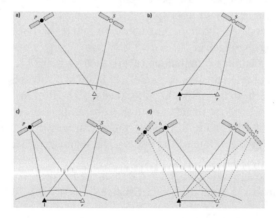

Fig. 3. Various differencing strategies: (a) between-satellite single differencing; (b) between-receiver single differencing; (c) double differencing; (d) triple differencing (Dennis Odijk, differenced positioning models, Springer handbook of global navigation satellite systems, 2017).

8 RTK GPS

The RTK is yet another name for the differential GPS by means of bearing phases, by which precision positioning is taken to the level of centimetres in real time over short distances. It is based on the following features: transfer of pseudolatitudes and bearing phases data from a referent station (base) to a receiver (rover) in real time, dealing with the 'on the way' or 'on the fly' (OTF) ambiguity on a receiver, and reliable determining of basic vector in real time.

One of major limiting factors of the RTK solution is a fact that errors increase as the distance from a referent station increases as well. The general rule for achieving accuracy is 10 mm +1 to 2 ppm for horizontal coordinates, and 15/20 mm +2 ppm as regards height component. In case of greater distances, more referent stations present a solution to the problem.

9   The Network of Reference Stations of High-Precision

By establishing referent stations providing data on bearing phases for precise DGPS applications in real time, distance-related error problems have become clear. Werrhen precision within 1 cm is required, a number of referent stations with adequate density would be unrealistically high, especially in periods of intense ionospheric interference. By connecting stations and error condition assessment in real time, the problem is overcome. In network solution, the error condition of the area is assessed and transferred to a receiver, where measurements are corrected in line with it.

As stated above, one of major drawbacks of the RTK of one base is a fact that the maximum distance between a referent station and a rover receiver must not exceed 10 to 20 in order for quick and reliable solutions of ambiguous bearing phases. This limitation is caused by partialities depending on distance, by ionosphere (refraction of signal), but by orbital error and tropospheric refraction as well. However, these errors can be precisely modeled through measurements along a series of GNSS referent stations surrounding the receiver.

Therefore, the solution to limitations of distance of the RTK of one station is in application of several referent stations, popularly known as the network RTK (NRTK). The distance between stations should not exceed 100–200 km in order for precise correction models of distance-related errors in to be realized in real time.

With regard to the Republic of Srpska, there is a network RTK service of positioning, providing for the whole of the entity. The major drawback of this method is related to two-way resolution of network observations, which usually requires a period of initialization of several minutes. Since the distance between referent stations across the Republic of Srpska is around 50 km, distance limit of 100–200 km should not be exceeded.

10   Conclusions

The Republic of Srpska has dated height data, which urged the update of Order II benchmarks and leveling lines in order for creating conditions for designing a geoid model which would serve, like in some European countries, as a basis for the replacement of the geometric leveling with the GNSS leveling. Consequently, the next step is the introduction of the Order III network. After this step has been completed, it is necessary for the Administration for Geodetic and Property Affairs of the Republic of

Srpska and its counterpart in the Federation of Bosnia and Herzegovina to make a list of identical benchmarks of Austro-Hungarian leveling, Order I and II leveling of the former Yugoslavia, and new leveling of Bosnia and Herzegovina.

By renewing the height reference system of the Republic of Srpska, there are going to be foundations laid for modern height basis of the Republic of Srpska, which can be further applied in deformation analysis and determining recent vertical movements of the soil in Bosnia and Herzegovina.

References

1. Zrinjski, M., Barković, Đ., Razumović, I.: Automatizacija ispitivanja preciznosti nivelira i umjeravanja invarnih nivelmanskih letvi. Geodetski list **64**(87), 4, 279–296 (2010)
2. Klak, S., Feil, L., Rožić, N.: Studija o sređivanju geometrijskog nivelmana na području Republike Hrvatske. Geodetski fakultet Sveučilišta u Zagrebu, Zagreb (1992)
3. Klak, S., Feil, L., Rožić, N·: Izjednačenje nivelmanskih mreža svih redova u II. Nivelmanskom poligonu II. NVT-a, Geodetski fakultet Sveučilišta u Zagrebu, Zagreb (1994)
4. Bilajbegović, A., Feil, L., Klak, S., Škeljo, L.: II Nivelman visoke tačnosti SR Bosne i Hercegovine, Crne Gore, Hrvatske, Slovenije i SAP Vojvodine, 1970–1973, Zbornik radova Geodetskog fakulteta sveučilišta u Zagrebu, Zagreb (1986)
5. Lachapelle, G., Alves, P., Paulo, L.F., Cannon, M.E.: DGPS RTK Positioning Using a Reference Network., Presented at ION GPS-00 (Session C3), Salt Lake City, 19–22 September 2000
6. Lachapelle, G., Cannon, M.E., Fortes, L.P.S., Alves, P.: Use of multiple reference GNSS stations for RTK positioning. In: Proceedings of World Congress of International Association of Institutes of Navigation, Institute of Navigation, Alexandria (2000)
7. Rezo, M., Markovinović, D., Šljivarić, M.: Analiza točnosti nivelmanskih mjerenja i jedinstveno izjednačenje II. NVT-a, Geodski list **1**, 1–25 (2015). Zagreb
8. Grgić, I., Lučić, M., Trifković, M.: Visinski sustavi u nekim europskim zemljama, Geodski list **2**, 79–96 (2015). Zagreb
9. Višnjić, R.: Nivelmanski radovi na teritoriji Republike Srpske, XII Međunarodna naučno-stručna konferencija, Savremena teorija i praksa u graditeljstvu, STEPGRAD, pp. 327–335, Banja Luka (2016)
10. Tucikešić, S., Jakovljević, G., Gučević, J.: Modelovanje razlike referentnih površi tijela Zemlje za rješavanje problema vertikalnog pozicioniranja, Naučno-stručni časopis iz oblasti tehničkih nauka i struka "Tehnika", Savez inženjera i tehničara (2016). ISNN 0040-2176
11. Tucikešić, S., Gučević, J.: A-priori accuracy of 1D coordinates in the network of combined levelling. In: INGEO2014, 6th International Conference on Engineering Surveying (2014)
12. Ristić, K., Tucikešić, S., Milinković, A., Božić, B., Jaćimović, S.: Uspostava geodetske mreže primjenom globalnih navigacionih satelitskih sistema, XII Međunarodna naučno - stručna konferencija "Savremena teorija i praksa u graditeljstvu", Banja Luka, 7–8 Decembar 2016
13. Ristić, K., Tucikešić, S., Milinković, A.: Infrastruktura kvaliteta GPS mjerenja, Naučno - stručni časopis iz oblasti tehničkih nauka i struka "Tehnika". Savez inženjera i tehničara Srbije **24**(2), 236–241 (2015). ISSN 0354-2300

Rutting Performance on Different Asphalt Mixtures

Čchajić Adnan[✉]

Master of Civil Engineering, Modus Projekt d.o.o., Alije Izetbegovića bb,
Kakanj, Bosnia and Herzegovina
adnan.ceh@gmail.com

Abstract. Rutting is main type of road pavement distress. Rutting longitudinal depressions in wheel path caused by repeated heavy traffic load, shear failure in asphalt layer or both. Prediction of rutting development is essential for efficient management of road pavements. This paper presents a rutting performance on different asphalt mixtures. Three asphalt mixtures are made and tested in wheel tracking test. Knowledge from these tests, in future, will be used for developing a model for prediction a rutting progression in asphalt layer of road pavement.

Keywords: Rutting · Asphalt mixtures · Wheel tracking test · Prediction

1 Introduction

The main types of damage that occur as a result of the exploitation of the road are cracks and permanent deformations (rutting). The prediction of the intensity of these damages is a key input for the effective maintenance of pavement structures.

Permanent deformations (rutting) represent plastic deformations of individual layers of pavement structures. Rutting directly affect the quality of traffic flow, safety, but what is most important, the phenomenon of the track directly affects the durability of the pavement structure. With the increase in traffic load, and especially heavy freight vehicles, they represent the dominant type of damage to pavement structures.

Rutting are manifested as longitudinal depression in the vehicle wheel passage zones. They are caused by the effects of the vehicle and predominantly occur as a result of the action of traffic loads. In addition, climatic conditions can also have a significant impact on the occurrence of the track, especially when the asphalt layers are exposed to the action of high temperatures [1].

This paper presents a rutting performance on different asphalt mixtures. Three asphalt mixtures are made and tested in wheel tracking test. Knowledge from these tests, in future, will be used for developing a model for prediction a rutting progression in asphalt layer of road pavement.

© Springer Nature Switzerland AG 2019
S. Avdaković (Ed.): IAT 2018, LNNS 60, pp. 224–229, 2019.
https://doi.org/10.1007/978-3-030-02577-9_22

2 Rutting Development Mechanism

Four primary types of rutting have been identified [2]:

- consolidation - occurs due to insufficient compaction during pavement construction,
- surface wear - occurs due to surface abrasion by chains and studded tires,
- plastic flow - occurs when there is insufficient stability in the hot-mix asphalt,
- mechanical deformation - results from insufficient structural capacity of the pavement (Figs. 1 and 2).

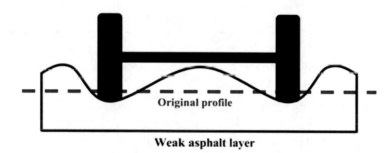

Fig. 1. Rutting in asphalt layer of road pavement

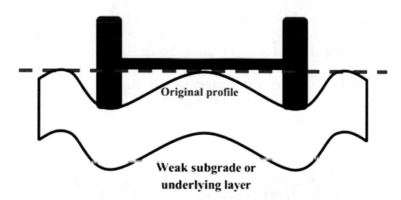

Fig. 2. Rutting due to subgrade failure

Rutting due to the accumulation of plastic strain in the asphalt pavement layers is the dominant type of rutting because they are in direct contact with the motor vehicle and the pressure of the same is directly manifested in the asphalt layers (Fig. 3).

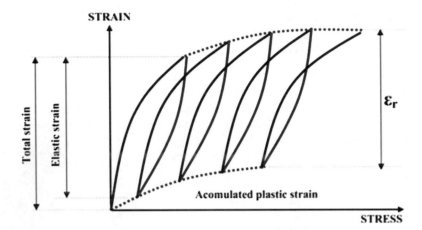

Fig. 3. Acomulated plastic strain in asphalt layer

3   Wheel Tracking Rest

Tests were performed on three types of asphalt mixtures such as: classic asphalt mixture, mix with polymer - modified bitumen and SMA mixture.

Testing of asphalt samples was carried out in the so-called Wheel tracking test.

The Wheel tracking test is a standard laboratory test of asphalt samples on the occurrence of permanent deformations. It is based on the principle that the asphalt sample is exposed to the action of the simulated traffic load, which is manifested under the constant force action with a certain number of passes (cycles). The test conditions are strictly controlled, and one of them is a test temperature of 60 °C. On each number of crossings of the standard load point, the total deformation of the treated asphalt sample is recorded, whereby the load time can be directly controlled, for example, 10,000 over-load loads or defining a critical deformation of the sample [4] (Figs. 4 and 5).

Fig. 4. Wheel tracking test

Fig. 5. Wheel tracking test asphalt sample

4   Results and Discussion

Observing the samples that have been tested it can be concluded that two areas of behavior are clearly visible. Namely, the behavior of the sample is clearly separated by up to 1,000 cycle of the wheel passage in the experiment and in the range of 1,000–10,000 cycles of the of the wheel passage.

The values of the rutting in the area up to 1,000 crossing points is an average of 50–60% of the maximum value of rutting obtained at the end of the experiment. As an illustration we provide a few examples of the values obtained from the experiment (Figs. 6, 7 and 8).

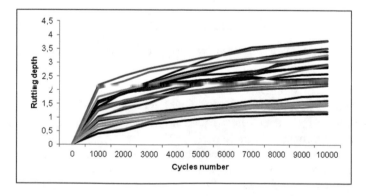

Fig. 6. Wheel tracking test result for classic asphalt mixtures

It should be emphasized that the experimental part of the research will be carried out on samples from the real asphalt pavement structure, rather than on the samples in the laboratory. This approach has been selected with the aim of the research to be based

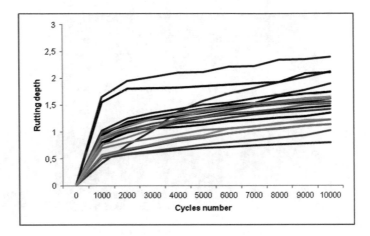

Fig. 7. Wheel tracking test result for polymer modified asphalt mixtures

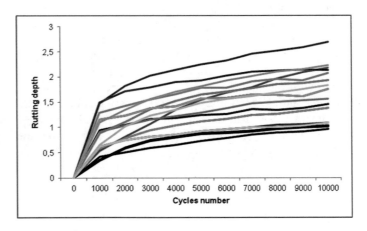

Fig. 8. Wheel tracking test result for SMA asphalt mixtures

on mixtures that have already been incorporated due to the understanding of the behavior of the mixtures and after the installation process, since it can be hypothesized that the size of the permanent deformation in the form of cartridges is influenced, for example, way of compacting and other.

The research is based on mixtures that have already been constructed due to the understanding of the behavior of the mixtures after the construction process.

Many previous research have also shown that there is a nonlinear relationship between the size of the rutting and the timing of the exploitation of the pavement structure. In accordance there are two phases of behavior. In the first phase, the value of the rutting is significantly increased as a result of compaction due to the effect of traffic load. In the second, the value of the rutting is rising, but not the same progression as in the first stage (Fig. 9) [5].

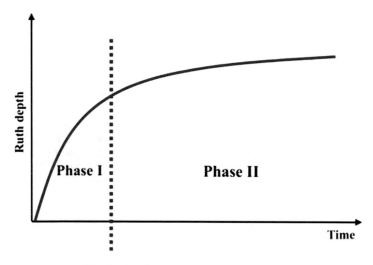

Fig. 9. Rutting versus time of exploitation

5   Conclusion

In the preliminary research, there are zones where rutting significant increases (even up to 50–60% of the maximum value), and after that the continual further increases to the maximum value (end of the test according to the standard). Research shows that the zones in which the treated sample behaves differently are clearly distinguished.

Accordingly, further research will be based on the separation of the area of development of the rutting in time (number of cycles) in two phases with observation of the influence of individual variables that characterize the properties of the asphalt mix with the aim of developing a model for the prediction of the value of rutting.

References

1. Organization for Economic Cooperation and Development (OECD) (1988): Heavy, trucks, climate and pavement damage, prepared by an OECD scientific experts group, Paris (1988)
2. Uzarowski, L.: The development of asphalt mix creep parameters and finite element modeling of asphalt rutting. Waterloo, Doctoral thesis, Ontario, Canada (2006)
3. Rabbira, G.: Permanent Deformation Properties of Asphalt Concrete, Doctoral thesis, Department of Road end Railway Engineering Norwegian University of Science and Technology (2002)
4. Čehajić, A., Pozder, M.: Uticaj vrste veziva na deformabilnost asfalta. IV kongres o cestama, Sarajevo (2014)
5. Baghaee, T., et al.: A review on fatigue and rutting performance of asphalt mixes. Sci. Res. Essays **6**(4), 670–682 (2011). Center for Transportation Research, University of Malaya, 50603 Kuala Lumpur, Malaysia

Monitoring of the Highway Construction by Hybrid Geodetic Measurements

Esad Vrce[✉], Medžida Mulić, Dževad Krdžalić,
and Džanina Omićević

Faculty of Civil Engineering, Department of Geodesy and Geoinformatics,
University of Sarajevo, Sarajevo, Bosnia and Herzegovina
{esad.vrce,medzida.mulic,dzevad.krdzalic,
dzanina.omicevic}@gf.unsa.ba

Abstract. Hybrid surveying measurement application is one of the most demanding deformation analysis tasks in a project of monitoring object. Incorrect resulting positions can lead to incorrect conclusions and thus lead to nasty consequences in the assessing the behavior of an object, which might cause human casualties and material damage.

This paper presents the monitoring of 3D damages caused by landslides, on the highway construction at the section A1 Lašva - Kakanj in Bosnia and Herzegovina. The zero series of measurements was made in May 2015. Since then, six series of measurements have been performed. Geodetic networks of the different types were set at monitored part of highway, such as: micro-triangulation, leveling and GNSS (Global Navigation Satellite Systems) networks, consisting of 60 points, of which 18 represent control points. In order to make the most of the advantages of satellite and classical terrestrial geodetic measurements, i.e. hybrid measurements, a software solution was developed. Determining the common factor variance was a central problem because of the heterogeneity of the measurement vector defined by: horizontal directions, distances, vertical angles, height differences, and GNSS vectors.

Based on the processing of non-homogeneous measurement vectors in different series, the displacements of the geodetic points on the object were determined. However, analyzing of the displacements, deformations were estimated. The standard deviations of the points position along the coordinate axes are <2 mm. Points position accuracy in hybrid networks is up to 40% higher than in the individual classical terrestrial networks.

Keywords: Object monitoring · Deformation analysis · Hybrid measurements Displacement · Deformation

1 Introduction

Hybrid surveying measurement application is one of the most demanding deformation analysis tasks in a project of the monitoring a facility. The most important deformation analysis tasks are: to describe reality in the most precise possible way, to define the conditions of the measurement that effect on the stability of the monitored facility, and to combine geodetic measurements into an integrating form, with the aim of obtaining

S. Avdaković (Ed.): IAT 2018, LNNS 60, pp. 230–240, 2019.
https://doi.org/10.1007/978-3-030-02577-9_23

reliable results. However, erroneous results can cause the inaccurate conclusions and thus lead to unimaginable consequences in assessing the behavior of the facility, what again could trigger losses of the human's life and material damage. We have witnessed recent damage and deformation on objects caused by landslides. It is possible to moderate or even prevent these consequences.

In the area of monitoring the behavior of construction facilities (roads, bridges, tunnels, dams, landslides, etc.), using modern technologies, we have entered in a completely new technological era. The most frequently used geodetic measurement systems for monitoring are: terrestrial (total stations) and satellite (GNSS). The third component of the 3D position (height above the sea level), is most often determined by geometric leveling, to get precision and reliability. Technological advances in the equipment development and measurement methods, as well as the methods of measurement processing and analyzing, are so big and fast that changes can only be monitored by skilled specialists. However, it is important that other geodetic experts involved in a monitoring the facilities are familiar with modern trends in this field.

This paper is focused to monitoring a landslide on the highway of Corridor Vc. The landslide stretches on a part of the slope, in the length of about 550 m, involves a part of the newly designed highway route in the length of approx. 900 m, and directly endangers the stability of the highway construction. At the foot of the landslide is the river Bosna which directly strikes the slope of the landslide. Due to the damage (Figs. 1 and 2), obvious progression and crack propagation, the urgent geodetic monitoring was necessary.

Fig. 1. Cracks on highway   **Fig. 2.** Crakcs on regional road near highway

Geodetic monitoring was carried out in the framework of investigative works, for detecting the causes of damage and finding solutions for landslide rehabilitation. The goal of geodetic measurements is to provide input data for determining the state of the existing pavement structure, as well as monitoring the condition after eventual remediation within a certain time period.

Control geodetic points are stabilized within the wider area of the landslide on the highway and the regional road on Corridor Vc. These points define a control network for geodetic monitoring. It is common that the network for determining displacement and deformations is divided into a reference network and a network composed of points on the object [1], what is done in this project as well. The exact location of the network points is determined with geologists. Twelve points were selected to be in a potentially unstable area (points on the object) and six points to be in a zone outside of the landslide and represent the reference network.

If we use combine measurements, we have two main computational problems to resolve:

- problem of different coordinate systems of measurements, and
- problem of weights of group of measurements.

Coordinates resulting from satellite measurements are presented in a global geocentric coordinate system. Classical terrestrial measurements are performed in a local astronomical coordinate system. Geodetic measurements are carried out in the real gravity field. In order to use classical terrestrial measurements (azimuths, directions, angles, lengths) on the surface of the Earth in the adjustment and computation in a 2D, 3D or in a cartographic projection, it is necessary to reduce them. Coordinate systems in which measurements are made should be linked in an appropriate way. If we want to use coordinates from one system in another system, it is necessary to perform 3D transformation of the coordinates or coordinate differences from one system to the other.

Vector of the measurements used for adjustment is, in this case, heterogeneous vector, composed by directions, distances, vertical angles height differences and vector of coordinate differences. So, there is a problem with estimation of common variance factor. Adjustment results are directly related with weights of measurements. We can expect good results of the adjustment using least square method, only if we have correct a-priori weights. Also, we can add here a problem of correlation in classical terrestrial measurements. In most cases, classical terrestrial measurements are considered as statistically uncorrelated, so the variance-covariance matrix and the weight matrix are usually taken as a diagonal matrix. But, in satellite measurements, the three components of the vector are correlated, so the variance-covariance matrix is 3×3.

2 Methodology

2.1 Measurements

The first series of measurements, zero epoch, is performed in May 2015. Afterwards, six more series of measurements were performed. So there was six series of measurements and 60 control points. Only two series of measurements with 18 points (points where 3D measurements are performed) are presented in this paper:

- on 15 points geometric leveling measurements carried out,
- on 12 points classical terrestrial measurements carried out,
- on 16 points GNSS static measurements carried out.

For the remaining 42 points, only vertical displacements and deformations were determined by the method of geometric leveling and they were not shown in here.

For determination of coordinates in global geocentric system (X, Y and Z) and local coordinates (north-n, east-e and up-u) it was necessary to do next:

- Levelling network, geometric levelling, (Fig. 3, blue color): - 15 points, - 4 polygons, - 20 level sides, - precise digital level Leica DNA 03, - invar barcode staff 3 m length,
- Terrestrial network (Fig. 3, red color): - 12 points, - 55 distances (Leica Di 2002), - 55 horizontal directions (Leica TC 1800) i - 55 vertical angles (Leica TC 1800),
- GNSS network (Fig. 4): - 16 points, - 48 GNSS vectors, - Trimble 4000SSI, Trimble R6 and Topcon Hiper II,

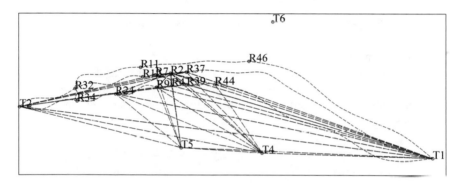

Fig. 3. Levelling (blue) i terrestrial 3D network (red)

The basic information about combined network: number of measurement 327, number of distances 55, number of vertical angles 55, number of horizontal directions 55, number of height differences 21, number of GNSS vectors 48, number of unknowns 64 (18 * 3 coordinates +6 stations +4 transformation parameters), geodetic datum defect 4, number of redundant measurements 267, the percentage of terrestrial measurements 56% and the percentage of GNSS measurements 64%.

2.2   Network Adjustment

GNSS vectors are processed using by Trimble Business Center software. Control network (all measurements together) is adjusted an results are given as coordinates X, Y, Z, which are then transformed to local coordinates n, e, u. Partial derivations – coefficients, equations f_o corrections, mathematical model of adjustment and deformation analysis are performed according to [1, 2, 4, 6, 8]. Analysis of accuracy, adjustment and deformation analysis are carried out in the same coordinate system. Network adjustment and deformation analysis are done with in house developed software at the Faculty of Civil Engineering University of Sarajevo.

Hybrid adjustment of terrestrial and GNSS measurements is possible only in same if measurements are in same coordinate system.

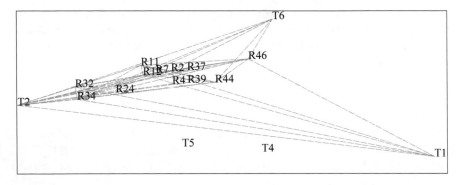

Fig. 4. GNSS network

It is possible (Fig. 5) to transform GNSS measurements to local astronomical coordinate system, and then these measurements can be adjusted in this system together with terrestrial measurements. The other possibility is to transform terrestrial measurements into global geocentric coordinate system and then adjust them together with GNSS measurements in this system. Third option is transform all measurements (terrestrial and GNSS) to new coordinate system, and then adjust them together in that new coordinate system.

local astronomical C.S.   global geocentric. C.S.

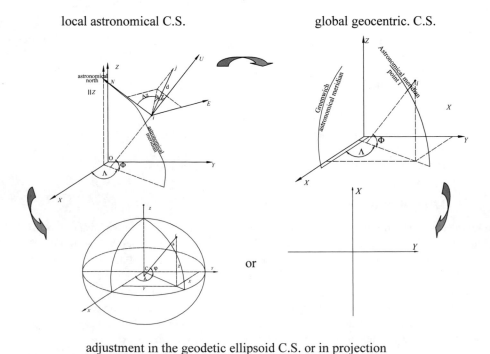

adjustment in the geodetic ellipsoid C.S. or in projection

Fig. 5. Different cases of hybrid adjustment of terrestrial and GNSS measurements

For combination of GNSS and classical terrestrial measurements it is necessary to take four transformation parameters (scale and three rotations) as unknowns in the adjustment.

2.3   Estimation of Variance Components

It is necessary to have one matrix of weights for all groups of measurements. Using estimation of variance components procedure, it is possible to determine individual variance component factor for each group of measurements, and common variance factor for all measurements together [1, 6, 7]. Computation of variance components is iterative procedure. It is repeated until all group and common variance factor is not a 1.

2.4   Deformation Analysis

The consistency between two epochs of measurements is usually checked by global congruency test. Stable points are those points that have not changed their position in a time interval between two epochs. The congruency of the network can be tested using test statistics calculated according to the formula [3, 5] and [9]:

$$F = \frac{\hat{\mathbf{d}}^T \mathbf{Q}_d^+ \hat{\mathbf{d}}}{hs^2} \qquad (1)$$

Where is: $\hat{\mathbf{d}} = \hat{\mathbf{x}}_2 - \hat{\mathbf{x}}_1$ - vector of coordinate differences, $h = rang(\mathbf{Q}_d)$ - rank of the cofactor matrix of coordinate difference, $\mathbf{Q}_d^+$ - pseudoinverse matrix of cofactors i s^2 - common variance factor.

When the value of test statistic F is lower than critical value $F_{1-\alpha,h,f}$:

$$F \leq F_{1-\alpha,h,f} \qquad (2)$$

then zero hypothesis is accepted, which is actually mean that coordinates from both epoch of measurements are congruent. In other words, all network points are stable.

Stable points on the object are those points that have not changed their position in a time interval between two epochs. Congruency of the points is checked with the test statistics which is calculated according to the formula:

$$F_i = \frac{\hat{\mathbf{d}}_i^T \mathbf{Q}_{d_i}^{-1} \hat{\mathbf{d}}_i}{3s^2} \qquad (3)$$

where is: $\hat{\mathbf{d}}_i = \hat{\mathbf{x}}_{2i} - \hat{\mathbf{x}}_{1i}$ - vector of coordinate differences of testing points, $\mathbf{Q}_{d_i}^{-1}$ inverse matrix of cofactors of coordinate differences.

If value of test statistic F_i is lower than critical value $F \leq F_{1-\alpha,h,f}$, zero hypothesis is accepted, which means that object point coordinates from both epochs are congruent. If zero hypothesis is not accepted, then we can claim that object point is moved with significance of $(1 - \alpha)$.

3 Results and Discussion

Based on the processing of a common vector of measurement and network adjustment, the final coordinates of the points are calculated, their differences - the displacements, and analyzes of the control points are performed. The emphasis was given to the importance of applying hybrid measurements to reduce standard deviations of the displacement determination and thus to increase their reliability. The estimation of the variance components of individual measurement groups and their impact on the final adjustment results were also in the focus of the processing. Below are given summary results, point's displacements, standard deviations, as well as the proportionality factors that need to be applied to variance components of individual groups of measurements. The coordinates of the points, the results of adjustment, the accuracy and other numerical quantities, are not shown here. Analysis of the results was done in three ways or three variants:

- *Hib* - variant, a combination of satellite and classic terrestrial measurements, hybrid measurements,
- *Ter* - variant, only classical terrestrial measurements and
- *TerN* - variant, classical terrestrial measurements including levelling.

Standard deviations of coordinates (X, Y, Z) of points are given on Figs. 6a, b and c.

It can be seen that standard deviations are determined on the basis of only terrestrial measurements without leveling, the variant *Ter*, (Fig. 6b) are significantly higher than standard deviations resulted from hybrid measurements, variant *Hib*, (Fig. 6a). Standard deviations of points T4, T5 i T6 (Fig. 6a) are somewhat larger (about 2 mm) than the other points (about 1 mm). The reason for the increase is that the mentioned points are determined only by GNSS measurements. Also, we see that standard deviations determined on the basis of only terrestrial measurements, including levelling, variant *TerN*, (Fig. 6c) are slightly higher than the standard deviations determined on the basis of hybrid measurements, but also less compared to the variant *Ter*.

The minimum, maximum and average values of standard deviations for different variants and measurement combinations are given in Table 1.

Table 1 show that almost all values are the least in the *Hib* variant. Almost all values are the largest in variant *Ter*, so this variant is omitted from further analysis. We retain variants *Hib* and *TerN*. The standard deviations of the points in local coordinate system (n, e, u) are given in Fig. 7(a) and (b). We can also note that standard deviations are determined on the basis of the *TerN* variant (Fig. 7b) are slightly higher than the standard deviations determined on the basis of the *Hib* variant (Fig. 7a). Standard deviations at points T4, T5 and T6 (Fig. 7a) are larger (especially u component, up to 3 mm) than the other points, because they are determined only from GNSS measurements without geometric leveling.

Standard deviations X, Y, Z i n, e, u, variant Hib, (Figs. 6a i 7a) are small values (mostly smaller than 2 mm, only few up to 3 mm). The values of standard deviations indicate that the applied adjustment models satisfy the necessary accuracy in the monitoring and determination of the deformation of the control points.

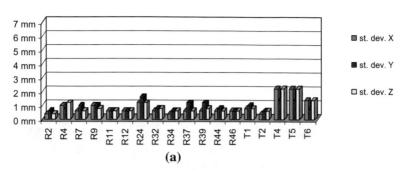

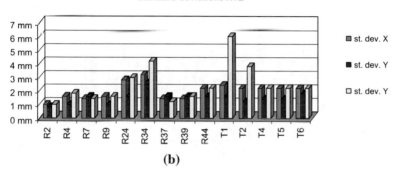

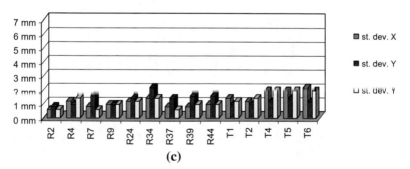

Fig. 6. (a) Standard deviations XYZ of points (all measurements together, variant *Hib*) (b) Standard deviations XYZ of points (terrestrial measurements only, variant *Ter*) (c) Standard deviations XYZ of points (terrestrial measurements with levelling, variant *TerN*)

Improvements of the standard deviations along axes X, Y, Z, *Hib* variant, compared to *TerN* expressed in percent, are 43%, 43% and 35% respectively, or 40% in average. A similar comment can also be made for the improvement of standard deviations

Table 1. Standard deviations by XYZ axes for various combinations of measurements (mm)

	Hybrid measurements (*Hib*)			Terrestrial measurements without levelling (*Ter*)			Terrestrial measurements with levelling (*TerN*)		
	X	Y	Z	X	Y	Z	X	Y	Z
Min	0.4	0.6	0.4	1.0	0.8	1.0	0.6	0.8	0.6
Max	2.2	1.6	2.2	3.2	2.6	6.0	2.2	2.2	2.0
Aver	0.9	0.9	0.9	2.0	1.5	2.5	1.3	1.3	1.2

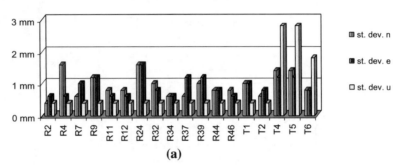

(a)

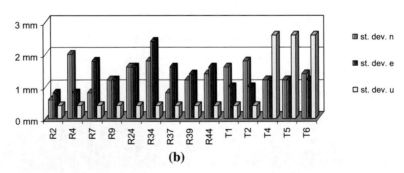

(b)

Fig. 7. (a) Standard deviations of the components *n, e, u* (variant *Hib*) (b) Standard deviations of the components n, e, u (variant *TerN*)

along axes *n, e, u*, variant *Hib*, in relation to *TerN*. The improvements expressed in percentage are in the order of 41%, 47% and 17% or, the average value of 38%. The least improvement is for the up component. This is also logical because the up component is determined from geometric levelling, which gives the highest possible precision. Geometric leveling is also included in the variant *Hib* and in the variant *TerN*.

Estimation of variance components of group of measurements leads to factor of proportionality (Table 2). Then, it is used for better estimation of a priori standard deviations of measurements.

Table 2. Results of estimation of variance components

Measurement	Factor of proportionality
Distances	0.92
Vertical angles	0.86
Heights	1.14
Horizontal directions	0.72
GNSS vectors	7.30

As the factor must be around the unit (if the a priori values are well determinate), we can conclude from the table that a priori values of standard deviations of terrestrial measurements are determined quite correctly, as opposed to GNSS measurements where factor is 7.30 (which is agree with the values commonly referred in the literature,

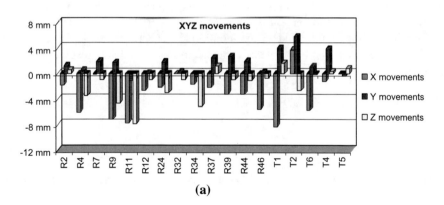

(a)

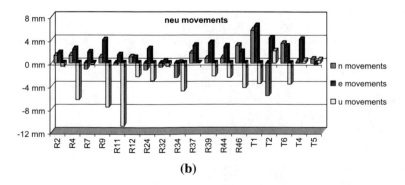

(b)

Fig. 8. (a) Point movements in X, Y, Z (b) Point movements in n, e, u

this coefficient is between 5 and 10). By implementing the method of estimation the variance components to the measurement groups, a positive effect on the final results has been obtained because some a priori values are not well-defined (for GNSS it is usually questionable).

Since the model with hybrid measurements gives better (more accurate and reliable results), it was chosen for further processing, and determination of the movement of the points.

The movements of the points determined on the basis of the variant *Hib* are given in Figs. 7(a) and (b) (Fig. 8).

4 Conclusion

This paper presented model of the adjustment and deformation analyses of the 3D hybrid control geodetic network established to monitor the landslide on the Vc highway. A synthesis of the results of modern satellite techniques and conventional terrestrial measurement methods are shown. Deformation analysis determines the points in which there have been significant displacements.

Analyzing standard deviations, we can conclude that the accuracy of determining the movements and deformations of the points is satisfactory. Standard deviations in hybrid networks are less compared to conventional terrestrial networks. The improvement expressed in percent is 40%. The values of standard deviations from the previous item indicate the need to use combined geodetic networks. The proposed models can be used in most of the projects to determine the displacements and deformations.

References

1. Caspary, W.F.: Concepts of Network and Deformation Analysis, School of Surveying, The University of New South Wales, Kensington (1987)
2. Ghilani, C.D., Wolf, P.R.: Adjustment Computations - Spatial Data Analysis, 4th edn. Wiley, Hoboken (2006)
3. Kuang, S.: Geodetic Network Analysis and Optimal Design: Concepts and Applications. Ann Arbor Press Inc., Chelsea (1996)
4. Leick, A.: GPS Satellite Surveying. Wiley, New York (1995)
5. Mihailović, K., Aleksić, I.: Deformaciona analiza geodetskih mreža, Građevinski fakultet Univerziteta u Beogradu, Institut za geodeziju, Belgrade (1994)
6. Niemeier, W.: Ausgleichungsrechnung. Walter de Gruyter, Berlin, New York (2002)
7. Perović, G.: Metod najmanjih kvadrata (Monografija). Građevinski fakultet, Belgrade (2005)
8. Strang, G., Borre, K.: Linear Algebra Geodesy, and GPS. Wellesley-Cambridge Press, Wellesley (1997)
9. Vrce, E.: Razvijanje optimalnog modela za izjednačenje trodimenzionalnih hibridnih mreža i prikladne deformacijske analize. Doktorska disertacija, Sarajevo (2016)

Accuracy of the Reflectorless Distance Measurements Investigation

Džanina Omićević[(⊠)], Dževad Krdžalić, and Esad Vrce

Faculty of Civil Engineering, Department of Geodesy, University of Sarajevo,
Sarajevo, Bosnia and Herzegovina
dzanina_omicevic@yahoo.com,
dzanina.omicevic@gf.unsa.ba,
dzevadkrdzalic@gmail.com, vrceesad@hotmail.com

Abstract. Reflectorless electronic distance measuring has found application in various surveying tasks. Accuracy of the distances measured using by instrument type GPT 7505 which has a built-in laser measuring device. Research is focused to surveying to the reflecting surfaces which are made of the five different material: metal sheet, fine and coarse facade, styrofoam and wood. Also, experiment conducted for three different angles of inclination (0°, 45°, 60°) to the reflecting surfaces. However, because of the limited manipulation capability, 55 × 55 cm signals have been created. Results show that better accuracy achieved to the rough (styrofoam and wood) than to smooth reflecting materials. The maximum range of 1400 m achieved if material rough reflecting, but the smallest distance of about 750 m to the sheet metal reflecting surface. The results also indicate that the inclination of the reflecting surface affects the range, as well, especially for the styrofoam and wood surfaces. In last scenario rang was shorter, and was about 400 m. With the increase in range, the expected accuracy of the measured rang was expected, as shown by the test results. For ranges between 150–750 m, the best accuracy is achieved if styrofoam was reflecting, and the worst for sheet metal. As a conclusion, it is difficult to define the trend of accuracy in relation to reflecting surface material, either, the inclination of the measuring rang to reflecting surface.

Keywords: Reflectorless distance measurement · Range · Accuracy
Different reflecting material · Angle of incidence

1 Introduction

New technology for the distance measuring without the reflector and the laser scanning was developed during last three decades. The application of this measuring technology is a very useful tool in geodetic practice. It has opened the door to meet most of society's and geodetic requirements. The ability to measure distances without a reflector is now available in most modern total station. Surveyors measure distance in this way and therefore, they should be sure that this measured distance is consistent with the ones measured to the reflecting prism, within the allowed tolerance. Distance measurement without a reflector found the application for determining the axes and tunnel profiles, polar set of building facades, the interior of the building, hardly

© Springer Nature Switzerland AG 2019
S. Avdaković (Ed.): IAT 2018, LNNS 60, pp. 241–249, 2019.
https://doi.org/10.1007/978-3-030-02577-9_24

accessible and non-accessible points of natural objects (stone-pits, hanging rocks), in measuring the position and dimensions of dangerous facility (heating system line, gas pipelines), for determining the displacement of specific objects, etc. In usual measurements, the target point is signaled with a reflector.

Measurements are only possible if the instrument is directed at the target point, so that the amount of reflected signal returned to the receiver is at least minimal. In the case of direct measurements to arbitrary objects, additional problems occur because reflection on diffuse objects is unreliable, because for this method of measurement there are characteristic large changes in the reflection [5].

In addition to the ability to measure up to hard-to-reach and inaccessible points, the security of operators is increased, for example, in the underground and mining measurements, at traffic intersections, or where finding secure access is not possible.

As mentioned earlier, this technology does not require the use of a reflector for distance measuring. Measurements are performed so that the instrument emits a beam of light towards a target, what is than reflected back into the instrument and then distance is calculated.

This measurement technology has some limitations that need to be investigated. Problems arising due to the effects of the type and quality of the materials from which the light beam is reflected, colors of the reflecting surface, distance and angle of light beam [1–7]. Therefore, it is very important to understand the possibilities and disadvantages of applying this technology. In order to measure the distance without using a reflector, the surface of the measuring object must act as a reflector. The largest distance, which can be measured without the use of a reflector, depends primarily on the type of reflective surface material, its structure, color and angle under which the light beam comes. Different effects such as spraying, transmission, absorption and weather conditions may affect the path, or, the wavelength of the reflecting signal. Most reflective surfaces and materials can be divided into two groups, reflective and diffuse. Reflective surfaces and materials are usually relatively smooth, such as metal, walls, concrete or some polished surfaces or coatings. In the case of diffuse ones, reflection will appear on rough surfaces and materials such as asphalt, bricks, stones, dirt and jagged surfaces [4].

2   Methodology

At the Geodesy Department of the Faculty of Civil Engineering, University of Sarajevo, within the diploma paper [4], the testing of the accuracy of the electronic distancemeter for measuring without a reflector was performed.

All measurements were performed with Topcon GPT 7505, and its technical specifications are given in Table 1.

The accuracy, range of measurement of the distance without the use of a reflector was investigated, depending on the type of material, distance to, and inclination of the reflecting surface, relative to the beam axis [6]. For testing the different materials are selected, which are commonly encountered in constructing and are surfaces at the buildings: sheet, fine and coarse facade, styrofoam and wood. The angle of the inclination of reflecting surfaces was selected to be three values: 0°, 45° and 60°. Because of the limited manipulation capability of these materials, 55 × 55 cm signals have

Table 1. Technical specifications of a tested electronic range finder

Topcon GPT - 7505		
Angle measurement		
Method	Absolute Reading	
Min. reading	$1''/5''$	
Accuracy	$5''$	
Compensation	Dual axis	
Compensation range	$\pm 6''$	
Telescope		
Magnification	$30\times$	
Min focus distance	1.3 m	
Prism mode distant measurement		
Distant 1/3/9 prism	3000 m/4000 m/5000 m	
Accuracy	$\pm (2 \text{ mm} + 2 \text{ ppm} \times D)$	
Non prism mode distant measurement	*Normal mod*	*Long mod*
Range	1.5 m – 250 m	5.0 m – 2000 m
Accuracy	$\pm (5 \text{ mm})$	$\pm (10 \text{ mm} + 10 \text{ ppm} \times D)$
- Fine	$\pm (10 \text{ mm})$	$\pm (20 \text{ mm} + 10 \text{ ppm} \times D)$
- Coarse	$\pm (10 \text{ mm}$	$\pm (100 \text{ mm})$
- Tracking		

been produced. In order to get the accuracy of measurements, each individual distance was measured 10 times. Measurement processing and the evaluation of the accuracy of the measured distance is derived by applying Eq. (1) and (2):

$$v_i = \bar{d} - d_i \tag{1}$$

$$s = \sqrt{\frac{[vv]}{n-1}} \tag{2}$$

where:

v_i deviation of ith observation,
$\bar{d}$ arithmetic mean of the observation distance,
d i^{th} distance,
n number of observations.

3   Data Processing and Results

3.1   Range of Electronic Distancemeter Topcon GTP 7505

In order to investigate the range of the distancemeter, the distance is increased as long as the measurement of length is not possible. During the research, the influence of five types of materials and an angle examined. Numerical values are summarized in Table 2 and additionally shown graphically (Fig. 1).

Table 2. Range of electronic distancemeter Topcon GTP 7505 depend into different materials and inclined angle of reflecting surface

Material	Range [m] dependent on angle of incidence [°]		
	0°	45°	60°
Sheet metal	1000.161	750.386	750.401
Wood	1400.113	1000.142	1000.145
Coarse facade	1350.234	1000.137	1000.138
Fine facade	1350.242	1000.139	1000.142
Styrofoam	1400.136	1000.158	1000.153

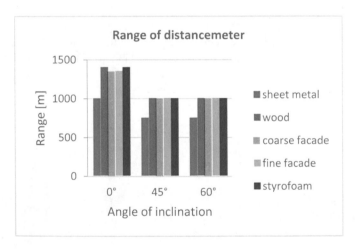

Fig. 1. Range electronic distancemeter Topcon GTP 7505 depend on different materials and inclination angle of the reflecting surface

Results in Table 2 and Fig. 1, show that the range of the electronic distance meter (EDM), influences not only by the material of which the reflective surface is made, but also by angle of beam relative to the surface.

The largest range is achieved in case of the reflection of the light beam from the wooden surfaces and styrofoam, 1400.136 m, while a somewhat shorter range is achieved by measuring to a fine and rough facade. The worst result, i.e. the smallest range achieved when measuring on the sheet metal, and measured the length of 750.401 m. It should be noted here that the sheet metal is very smooth. These measurement results were achieved when the reflecting beam was perpendicular to the reflecting surface to which the measurement was performed.

Table 2 shows that, as expected, the result of the measurement of length was influenced by the inclination of the reflecting surface, and a significantly shorter range of measured length was achieved. It is interesting to note that the same results were achieved at the inclination of 45° and 60° for all the materials used.

Based on the results of the test, it was shown that when measuring the distance without the use of the reflector, better results give coarser than the smooth materials. In this case, materials such as styrofoam, fine facade, coarse facade and wood have shown similar reflective properties.

3.2   Accuracy the Reflectorless Distance Measurement

The accuracy of the measurement of the distance by the tested EDM was estimated based on a series of measurement of deviations from the arithmetic mean. Internal accuracy was determined of the different distances and slopes. The standard deviation of the measurements of different distances to the different types of reflective surfaces were presented numerically in Table 3 and graphically in Fig. 2.

Table 3. Standard deviation of distance for different materials when angle of incidence is 0°

Distance [m]	Sheet metal	Wood	Coarse facade	Fine facade	Styrofoam	Distance [m]
23.953	0.001	0.005	0.008	0.003	0.006	0.007
71.892	0.000	0.002	0.002	0.001	0.004	0.009
95.913	0.001	0.001	0.001	0.002	0.004	0.010
143.924	0.001	0.006	0.001	0.005	0.006	0.008
250.305	0.000	0.032	0.033	0.031	0.034	0.025
500.24	0.001	0.035	0.034	0.044	0.045	0.032
750.424	0.001	0.054	0.044	0.042	0.047	0.032
1000.16	0.001	0.015	0.007	0.016	0.031	0.025
1330.24	0.001	–	0.014	0.015	0.021	0.026
1400.13	0.000	–	0.024	–	–	0.007

From the previous table and Fig. 2 we can notice large oscillations in the accuracy of measuring the length depending on the type of material and the distance being measured. The chart shows the standard deviations of the measured using the reflector.

When measuring the distance with the use of a reflector, regardless of distance, an excellent accuracy has been achieved, and its value is 1 mm. Based on the chart lines, we note that the standard deviation is within 1 cm for the distances up to 150 m for all types of materials. The accuracy achieved for all types of materials is similar, with the exception of styrofoam which has a slightly lower accuracy of 4 to 6 mm. It is

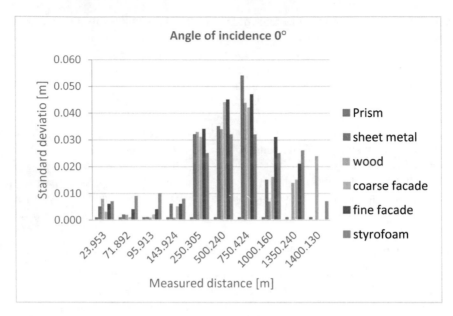

Fig. 2. Standard deviation of distance for different materials - angle of inclination 0°

interesting to note that some materials, as the wood, and fine facade have shown a worse result in measuring the shortest distance.

If the distance increase, the accuracy of the measured distance was expected to decrease. For a range of 150 m–750 m, the standard deviation reached a value of 5.4 cm when measured to the sheet metal. In this range of measured distances, styrofoam shown best results, and the worst accuracy was achieved by measuring to the sheet. When measuring distances of 1000 m or more, there was an unexpected increase of the accuracy. Each of the materials used showed different accuracy, but on the basis of the obtained results it is difficult to determine the trend. Also, for all five materials used: styrofoam, fine facade, coarse facade, wood and sheet metal, standard deviation of measured distances with slopes of 45° and 60° reflective surface were determined. Numerical values are presented in table and graphically (Table 4).

Table 4. Standard deviation of distances for different materials, when angle of inclination is 45°

Distance [m]	Sheet metal	Wood	Coarse facade	Fine facade	Styrofoam
23.953	0.010	0.008	0.005	0.004	0.002
71.892	0.002	0.005	0.008	0.007	0.005
95.913	0.012	0.005	0.003	0.001	0.005
143.924	0.014	0.002	0.005	0.007	0.006
250.305	0.025	0.030	0.027	0.028	0.023
500.24	0.026	0.027	0.048	0.035	0.033
750.424	0.043	0.051	0.054	0.051	0.051
1000.16	–	0.017	0.025	0.018	0.030

In Fig. 3, the standard deviations of the measured distances for the inclined 45° reflective surface are shown. Lines representing a standard deviation are broken down, and the changes are such that it is difficult to draw some trend. For measured distances up to 150 m, standard deviations are within 1 cm, except for sheet metal when it is 1.4 cm. At greater distances the accuracy drops, the maximum standard deviation is 5.4 cm in a rough facade. For a measured distance of 1000 m, for all types of materials, there is an increase in the accuracy, with the exception of sheet metal, when this distance could not be measured. It is difficult to find a logical explanation for this accuracy increase.

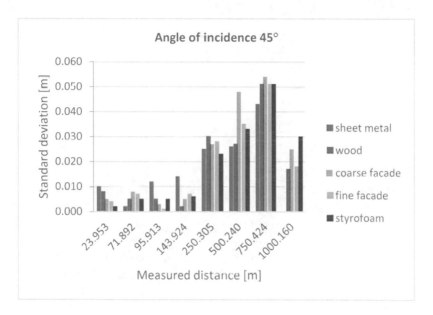

Fig. 3. Standard deviation of distances for different materials - angle of inclination 45°

If we compare the numerical values of the results in Tables 2 and 3, it can be said that due to the inclination of the reflecting surface there was no significant impact on the accuracy.

In the slope of the 60° standard deviations of the measured distances shown in Fig. 4, there are no significant changes in relation to the standard deviation at 45° inclination. The accuracy is somewhat smaller and for the distance of 150 m, the standard deviation is 2 cm for the sheet metal. With an increase in length, the accuracy decreases, and at a length of 750 m the standard deviation value, as presented in Table 2, is 6.1 cm for sheet metal. As in all previous cases, when measuring lengths of 1000 m, accuracy increases for all materials (Table 5).

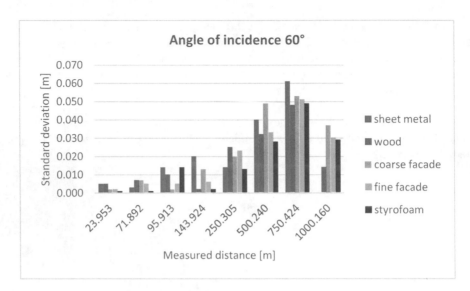

Fig. 4. Standard deviation of distance for different materials - angle of inclination is 60°

Table 5. Standard deviation of distances for different materials when angle of incidence is 60°

Distance [m]	Sheet metal	Wood	Coarse facade	Fine facade	Styrofoam
23.953	0.005	0.005	0.002	0.002	0.001
71.892	0.003	0.007	0.007	0.005	0.001
95.913	0.014	0.010	0.002	0.005	0.014
143.924	0.020	0.002	0.013	0.006	0.002
250.305	0.014	0.025	0.020	0.023	0.013
500.24	0.040	0.032	0.049	0.033	0.028
750.424	0.061	0.048	0.053	0.051	0.049
1000.16	–	0.014	0.037	0.030	0.029

4   Conclusion

This paper present results of an experiment with the distance measuring without the reflector. After the testing performed by the electronic distance meter, a range of 2000 m has not been reached, what is specified in the manufacturer's specification. A range of 2000 m probably not achieved due to the dimension of a signal of 55 × 55 cm. These signal dimensions have been selected due to limited manipulation capability. The tests have shown that there is a dependence of the range on the type of material and the inclination of the reflecting surface. The largest range is achieved in the reflection of light beam from wooden surfaces and styrofoam of some 1400 m, while the worst result achieved when measuring on the surface of the sheet of 750 m, which is very smooth. The slope influence is greatest at the styrofoam and wood surface, where the range of the range between the maximum and minimum value is

400 m, reduced by some 40%. Less impact on the fine and rough facades of 350 m (35% reduction). The smallest impact is of the sheet metal, where the minimum and maximum range difference is 250 m (25% reduction). Based on the results of the test, it has been shown that measuring the distances without a reflector, better results give coarser than the smooth materials. In this case, the materials such as styrofoam, fine facade, coarse facade and wood show similar reflective properties. The test has confirmed that there is a significant impact of the inclination of the reflecting surface and range that can be achieved when measuring the distance without a reflector.

The accuracy of the tested total station depends on the measured distance. Increasing the measured distance reduced the accuracy, what was expected. The accuracy of the measured the distances without a reflector is of the level of 3 to 6 cm. On the basis of the obtained results, it is difficult to determine the trend of the influence of the type of material and the inclination on the accuracy of the length measurements.

References

1. Alan, H.: Investigation into the effectiveness of reflectorless technologies on structural surveillance monitoring. Diploma thesis, University of Southern Queensland, Australia (2009)
2. Amezene, R., Bekele, B.: Accuracy analysis and calibration of total station based on the reflectorless distance measurement. Master of Science Thesis in Geodesy, School of Architecture and the Built Environment, Royal Institute of Technology Stockholm (2012)
3. Fawzy, H.E.-D.: Evaluate the accuracy of reflector-less total station. Int. J. Civ. Eng. Technol. (IJCIET) 6(3), 23–30 (2015)
4. Jones, J.: Comparison of robotic total stations for scanning of volumes or structures. Diploma thesis, Southern Queensland, Faculty of Engineering and Surveying (2009)
5. Kogoj, D.: Fähigkeiten elektronischer Distanzmesser bei reflektorloser Distanzmessung, AVN 186-190 (2001)
6. Latić, B.: Testing the accuracy of length measurements without reflectors. Diploma thesis, University of Sarajevo, Bosnia and Hercegovina (2012)
7. Záměþníková, M., Neuner, H., Pegritz, S.: Influence of the incidence angle on the reflectorless distance measurement in close range. In: INGEO 2014 – 6th International Conference on Engineering Surveying Prague, 3–4 April 2014

Impact of the Heterogeneous Vector of Surveying Measurements to the Estimation of the Posteriori Stochastic Model

Dzanina Omicevic[✉]

Faculty of Civil Engineering, Department of Geodesy, University of Sarajevo,
Sarajevo, Bosnia and Herzegovina
dzanina_omicevic@yahoo.com,
dzanina.omicevic@gf.unsa.ba

Abstract. Heterogeneous vector of the surveying measurements processing usually apply method of least squares. It requires that stochastic model to be defined appropriately. The focus of this paper is to research the stochastic model application in 3D parametric adjustment of the geodetic networks. The study was conducted on the sample of two test surveying networks: network "Cube" dealing with simulated measurements, and network "FCESA" with real measurements. Simulated 3D geodetic networks are formed out of the eight points, and "FCESA" of ten points. Heterogeneous measurement vectors are defined by the: horizontal directions p, distances d, zenith angles z, heights differences Δh, and coordinate differences Δy, Δx and ΔH. Same variance factor applies to each group of measurements to be involved in the adjustment. There are different methods that estimate weights of unknown parameters, simultaneously while estimating unknown parameters. The tests were conducted using different methods: Helmert, MINQUE, Förstner, Ebner, and AUE. The results clearly show that different types of measurements and their combinations, which constitute a heterogeneous vector, have a significant impact to a posteriori estimate of standard deviation. The most significant change in standard deviation occurs if horizontal directions are left out of the non-homogeneous vector.

Keywords: Heterogeneous measurements vector · Variance factor
Variance · A posteriori estimation · Stochastic model · Standard deviation

1 Introduction

In order to solve complex geodetic task, different combinations of measurement techniques are used. The observation vector in these situations is heterogeneous because it contains various types of geodetic measurements.

Processing of the heterogeneous vector of the surveying measurements in the most geodetic applications commonly use method of least squares that requires the stochastic model to be defined appropriately. The stochastic model describes the accuracy of the measured values and their relations. The covariance matrix is generally not known. Its appearance therefore needs to be priori assessed.

Same variance factor applies to all groups of measurement if heterogeneous measurements to be involved in the adjustment procedure, its determination is very

© Springer Nature Switzerland AG 2019
S. Avdaković (Ed.): IAT 2018, LNNS 60, pp. 250–260, 2019.
https://doi.org/10.1007/978-3-030-02577-9_25

complex, because it requires knowledge of the internal statistical structural relation between individual groups of measurements [1]. There are different methods for estimating the variance components, and differ in the use of the estimated principle of estimation and the assumed distribution.

Unreliable a priori determination of the weight matrix causes an inadequate evaluation of the measurement vector of residuals. The consequence of this is the non-objective evaluation of the accuracy of the measurement and the final results. An even bigger problem is that estimates of weight depend directly on the estimation of the required values, most often the coordinates of unknown points.

The solution of this complex situation is to use methods that estimate weights of unknown parameters, simultaneously while estimating these unknown parameters. The most common way of solving the problem is to form a group of observations in which the relations of accuracy are known, and each group of measurements assigns a specific corresponding variance factor. A posteriori estimated value of the variance of the single-group factor then determines the new weight of groups, which are used a basis for in the next step of iterative procedure.

2 Iterative Estimation Procedure

The process of adjusting of geodetic data is based on the observation model and consists mainly of two mathematical models. On the one hand, it is a functional model that defines mathematical relationships between unknown parameters and observations, together with properly modeled systemic influences. The second component is the definition of a real stochastic model in the form of variance - covariance matrix, which describes functional model errors that are random variables. Stochastic model adjustment 3D of heterogeneous vector of the surveying measurements, which can be contain horizontal directions p, distances d_{TPS}, distances from GNSS observations d_{GNSS}, zenith angles z, height differences Δh measured by trigonometric leveling method and coordinate differences Δy, Δx and ΔH is given by expression in Eq. 1 [2]:

$$E\{vv^T\} = C_l = \sigma_0^2 Q = 1 \cdot Q = P^{-1} \tag{1}$$

where is C_l generally variance covariance matrix. Matrix P is block diagonal matrix of weight and it is inversion of symmetric positive-definitive cofactor matrix Q. Variance factor σ_0^2 is a priori selected and usually is one.

Covariance matrix of observations C_l it must be at least approximately known for applying algorithms for estimating variance components. This matrix is performed from initial values $\hat{\theta}_0$ unknown variance components for start iteration procedure. Hence, different assessment methods are only the locally best estimators, which provide variance components dependent on the a priori choice of variance factors.

This problem can be solved by applying iterative procedures (see Fig. 1). They provide the global best estimator, less dependent on the initial assessment of unknown variance components. This is especially important if the initial estimate is unreliable or uncertain.

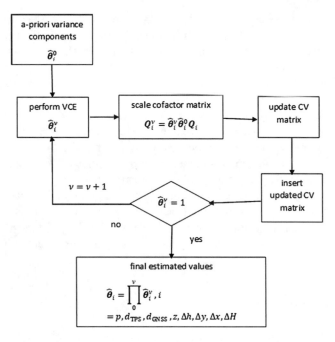

Fig. 1. Iterative variance component estimation computational scheme (according [8])

A priori component of variance $\hat{\theta}_i^0 = \sigma_i^2$ the initial values are specified for each measurement groups $i = p, d_{\text{TPS}}, d_{\text{GNSS}}, z, \Delta h, \Delta x, \Delta y, \Delta H$ or coord. Initial approximate values are used in the first (initial) iteration and use to derive the first variance component $\hat{\theta}_i^v$ for each measurement groups, which was applied to the appropriate cofactor matrix. This will update the matrix cofactor.

In the next step iteration $v = v + 1$, the variance component estimation is repeated by defining the results of the estimation from the previous iteration as a priori values, and a new set of variance components $\hat{\theta}_i^{v+1}$ can be obtained.

The convergence criterion of the iteration process a posteriori estimation of the variance component of the heterogeneous vector of measurement is given by expression in Eq. (2):

$$\hat{\boldsymbol{\theta}}^v = \begin{bmatrix} \hat{\theta}_p^v \\ \hat{\theta}_{d_{\text{TPS}}}^v \\ \hat{\theta}_z^v \\ \hat{\theta}_{\Delta h}^v \\ \hat{\theta}_{\Delta x}^v \\ \hat{\theta}_{\Delta y}^v \\ \hat{\theta}_{\Delta H}^v \end{bmatrix} \rightarrow \begin{bmatrix} 1 \\ 1 \\ 1 \\ 1 \\ 1 \\ 1 \\ 1 \end{bmatrix} \qquad (2)$$

The convergence criterion can be achieved if the difference between the two adjacent iterations is smaller than the defined value in the Eq. (3), e.g. in this case is selected 0, 0001:

$$\left| \boldsymbol{\theta}^{v} - \boldsymbol{\theta}^{v-1} \right| < 0,0001 \tag{3}$$

The conditional Eq. (3) is tested for all components to ensure that the selected criterion is satisfied because the convergence rate is not the same for all variance components. The final estimated variance component values for each group of measurements in n. iterations is given by expression in Eq. (4) (Fotopoulos 2007):

$$\hat{\sigma}_i^2 = \hat{\theta}_i - \prod_{v=0}^{n} \hat{\theta}_i^v,$$
$$i \in [p, d_{\text{TPS}}, d_{\text{GNSS}}, z, \Delta h, \Delta y, \Delta x, \Delta H] \tag{4}$$

3 Testing

In order to investigate the impact of the number of groups on a posteriori estimation of standard deviation tests were performed on two geodetic networks. Theoretic geodetic network "Cube" (see Fig. 2), where heterogeneous measurement vector is formed from seven groups simulated observations (horizontal directions p, distance d, zenith angles z, difference heights Δh and coordinate differences Δy, Δx. and ΔH.). Each group formed 48 measurements.

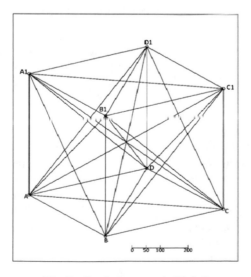

Fig. 2. Geodetic network "Cube"

Geodetic network "FCESA" (see Fig. 3) is a real geodetic network which has a 10 points. Heterogeneous measurement vector is formed from five groups: 34 horizontal directions p, 17 distance TPS d_{TPS}, 15 GNSS distance d_{GNSS}, 20 zenith angles z, and 24 height differences Δh measured by trigonometric leveling method.

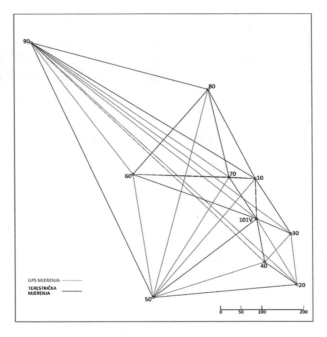

Fig. 3. Geodetic network "FCESA"

The tests were conducted using strongly methods Helmert [7] and MINQUE [9], and simply methods Förstner [4] and Ebner [3]. As there is a potential for a negative evaluation of the variation component estimation, it was selected AUE [1] as an alternative MINQUE method, which is roughly unbiased and gives no negative estimates.

In order to determine the influence of the number of measurement groups on the a posterior assessment of the standard deviation, the number of groups is reduced, simply throw out a set of measurements, but were formed various combinations of the group of measurement. The aim of e research was to determine whether and how the value of a posteriori estimation of the standard deviation of individual measurement groups are changed, whether these deviations are significant. Which of the applied methods is robust, i.e. whether it is stable. Also, it was very important to find out if there are problematic combinations of measurements and what is a minimum number of group that estimation is a possible.

3.1 Results in Geodetic Network "Cube"

In the geodetic network "Cube" to investigate the influence of group numbers on a posterior assessment of standard deviation, were made 36 combinations; of which seven are combinations with six groups of measurements, 13 combinations with five groups of measurements, five combinations with four measurement groups, seven combinations with three measurement groups, and four combinations with two measurement groups. The results of numerical research are graphically presented.

In the case, when the vector is made the six group of measurement MINQUE and AUE are given an identical result. Similarly, Helmert's and Ebner's methods yielded mutually identical results, while Förstner's method given a slightly different result. By changing the combination of the types of geodetic measurements, the final values of the estimated standard deviations were slightly chaneged.

Similar estimation results are obtained when the vector makes five groups of measurements. With Fig. 4 it can be clearly seen that there is no estimation of standard deviation by the Förstner's method when the measurement vector is: horizontal

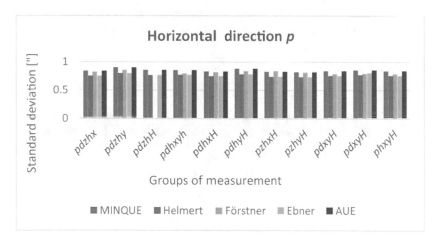

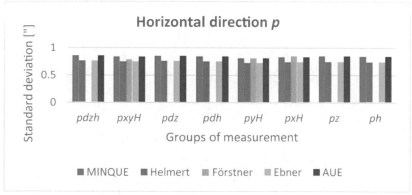

Fig. 4. Estimation of standard deviation of horizontal direction in the geodetic network "Cube" [8]

directions p, distance d, zenith angles z, height differences Δh i coordinate differences ΔH $(pdzhH)$. During the iteration process, the estimation of standard derivation for horizontal directions s_p is negative value.

As the number of measurement groups reduced, the same trend was maintained. However, combinations of measurements have appeared where Förstner's method during the iteration process yields negative values. Critical measurement vectors are the following combinations of geodetic measurements:

1. Horizontal directions p, distances d, zenith angles z i height difference $s\Delta h$ $(pdzh)$,
2. horizontal directions p, distances d, zenith angles z (pdz),
3. horizontal directions p, distances d and height differences Δh (pdh) i
4. horizontal directions p i height differences Δh (ph).

Final values of the standard deviation of the group measurement s of distance s_d in the case when a measurement vector is a combination of six types of geodetic measurements in a group, all methods are slightly different, regardless of the selected combination of measurement groups involved in adjustment.

As the number of measurement groups is reduced, Fig. 5 shows significant deviation from Förstner's method of a posteriori estimation. This is the case with the above mentioned critical measurement vectors:$pdzhH$, $pdzh$, pdz, pdh i pz.

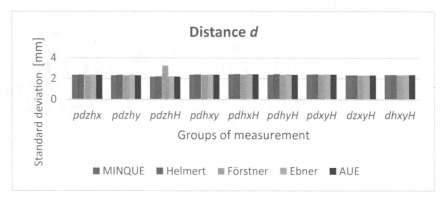

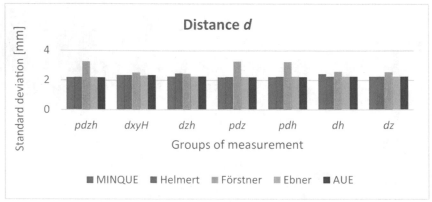

Fig. 5. Estimation of standard deviation of distance in the geodetic network "Cube" [8]

A posteriori estimation standard deviation of distance s_d, standard deviation of zenith angle s_z and standard deviation of height difference $s_{\Delta h}$ finally value of estimation are identical for MINQUE, Helmert's method, Ebner and AUE method, respectively.

The results of estimation standard deviations of coordinate differences Δx, Δy and ΔH are identical for MINQUE, AUE and Förstner's method.

3.2   Results in Geodetic Network "FCESA"

Similar research was carried out in the real geodesic network "FCESA", where 15 combinations were performed; of which five combinations with four measurement groups, six combinations with three measurement groups and four combinations with two measurement groups. If a posteriori standard deviation horizontal direction s_p is considered in Fig. 6 can be see it shows the sensitivity to the number of group measurements, as well as the combination of measurements that makes the vector. The critical vector of measurement is combination horizontal directions p and zenith angles z. In this case, Förstner's method a posteriori estimation is given a negative values of standard deviation. For the measurement vector horizontal directions p and height differences Δh Förstner's method a posteriori estimates give a significantly different value to the other methods.

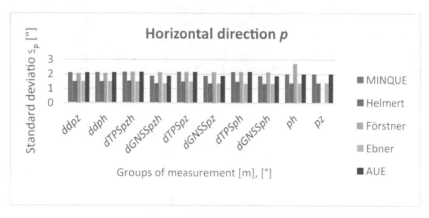

Fig. 6. Estimation of standard deviation of horizontal direction in the geodetic network "FCESA" [8]

It is also very interesting that Helmert's and Ebner's methods give the same results for all combinations, and the results differ from the finite values of other methods, including similarities, with the exception of two combinations of ph and pz, where Förstner's methods significantly deviate. Figure 7 it can be seen that a significant change of the value of a posteriori estimation of the standard deviation $s_{d_{TPS}}$ occurred when the horizontal direction p.

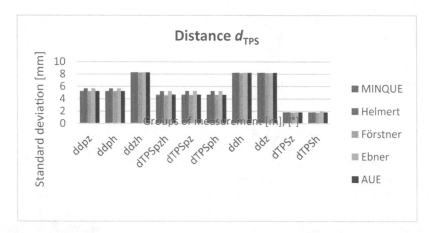

Fig. 7. Estimation of standard deviation of distance d_{TPS} in geodetic network "FCESA" [8]

Methods of estimation MINQUE and AUE are given identical of results, and the other methods give very close results. The identical evaluation result is obtained in the case when group of distance (d_{TPS} and d_{GNSS}) are added group of measurements that allow determination of height differences ($ddzh$, ddh i ddz).

With Fig. 8 it is clear that a significant change in the value of a posteriori estimates of standard deviation d_{GNSS} comes in the case when the measurement vector does not contain horizontal direction p. In other cases (four types of geodetic measurements $ddzh$, than combination of tree types of measurements ddh and ddz) estimation of standard deviation $s_{d_{GNSS}}$ is too optimistic in comparison to other cases, which is a practical experience unrealistic. Methods of a posteriori estimation MINQUE and AUE give the same result.

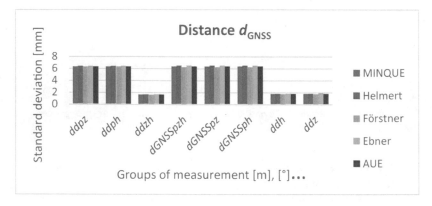

Fig. 8. Estimation of standard deviation of distance d_{GNSS} in geodetic network "FCESA" [8]

Analysis of results with Fig. 9, it is clear that a change of the a posteriori value of the standard deviation of the zenith distance z comes when the horizontal distances p are excluded from the measurement vector. Estimation of standard deviation zenith angle s_z is excited, the value achieved by the Förstner's method in case we have a combination pzh with distances (d_{TPS} or d_{GNSS}). The critical vector of measurements is a combination of p and zenith angle z. Förstner's method a posteriori estimation is given negative value of standard deviation of horizontal distance s_p. With the exception of these two cases, in all other combinations of geodetic measurements all methods

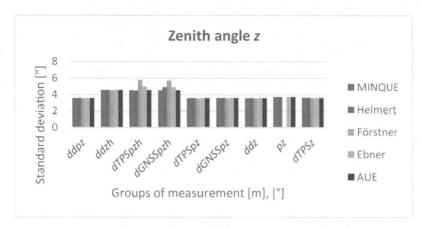

Fig. 9. Estimation of standard deviation of zenith angle z in geodetic network "FCESA" [8]

give the same or similar estimation value.

4   Conclusion

Results clearly show that the number of groups of measurement, and types of measurement forming the observation vector, have an impact to a posteriori estimates of the standard deviation. It should be noted that there is significant change in the posteriori estimation of the standard deviation if the horizontal direction p is omitted from the vector of measurement group. In theoretical geodetic network "Cube" MINQUE and AUE are given identical result, than Helmert and Ebner method, respectively.

Significant change of the value of a posteriori estimation of the standard deviation in the geodetic network "FCESA" occurred when a group of measurements of horizontal directions p were cut out from the measurement vector. This is especially emphasized for the measurement vector, containing distances d_{TPS} and d_{GNSS}, the zenith angle z and the height difference of Δh (ddzh), although it is noteworthy that four measurement groups are included. Other methods give close results, noticeable disagreement between the third and fourth decimals behind the comma.

In this investigation, Förstner 's method has been shown to be sensitive to changing the number of measurement groups as well as a combination of measurements which contain the heterogenic vector. Particularly critical combinations were: $pdzhH$, $pdzh$, pdz, pdh and ph in the theoretical network "Cube".

References

1. Amiri-Simkooei, A.: Least - squares variance component estimation: theory and GPS applications. Ph.D. thesis, Netherland Geodetic Commission, Delft (2007)
2. Bähr, H., Altamimi, Z., Heck, B.: Variance Component Estimation for Combination of Terrestrial Reference Frame. University Karlsruhe, Karlsruhe (2007)
3. Ebner, H.: A posteriori Variancschätcungen für die Koordinaten unabhängiger Modelle. Zfv Nr. 4, 166–172 (1972)
4. Förstner, W.: Ein Verfaren zur Schatzung von Varianz und Kovarianzkomponenten. AVN 11-12, 446–453 (1979)
5. Fotopoulos, G.: An analysis on the optimal combination of geoid orthometric and ellipsoidal height data. Ph.D. thesis, Department of Geomatics Engineering, University of Calgary, Calgary (2003)
6. Ghilani, C.D., Wolf, P.R.: Adjustment Computations Spatial Data Analysis. Wiley, Hoboken (2011)
7. Grafarend, E.W.: An introduction of variance-covariance estimation of Helmert type. Zfv 4, 161–180 (1980)
8. Omicevic, D.: Homogenization of weight of surveying measurements. Ph.D. thesis. University of Sarajevo, Bosnian and Hercegovina (2017)
9. Satirapod, C., Wang, J., Rizos, C.: A simplified MINQUE procedure for the estimation of variance-covariance components of GPS observables. Surv. Rev. **36**, 582–590 (2002)

GNSS Reference Network - Accuracy Under Different Parameters Variation

Medžida Mulić[1](✉) and Asim Bilajbegović[2]

[1] Faculty of Civil Engineering, Department of Geodesy and Geoinformatics,
University of Sarajevo, Sarajevo, Bosnia and Herzegovina
medzida.mulic@gf.unsa.ba
[2] Technical University of Dresden, Dresden, Germany
bilajbegov@gmail.com

Abstract. Global Navigation Satellite Systems-GNSS, such as, American GPS, European Galileo, Russian GLONASS or Chinese BeiDou, allow 3D coordinates of the geodetic reference networks to be determined in Global Geodetic Reference System, with an accuracy and economic efficiency that could not be imagined a half of century ago. However, the accuracy of GNSS positioning is affected by a very large number of errors, to mention only some: satellites orbit errors, satellite and receiver's clock errors, satellite and receiver antenna' errors, poor geometry distribution of the satellite, the influence of propagation media, the environment of the observing station, interference with other radio signals, etc.

In order to obtain high accuracy of the coordinates in a geodetic network, it is necessary to pay particular attention to the selection of appropriate equipment, techniques and procedures for data collection, as well as the location of observing reference sites. Nothing less important is to choose data processing methods. Applying carefully selected strategy for data processing, a large number of errors can be eliminated or significantly reduced. This paper discusses the impact of various parameters on the accuracy of the reference network of Bosnia and Herzegovina, such as: the accuracy of the applied global reference frame, the effects of the individual reference stations selected as datum stations, period of time and number of observation sessions, the accuracy of satellite orbit, the impact of ionospheric and tropospheric refraction, etc.

Keywords: GNSS · Global geodetic reference system · Reference network
Coordinates accuracy · Data processing · Error sources

1 Introduction

Global Navigation Satellite Systems (GNSS), is a term describing a group of satellite systems that provide services such as navigation, positioning and time measurement in real time, for the different users, military and civilian as well, located all over the globe, for the 24 h a day, seven days a week, under any weather conditions [1]. There are four operational global satellite systems today, but with varying degrees of development, such as, American Global Positioning System (GPS), European Galileo, Russian Global Navigation Satellite System (or Globalnaya Navigazionnaya Sputnikovaya

© Springer Nature Switzerland AG 2019
S. Avdaković (Ed.): IAT 2018, LNNS 60, pp. 261–272, 2019.
https://doi.org/10.1007/978-3-030-02577-9_26

Sistema - GLONASS), or Chinese BeiDou System (BDS). In addition to these global ones, there are regional navigation systems as well: Indian Regional Navigation Satellite System (IRNSS), which is fully operational, and Japanese regional developing system, Quasi-Zenith Satellite System (QZSS).

GNSS allow 3D coordinates of the geodetic reference networks to be determined in Global Geodetic Reference System [2], which gives advantages in different segments: economic, scientific and practical applications. The accuracy and economic efficiency of the positioning, navigation and timing (PNT) that possible to achieve today with GNSS technology, could not be imagined just half a century ago. The first GPS satellite was launched only 40 years ago [3]. However, the accuracy of GNSS positioning is affected by a very large number of errors, to mention only the main: satellites orbit errors, satellite and receiver clock's errors, satellite and receiver antenna's errors, poor geometry distribution of the satellite, known as the term Dilution of Precision (DOP), the influence of propagation media, such as ionosphere and troposphere refractions, the impact of the environment of the observing station, known as multipath effect, interference with other radio signals, etc.

So, user who are interested in the high-accuracy applications i.e. the 3D positions of the observed points, must be very skilled in the strategy of these errors reductions or elimination [4]. In order to obtain high accuracy of the coordinates in a geodetic reference network, it is necessary to pay particular attention to the selection of appropriate equipment, (i.e. receivers and software), to apply appropriate techniques and procedures for data collection, as well as, to choose a convenient location of the observing reference sites [5]. Nothing less important is to choose data processing methods. Applying carefully selected strategy for data processing, a large number of errors can be eliminated or significantly reduced [6].

This paper discusses the impact of various parameters on the accuracy of the geodetic reference networks of Bosnia and Herzegovina. Bosnian modern reference frames are connected to the global geodetic reference networks, such as ITRF (International Terrestrial Reference Frame), what is the realization of the ITRS (International Terrestrial Reference Systems). ITRS is theoretical definitions of an ideal global geodetic reference systems. It is consisted of a set of mathematical and physical prescriptions and conventions together with the modelling required to define origin, scale, orientation and time evolution of a Conventional Terrestrial Reference System (CTRS) [7, 8]. ITRS is recommended by International Union of Geodesy and Geophysics (IUGG) [9], by the resolution no. 2 adopted in Vienna, 1992.

So, to meet adopted international conventions and recommendations, Bosnian modern geodetic reference networks has to respect highest accuracy standards. The high-accuracy coordinates od reference network depend of many parameters: the accuracy of the applied global reference frame in the data processing, the effects of the individual reference stations selected as datum stations, period of time and number of observation sessions, the accuracy of satellite orbit used, the impact of ionospheric and tropospheric refraction, etc.

2   Data and Methodology

Data are collected during three GPS observing campaigns, organized in the territory of Bosnia and Herzegovina, in following years: 1998, 2000, and 2005. GPS two-frequencies receivers used in campaigns were satisfying the highest requested criteria for geodetic networks. The first and second GPS campaigns carried out for the reason of establishing a modern geodetic reference network in Bosnia and Herzegovina and to include it to the European Reference Frame (EUREF). The first one, named Balkan'98 GPS campaign was supported by Federal Agency for Cartography and Geodesy from Frankfurt am Main, Germany. It was regional GPS campaign, since synchronized GPS observation organized in Albania, Bosnia and Herzegovina, SR Yugoslavia, but today three states exist: Serbia, Montenegro, Kosovo. In Bosnia and Herzegovina 13 stations observed for five sessions of 24 h each. Results of data processing satisfied requested criteria and Bosnian reference network accepted as the part of EUREF network at the EUREF Symposium, held in Praha 1999 [10]. That network represents passive GPS reference network of Bosnia and Herzegovina with centimeter accuracy and is named BIHREF (B&H Reference Frame) 98.

Second GPS campaign carried out to densify BIHREF 98 network and was carried out in the summer 2000 with support of the Geodetic administrations of B&H [11]. About 30 stations observed for the two 24 h session but five stations from BIHREF observed for five days, as well as station SRJV in Sarajevo, as the first permanent EPN station in the region (Fig. 1).

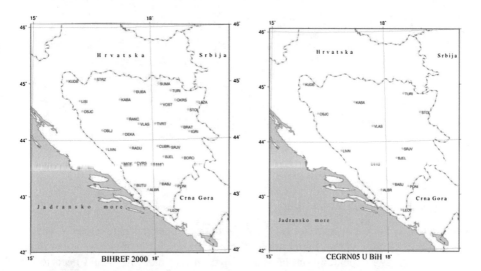

Fig. 1. Map of distribution of the observed stations in the GPS campaigns BIHREF2000 and CEGRN05 in the territory of Bosnia and Herzegovina [4]

Third GPS campaign organized in summer 2005, in the frame of the scientific project named Central European Research Geodynamical Project, with acronym

CERGOP/2 Environment, supported by EU FP5 program. GPS Campaign was named CEGRN (Central European Geodynamical Reference Network). During CEGRN 2005 GPS campaign, in Bosnia and Herzegovina, 16 stations were observed (Fig. 1 rights), for five to seven days in 24 h sessions [12, 13]. Beside these GPS campaign observing data, data from SRJV station in Sarajevo, located at the roof of Department of Geodesy and Geoinformation at Faculty of Civil Engineering University of Sarajevo were processed and analyzed [13, 14].

Some stations from IGS (International GNSS Service) network used as the reference, Fig. 2, during data processing of Bosnian GPS campaigns, and their observing data were downloaded from IGS web site. However, precise efemeride i.e., data of satellite positions in their orbits, are used from IGS, as well.

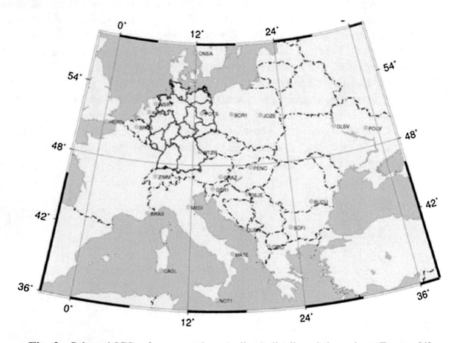

Fig. 2. Selected IGS reference stations (yellow) distributed throughout Europe [4]

2.1 Method of Data Processing

There are two basic concepts for GPS/GNSS data processing: parameters estimation, and parameters elimination. Furthermore, the procedures for estimating the parameters differ significantly depending on the type of parameters that one would need to evaluate. Thus, the parameters that can be estimated can be divided into three groups:

a. The parameters that describe the geocentric position and movement of GNSS observing station (coordinates and velocity vectors and some others parameters not relevant for this paper),

b. Parameters that describe the movement of satellites, and

c. Parameters that directly affect the motion of satellites, such as: atmospheric parameters, clock parameters, signal propagation delays.

Therefore, it follows from the foregoing that it is not possible to determine all the parameters at the same time. Certain conditions must be introduced in order to avoid the singularities of the equation system. In principle, in the process of processing one parameter group, it is necessary to consider other parameters known. For example, in the process of determining the coordinates of the GNSS stations, the orbits of the satellites are considered known. Methods of estimating atmospheric parameters (troposphere and ionosphere), clock parameters, or signal delays are usually carried out in the pre-processing.

The other concept of GNSS observing data processing, instead of evaluating the parameters in the adjustment procedure (using the least squares method), parameters are eliminated, with a suitable combination of observations equations. Thus, the specific parameters are being annulled when simultaneous observations are defined at different GNSS points. These techniques are often used to evaluate atmospheric parameters, clock parameters, or signal delay [4, 15].

To evaluate the coordinates of the GNSS observations, precise IGS orbits (precise ephemerides) are available, which leaves no space for speculation on the topic of orbit estimates as unknown parameters, in the procedures of the processing of observation data in networks such as BIHREF98, BIHREF 2000, CEGRN05, or the like.

It is also possible to apply different strategies in the procedures for estimating the coordinates of the geodetic reference network. Previous research has shown that the application of different processing strategies yields significantly different values of the estimated coordinates [4, 16, 17].

GPS data processing for campaigns Balkan 98, BIHREF 2000, CEGRN05, followed, in general, EUREF criteria and recommendations, but it was not same procedure for all campaigns. It is so, because of the new findings that the scientific community collects through various research projects over time suggests that, the strategy should be changed or updated, in order to achieve better accuracy and reliability of the results.

In addition to the above, research [4] has shown that there is a need for the repeated processing of old GPS observation data, because of new available data, such as:

a. New reference frames, (ITRF2000, ITRF2005, ITRF2008, ITRF20014) which are better defined and have better accuracy than the frames that preceded to them.
b. Re-processed satellite orbits, which are better accuracy due to the application of more accurate reference frames and calibration of the antennas,
c. Available data for the absolute calibration of the receiver's antennas (for IGS stations used as reference) but also the calibration data available of the satellite's antennas.

Benefit of the improved data of new reference frames, reprocessed orbits and calibration of the antennas applied in the strategy processing for two campaigns BIHREF2000 and CEGRN2005, are showed at Fig. 3.

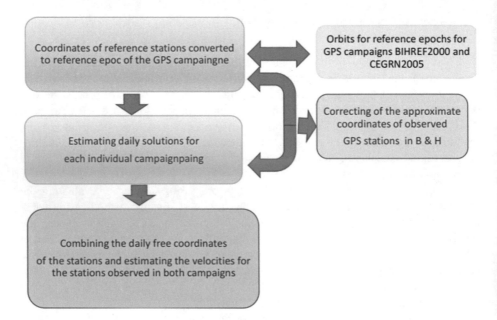

Fig. 3. Processing strategy for GPS campaigns BIHREF2000 and CEGRN2005

3 Results

Although EUREF criteria [18, 19] were applied during the processing of GPS data in B&H, Balkan 98, BIHREF 2000, CEGRN05, the resulting coordinates and their standard deviations varied depending on different factors. Variations were noticed because different ITRF realizations applied, the significant systematic impact of some reference IGS stations noticed, meteorological conditions in B&H were not the same during all campaigns, the number of 24 h sessions was different, orbits of different accuracy were used, interference with other radio signals noticed, significant effect of multipath effects on some stations detected, etc.

When processing GPS Balkan 98 campaign two different ITRF realisation applied, ITRF97 and ITRF2000. It was for the investigating resulting coordinates accuracy. The coordinates 13 stations of the B&H reference network show systematic differences in Nord component of the positions for about 20 mm, since ITRF2000 frame has that updated accuracy relating to realisation ITRF97, what is described in [11] and [17]. In the same research strong bias in component E of the positions of B&H network noticed, and explained as systematic influence of the reference station SOFI, in Sofia, Bulgaria.

Next example of the impact of the reference frames with improved accuracy is described in [4]. GPS campaign BIHREF2000 was processed applying ITRF2000 and ITRF2005 and resulting coordinates showed better accuracy of the stations' coordinates in B&H since ITRF2005 get benefit of the availability data of the absolute calibration antenna. But, even better coordinate accuracy resulted if re-processed orbit applied.

Figure 4 show good accuracy of the coordinates and horizontal "movement" of the stations in B&H, in relation to the fixed datum stations BOR1, GRAZ, WTZR, and ZIMM, which are located northern of the B&H network. Errors of horizontal velocity were determined with a probability of 95%. The red line shows the direction of displacement, and the ellipses of the errors are in blue. Random red line orientation shows that there is no systematic impact, but there are probably random errors in calculating the coordinates, and the velocities (Fig. 5).

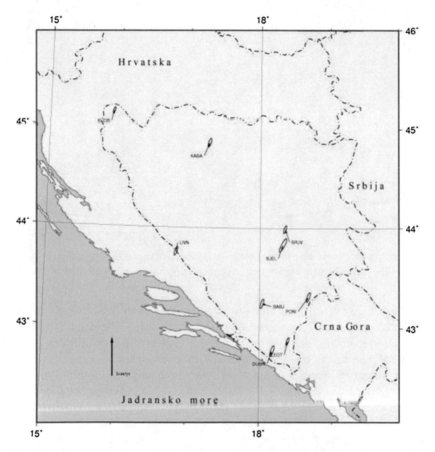

Fig. 4. Accuracy of the estimated coordinates and the velocities of the reference stations in B&H, which are observed twice, in BIHREF2000 and CEGRN2005 [4].

Velocities estimated only for those stations which are observed twice, in BIH-REF2000 and CEGRN2005 campaigns. Permanent station SRJV in Sarajevo got smallest ellipse of error, but BJEL, station at Bjelašnica mountain got the biggest ellipse of errors. It is explained with detected multipath effect caused by GPS signals reflecting from the building near pillar of reference station at Bjelašnica, what reported

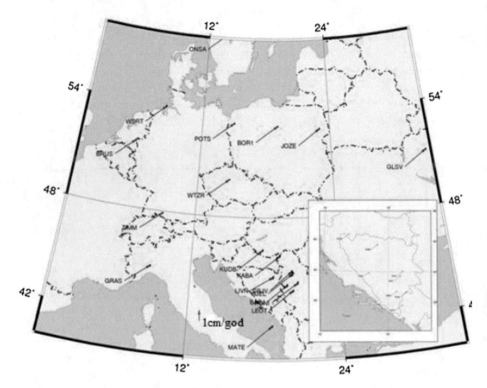

Fig. 5. Velocities of some reference stations in B&H. Behavior of B&H station indicate same direction as the EUREF network, i.e. northeast movement of the Eurasian tectonic plate [4].

in [20], (Fig. 6). Another reason for low accuracy of station BJEL is only two day of the observation, but also bad meteorology conditions during campaign BIHREF2000.

Another systematic impact at the reduced accuracy of the coordinates of the reference B&H network noticed at station CVRS at mountain Čvrsnica. During GPS campaign Balkan 98, there was a competition of the radio amateurs, in the building near pillar of the reference station CVRS. Standard deviation of that station was twice bigger that average. It is example of the interference of the another radio signal to the reduction of the accuracy of the reference station coordinates by GPS methods.

The research of the ionosphere condition and space weather based on GNSS data represent important area of GNSS application. During investigation of the atmospheric parameters

The research of the ionospheric effect on the accuracy of the coordinates of the reference station SRJV for March 15, 2015, an event known as the "St. Patricks' Day storm" carried out [21]. The most powerful space weather geomagnetic storm occurred in the 24 solar cycle on that day. This research showed that the ionospheric free combination L3 method is able to eliminate the strong ionospheric effect to the

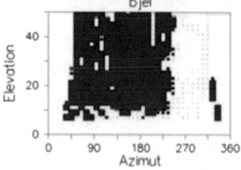

Fig. 6. Map of the detected multipath effect at reference station BJEL at Bjelašnica mountain. Photo above the map show the pillar of reference station is very closed to "ski" station at the top of the Olympic mountain [4, 20].

coordinates of the SRJV station. However, in the same research, it was found that the coordinates at the beginning of March 2015 compared to the EUREF weekly solution, showed significant deviations in the position, as shown in Figs. 7 and 8. A more detailed investigation found that this deviation of the SRJV coordinates probably happened due to the heavy snow at the beginning of March 2015 in Sarajevo. The deviation of the coordinates in all tree positioning components explained by the fact that the tropospheric model applied to estimate the tropospheric parameters during the processing of GNSS data, was not able to give an adequate accuracy of tropospheric parameters. Probably, such a high snow is not in accordance with the troposphere model for that part of the year for the latitude and height of Sarajevo.

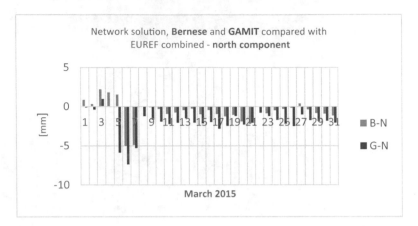

Fig. 7. SRJV station coordinate deviation (6–7 mm for March 5–7, 2015) in north direction, if compared with EUREF weekly solution.

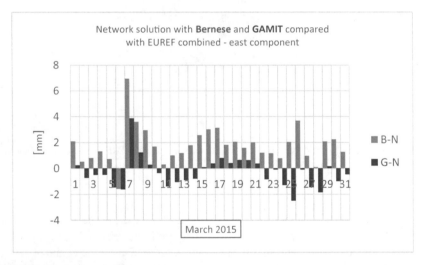

Fig. 8. SRJV station coordinate deviation (7 mm for March 7, 2015) in east direction, if compared with EUREF weekly solution.

4   Conclusion

This paper considers impact of the different parameters variation to the accuracy of the coordinates of the modern geodetic reference network of Bosnia and Herzegovina which determinates by GNSS techniques. Three GPS campaigns considered: Balkan 98, BIHREF2000 and CEGRN2005.

It can be concluded from the presented results that the determination of high accuracy coordinates is a very sensitive area of geodetic science and practice. Many factors are perceived as important parameters in procedure of processing the GNSS observations to estimate the high-precise coordinates and the stations' velocities, which outline the following: good observation data, carefully thought-out strategy of data processing, accuracy of the used ITRS realization, the effects of the individual reference stations selected as datum stations, period of time and number of observation sessions, the accuracy of satellite orbit, the impact of ionospheric and tropospheric refraction, interference with the radio signals from other sources, multipath effect at observing stations, etc.

So, processing of the GNSS observing data is very sophisticated geodetic task. It is never ending work and reference networks should be re-processed in accordance with available parameters: re-processed satellite's orbit, new published ITRF realizations, re-observed network's GNSS data if available, but always, careful strategy of the processing should be applied.

References

1. Hofmann-Wellenhof, B.: GNSS-Global Navigation Satellite System. GPS GLONASS, Galileo, & More. Springer, New York (2008). ISBN 978-3-2011-73012-6
2. Mulić, M.: Geodetic Reference Systems (in Bosnian language), Faculty of Civil Engineering University of Sarajevo (2018). ISBN 978-9958-638-50-3
3. Leick, A., Rapoport, L., Tatarnikov, D.: GPS Satellite Surveying, 4th edn. Wiley, NewYork (2015). ISBN 978-1-118-67557-1
4. Mulić, M.: Investigation of the impact of the ITRF realisation on the coordinates, their accuracy and the determination of the velocity vectors of the GPS points in the territory of B&H. University of Sarajevo, Ph.D. thesis (2012)
5. Mulić, M., Bilajbegović, A.: Improved accuracy of the reference frame of Bosnia and Herzegovina, Geophysical Research Abstracts Vol. 14, EGU2012-13406, 2012 EGU General Assembly (2012)
6. Grewal, M.S., Andrews, A.P., Bartone, C.G.: Global Navigation Satellite Systems, Inertial Navigation and Integration, 3rd edn. Wiley, Hoboken (2013). ISBN 978-1-118-44700-0
7. International Earth Rotation Services. https://www.iers.org/IERS/EN/DataProducts/data.html. Accessed 14 May 2018
8. International Earth Rotation Services. https://www.iers.org/IERS/EN/Science/Recommendations/resolutionCTRS.html. Accessed 14 May 2018
9. International Union of Geodesy and Geophysics-International association of Geodesy. http://www.iugg.org/associations/iag.php. Accessed 14 May 2018
10. Altiner, Y., Schluter, W., Seeger, H.: Results of the Balkan 98 GPS campaigns in Albania, Bosnia and Herzegovina, and Yougoslavia. In: Gubler, E., Torres, J.O., Hornik, H. (eds.) Report on the Symposium of the IAG Subcommision for Europe (EUREF) held in Prague, 2–5 June 1999. Publication No. 8, pp. 106–113 (1999)
11. Mulić, M., Bilajbegović, A., Altiner, Y.: Processing strategy variations on GPS position estimates. In: Geodynamics of the Balkan Peninsula, Report on Geodesy, No. 5(80)2006. Monograph Published in the Frame of Project CERGOP/2 Environment. Warsaw University of Technology, Institute of Geodesy and Astronomy (2006)

12. Caporali, A., Aichhorn, C., Becker, M., Fejes, I., Gerhatova, L., Ghitau, D., Grenerczy, G., Hefty, J., Krauss, S., Medac, D., Milev, G., Mojzes, M., Mulic, M., Nardo, A., Pesec, P., Rus, T., Simek, J., Sledzinski, J., Solaric, M., Stangl, G., Vespe, F., Virag, G., Vodopivec, F., Zablotskyi, F.: Geokinematics of central Europe: new insights from the CERGOP 2/environment project. J. Geodynamics (2007). https://doi.org/10.1016/j.jog.2008.01.004

13. Caporali, A., Aichhorn, C., Becker, M., Fejes, I., Gerhatova, L., Ghitau, D., Grenerczy, G., Hefty, J., Krauss, S., Medak, D., Milev, G., Mojzes, M., Mulic, M., Nardo, A., Pesec, P., Rus, T., Simek, J., Sledzinski, J., Solaric, M, Stangl, G., Stopar, B., Vespe, F., Virag, G.: Surface kinematics in the Alpine-Carpathian-Dinaric and Balkan region inferred from a new multi-network GPS combination solution. Tectonophysics **474**, 295–321 (2009). ISSN: 0040-1951. https://doi.org/10.1016/j.tecto.2009.04.035

14. Mulić, M., Natraš, R.: Ionosphere TEC variations over Bosnia and Herzegovina using GNSS data. In: Cefalo, R., Zielinski, J.B., Barbarella, M. (eds.) New Advanced GNSS and 3D Spatial Techniques, Applications to Civil and Environmental Engineering, Geophysics, Architecture, Archeology and Cultural Heritage. Lecture Notes in Geoinformation and Cartography, pp. 271–283. Springer, Heidelberg (2008). https://doi.org/10.1007/978-3-319-56218-6_22

15. Seeber, H.: Satellite Geodesy, 2nd edn. Walter de Gruyter Berlin, New York (2003)

16. Mulić, M., Bilajbegović, A., Altiner, Y.: Untersuchung der Einflussfaktoren bei der Koordinatenbestimmung nach EUREF-Kriterien. Allegemeine Vermessungs-Nachrichten-AVN, No. 2/2006, pp. 49–55 (2006)

17. Mulić, M., Bilajbegović, A., Altiner, Y.: Processing strategy variations on GPS position estimates. geodynamics of the Balkan Peninsula, report on Geodesy, No. 5(80)2006. Monograph published in the frame of Project CERGOP/2 Environment. Warsaw University of Technology, Institute of Geodesy and Astronomy (2006)

18. Boucher, C., Altamini, Z.: Specifications for reference frame fixing in the analysis of a EUREF GPS campaign (2001). http://etrs89.ensg.ign.fr/memo-V8.pdf. Accessed 15 May 2018

19. Bruyninx, C., Altamimi, Z., Caporali, A., Kenyeres, A., Stangl, G., Torres, J.A.: Guidelines for EUREF densifications (2013). vers.5. ftp://epncb.oma.be/pub/general/Guidelines_for_EUREF_Densifications.pdf. Accessed 15 May 2018

20. Bilajbegović, A., Mulić, M.: Multipath effect investigation for the improvement of BIHPOS. In: BALGEOS Workshop, Sofia, 20 November 2008

21. Krdzalić, Dž., Tabaković, A., Horozović, Dz., Natraš, R., Mulić, M.: Coordinates estimate under disturbed ionosphere conditions: case study SRJV GNSS. In: 2nd Conference of West Balkan Geodetic Forum, Mostar, October 2017

Determination of Aiming Error
with Automatic Theodolites

Stefan Miljković[1], Vukan Ogrizović[2(✉)], Siniša Delčev[2],
and Jelena Gučević[2]

[1] Faculty of Civil Engineering, University of Belgrade, Belgrade, Serbia
stefan.miljkovic@grf.bg.ac.rs
[2] Faculty of Civil Engineering, University of Novi Sad, Subotica, Serbia
{vukan,delcevs,jgucevic}@gf.uns.ac.rs

Abstract. Automatic theodolites use the systems for automatic aiming, eliminating the operators' error. This error is dominant, meaning that it remains in the error budget. For that reason, the automation of this process reduces the total error of angle measurements. However, due to their construction and performance, those systems introduce the error acting like the aiming error. In our study case, we tested two classes of instruments with the automatic aiming systems, with 0.5″ and 3″ accuracy, respectively. The experiment comprises measurements of a number of horizontal and vertical directions, variating types of the prisms and distances to them, in order to randomize other error sources. Comparison of the measurements taken in real conditions with the values given by the manufacturers, after the statistical analysis, show that the error of automatic aiming depends on the environmental conditions and the prism type. In the case of high accuracy demands, further attention should be paid on the preparation of a measuring campaign, in order to reduce this effect.

Keywords: Theodolite · Autolock · Error · Trimble · Total station

1 Introduction

One of preconditions to quality assurance of surveying measurements is a metrological assurance of instruments and accessories, including the analysis of measuring method and determination of certain error sources that can occur during measurements. Historically, procedures for determination of common error sources for a number of surveying instruments are developed, making determination of the total measurement error possible. However, advances in technology and involving contemporary approaches into construction of surveying instruments, principles of certain elements functioning change basically, which led to need to treat some error sources in different way, because the technological concept of the measurement itself changes.

Angle measurement with theodolites, using the modern equipment, remains the same as with classical instrumentation, but some modifications to the equipment was made. Developing the concept of automatic signal search automatize the measuring process itself. Also, one of the error sources – aiming, is eliminated. Due to imperfection of the signal search system, another error source occured. In can be treated as

© Springer Nature Switzerland AG 2019
S. Avdaković (Ed.): IAT 2018, LNNS 60, pp. 273–284, 2019.
https://doi.org/10.1007/978-3-030-02577-9_27

the aiming error with classical theodolites. Determination of this error is important, because it is one of the dominant errors, which affect significantly the measurement accuracy.

This research is directed to determination of the practical aiming error of the automatic aiming system with Trimble instruments, in real conditions. In the experimental part, we measured more horizontal and vertical directions in more series, in different combinations (different prisms, sight lengths, etc.). Two instruments are tested, both of Trimble series "S".

2 Methodology and Background

Measuring method of a quantity assumes the whole process from the primary standard to the value of the measurand and its error estimation [2]. The need for the measuring method analysis occurs when the objective accuracy of measurement of the quantity, applying the given method, should be determined. For that reason, all error sources of the certain method should be discovered, if they can impact the measurement accuracy. Also, one should determine the character and value of the measurement quantity error, which errors influence the measurement accuracy, and which of them can be made insignificant. Finally, the analysis contains development of the measurement error equations, error conditions equations, and the equations of the error sources that could be made insignificant to the measurement accuracy [3].

One of the most dominant errors of the angle measurement method is the aiming error. Aiming is a procedure for aligning the line sight of the theodolite with the signal symmetry line. This is a direct measurement, the procedure with an operator, the theodolite, and the signal. In the classical concept, the aiming error depends on the precision and the quality of the theodolite, signal type and capability of the operator. With the modern instruments, the operator role is completely switched with the systems for automatic aiming. However, the error still remains, only with different sources, due to imperfection of the automating aiming system.

2.1 Determination of the Aiming Error of the Automatic Theodolites

The basic measuring quantity in the angle measurement procedure is a direction (horizontal or vertical). The error of the measured direction can be determined a priori, based on the known values of all influences, applying the expression for the previous accuracy estimation. If the error sources are unknown, then the error can be calculated from the measurement results, experimentally, by repeating measurement of the same direction. The variance of the measured direction is obtained as the sum of all dominant variances, that exist during the measurement (classical approach) [3]:

$$\sigma_p^2 = \sigma_{\tau_{RM}}^2 + \sigma_{v_z}^2 + \sigma_k^2 + \sigma_r^2 + \sigma_z^2 \tag{1}$$

with:

σ_p - total error of the measured direction,

$\sigma_{\tau_{RM}}$ - total error of the equipment,

σ_{v_z} - aiming error,
σ_k - coincide error,
σ_r - refraction error, and
σ_z - reading rounding error.

If the values of the error impacting the measured direction are unknown, than the direction error can be empirically represented as the repeatability error, i.e., dispersion of the same direction in more series:

$$\sigma_p^2 = S_p^2 = \sigma_{\tau_{RM}}^2 + \sigma_{v_z}^2 + \sigma_k^2 + \sigma_r^2 + \sigma_z^2 \tag{2}$$

with:
S_p - repeatability error (dispersion).

Repeatability error of the measurement set is represented by standard deviation of all performed measurement at the given point:

$$S_p^2 = \frac{\sum\limits_1^n (p_i - \bar{p})^2}{n - 1} \tag{3}$$

where:
$\bar{p}$ - average value of the measured direction,
p_i - i^{th} measured direction, and
n - number of repeats.

After analysing some error sources from (1), the following conclusions are drawn:

- Each repeated measurement is performed at the same place on the limb, which means that the instrument error is the same in each series and, therefore, does not affect the measurement dispersion.
- If electronic theodolites are used, than there is no coinciding, and, consequently, there is no coincide error.
- There is no impact of the refraction error, because the lengths of the line sights are the same, the measurements last for a short time, and the refraction error influence all measurements equally.
- The reading error of the modern electronic instruments construction are insignificant to the accuracy class of the instrument.

When certain error sources, after the drawn conclusions, are removed from (2), the fact remains, that the repeatability error depends mostly on the aiming error:

$$S_p^2 = \sigma_{v_z}^2 \tag{4}$$

It can be concluded from (3) that the aiming error can be obtained as the standard deviation, i.e., the repeatability error of the same measurement in more series:

$$\sigma_{vz}^2 = \frac{\sum\limits_{1}^{n} (p_i - \bar{p})^2}{n} \tag{5}$$

If there are measurements, at each point, in two faces with n repeats in each face, the errors for both faces can be calculated from the obtained results. Since this is the same quantity obtained from two measurement sets, the final aiming error can be expressed as:

$$\sigma_{VZ}^2 = \frac{\left(\sigma_{vz}^2\right)_I + \left(\sigma_{vz}^2\right)_{II}}{2} \tag{6}$$

where:

σ_{VZ} - total aiming error,

$\left(\sigma_{vz}^2\right)_I$ - aiming error from the face I,

$\left(\sigma_{vz}^2\right)_{II}$ - aiming error from the face II.

If the measurement is performed with few different lengths of line of sight, the total aiming error for the certain instrument is obtained from:

$$\sigma_V^2 = \frac{\sum\limits_{1}^{n} \left(\sigma_{VZ}^2\right)_i}{n} \tag{7}$$

with:

σ_V - total aiming error,

$\left(\sigma_{VZ}^2\right)_i$ - total aiming error at one signal, with i^{th} line of sight,

n - number of the markers used for measuring.

From the measurement results, the aiming error is calculated in this way, also, when the operator aims manually, using the instruments with no automatic aiming systems. With manual aiming, the error depends on the shape and the size of the marker, quality of the instrument, and, mostly, from the subjective operator error. A part of the error, related to the marker and the instrument, can be partially standardized. However, the most of the error still remains to the capability of the operator. In the case of the instruments which own the automatic aiming systems, the aiming error depends on the marker type, and on the accuracy of the automatic aiming system.

2.2 Analysis of the Quality of the System for Automatic Target Recognition

Each instrument owns a declared accuracy of the automatic aiming, which is prescribed by the manufacturer. Using the procedure for determining the aiming error experimentally, the empiric error of the system in the real conditions can be calculated. Statistical tests are used for testing the congruency between the declared and the experimental system errors.

Testing can be performed by Fisher test of the congruency of the declared standard instrument error, and the standard error obtained from the measurement results. During testing, it is assumed that the measurement follow Fisher distribution with the probability 95%. The test statistics is based on:

$$\sigma_V^2 < F_{gran}^2 \tag{8}$$

where:

$F_{gran} = \sqrt{\sigma_{av}^2 \cdot F_{(\alpha,f,\infty)}}$ - border value of the statistical congruency for given declared accuracy of the automatic aiming system,

σ_V - aiming error obtained from the measurement results,

σ_{av} - declared aiming error of the automatic aiming system,

$F_{(0.05,f,\infty)}$ - quantile of Fisher distribution,

f - number of performed measurements in both faces,

α - probability.

Statistical results come from:

$\sigma_V^2 < F_{gran}^2$ - aiming error obtained from the measurement results in congruent with the error declared by the instrument manufacturer.

$\sigma_V^2 > F_{gran}^2$ - aiming error obtained from the measurement results is not congruent with the error declared by the instrument manufacturer.

3   Experiment and Results

The goal of this research is to find out the amount of the error of the robotized total station, i.e., their automatic theodolites, during the measurement of the horizontal and vertical angles. The biggest revolution in technological development of this kind of the instruments caused resolving the problem of automatic aiming. We intended to find out whether the real aiming error is congruent to the error declared by the manufacturer, performing the experiment in real conditions, in field.

3.1   Polygon for the Experiment Performance

Based on empirical experiences, we took into account the fact that the aiming error, besides the instrument quality, shape and the size of the marker, and accuracy of the automatic tracking system, indirectly depends on the distance from the prism. If, during aiming, the same instrument, same tracking system for the same prism are used, the only factor that can have the impact to the aiming error is the distance from the prism, i.e., the length of the line of sight. According to that, we designed the polygon with several points, placed in the characteristic distances which the users mostly find in practice: 10 m, 25 m, 50 m, 100 m, 200 m, and 300 m. After 300 m, the points are places at each 200 m, to the upper limit of the system for automatic detection of the prism (Fig. 1).

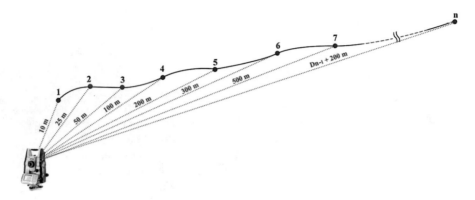

Fig. 1. Polygon for the experiment realization

3.2   Instrumentation and Accessories

The instruments used in this research are the automatized robotized total stations. We used two instruments, Trimble S6 and Trimble S8.

Trimble S6 DR300+ (Fig. 2) has the declared accuracy of horizontal and vertical angles 3″. Accuracy of distance measurements is 2 mm + 2 ppm with prism, and 3 mm + 2 ppm, in reflectorless mode. Maximum reach of the Autolock system is up to 700 m, depending on the prism type, while its declared accuracy is 2″. We used three prisms in combination with this instrument: Trimble 360° prism, Trimble 360° mini prism, and universal standard prism.

Trimble S8 is the first 1″ automatic total station produced by the Trimble company. It is dedicated to the most challenging works in surveying and in complementary disciplines. The instrument used for this research is Trimble S8 0.5 HP (Fig. 3).

The total station has the declared accuracy of horizontal and vertical angles 0.5″. The accuracy of length measurements is 0.8 mm + 1 ppm with the prism, and 3 mm + 2 ppm in the reflectorless mode. The maximum range of the Autolock system is up to 700 m, depending on the prism type, while the declared accuracy of the Autolock system, in the Finelock mode, 0.68″. A Trimble 360° and a SECO universal standard prism are used as the auxiliary measurement equipment.

3.3   Measurement Realization

We measured at two different locations. The instruments and the prisms are placed onto massive tripods, following the distribution explained in Fig. 1. During the measurements, environmental conditions were monitored. The instruments were exposed to external temperature minimum 30 min before measurements. The measurements performed during the extensive sunshine requested the sunshade, in order to avoid the de-

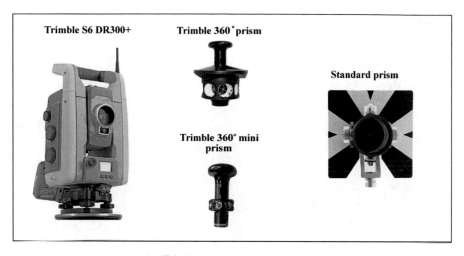

Fig. 2. Trimble S6 DR300+ with accessories

Fig. 3. Trimble S8 0.5 HP with accessories

rectification and tripod twist. During the measurements, the operator remained as steady as possible, to exclude any disturbances of the tripod. The instrument was controlled remotely, using a wireless controller. The measurements were logged into the instrument internal memory.

3.4   The Experiment Results

The results are shown in Tables 1 and 2.

Table 1. Aiming errors of horizontal directions for Trimble S6 DR300+

Point number	Length of line of sight [m]	Prism types		
		Trimble 360°	Trimble 360° mini	Standard universal prism
		σ_{vz} ["]	σ_{vz} ["]	σ_{vz} ["]
1	10.41	4.57	1.34	4.62
2	25.13	1.29	3.18	3.13
3	50.31	4.34	4.68	1.07
4	99.53	2.28	2.87	-
5	200.69	-	2.36	2.50
6	299.33	2.76	-	2.42
Total aiming error		**3.29**	**3.09**	**2.75**

Table 2. Aiming errors of vertical directions for Trimble S6 DR300+

Point number	Length of line of sight [m]	Prism types		
		Trimble 360°	Trimble 360° mini	Standard universal prism
		σ_{vz} ["]	σ_{vz} ["]	σ_{vz} ["]
1	10.41	1.79	1.39	1.93
2	25.13	1.03	1.56	1.97
3	50.31	1.88	1.85	1.47
4	99.53	0.97	1.51	-
5	200.69	-	1.00	1.57
6	299.33	1.36	-	0.93
Total aiming error		**1.45**	**1.49**	**1.62**

Graphical presentations of the research obtained by Trimble S6 DR300+ are given in Figs. 4 and 5.

The declared accuracy of the Autolock system with Trimble S6 DR300+ is 2″. If a test statistics is applied, it is calculated that the border statistical congruence, with probability 95%, equals $F_{gran} = 2.42″$. The comparative analysis of the experiment results with the statistical analysis for Trimble S6 DR300+ is given in Table 3.

The results of the experimental research, i.e., the aiming errors for the horizontal and vertical directions obtained by Trimble S8 0.5 HP are depicted in Tables 4 and 5.

Graphical representations of the results obtained by Trimble S8 0.5 HP are given in Figs. 6 and 7.

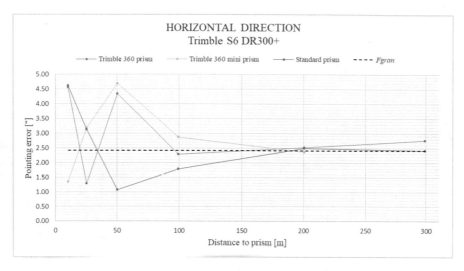

Fig. 4. Graphical representation of the aiming error of the horizontal directions for Trimble S6 DR300+

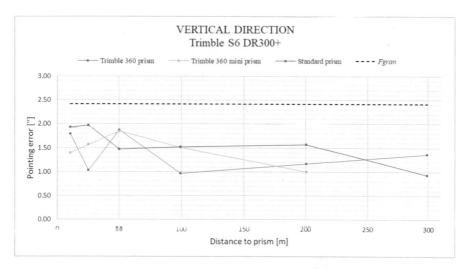

Fig. 5. Graphical representation of the aiming error of the vertical directions for Trimble S6 DR300+

The declared accuracy of the Autolock system of the Trimble S8 0.5 HP in the Finelock mode is 0.68″. By applying the test statistics, it is obtained that the vorder statistical congruency, with probability 95% is 0.82″. Comparative analysis of the experimental results with statistical analysis for Trimble S8 0.5 HP is shown in Table 6.

Table 3. Comparative analysis - Trimble S6 DR300+

Trimble S6 DR300+		σ_{VZ}^{min}	σ_{VZ}^{max}	σ_V	F_{gran}
Horizontal directions	Trimble 360°	1.29″	4.57″	3.29″	2.42″
	Trimble 360° mini	1.34″	4.68″	3.09″	2.42″
	Standard prism	1.07″	4.62″	2.75″	2.42″
Vertical directions	Trimble 360°	0.97″	1.88″	1.45″	2.42″
	Trimble 360° mini	1.00″	1.85″	1.49″	2.42″
	Standard prism	0.93″	1.97″	1.62″	2.42″

Table 4. Aiming errors of the horizonlta directions - Trimble S8 0.5 HP

Point number	Length of line of sight [m]	Prism types	
		Standard SECO prism	Trimble 360°
		σ_v [″]	σ_v [″]
1	10.14	1.79	1.17
2	25.64	1.52	0.76
3	50.41	1.38	0.99
4	100.36	1.36	0.50
5	199.79	1.40	-
6	300.27	1.48	-
7	500.92	1.54	-
8	672.48	1.80	-
Total aiming error		**1.54**	**0.89**

Table 5. Aiming errors of the vertical directions - Trimble S8 0.5 HP

Point number	Length of line of sight [m]	Prism types	
		Standard SECO prism	Trimble 360°
		σ_{vz} [″]	σ_{vz} [″]
1	10.14	1.16	1.02
2	25.64	0.49	0.69
3	50.41	0.58	0.54
4	100.36	0.54	0.58
5	199.79	1.56	-
6	300.27	1.23	-
7	500.92	1.21	-
8	672.48	1.35	-
Total aiming error		**1.09**	**0.73**

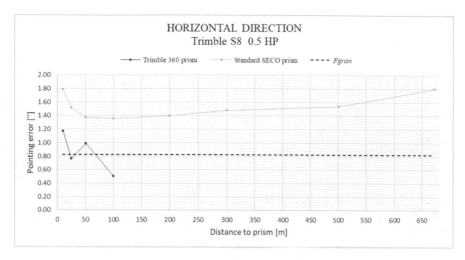

Fig. 6. Graphical representation of the aiming error of the horizontal distances for Trimble S8 0.5 HP

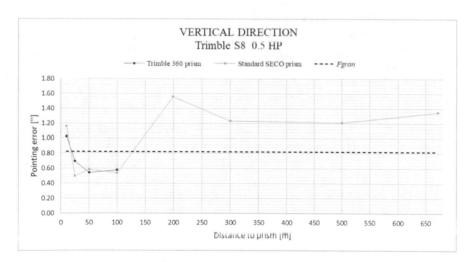

Fig. 7. Graphical representation of the aiming error of the vertical directions for Trimble S8 0.5 HP

Table 6. Comparative analysis - Trimble S8 0.5 HP

Trimble S8 0.5 HP	σ_{VZ}^{min}		σ_{VZ}^{max}	σ_V	F_{gran}	
Horizontal directions	Trimble 360°		0.50″	1.17″	0.89″	0.82″
	Standard SECO prism		1.36″	1.80″	1.54″	0.82″
Vertical directions	Trimble 360° prism		0.54″	1.02″	0.73″	0.82″
	Standard SECO prism		0.49″	1.56″	1.09″	0.82″

4 Conclusion and Remarks

We calculated the aiming errors obtained using the systems for automatic aiming in field conditions. For Trimble S6 DR300+, the aiming error of the horizontal directions corresponds to the interval 1.07″–4.68″, while the aiming error for the vertical directions belongs to 0.93″–1.97″ interval, depending on the prism type and length of the line of sight during the measurements. For Trimble S8 0.5 HP, the aiming error for the horizontal distances has the values 0.50″–1.80″, and the aiming error for the vertical directions falls into the interval 0.49″–1.56″.

Statistical testing of the obtained results, shown in Tables 3 and 6, leads to the conclusion that only in the case of the vertical directions measurements with Trimble S6 DR300+, declared accuracy is obtained throughout the polygon. In all other cases, there are errors incongruent with the declared values. Besides that, the prism type does not affect the aiming accuracy significantly, except for Trimble S8 0.5 HP, where it affects the horizontal directions.

Analysing Figs. 4, 5, 6 and 7, it can be concluded that the obtained results disperse more on distances less than 100 m, while the oscillations at longer distances get smaller.

The overall impression, according to the measurement results of this research, is that none of the instruments completely reaches the declared standard of automatic aiming. However, further investigation of this effect is needed. To improve the objectiveness of the experiment, the tripods should be switched with concrete pillars. The measurements should be performed in more stable weather conditions, without strong sunshine or high humidity, to reduce those influences.

References

1. Leica Geosystems. Leica TPS1200+: White Paper. Heerbrugg (Switzerland) (2009)
2. Синиша Делчев. *Геодетска метрологија.* Академска мисао, Београд, ИСБН 978-86-7466-640-1 (2016)
3. Чинкловић Никола. Анализа и претходна оцена тачности метода геодетских мерења. Геокарта, Београд (1978)
4. http://tknsc.trimble.com/Default.aspx?ip_server=http://tknbe.trimble.com:8080&qtype=start - Trimble Knowledge Network, website
5. www.leica-geosystems.com - Leica Geosystems, official website
6. www.trimble.com - Trimble, official website

Performance Analysis of Main Road Section in Bosnia and Herzegovina in Terms of Achieved Average Speeds

Sanjin Albinovic[(✉)], Ammar Saric, and Mirza Pozder

Faculty of Civil Engineering, Department of Roads and Transportation,
University of Sarajevo, Sarajevo, Bosnia and Herzegovina
sanjin.albinovic@gmail.com,
ammar.saric@hotmail.com, pozder.mirza@hotmail.com

Abstract. Although recently has been an expansion in building of highways in Bosnia and Herzegovina, two-way roads are still a very important part of traffic network. They continue to be the main link between cities, with significant traffic loads on particular sections. Accordingly, these roads should meet some of the basic requirements in terms of operational performances and safety. In this paper, performance characteristics of one section of the main road has been analyzed from the aspect of average speed of traffic flow. The analysis was carried out with different methodologies that are in use in the world and in Bosnia and Herzegovina. The research is focused on comparing all results obtained with different methodologies in order to find the most applicable methodology for traffic conditions in B&H.

Keywords: Two-way roads · Speed · Traffic-flow

1 Introduction

Two-way highway form about 90% of the road transport network in B&H and connect all major cities.

On some main roads sections, especially near the towns or populated areas, the traffic flows is significantly high.

In these situations quality of the traffic flow is getting worse as well as the safety of highway users (drivers).

From the aspect of their functionality, this question arises: whether these roads fulfill the basic conditions and the purpose for which they were built?

Therefore, the roads that were once planned, designed and built to connect intercity centers and to fulfill certain conditions regarding the travel speed, due to building a settlement near the road (the problem of corridors settlements) over time become part of the traffic network of the populated place. Consequently, there is a significant deterioration of their performance characteristics.

In order to solve these problems, road infrastructure managers are forced to design and build bypass around settlements. On the other hand, on sections of two-way highways in rural area, the problem are parts with a slightly larger longitudinal slope.

© Springer Nature Switzerland AG 2019
S. Avdaković (Ed.): IAT 2018, LNNS 60, pp. 285–300, 2019.
https://doi.org/10.1007/978-3-030-02577-9_28

At these sections, there are a significant decrease in the speed of heavy vehicles and consequently other traffic participants. In this case, as a solution to the problems imposed by the construction of lane for the slow vehicles (i.e. climbing lane) or possibly base tunnels (which is more expensive solution).

In order to make an optimal solution for these problems, it is necessary to conduct a capacity analysis as a basic indicator of the quality of traffic flow in the first phase and in some cases it is necessary to make an analysis from the aspect of traffic safety. Based on the results of these analyzes or knowledge of the quality of traffic conditions on a certain section of the road, a decision should be made on the implementation and selection of appropriate measures in order to improve the functionality of the analyzed road.

There are various methodologies for capacity analysis and some of the most famous are: Highway Capacity Manual (HCM) [1], Handbuch für die Bemessung von Straßenverkehrsanlagen (HBS), Ausgabe 2015 [2]. These methodologies are applied for the analysis of two-way two lane highways in rural (suburban) areas with assumption of uninterrupted flow.

In these traffic conditions, vehicles carrying out the passing maneuvers of slower vehicles must use a lane for the traffic in opposite direction. In doing so, they must use time gaps of corresponding sizes in flow in opposite direction as well as sections with appropriate geometric characteristics. This often leads to the impossibility of passing maneuvers of slower vehicles and creating a (column-following) platoons of vehicles in which decreases the flow speed and increases of density.

Capacitive analysis consists of two main steps. One is the determination of the capacity (c) of the sections (i.e. the degree of saturation $x = q/c$) as quantitative measures and the levels of service (LOS) describing the quality of traffic flows on the analyzed sections of highways (roads).

According to the above methodologies, performance characteristics of one section of the main road has been analyzed from the aspect of average speed of traffic flow. The results of the analysis has been compared with the results of in-situ research. In this way, we determined the applicability of these methodologies on the main roads and local conditions in B&H.

1.1 HCM Methodology

An operational analysis is conducted to observe both directions at the same time. Two-way segments may include more sections of two-lane highway with a homogeneous cross section and relatively constant traffic load and structure along the entire segment. They have to be in level or rolling terrain.

Roads in the mountainous terrain or with a longitudinal slope greater than or equal to 3% over a length greater than or equal to 1 km have to be analyzed as directional segments on specific grades or directional segments including passing and truck climbing lanes. But, if only one direction is analyzed, an analysis of the directional segment in the level or rolling terrain also can be applied.

In this methodology, three classes of roads are defined:

1. *Class I* are two-way highway which connecting larger centres and represent primary connectors in the state highway network. They serve for long distance trips and on them drivers expect to travel at relatively high speeds.
2. *Class II* are two-way highway which serve for short distance trips, may be the initial and final part of long distance trip or serve for recreational purposes and theirs function in highway network should be to connectors for Class I highways. Driving speed is not primary and drivers don't expect to travel at relatively high speeds.
3. *Class III* are two-way highway that can be parts of the class I and/or class II highway that pass through small towns or developed recreational zones. They have a significant local traffic and a large number of access points. The travel speed of is generally reduced due to the posted speed limits.

To describe the level of service on these roads, three measures of efficiency of traffic flows are used:

1. The average travel speed (ATS) is a measure of mobility on the road. It is obtained If the length of the section is divided by the average traveling time of all vehicles exceeding the subject length in both directions at a given interval.
2. Percent time spent following (PTSF) represents the freedom of manoeuvring and comfort of the trip. It is the average percentage of travel time that vehicles have to spend in platoons behind a slower vehicle due to inability to pass.
3. Percent of free flow speed represents the ability of vehicles to travel with speed at or near posted speed limits

The traffic flow values increases with correction factors for all deviations in the environmental, traffic and geometric characteristics of the considered section of the values defined for the ideal (base) conditions[1].

The capacity of the two-lane rural highway under ideal (base) conditions is 1700 pc/h in one direction, i.e. 3200 pc/h in both directions.

In two-lane highways of Class I for determining the level (LOS) of service, both measures of efficiency (ATS and PTSF) use (Fig. 1). For highways of Class II level of service is determined only according to the PTSF criteria, while for Class III only according to the PFFS criteria (Fig. 1). The Level of Service (LOS) criteria for two-lane highways are shown in Table 1.

Capacity analysis process for the two-lane highway according to IICM methodology in the form of flowchart is shown in Fig. 1.

1.2   HBS Methodology

HBS is a German Highway Capacity Manual (first edition HBS 2001 is publicized in January 2002 from FGSV) and in principle follows the ideas of the American Highway Capacity Manual with adaptation to local standards and traffic conditions based on the results of numerous researches carried out in Germany (Brillon, Großmann, Blanke).

---

[1] Lane width $\geq$ 3,6 m; Clear shoulders $\geq$ 1,8 m; No no-passing zones; Only passenger cars in the traffic flow; Level terrain and No impediments to through traffic (e.g., traffic signals, turning vehicles).

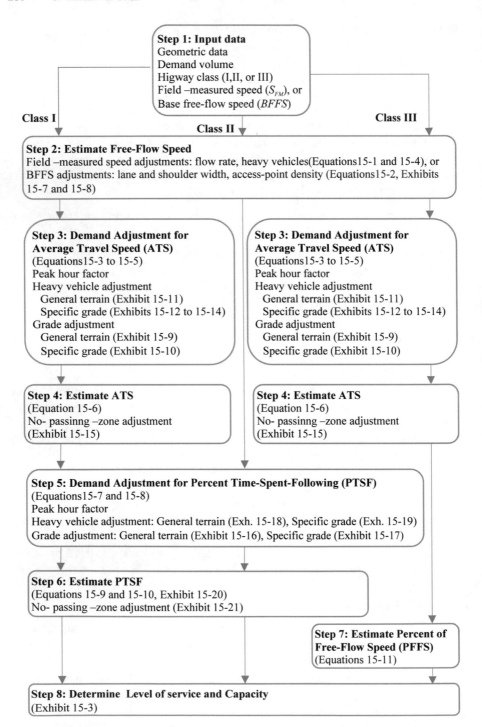

Fig. 1. Flowchart of the Two-lane highway methodology (HCM 2010) [1]

Table 1. Level of service criteria for two-lane highways

LOS	Class I highways		PTSF (%)	Class II highways PTSF (%)	Class III highways PFFS (%)
	ATS (km/h)	PTSF (%)	PTSF (%)	PFFS (%)
A	>88.5	≤ 35	≤ 40	>91.7
B	>80.46 ≤ 88.5	>35 ≤ 50	>40 ≤ 55	>83.3 ≤ 91.7
C	>72.42 ≤ 80.46	>50 ≤ 65	>55 ≤ 70	>75.0 ≤ 83.3
D	> 64.37 ≤ 72.42	>65 ≤ 80	>70 ≤ 85	>66.7 ≤ 75.0
E	≤ 64.37	>80	>85	≤ 66.7

The procedure applies to calculation the capacity and traffic quality conditions of two lane rural highways (Brilon 1998 and Weiser 2003) for sections that do not lie in the area of influence of adjacent nodal points. It is valid for 2-lane rural highways of category A with coupling function degrees I to III by RAS-N, 1988. Also it is applied for cross sections RQ 10.5 (2 × 3.5 m + 2 × 0.25 m + 2 × 1.5 m) and RQ 9.5 (2 × 3.0 m + 2 × 0.25 m + 2 × 1.5 m) according to RAS-Q, 1996 and similar cross sections.

The methodology implies that different widths of the traffic lanes mentioned cross sections and slight deviations from these widths do not have a significant impact on the capacity and level of service i.e. on the quality of traffic flows.

Road sections with variable influencing parameters (e.g. longitudinal slope and the bendiness) are divided into segments.

The segment begins with the change of some of the influencing parameters and should be long at least 300 m. The usual procedure is that both lanes or directions are viewed at the same time.

The parameters that affect the flow of traffic and thus the operational speed are called influencing parameters, and the process of calculating the capacity and the quality of the conditions of the traffic flow begins with the determination of the first influencing parameter, and that is the longitudinal slope. On longitudinal slopes greater than 2%, the travel speed of all vehicles, especially heavy vehicles, is reduced. Therefore, these segments are taken into calculation with the size of longitudinal slope and its length. The impact of the longitudinal slope is transferred to the upgrade class from 1 to 5.

Stepping into these classes is done for a segment of a certain length and slope:

– based on the constant speed of a metering heavy vehicle that can be held on the part of the section,
– or on the basis of medium-speed measurement of heavy vehicles on the section of which is calculated as the arithmetic average of the beginning and end of the slope (climb).

The speed of the metering heavy vehicle is determined using the diagram in Fig. 2 separately for both directions, and less value is relevant.

After determination of relevant speed for heavy vehicle according to the criteria in Table 2, road section is classified into one slope category.

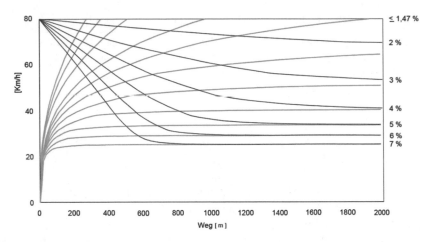

Fig. 2. Speed diagram for a metering heavy vehicle with different longitudinal slopes and its lengths [2]

Table 2. Capacities on two-lane rural highway (both directions)

Climbing class	Curvature (grad/km)	Capacity [veh./h]					
		Heavy vehicles [%]					
		0	5	10	15	20	25
1	0–75	2500	2490	2370	2290	2255	2215
	75–150	2075	2075	2065	2060	2060	2060
	150–225	1935	1875	1840	1815	1800	1780
	>225	1855	1805	1770	1745	1740	1720
2	0–75	2500	2420	2295	2195	2155	2100
	75–150	2070	2070	2065	2060	2050	2045
	150–225	1930	1870	1830	1810	1795	1780
	>225	1855	1795	1760	1735	1715	1700
3	0–75	2500	2115	1965	1865	1795	1750
	75–150	2000	1975	1925	1865	1795	1750
	150–225	1930	1840	1795	1755	1735	1720
	>225	1855	1780	1740	1705	1680	1675
4	0–75	2400	1735	1590	1510	1445	1405
	75–150	2000	1680	1580	1510	1445	1405
	150–225	1930	1665	1570	1510	1445	1405
	>225	1855	1650	1570	1510	1445	1405
5	0–75	2000	1400	1230	1140	1055	950
	75–150	1800	1385	1230	1140	1045	950
	150–225	1800	1370	12300	1140	1045	950
	>225	1795	1360	1230	1140	1040	940

The following two influencing parameters that are determined and which have a significant impact on the operational speed on two-lane roads are relations of the curve of a road and the possibility of passing through the vehicle. Both these influencing parameters are expressed through the curvature in grad/km. The sum of the curvature of route and the addition based of the sections with no passing zone, affect the overall influencing parameters in determination of quality of traffic flow.

The impact of heavy traffic on the medium speed of a passenger car and capacity is taken into account and it is given graphically in the "q-V" diagrams for the five climbing classes and within each class of climb for four groups the combined influence of curvature and impossibility passing (no-passing zone).

The impact of speed limits is not taken into account in the capacity analysis under this methodology and this may constitute a certain limitation in the application of this methodology in areas where there are settlements along the two way highway.

In this methodology the capacities of subject roads section are given in tables (Table 2) depending on of previously defined influential parameters.

Practically dependent on the given conditions of road geometry (horizontal, vertical and defined cross sections) and traffic flow structure.

The level of service is determined by the density of the traffic flow. For simplicity, it is calculated with a single fictitious traffic flow density, where all motor vehicles are present in the traffic flow value q, while the average speed of travel relates only to passenger cars.

Each traffic density k belongs, depending on the characteristics of the section and the traffic boundary conditions, the average speed of the passenger vehicle VR, which is obtained from the diagram (Fig. 3) and the quality of the traffic flow, i.e. the level of service A to F, is determined according to the traffic flow density limits given in Table 3.

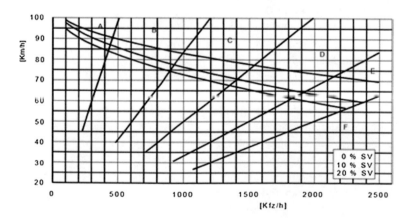

Fig. 3. The average travel speed of a passenger car depending on the traffic volume (climbing class 1) for curvature in the range KU = 0–75 city/km [2]

Table 3. Limits of traffic flow density as a Level of Service criteria

LOS	Traffic flow density[a]
	pc/km
A	≤ 5
B	≤ 12
C	≤ 20
D	≤ 30
E	≤ 40
F	>40

[a]The traffic flow density refers to vehicles of both directions

Capacity analysis process for the two-lane highway according to HBS methodology is shown in Table 4.

Table 4. Capacity analysis process for the two-lane highway according to HBS methodology

	Form 1: The quality of the traffic flow on the analyzed section of two-way rural highway
	The analyzed section between:
	Part of section no:
1.	Road category (RAS-N)
2.	Desired travel speed V_B [km/h]
3.	Measuring traffic volume q_B [Kfz/h]
4.	Percent of heavy traffic b_{SV} [%]
5.	Cross-section (RAS-Q)
6.	Desired grade of quality (Table 5-3) [1] QSVi [-]
7.	The length of section two-way rural highway Li [m]
8.	Longitudinal slope s_i [%]
9.	Minimum average speed of heavy vehicle V [km/h]
10.	Climbing Class (Table 5-1) [1] [-]
11.	Horizontal curvature (Eqs. 5-1) [1] KU [grad/km]
12.	No passing zone [%]
13.	Addition for curvature (Table 5-2) [1] [grad/km]
14.	Total curvature (horizontal + addition) [grad/km]
15.	The achieved travel speed of the passenger car (pic. 5-2–5-6) $V_{R,i}$ [km/h]
16.	Traffic flow density (=$q_{B,i}/V_{R,i}$) (Eq. 5-2) k_i [Kfz/km]
17.	Degree of the quality a part of the analyzed section (Table 5-3 or Figs. 5-2 to 5-6) QSV$_i$ [-]
18.	Average travel speed of a passenger car (Eq. 5-4) V_R [km/h]
19.	Average traffic density (Eq. 5-5) k [Kfz/km]
20.	Degree of traffic flow quality (Table 5-3) QSVGes [-]

1.3   Interactive Highway Safety Design Model (IHSDM)

Interactive Highway Safety Design Model (IHSDM) is software for analyzing the safety and impact assessment of designed geometric elements to road safety. There are several modules implemented in this software: Crash Prediction Module, Design Consistency Module, Intersection Review Module, Policy Review Module, Traffic Analysis Module and Driver/Vehicle Module that treat different areas in order to achieve better analysis of the highways from the aspect of the safety of highways users.

Design Consistency Module (DCM), or the module of consistency or alignment of project elements, is useful for determining possible security problems in horizontal curves. Expectations of drivers when driving on rural two-way highways is that they will be able to maintain a uniform speed, which is not possible in most cases. DCM uses a speed profile model that estimates the speed of 85% of freeway passenger cars at each point along the road (hereinafter V85).

The speed-profile model combines estimated 85th percentile speeds on curves (horizontal, vertical, and horizontal vertical combinations), desired speeds on long tangents, acceleration and deceleration rates exiting and entering curves, and an algorithm for estimating speeds on vertical grades.

The module identifies two potential consistency issues: (1) large differences between the assumed design speed and estimated 85th percentile speed, and (2) large changes in 85th percentile speeds from an approach tangent to a horizontal curve (Fig. 4).

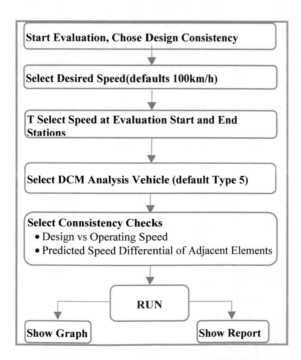

Fig. 4. DCM evaluation procedural chart.

TWOPAS model is used to simulate traffic operations on two-lane highway in the DCM module.

TWOPAS is a microscopic model that simulates traffic operations on two-lane highways, checking the position, speed and acceleration of each individual vehicle on the simulated road at intervals of one second along the highway.

The model takes into account the effects of traffic operations from highway geometry, traffic control, driver preferences, vehicle size and performance in both directions at any time. The model incorporates realistic passing and passing attempts at approved places for passing on two-lane highways. At large longitudinal slopes the speed of passenger cars is reduce. The TWOPAS model contains equations that can present the effects of slope on the speed of passenger cars. The result is the operating speed profile (Fig. 5.) for the selected vehicle type, depending on the effects of the vertical elements [3].

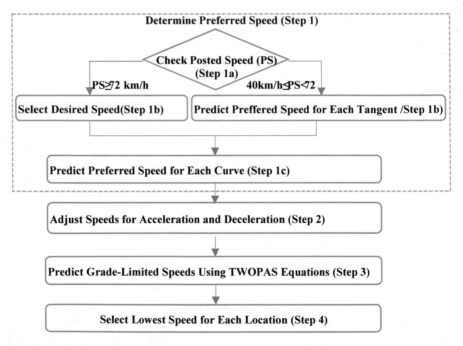

Fig. 5. Speed profile model procedure flowchart.

The procedures shown in the DCM module were used for the analysis of a section of the main road in BiH and will be presented in the text below.

1.4   Performance Analysis of Main Road Section in B&H

As part of the master thesis [4], the analysis of the performance of the section of main two-lane highway M5 (Lašva - Bihać) in B&H was carried out.

In order to perform the analysis according to the above methods, the M5 highways is divided into 35 segments depending on the AADT occurring on the analyzed road, and the length of the longitudinal slope.

Considering the large length of the analyzed road section (L ≈ 243 km), this paper will show the results for only one section M5 -1 to M5-7 (L ≈ 52 km).

1.5   Input Data for Performance Analysis

The input data needed for road performance analysis are collected from existing databases, studies and projects, as well as counting of traffic made for the analyzed road and are shown in the Table 5.

As described in this work, analyzing data in IHSDM, or in its DCM (Design Consistency Module), V85 speed was obtained (Fig. 6). DCM reports are extensive, and accordingly, the graphs show the change of V85 speeds for individual parts (Fig. 7).

The average values of the V85 speeds by certain segments are shown in the following Table 6 (as the mean value for the same segments).

Using HCS (Highway Capacity Software) that operates according to the described HCM methodology, data analysis and the values obtained are shown in Table 7. For input data of BFFS (Determination of the base free-flow speed), the V85 values obtained by analysis in IHSDM.

All other input data are given in Table 5.

The results of the calculation according to the HBS methodology are presented in the following Table 8:

Also, an analysis of the considered section was performed in order to determine the free-flow speed (according the following equation):

$$V_{ffs} = 38.38 - 0.034 KK - 1.461 UN + 12.172 ST \qquad (1)$$

where are:

KK   - curvature,
UN   - longitudinal slope and
ST   - lane width.

This equation was obtained on the basis of research conducted on the main two-way roads in B&H in order to prepare of the PhD dissertation [5] (Table 9).

Data collection was carried out in September 2008 using the mobile observer method (MoM). Driving was done in two directions. This method, among others, includes data on the speed limit, the length of the limitation and the driving time per segment. Based on the collected data, the traffic flow speed is calculated based on the governing traffic conditions (road and traffic conditions).

The following Table 10 shows the measured traffic speeds. [6]

After analyzing and calculating the speed using different methods, a comparison was made.

In order to make comparison across the unique segments, those taken from field research were taken as the basic segments.

Table 5. The size of limit values of parameters of track geometry [5]

Section M5 Station		1-1	1-2	1-3	1-4	1-5	1-6	1-7
		14 + 000.00	22 + 665.00	28 + 258.80	45 + 478.80	46 + 995.00	47 + 712.50	50 + 140.00
		22 + 665.00	28 + 258.80	45 + 478.80	46 + 995.00	47 + 712.50	50 + 140.00	66 + 740.00
1.	Segment length (m)	8665.00	5593.80	17220.00	1516.20	717.50	2427.50	16600.00
2.	Traffic volume (veh/h)	277	125	125	125	125	125	125
3.	Traffic volume per lane (veh/h)	138	62	62	62	62	62	62
4.	Heavy vehicles (%)	10.10	16.60	16.60	16.60	16.60	16.60	16.60
5.	Trucks (%)	3.90	7.10	7.10	7.10	7.10	7.10	7.10
6.	Average long. slope (%)	0.16	4.10	0.51	−4.30	−0.75	−2.95	0.00
7.	Curvature (°/ km)	46.97	180.20	35.42	29.02	66.90	19.77	21.63
8.	No-passing zone (%)	48.12	88.76	38.26	44.93	0.00	54.79	26.63
9.	Access points per km	4.62	1.43	0.70	1.98	1.39	2.06	0.96
10.	Road width (m)	7.00	7.00	7.00	7.00	7.00	7.00	7.00
11.	Lane width (m)	3.50	3.50	3.50	3.50	3.50	3.50	3.50
12.	Shoulders width (m)	1.00	0.50	0.50	1.00	1.00	1.00	0.50
13.	Average speed limits (km/h)	58.39	59.82	78.79	80.00	80.00	69.95	80.00

The speeds obtained by other methods on the 35 mentioned segments were translated into the six segments shown (Table 11).

This is done by the equation:

$$V = \frac{\sum_{i=1}^{n} L_i}{\sum_{i=1}^{n} L_i \cdot V_1} \tag{2}$$

Where are: L_i the segment length (m) and V_i speed in a segment (km/h) (Fig. 8).

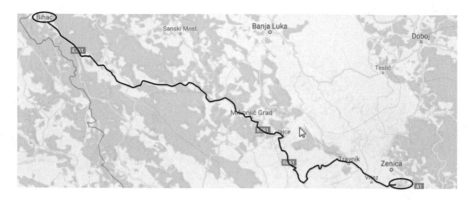

Fig. 6. Analyzed main road section in B&H

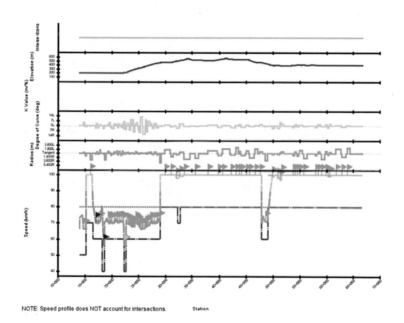

NOTE: Speed profile does NOT account for intersections.   Station

Fig. 7. IHSDM – DCM report for obtained section of road

Table 6. The average values of the speed V85 by certain segments [4]

Section M5 Station	1-1	1-2	1-3	1-4	1-5	1-6	1-7
	14 + 000.00	22 + 665.00	28 + 258.80	45 + 478.80	46 + 995.00	47 + 712.50	50 + 140.00
	22 + 665.00	28 + 258.80	45 + 478.80	46 + 995.00	47 + 712.50	50 + 140.00	66 + 740.00
1. V85 (km/h)	70.51	69.37	95.38	97.50	100.00	82.90	99.59

Table 7. Level of service and average travel speed by certain segments – HCM methodology [4]

Section M5 Station		1-1	1-2	1-3	1-4	1-5	1-6	1-7
		14 + 000.00	22 + 665.00	28 + 258.80	45 + 478.80	46 + 995.00	47 + 712.50	50 + 140.00
		22 + 665.00	28 + 258.80	45 + 478.80	46 + 995.00	47 + 712.50	50 + 140.00	66 + 740.00
1.	BFFS V85 (km/h)	70.50	69.40	95.40	97.50	100.00	82.90	99.60
2.	FFS (km/h)	62.3	61.8	87.2	91.3	94.4	76.7	91.4
3.	ATS (km/h)	53.8	54.4	81.4	83.3	92.4	69.4	88.3
4.	PTSF (km/h)	44.0	55.9	33.4	35.7	11.9	40.0	23.9
5.	v/c	0.11	0.13	0.08	0.05	0.05	0.05	0.05
6.	LOS	E	E	B	B	A	D	B

Table 8. Level of service and average travel speed by certain segments – HBS methodology [4]

Section M5 Station		1-1	1-2	1-3	1-4	1-5	1-6	1-7
		14 + 000.00	22 + 665.00	28 + 258.80	45 + 478.80	46 + 995.00	47 + 712.50	50 + 140.00
		22 + 665.00	28 + 258.80	45 + 478.80	46 + 995.00	47 + 712.50	50 + 140.00	66 + 740.00
1.	Climbing class	1	3	1	4	1	3	1
2.	q_m/C	0.15	0.07	0.07	0.08	0.05	0.07	0.07
3.	V_r (km/h)	65.0	62.0	68.0	62.0	95.0	66.4	68.0
4.	Density (veh/km)	4.26	2.01	1.83	2.01	1.31	1.89	1.83
5.	LOS	A	A	A	A	A	A	A

Table 9. The average values of the FFS speed by certain segments [5]

Section M5 Station	1-1	1-2	1-3	1-4	1-5	1-6	1-7
	14 + 000.00	22 + 665.00	28 + 258.80	45 + 478.80	46 + 995.00	47 + 712.50	50 + 140.00
	22 + 665.00	28 + 258.80	45 + 478.80	46 + 995.00	47 + 712.50	50 + 140.00	66 + 740.00
1. V_{ffs} (km/h)	79.15	68.86	79.03	73.72	77.61	75.99	80.24

Table 10. The average values of the measured traffic speed by certain segments [6]

Section M5 Station	1-1 to 2-9	2-10 to 13	3-1	3-2 to 3-10	3-11	3-12 to 3-15
	14 + 000.00	87 + 330.00	0 + 000.00	33 + 710.00	68 + 675.00	83 + 405.00
	44 + 240.40	96 + 760.00	33 + 710.00	66 + 655.00	83 + 405.00	94 + 160.00
1. V_{MoM} (km/h)	61.00	40.29	48.97	49.76	37.64	37.20

Table 11. The average values of speed by certain segments according to different methodology

Section M5 Station	1-1 to 2-9	2-10 to 13	3-1	3-2 to 3-10	3-11	3-12 to 3-15
	14 + 000.00	87 + 330.00	0 + 000.00	33 + 710.00	68 + 675.00	83 + 405.00
	44 + 240.40	96 + 760.00	33 + 710.00	66 + 655.00	83 + 405.00	94 + 160.00
1. HBS 2001	65.91	59.79	61.00	59.48	52.00	51.04
2. FFS [5]	76.98	68.78	69.31	66.21	72.98	67.50
3. HCM	73.55	52.27	60.90	57.83	38.70	43.09
4. MoM [6]	61.00	40.29	48.97	49.76	37.64	37.20
5. Speed Limit[a]	72.38	58.35	59.68	61.53	57.27	64.49

[a]Average speed limit [6]

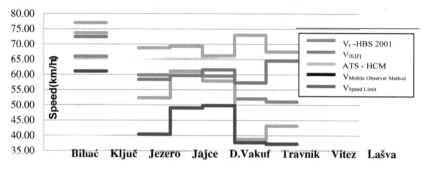

Fig. 8. Graphical representation of average speeds for each segments

2 Conclusion

From the previously shown results, it can be noticed that the average travel speed obtained by the HCM method, has at least deviations from the measured velocity by the mobile observer method, and that the speed change graphs have a similar shape. Also, the results of HBS 2001 method do not many deviate of the two already mentioned methods, although that results are too uniform in segments and do not show too much difference in speeds where road elements are very limiting in terms of travel speed. The V85 speed obtained by the IHSDM (TWOPAS method) are slightly higher than the free-flow speed obtained by the Eq. 1, which was obtained on the basis of studies conducted on roads of similar characteristics in B&H.

Generally, it can be concluded that there are significant differences in speeds and that more detailed research on a larger sample is required in order to determine the most acceptable methodology for existing road conditions in B&H.

References

1. Highway Capacity Manual (HCM 2010): Chapter 15 –Two-lane highways, pp. 15-1–15-64. The Transportation Research Board's (TRB), Washington, DC 20001 (2010)
2. Handbuch für die Bemessung von Straßenverkehrsanlagen (HBS), Forschungsgesellschaft für Straßen- und Verkehrswesen(FGSV), Ausgabe 2015
3. Interactive Highway Safety Design Model (IHSDM)-Version 13.0.0. Design Consistency Module Engineer's Manual. Federal Highway Administration, Office of Safety Research and Development, 6300 Georgetown Pike, McLean, VA 22101, September 2017
4. Redzic, S.: Istraživanje funkcionalnih karakterstika magistralne ceste M5: Lašva - Bihać, Master thesis, Građevinski fakultet Univerziteta u Sarajevu, Sarajevo (2014)
5. Lovric, I.: Modeli brzine prometnog toka izvangradskih dvotračnih cesta, Ph.D. dissertation, Građevinski fakultet Sveučilišta u Mostaru, Mostar (2007)
6. Metodologija za rangiranje prioriteta intervencija na magistralnim cestama FBIH, Sarajevo (2009)

Robotics and Biomedical Engineering

Torsional Vibration of Shafts Connected Through Pair of Gears

Ermin Husak$^{(\boxtimes)}$ and Erzad Haskić

Technical Faculty, University of Bihać, Bihać, Bosnia and Herzegovina
erminhusak@yahoo.com

Abstract. Analyzing the torsional vibration of the shafts is of great importance in the process of developing the mechanical systems. It is particularly important to know the values of natural frequencies and mode shapes of such systems to avoid the effect of resonance. These values are derived from models that describe the motion of shafts during vibration. Torsional shaft vibrations can be modelled as discrete and continuous systems. A discrete system can be modelled as systems with one, two, three or more degrees of freedom. In this paper an analysis of two shafts connected through pair of gears is carried out. The analysis has been performed analytically as a three degree of freedom system and numerically with the use of ANSYS software.

Keywords: Torsional vibration · Shaft · Natural frequency
Degree of freedom

1 Introduction

Shafts are one of the basic structural elements in the machine. They are used in various mechanisms for the movement of rotary motion. Detailed shaft analysis, as well as the systems in which it participates, allows reducing unwanted consequences. In addition to the strength analysis, the calculation of true values of the natural frequencies of the torsional vibration of the shaft and the system allows to preventively affect the possible occurrence of the resonance. As is known, resonance is the case when the frequencies of some excitation force coincide with some of the natural frequencies of the elastic system. Resonance causes large amplitudes that inevitably cause the system to fall. Possibility of manipulating the values of natural frequencies during design of the system and with the most commonly known forced frequency we are able to avoid resonance or at least a longer period of its operation.

Referring to a concrete example of the torsional vibration of the gear unit, an approach to solving such problems is illustrated, namely the introduction of a torsional vibration of the gearbox to a free torsion system, whose solution would not be a problem [1–4]. In addition to analytical solving and calculation of frequencies, ANSYS simulations were performed for the same purpose.

© Springer Nature Switzerland AG 2019
S. Avdaković (Ed.): IAT 2018, LNNS 60, pp. 303–308, 2019.
https://doi.org/10.1007/978-3-030-02577-9_29

2 Torsional Vibration of Shafts in Gear Units

The torsional vibration of shafts in gear units is one of the most common problems of torsional shaft vibration. Reduction can be made with two, three, and more gears, and the problem of torsional vibration of the gear unit is reduced to the problem of torsional system transforming the pair of gears into one rotor.

To find out more about the torsional vibration of the shaft of the gear unit, we will solve the problem of torsional vibration of the gear unit with two geared gears shown in the following Fig. 1.

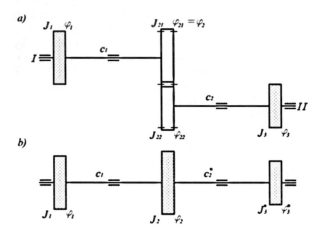

Fig. 1. (a) Gear unit, (b) Free torsional system

The gear unit shown in Fig. 1a consists of two shafts (I and II). On each shaft, a rotor and gear are inserted. The moment of inertia for the longitudinal axis of the rotor on the first shaft is J_1, and the moment of inertia of the rotor on the second shaft is J_3. The moments of inertia of gears for the longitudinal shaft axis are J_{21}- for gear on shaft I, and J_{22}- for gear on shaft II [5].

It will be assumed that the stiffness of the shaft I and II are c_1 and c_2. The torsion angle of the shaft I at the rotor position is φ_1 and in the gear position φ_{21}. Similarly, the torsion angle of the shaft II, at the disc position is φ_3 and at the gear position is φ_{22}.

Gears on the shafts I and II have diameters $D_1 \, i \, D_2$, radii $R_1 \, i \, R_2$, number of teeth $z_1 \, i \, z_2$, and angular velocities $\omega_1 \, i \, \omega_2$. The gear ratio is

$$i = \frac{D_2}{D_1} = \frac{R_2}{R_1} = \frac{z_2}{z_1} = -\frac{\omega_1}{\omega_2} = -\frac{\varphi_1}{\varphi_2} = \frac{1}{k}, \tag{1}$$

where k – reduction coefficient.

The kinetic and potential energy of the gear unit according to Fig. 1 are

$$E_k = \frac{1}{2}\left(J_1\,\dot{\varphi_1}^2 + J_{21}\,\dot{\varphi_{21}}^2 + J_{22}\,\dot{\varphi_{22}}^2 + J_3\,\dot{\varphi_3}^2 \right),$$

$$E_p = \frac{1}{2}\left[c_1(\varphi_{21} - \varphi_1)^2 + c_2(\varphi_3 - \varphi_{22})^2 \right]. \tag{2}$$

Using the Eq. (1) it can be written $\varphi_{22} = -k\varphi_{21}$. Also can be written $\varphi_{21} = \varphi_2$, where $\varphi_{22} = -k\varphi_2$. Returning these equations to kinetic and potential energy following equation are obtained

$$E_k = \frac{1}{2}\left[J_1\,\dot{\varphi_1}^2 + (J_{21} + k^2 J_{22})\,\dot{\varphi_2}^2 + J_3\,\dot{\varphi_3}^2 \right],$$

$$E_p = \frac{1}{2}\left[c_1(\varphi_2 - \varphi_1)^2 + c_2(\varphi_3 + k \cdot \varphi_2)^2 \right]. \tag{3}$$

The torsional vibration of the gear unit shown in Fig. 1a can be reduced to the torsional vibration of the free torsional system with three rotors, as shown in Fig. 1b, while ensuring that the kinetic and potential energy of the reduced part of the system remains unchanged.

Reduction is done in such a way that the moment of inertia of the second rotor J_3, torsion angle φ_3, and stiffness c_2 of shaft II transform to shaft I with new symbols J_3^*, φ_3^* i c_2^*. Taking into account that the kinetic and potential energy of the reduced part of the system remain unchanged due to reduction, it can be written

$$E_k^* = \frac{1}{2}J_3\,\dot{\varphi_3}^2 = \frac{1}{2}J_3^*\left(\dot{\varphi_3}^*\right)^2,$$

$$E_p^* = \frac{1}{2}c_2\varphi_3^2 = \frac{1}{2}c_2^*\left(\varphi_3^*\right)^2. \tag{4}$$

Using the Eq. (1) it can be written

$$\varphi_3^* = -\frac{1}{k}\varphi_3, \tag{5}$$

and in order to satisfy the condition (3) the following equality must be satisfied

$$J_3^* = k^2 J_3,$$

$$c_2^* = k^2 c_2. \tag{6}$$

Moment of inertia for both gears on shaft I is

$$J_2 = J_{21} + k^2 J_{22}. \tag{7}$$

It is now possible to write the terms for the kinetic and potential energy of the gear unit transformed into free torsional system

$$E_k = \frac{1}{2} \left[J_1 \, \dot{\varphi}_1{}^2 + J_2 \, \dot{\varphi}_2{}^2 + J_3^* \left(\dot{\varphi}_3{}^* \right)^2 \right],$$

$$E_p = \frac{1}{2} \left[c_1 (\varphi_2 - \varphi_1)^2 + c_2^* \left(\varphi_3^* - \varphi_2 \right)^2 \right]. \tag{8}$$

The free torsional system shown on Fig. 1b has three degrees of freedom and its motion is described by means of three generalized coordinates $(\varphi_1, \varphi_2, \varphi_3^*)$. Differential equations of motion for free torsional system

$$\begin{aligned} J_1 \, \ddot{\varphi}_1 - c_1 (\varphi_2 - \varphi_1) &= 0, \\ J_2 \, \ddot{\varphi}_2 + c_1 (\varphi_2 - \varphi_1) - c_2^* \left(\varphi_3^* - \varphi_2 \right) &= 0, \\ J_3^* \, \ddot{\varphi}_3{}^* + c_2^* \left(\varphi_3^* - \varphi_2 \right) &= 0. \end{aligned} \tag{9}$$

There are nontrivial solutions if the determinant of the system is zero, i.e. if it is

$$\Delta(\omega^2) = \begin{vmatrix} (c_1 - J_1 \omega^2) & -c_1 & 0 \\ -c_1 & (c_1 + c_2^* - J_2 \omega^2) & -c_2^* \\ 0 & -c_2^* & (c_2^* - J_3^* \omega^2) \end{vmatrix} = 0, \tag{10}$$

or

$$\Delta(\omega^2) = \omega^4 - \left(\frac{J_1 + J_2}{J_1 J_2} c_1 + \frac{J_2 + J_3^*}{J_2 J_3^*} c_2^* \right) \omega^2 + \frac{J_1 + J_2 + J_3^*}{J_1 J_2 J_3^*} c_1 c_2^* = 0. \tag{11}$$

The solution of the frequent equation shows that it is $\omega_1 = 0$ which tells us that in this case the whole system is rotating, while ω_2 and ω_3 shows vibration of the system.

3   Analysis of Vibrations in ANSYS

If we choose a system whose values are $J_1 = 9000$ kgm^2, $J_2 = 1310$ kgm^2, $J_3^* = 8000$ kgm^2, $c_1 = 981747$ Nm/rad and $c_2 = 15904312$ Nm/rad and put them in Eq. (11) we get the following natural frequencies solutions $\omega_1 = 0$, $\omega_2 = 14.32$ rad/s and $\omega_3 = 121.57$ rad/s. When we generate the geometry and define the material in ANSYS so that moments of inertia of rotors and stiffness of shafts correspond, we can obtain the results of natural frequencies and mode shapes. It is necessary to set the appropriate boundary conditions and discredit the system to the finite elements.

The first mode shape is pure rotation of the system because the system is free (Fig. 2).

In Fig. 3 is shown second mode shape at natural frequency of 2.77 Hz which corresponds to 17.40 rad/s.

Fig. 2. First mode shape

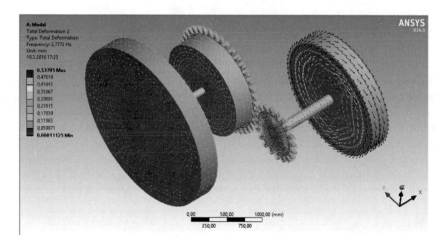

Fig. 3. Second mode shape

Figure 4 shows the third mode shape at the natural frequency 20.34 Hz which corresponds 127.73 rad/s. As can be seen from Figs. 3 and 4, the amplitude of element 1 is almost negligible. The reason for this is a great moment of inertia J_1 in relation to moments of inertia of other elements so that it can be fixed and system analyzed as a system with two degrees of freedom [6].

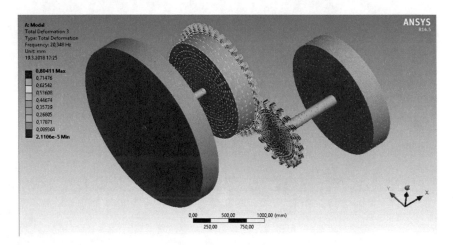

Fig. 4. Third mode shape

4   Conclusion

By conducting the analysis of torsional vibrations in this paper, we have pointed out the methods of solving specific problems with torsional vibrations. Since the rotational motion contained in the majority of drives, vibration torsion occurs in the rotors, axis, shafts, gears units, tools, jacks, and other structure elements, so studying torsional vibrations is of great significance. Today's great advantage is a diverse approach to software that supports static and dynamic analysis of simulated models. ANSYS belongs to a group of such software, and access to the analysis carried out within it is similar to other programs that support simulation of the process.

References

1. Tongue, B.H.: Principles of Vibration, 2nd edn. Oxford University Press, Oxford (2002)
2. Karabegović, I., Husak, E., Pašić S.: Vibration analysis of tool holder during turning process. In: 14th International Conference Mechanika, Kaunas, 2–3 April 2009, pp. 200–204 (2009)
3. Özkal, F.M., Cakir, F., Arkun, A.: Finite element method for optimum design selection of carport structures under multiple load cases. Adv. Prod. Eng. Manag. **11**(4), 287–298 (2016)
4. Karabegović, I., Novkinić, B., Husak, E.: Experimental identification of tool holder acceleration in the process of longitudinal turning. J. FME Trans. **43**(2), 131–137 (2015). Faculty of Mechanical Engineering Beograd
5. Radosavljevic, G.B.: Theory of Oscillation. Mašinski fakultet Beograd (1972)
6. Husak, E., Kovačević, A., Rane, S.: Numerical analysis of screw compressor rotor and casing deformations. In: Advanced Technologies, Systems, and Application II. Lecture Notes in Networks and Systems, pp. 933–940. Springer, Heidelberg (2018)

Conceptual Approaches to Seamless Integration of Enterprise Information Systems

Vladimir Barabanov, Semen Podvalny(✉), Anatoliy Povalyaev,
Vitaliy Safronov, and Alexander Achkasov

Voronezh State Technical University, Moskovsky pr., 14, 394026 Voronezh,
Russian Federation
spodvalny2@mail.ru

Abstract. In this article the methods of specialized software systems integration are analyzed and the concept of seamless integration of production decisions is offered. In view of this concept developed structural and functional schemes of the specialized software are shown. The proposed schemes and models are improved for a typical machine-building enterprise.

Keywords: Seamless integration · Enterprise systems integration
Translation

1 Introduction

Quite often it happens when a lot of different software products are used by an industrial plant. This may be caused by integration of companies using different software, or simply by historical factors of development. There are several reasons for the simultaneous application of a variety of specialized software systems:

- The high complexity of today's products;
- Manufacturers are transformed into transnational corporations, and for the organization of their operation data replication is required;
- Assimilation of existing software infrastructures to maintain data integrity in mergers and take-over.

A company often first purchases the required software, and then the problem of how to integrate it with its existing information systems is solved. In this regard, analysis and integration of data from different software systems, the creation of joint documents is not only difficult, but also costly for the company. It is important that each organizational unit operates its information and processes it in its own manner. That is why in the course of corporate systems implementation special corporate standards are introduced for data exchange formats. Often companies when performing certain production tasks use software solutions from different vendors. Their integration is ensured by means of data conversion from one format to another, which often causes errors and degrades the information quality.

To prevent this, it is necessary to introduce single vendor software solutions, which will save on software integration and updating. However few providers offer a full

© Springer Nature Switzerland AG 2019
S. Avdaković (Ed.): IAT 2018, LNNS 60, pp. 309–322, 2019.
https://doi.org/10.1007/978-3-030-02577-9_30

range of management and project funds and companies are not always ready to simultaneously change the manufacturing process. Among the shortcomings of the existing information systems developed by most design companies are:

(1) Narrow specialization of design and calculation causing incorrect problem statement and insufficient complete analysis of the results. At the same time there is no possibility for designers to carry out their own preliminary calculations of developed product, which result in complexity and timing increase in the project work as a whole.

A large number of paper documents which has the following disadvantages: slow retrieval of documents; difficulty in document tracking at all stages of its life cycle; the duration of the timing and coordination of documents; the likelihood of errors in the processing and transmission of information collection increases; increase in the information processing terms.

(2) A set of specialized and poorly integrated techniques implemented in the Microsoft Excel environment is usually used in calculations.

(3) Also, one of the drawbacks of currently used methods is the considerable length of the calculations caused by the need to perform time-consuming procedures of phased iterative calculation and parameters associating.

The outdated calculation methods usage and their manual implementation increases the complexity of the primary engineering analysis [1].

2   Justification of the Need to Improve Engineering Enterprise's Information System

Information systems of most engineering enterprises (for example, a company producing high-tech main oil pumps) as a rule is not optimal and does not provide the possibility of carrying out all the necessary calculations for the company's specialists. It requires the involvement of third parties. Existing hardware and software are often outdated and are able to perform only standard computing tasks on existing engineering methods or special techniques developed by the company's specialists. The computerization of traditional methods to solve engineering design problems at the machine-building enterprises is not systematized. This is mainly carried out by means of in-house design applications providing only particular design problems solutions.

Currently a full design study of the object characteristics with the required parameters of accuracy and turnaround time can be achieved through the creation of automated software and hardware systems that are based on a computer analysis using physical and mathematical models that describe the hydrodynamic, thermal, and other processes occurring in created product. The method of defect elimination and improving the existing engineering techniques is the development of comprehensive automated software and hardware modules to be integrated into a single computational environment for managing and sharing data between software applications within a single company's information environment.

3   Methods of Corporate Software Support Systems for Product Life Cycle Maintenance Integrating

Integration provides a solution of a data mismatch problems in two or more systems used in the design organization and the construction of the organization's IT infrastructure. Technical problems to be solved in the course of work on the systems integration include:

- The semantic data reconciliation - bringing data in different systems for "total solution".
- Construction of single classifiers and directories - building a one-one correspondence between the elements of directories in different systems and fixing in the additional structures which exercise the "translation" functions.
- Creation of software interfaces of integrable systems for data transfer and call system functions on external events.
- Development of converters for data transmission from one system to another and output data formats for transmission including the realtime.
- Logical systems binding - building algorithm which enables to display one system "events" to other systems.
- Designing mechanisms for remote synchronization (replication) of data and their distributed development.
- Designing interfaces that enable one to control the data flow, logic transformation and structures, define a single access rights and mechanisms for working together with data, etc.
- Development of additional means of access, analysis, and collaboration in data processing.

The information system is usually a combination of several components. Thus, the integration of information systems should be considered as the integration of their components.

The information system comprises the following components:

- The platform on which the other system components, including the hardware and system software, operate.
- The data that the system works with consist of DBMS and database.
- The applications that implement the business logic of the data system. They consist of the components of the business logic, user interface, auxiliary components and the application server which provides storage and access to application components.
- The business processes which are a scenario of the user interaction with the system.

It is believed that the integration of information systems is the integration of one or more components of integrable information systems.

The objectives of the platforms integration are:

- Ensuring interoperability between applications running on different hardware and software platforms.
- Enabling applications developed for a specified software and hardware platform operate on platforms.

There are several ways to achieve these goals. Within each approach, there are different technologies:

- Remote Procedure Call (RPC),
- Middleware,
- Virtualization.

RPC (Remote Procedure Call) technology enables to publish the procedure and to call it for applications running on other platforms. The elements of such technologies are common to all platforms description interface procedures language (IDL, WSDL), procedure's "adapter" which translates external calls into internal ones and transmits the results back, and managers responsible for the delivery of enqueries and results between platforms in the network.

The middleware ideology is to develop application software without using particular operating system services by means of middleware services. The middleware developers create it for implementation in different operating systems, which reorganize respective functions of framework calls to the corresponding operating system calls.

"Virtualization" is the newest concept of platforms integration, as it greatly simplifies the use of different platforms and, accordingly, the use of systems demanding the presence of specific platforms for their functioning.

Information system works with data and is composed of a database for storage. Integration at the data level suggests the data sharing from different systems. Often it turns out to be easier than the applications integration, as industrial databases which store data information systems have advanced programmatic access capabilities to stored data from other applications.

Approaches to Data Integration:

- Universal access to data,
- The data warehouse (DW).

The universal data access technology provides uniform access to data of different DBMS through a dedicated driver [2]. The concept of data warehousing is to create a corporate data warehouse. Data warehouse is a database that stores the data collected from various information systems databases for further analysis. OLAP technology is used to create a data warehouse, other than the operational technologies DB-OLTP creation. Approaches to the creation and filling of data warehouses are reflected in the ETL paradigm (extraction, transformation, loading) [3].

The application level integration is put into practise through the use of ready-made application functions by other applications.

Existing approaches to application integration are:

- Applied Programming Interfaces (API),
- Messaging,
- Service-Oriented Architecture (SOA),
- User interfaces Integration.

Application Programming Interface of a particular system is a "declared" system's functionality which can be used outside. Functional is published as a set of functions or an object model. Service-oriented architecture (SOA) is a modern and popular paradigm. It is a logical continuation of the Web-services concept which is to publish the functional blocks of an application form permitting other applications to get access to them by the Web. Web-service is a small add-on software application functionality that converts calls obtainable via the Web into the internal application function calls and returns the results back. The main ideas of SOA are:

- Publication of functional enterprise applications as Web-services. Ordering published services as a catalog.
- Construction new applications on the basis of Web-based services through their combination.

Integration at the level of enterprise applications (EAI, Enterprise Application Integration) means the sharing executable code, not the internal application data (Fig. 1). Programs are divided into components that are integrated via standardized programming interfaces and special communication software. This approach of these components means to create a universal software kernel, which is used by all applications.

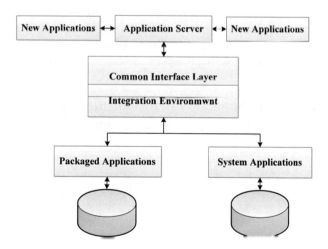

Fig. 1. Integration at the level of enterprise applications

Integration at the level of user interface enables applications interconnection by means of special user interface tools. The most comprehensive systems integration is the integration at the level of business processes. It provides application integration and data integration. Business Integration is "natural" for companies, since their work is based on business processes rather than applications, databases and platforms.

Corporate systems are complex software solutions. So it is impossible to use a single method for their integration. In order to provide specialized software integration, we suggest an approach structurally based on the individual functional features of the following methods of integration:

- Platforms integration (remote procedure call, the use of middleware);
- Data integration (the formation of a unified database as part of enterprise information systems);
- Integration of applications (using the API, work with a service-oriented architecture);
- Integration at the level of enterprise applications;
- Integration at the user interface level (creating cross-platform and inter-system interfaces, interaction of software systems);
- Integration of business processes.

It is significant that the joint use of the various methods individual elements is not contrary to the basic requirements for the enterprise software interaction and operation organization. This solution allows to create a single solutions database for support, maintenance, and planning of the product life cycle, and for ensuring full interoperability between them. Some aspects of integration will ensure the software interaction without user's direct participation which is not typical for any of the original methods. Thus, the use of the composite system integration will allow seamless integration with the formation of a single, integrated database [4].

Seamless integration is ensuring the interaction of two or more software systems with a "simplification" of the user influence on the data exchange between systems, due to the formation of a structured shared database; "embedding" translationing devices, conversion and transmission of data into the original software solution while maintaining its integrity and stability; creation and use of inter-module interfaces.

4 Structural and Functional Schemes of the Seamless Enterprise Software Systems Integration

Throughout the product lifecycle the same information is treated by different life cycle support software systems, but each system operates formed and recorded in an electronic database data and unique information generated by this process.

Paperless technology strategy is to create a single information space for all participants throughout the product lifecycle with the creation of the EPD (Electronic Product Definition). Accordingly, the use of a software system in a single information space should be methodologically compatible with paperless technology. In Fig. 2 there is a block diagram of an integrated information system based on paperless technology.

Software system in a single information space (SIS) provides:

- Joint development of an interactive environment;
- Structured electronic product description;
- Data protection and access to information about the product;
- Change management in an integrated database.

Without the formation of the SIS it is not possible to provide a functional, technological, information and logical compatibility and harmonization of the automated system design and technological complex with other software systems it interacts with.

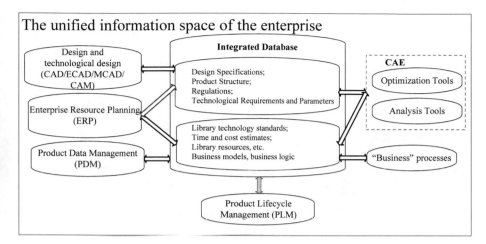

Fig. 3. Structure of the interaction of an integrated information system based on the paperless ideology

This means that the common principles and general rules for the formation and maintenance of information resources and information and telecommunication systems installed in the ERC should be respected by all actors of information relations, carrying out their activities and exchange of information within the framework of the company's SIS.

The main elements of the SIS are:

- Information Resources recorded on data carriers;
- Organisational structures for the operation and development of the SIS;
- Means of information interaction of subjects of information relations in the SIS to ensure regulated access to information resources on the basis of appropriate information technologies, including software, hardware, and legal documents.

The software as a part of corporate solutions is a set of highly specialized software products. It leads to a number of problems, such as their interaction and integration of software from different manufacturers. To resolve those problems, in accordance with the concept of seamless integration based on the structure of the interaction of an integrated information system in accordance with the paperless ideology we formed the scheme of functional interaction with support systems, planning and support of the product life cycle shown in Fig. 3. Interaction with solutions support planning and tracking the life cycle involves the use of structural typed datasets, the introduction of the unification of data on the interaction of elements for all kinds of descriptions of Digital Prototyping allows us to develop cross-cutting process which provides a seamless integration with external software.

The Feature of seamless integration software solution is the organization of inter-action with making support, maintenance and life cycle planning with a single information systems integrated database by means of program interfaces inter-module integration (Fig. 4).

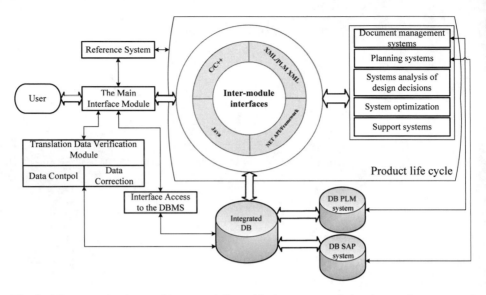

Fig. 3. The general scheme of interoperability with the systems of planning, maintenance and life cycle support

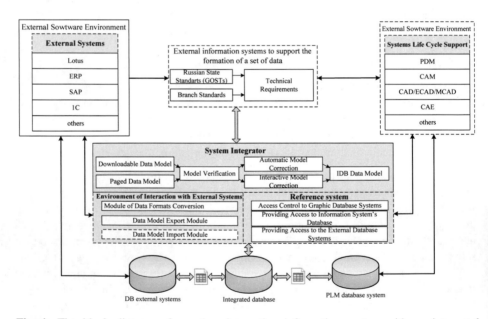

Fig. 4. The block diagram of seamless integration information system with an integrated database

On the basis of structural and functional scheme of software solutions a model of inter-module interfaces integration based on project data management with external specialized software was synthesized, Fig. 5. The advantage of this is the formation of a seamless integration and integration of used software solutions in a single information space with an integrated data management system project.

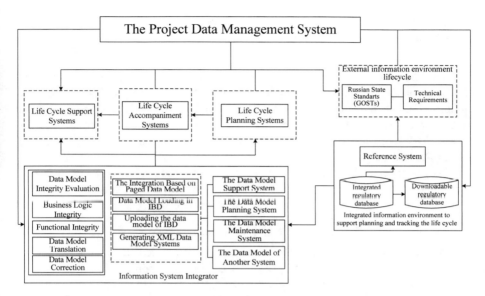

Fig. 5. The model of intermodule interfaces integration based on the project data system management

The interaction with the database enables to solve the problem of modules consistency and create a common information data space. The structure of the information system includes an external plug-in program graphics system, external information life cycle supporting environment, built-in reference system, external information design support environment, built-in projects integration environment, built-in information recovery data model that includes modules for interfacing with PLM, PDM, and ERP systems, modules of recovery data models, business model verification and correction modules. The structure and composition of the proposed software is not limited to the following list of modules, and it may be changed depending on a task.

Thus, the proposed model of inter-module interfaces integration of enterprise software, modular structure of the information system and integrated graphic elements database provide a seamless integration of software systems and a complete solution to a wide range of tasks, taking into account the specificity of the subject area.

5 Decisions on System and Subsystems Structure

As an example, the realization of the automated system of a design and technological complex (AS of DTC) consisting of a set of the firmware modules (FM) is considered: "Construction"; "Durability"; "Hydraulics"; "Warmth"; "Archive"; "Mechanics"; "Moulding"; "Assembly"; "3D Measurement"; "EIS", "Optimization", etc. The AS of

DTC has the flexible organization and possibility of adaptation under the changing external factors (such as changes in the organization of business processes, the current legislation, etc.) and provides:

- capability by number of users and the processed information;
- archival storage of information – according to the legislation of the Russian Federation and the existing nomenclature of the engineering company affairs.

Each FM provides carrying out all necessary types of calculations and has a possibility of adaptation under the changing external factors. The offered subsystems allow capturing fully activity of all projecting divisions and technological preparation of production divisions, executing integration of firmware modules into common information space, adjusting electronic document circulation and creating archive of electronic documentation.

The modules, making AS of DTC part are also intended for ensuring seamless input and output data integration. The data exchange module of electronic manufacturing techniques of products (EMTP) and standard reference information are intended for data exchange with external systems.

In PDM system, seamless integration of the input and output data of design and technological documentation preparation modules is provided by a set of the means allowing to make the automated formation of EMTP descriptions on the basis of data from the electronic structure of the product (ESP). The structure of intermodular interaction taking into account a choice as the interface of integration of PDM system is shown in Fig. 6.

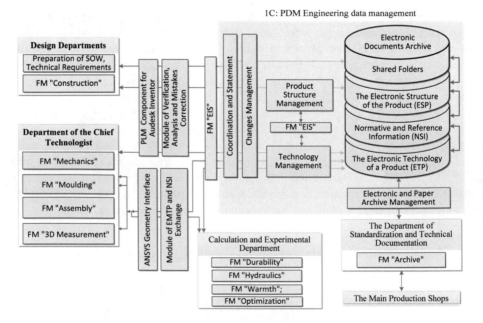

Fig. 6. The inter-module FM interaction

All means of information exchange can be divided into two parts:

- means of the internal information flows organization;
- means of interaction with external information systems.

The AS of DTC provides the centralized storage, processing and control of data received from various design systems in the central database realized by means of PDM. For the solution of this task the following functional subsystems are allocated:

- design data formation subsystem;
- subsystem of the automated control and design data updating;
- subsystem of graphic design data translation;
- external information environment of design support;
- life cycle support maintenance subsystem;
- business management subsystem;
- maintenance and support subsystem.

The version of the system's block diagram is shown in Fig. 7. The Basic purpose of FM "EIS" in the AS of DTC structure is creation of a single information space taking into account maintaining electronic structure of the product, the organization of electronic document circulation of all types of the design documents prepared in AS of DTC subsystems and the solution of several tasks on the organization of the top level developed design system management functions.

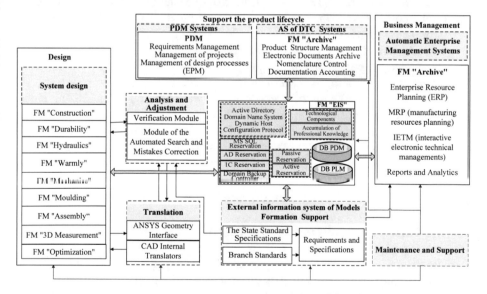

Fig. 7. The block diagram of AS of DTC

One of the functions of FM "EIS" in AS of DTC structure is giving program and information means for design-engineering preparation of the production [5]. Today PDM-systems are widely used as an integration tool of the automated design and technological preparation of production systems. Also they are used for the electronic document flow organization, creation of common information space at the enterprises engaged in design and production of difficult technical products; the organizational and technical systems providing management of all product information. For integration of firmware modules in the structure of AS of DTC, also for the solution of problems of management of engineering data, management of information on a product, management of specifications, maintaining of electronic archive of documentation it is possible to use system "1C:PDM. Engineering Data Management".

6 Means and Ways of Interaction for Information Exchange Between System Components

All means of information exchange can be divided into two parts:

- means of the internal data streams organization [6];
- means of interaction with external information system.

Taking into account the requirements and procedural restrictions, the decision is made:

(1) AS of DTC should be designed with formation of a single information space on the basis of local area network (LAN) with newly formed FM usage.
(2) The expedient type of LAN organization is the multiuser client - server architecture constructed on "multicascade star" topology in the form of an independent information cluster [7].
(3) The structure of a cluster has to include client FM, servers, calculators, switching equipment, peripheral equipment.
(4) It is expedient to arrange cluster compactly.
(5) It is expedient to license the specialized software used on client AM on client-server architecture (for the most effective use of software).
(6) Construction and modernization of the business scheme of projects passing stages (design, modeling, statement, return on completion, creation of modifications, release of the product, delivery in archive, transfer of technical documentation to the consumer) is carried out in the process of input FM in operation.

7 The Basic Methodological Provisions

A single information space is the integrated set of the spatially distributed bases and databanks of the systems, technologies of their maintaining and use, information and telecommunication systems and networks functioning on the basis of the uniform principles and common rules, providing information exchange of all participants. AS of DTC has hierarchical and multilevel structure of information resources:

The 1st level is the database of primary, most detailed, reliable and actual information about objects stored and used at the shop level in system of integrated management [8];

The 2nd level includes the databases of the aggregated indicators characterizing condition of the subjects, objects and processes controlled by governing bodies and the enterprise. The corporate network of data transmission for providing information exchange of AS of DTC with remote sources of information needs to be developed according to the following principles:

- possibility of further integration and use of the existing infrastructure of communication and telecommunication;
- priority investment into such infrastructure elements which will allow to achieve the solution of problems with optimum expenses and in the shortest possible time, also will provide possibility of long operation and networks modernization without their essential reconstruction;
- choice the specific telecommunications operator in accordance with the established procedure for carrying out management and administrative functions on the organization of users work in the corporate network;
- specific technological decisions and organizational forms of cooperation with the telecommunications operator striving for providing optimum operational costs and expenses.

Functioning of the software and hardware complex of AS of DTC and its interfacing to territorially remote local area networks is carried out within a uniform corporate area network of machine-building enterprise. Lately Enterprise Information Systems have been shifting toward cloud based environment [9] which enables complex systems to be supported with greater computing resources and achieve higher security [10] and access control [11], ease of information retrieval [12, 13].

8 Conclusion

The offered block and function schemes of the specialized software are developed on the basis of the concept of seamless integration with support, planning and life cycle maintenance systems. They are oriented to preservation of functional client-server model integrity. The software is realized in the form of a series of specialized program intermodular interfaces. The offered means are aimed at providing interaction of planning, maintenance, support systems and problem-solving in the sphere of production life cycle management.

Acknowledgments. The project was executed under the contract number 1450/300-13 dated February 24 between JSC "Turbonasos" and "Voronezh State Technical University" as a part of the "Development of the new generation main oil pumps production using the methods of multicriteria optimization and unique experimental base" project (The Russian Federation Government Resolution no. 218 from 9.04.2010).

References

1. Kenin, S.L., Barabanov, V.F., Nuzhny, A.M., Grebennikova, N.I.: Problems of graphic data translation in CAD-systems. Bull. Voronezh State Tech. Univ. **9**(3), 4–8 (2013)
2. Nuzhny, A.M., Safronov, V.V., Barabanov, A.V., Gaganov, A.V.: Creating an electronic archive by means of PDM-systems. Bull. Voronezh State Tech. Univ. **9**(6), 23–27 (2013)
3. Nuzhny, A.M., Grebennikova, N.I., Barabanov, A.V., Povalyaev, A.V.: Analysis of the factors of data management system selection. Bull. Voronezh State Tech. Univ. **9**(6), 25–31 (2013)
4. Safronov, V.V., Barabanov, V.F., Kenin, S.L., Pitolin, V.M.: Conceptual approach to seamless integration of management systems. Control Syst. Inf. Technol. **3**(53), 95–99 (2013)
5. Barabanov, V.F., Nuzhny, A.M., Grebennikov, N.I., Kovalenko, S.A.: Development of a universal technological data exchange module for 1C: PDM. Bull. Voronezh State Tech. Univ. **11**(6), 54–56 (2015)
6. Podvalny, S., Kravets, O., Barabanov, V.: Search engine features in gradient optimization of complex objects using adjoint systems. Autom. Remote Control. **75**(12), 2225–2230 (2014)
7. Podvalny, S.L., Vasiljev, E.V., Barabanov, V.F.: Models of multi-alternative control and decision-making in complex systems. Autom. Remote Control **75**(10), 1886–1891 (2014)
8. Ivaschenko, A.V., Barabanov, V.F., Podvalny, E.S.: Conditional management technology for multiagent interaction. Autom. Remote Control **76**(6), 1081–1087 (2015)
9. Şener, U., Gökalp, E., Eren, P.E.: Cloud-based enterprise information systems: determinants of adoption in the context of organizations. Commun. Comput. Inf. Sci. **639**, 53–66 (2016)
10. Chaudhry, P.E., Chaudhry, S.S., Reese, R., Jones, D.S.: Enterprise information systems security: a conceptual framework. Lecture Notes in Business Information Processing, LNBIP, vol. 105, pp. 118–128 (2012)
11. Dašić, P., Dašić, J., Crvenković, B.: Applications of access control as a service for software security. Int. J. Indus. Eng. Manag. (IJIEM) **7**(3), 111–116 (2016)
12. Dašić, P., Dašić, J., Crvenković, B.: Service models for cloud computing: search as a service (SaaS). Int. J. Eng. Technol. (IJET) **8**(5), 2366–2373 (2016)
13. Dašić, P., Dašić, J., Crvenković, B.: Applications of the search as a service (SaaS). Bull. Transilv. Univ. Bras. Ser. I: Eng. Sci. **9**(2), 91–98 (2016)

Microforming Processes

Edina Karabegović[1(✉)], Mehmed Mahmić[1], and Edin Šemić[2]

[1] Technical Faculty Bihać, University of Bihać, Dr. Irfana Ljubijankića bb,
77 000 Bihać, Bosnia and Herzegovina
edina-karabeg@hotmail.com, mmahmic@gmail.com
[2] Faculty of Mechanical Engineering, University "Džemal Bijedić" Mostar,
Maršala Tita bb, 88104 Mostar, Bosnia and Herzegovina
edin.semic@bih.net.ba

Abstract. Modern trends in process development are mainly based on the needs of the market for production that is characterized by the technical and economic advantages as compared to conventional production. Although the development of microsystems (MST) and microelectromechanical systems (MEMS) exists for more than 30 years, due to increased application in communication, automated production, transport, health care, and defense systems, there has been advancement of research related to the development and expansion of techniques for production of small size parts (from micro to nano sizes) in the recent years. The paper lists some of the techniques of microforming process with basic characteristics.

Keywords: Microforming · Volume microforming · Sheet microforming
Surface microforming

1 Introduction

Special requirements in the branches of the military, automotive and aerospace industries, telecommunications, medicine, and small-sized products (<1 mm) have led to the development of processes and systems for the production of required products.

The principles of microforming processes are based on the existing conventional metal and alloy processing (cutting processes: milling, turning, grinding, polishing, EDM, ECM, plastic forming processes such as compression, bending, deep drawing, forging, extrusion, hydroforming, incremental shaping, superplastic shaping, then laser welding processes, etc.). In addition to the cold microforming process, the research is focused on the processes of hot microforming.

Microforming is conducted for sheet forming and volumetric forming. The applied microforming processes meet requirements for high precision, short production time, high productivity, low cost, and other [1].

2 Microforming

Microforming presents forming of a part whose at least two dimensions have a size smaller than 1 mm.

© Springer Nature Switzerland AG 2019
S. Avdaković (Ed.): IAT 2018, LNNS 60, pp. 323–327, 2019.
https://doi.org/10.1007/978-3-030-02577-9_31

Figure 1 gives examples of parts obtained by the microforming process.

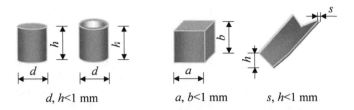

d, h<1 mm   a, b<1 mm   s, h<1 mm

Fig. 1. Parts obtained by microforming process [2]

The design of the microforming process requires a different approach compared to conventional plastic forming technologies, due to the so-called "size effect" that reflects the strength of the material, the lubrication and strain of the material. The reduction in size, from macro to micro size, does not change the material structure (grain size) and surface topography (roughness). The success of the process depends on the technique, forming conditions and material requirements. The materials that are formed in the microforming processes are metals, alloys and non-metals (plastics). The tools for performing the microforming process are more demanding than the ones for conventional forming. Particular attention is paid to the choice of materials and technology for its production in order to achieve high accuracy. High precision in design of tools is achieved by applying some of the advanced technologies or new energy (EDM-*Electric Discharge Machining*, laser). Tools for hot microforming are made of ceramic. The processing is performed on machining systems with automatic control. Microforming processes are divided into volumetric forming and sheet forming [3].

2.1   Volumetric Microforming

The process of volumetric microforming includes extrusion, forging, surface forging and other.

2.1.1   Microforming by Extrusion

The microforming process by extrusion is a simple process whereby the specimen is obtained from the wire cut into small pieces. Shaped parts have small dimensions up to several tens of micrometers, which makes them difficult to handle.

Figure 2 presents the tool for extrusion of metals in experimental conditions [4] and samples of different grain sizes in the material structure [5].

The grain size affects the properties of the material during forming, and the topography of the surface influences the tribological properties of the material. This certainly has an influence on the plastic yield stress and strain of the material.

For example, in conventional plastic forming, plastic yield stress of metal decreases with the increase of the grain in the material structure, whereas in microforming its increases. In microforming conditions, metal strain is lower, due to the stronger friction, resulting in difficult material yield stress [5].

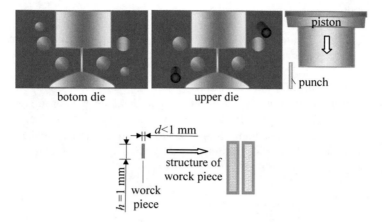

Fig. 2. Extrusion tool in experimental conditions and samples of different grains sizes

2.1.2   Shallow Engraving Forging

The shallow engraving forging is a simple process in which the workpiece is shaped by strain, and the shape is determined by engraving of the forging die. They are most commonly used for making coins, medals, jewelry, inscriptions, etc. Material behavior in the shallow engraving forging process of corresponds to the material behavior under micro-forging conditions. Figure 3 shows the comparison patterns of the products obtained by shallow engraving forging and shallow engraving micro-forging.

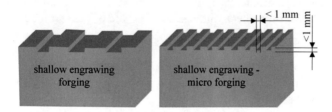

Fig. 3. The scheme of products obtained by forging

Investigations [6] were related to the influence of the forming force and crystalline grain size on total deformation and elastic springback during open forging process and coining (with limited material flow). The results of open forging process have shown that, with the same value of deformation force, the total deformation of the workpiece increases for the material of larger crystalline grain size, whereas the size of elastic springback at the end of deformation is smaller. This size ratio is changing during the process of coining, where the same deformation force in the crystalline grain growth decreases total deformation, whereas the elastic springback of the workpiece material increases.

2.2   Sheet Microforming

Several techniques for sheet forming can be used, including free sheet bending, laser bending, sheet deep drawing, incremental microforming, shearing and others.

2.2.1   Microforming by Sheet Bending

The bending of the sheet under microforming conditions is an area of interest in many studies. Research and analyzes relate to the influence of sample size, grain size, direction of sheet rolling, bending angle, bending radius, material behavior in the forming conditions, reduction of bending force, springback characteristic, etc.

Figure 4 gives an example of bending process.

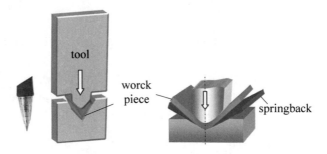

Fig. 4. Microforming by sheet bending [7]

The research [7] states that the thickness of the sheet, the bending force applied, the holding time in the tool, and etc., influenced the value of the springback of the shaped part.

2.2.2   Microforming by Deep Drawing of Sheet

The basic characteristics of the micro-deep drawing process are determined by micro-size and material behavior, as shown in Fig. 5. One of the essential parameters for material behavior in micro-conditions is friction, which requires good knowledge of lubricant influence on the process flow. The principle of classical lubrication in macro-conditions is not so acceptable in microforming conditions. Because of its small dimensions, the workpiece is difficult to clean after forming, and it is recommended to apply the appropriate tool coatings, instead of using mineral oils. The deep drawing process of micro-forming is performed more in dry condition.

The research in the paper [9] relate to the analysis of possible ways of friction protection. The results have shown that tools with coating have more advantages, as well as that friction is lower with application of diamond-like-carbon (DLC) coating tool compared to TiN coating.

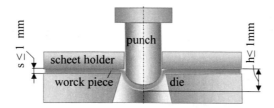

Fig. 5. The scheme of sheet microforming process by deep drawing [8]

3 Conclusion

The development of the material forming process is constantly increasing. The production of parts with micro dimension has certain requirements compared to the conventional forming of the macro dimensions. The main objective of the research mentioned in the paper was to achieve the best processing conditions, which are associated with greater accuracy and quality, lower stress, improved stress and more. Also, analysis and research results in micro forming conditions can serve in future researches that are being carried out today for nano-scale processes.

References

1. Karabegović, E., Brezočnik, M., Mahmić, M.: Nove tehnologije u proizvodnim procesima (Razvoj i primjena), Mašinski fakultet Mostar, Univerzitet Mostar, Bosna i Hercegovina, str. 70–72 (2014). ISBN 978-9958-058-02-8, COBISS.BH-ID 21640454
2. http://docplayer.net/docs-images/40/1595105/images/22-0.png
3. Jeswiet, J., Geiger, M., Engel, U., Kleiner, M., Schikorra, M., Duflou, J., Neugebauer, R., Bariani, P., Bruschi, S.: Metal forming progress since 2000. CIRP J. Manufact. Sci. Technol. **1**, 2–17 (2008). https://doi.org/10.1016/j.cirpj.2008.06.005
4. Piwnik, J., Mogielnicki, K., Gabrylewski, M., Baranowski, P.: The experimental tool for micro-extrusion of metals. Arch. Foundry Eng. (AFE) **11**(2), 195–198 (2011). ISSN (1897-3310)
5. Mogielnicki, K.: Numerical simulation in microforming for very small metal elements. In: Modelling and Simulation in Engineering Sciences. INTECH. http://dx.doi.org/10.5772/64275
6. Keran, Z.: Plitko gravurno kovanje s aspekta mikrooblikovanja, Doktorski rad, Sveučilište u Zagrebu, Fakultet strojarstva i brodogradnje, Zagreb, str. 21–26 (2010). http://repozitorij.fsb.hr/1098/1/27_10_2010_FORMA_KONACNA.pdf
7. Wan-Nawang, W.A., Qin, Y., Liu, X.: An experimental study on the springback in bending of w-shaped micro sheet-metal parts. In: MATEC Web of Conferences. https://doi.org/10.1051/matecconf/20152109015
8. Amin, T., Milad, A., Sorooshian, S., Lori, E.S.: A review on micro formings. Mod. Appl. Sci. **9**(9), 230–239 (2015). https://doi.org/10.5539/mas.v9n9p230. Published by Canadian Center of Science and Education
9. Hu, Z., Wielage, H., Vollertsen, F.: Economic micro forming using DLC- and TiN-coated tools. J. Technol. Plast. **36**(2), 51–59 (2011). https://doi.org/10.2478/v10211-011-0006-z

Matlab Simulation of Robust Control for Active Above-Knee Prosthetic Device

Zlata Jelačić[1]([✉]), Remzo Dedić[2], Safet Isić[3], and Želimir Husnić[4]

[1] Faculty of Mechanical Engineering, University of Sarajevo, Sarajevo
Bosnia and Herzegovina
jelacic@mef.unsa.ba
[2] Faculty of Mechanical, Electrical and Computer Engineering, University of
Mostar, Mostar, Bosnia and Herzegovina
[3] Faculty of Mechanical Engineering, University of Mostar "Džemal Bijedić",
Mostar, Bosnia and Herzegovina
[4] The Boeing Company, Ridley Park, PA, USA

Abstract. The locomotion of people with amputation is slower, less stable and requires more metabolic energy than the locomotion of physically fit individuals. Individuals with amputation of the lower extremities fall more often than able individuals and often have difficulty moving on uneven terrain and stairs. These challenges can mostly be attributed to the use of passive mechanical prosthetic legs that do not react actively to perturbations. Latest submitted solutions for active prosthetic devices of the lower extremities can significantly improve mobility and quality of life for millions of people with lower limb amputation, but challenges in control mechanisms of such devices are currently limiting their clinical viability.

Keywords: Above-knee prosthesis · Robust control · Tracking

1 Introduction

Dynamic equations of robotic manipulators present a complex, nonlinear and multi-variable system. One of the first methods of controlling such systems was inverse dynamics which is also known as a special case of the method of feedback linearization. However, plant variability and uncertainty are obstacles to exact dynamic inversion. Therefore, inverse dynamic control has limited practical validation.

Control of the variable impedance is one of the most popular prosthetic controls because of the independence of the system model. However, impedance controls are missing optimality and robustness due to several shortcomings: time consuming estimation of impedance parameters (unique for each amputee), difficulties in detection of sub-phase in one step, lack of feedback and passivity [1]. There have been several attempts to solve the limitations of ordinary impedance control [2, 3]. However, the above controls are independent of the system model and are missing mathematical proof of stability and robustness in the presence of system uncertainty, un-modelled dynamics and disorders.

© Springer Nature Switzerland AG 2019
S. Avdaković (Ed.): IAT 2018, LNNS 60, pp. 328–332, 2019.
https://doi.org/10.1007/978-3-030-02577-9_32

In order to overcome these difficulties, motion control techniques based on the passivity property of Euler-Lagrange equations are considered. Especially for the robust and adaptive control problems, the passivity-based approach shows great advantage over inverse dynamic method. Therefore, robust passivity-based control (RPBC) gained attention as a powerful nonlinear control law that can guarantee stability and tracking of arbitrary trajectories efficiently, despite uncertainties in plant model parameters.

2 Robust Passivity Based Control

2.1 Problem Description

This section describes the control of the above-knee prosthesis with actuated knee and ankle joints using robust passivity based controller. Above-knee controller receives input information $S_k = \{q_h, q_k^z\}$ from the combined system human prosthesis. Using linear transformation, set $S_p = \{q_p, q_p^z\}$ is generated from S_k, where q_p^z is the desired path for q_p. Controllers use S_p to generate prosthetic knee moment during the swing phase and a period of reliance, enabling the combined prosthetic system to mimic human movement, i.e. $q_p \rightarrow q_p^z \Rightarrow q_c \rightarrow q_c^z \Rightarrow y_u \rightarrow y_z$ with limited tracking trajectory errors.

Controllers use only the coordinates of the body and the reference trajectories of a healthy leg, without any dynamic information of a healthy body, in order generate the prosthetic angular momentum of the knee, which makes it possible for the combined human-prosthesis system to mimic the movements of a person without amputation. The proposed passivity-based robust controller is not only robust in relation to parametric uncertainties and unmodelled dynamics of the prosthesis, but also to different subjects with amputations.

2.2 RPBC for Active Above-Knee Prohesis

For the purposes of simulation, the system of the active above-knee prosthesis can be considered as a robotic manipulator consisting of three links with two one-axis joints. The coordinate system is set according to the standard Denavit-Hartenberg convention. The coordinate q_1 represents the angle in the hip in relation to the vertical axis. Coordinates q_2 and q_3 represent the angles in the knee and ankle joint, respectively (Fig. 1).

The flexible foot is attached to the ankle at an angle over the pylon. During the experiments, the foot will be placed on the Zebris plate to measure the vertical force of the reaction of the ground. This data is implemented in the control algorithm as an external non-conservative force because it can play an important role in the feedback part of the control algorithm.

In this case, the system can be considered as an active three-link planar robot, because movement is only observed in the sagittal plane. A robotic dynamic model in joint coordinates can be written as:

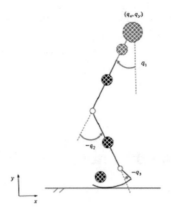

Fig. 1. Model of active above-knee prosthesis

$$D(q)\ddot{q} + C(q,\dot{q})\dot{q} + J_e^T F_e + g(q) = F_a \tag{1}$$

where $q = [q_1 \quad q_2 \quad q_3]^T$ is vector of joint angles, $D(q)$ is matrix of inertia, $C(q,\dot{q})$ is Centripetal and Coriolis matrix, J_e is the kinematic Jacobian of the point where external force acts, $g(q)$ is the gravitational vector and F_a is the vector of combined actuator inputs, where the effects of inertia and friction are incorporated.

As the position vector of the point where the maximum value of the reaction force of the ground is known, its global location can be calculated using the transformation matrix:

$$Z_{LC} = q_1 - l_{cy}cos(q_2 + q_3) + (c_3 + l_{cx})sin(q_2 + q_3) + l_2 sin(q_2) \tag{2}$$

Jacobian at the location of the maximum ground reaction force is given by:

$$J_e(1,1) = 0$$
$$J_e(1,2) = -(c_3 + l_{cx})sin(q_2 + q_3) + l_{cy}sin(q_2 + q_3) - l_2 sin(q_2)$$
$$J_e(1,3) = -(c_3 + l_{cx})sin(q_2 + q_3) + l_{cy}sin(q_2 + q_3)$$
$$J_e(2,1) = J_e(2,2) = J_e(2,3) = 0 \tag{3}$$
$$J_e(3,1) = 1$$
$$J_e(3,2) = (c_3 + l_{cx})cos(q_2 + q_3) + l_{cy}sin(q_2 + q_3) + l_2 cos(q_2)$$
$$J_e(3,3) = (c_3 + l_{cx})cos(q_2 + q_3) + l_{cy}sin(q_2 + q_3)$$

Horizontal component of the foot velocity V_f can be obtained from the Jacobian above, so the horizontal friction force F_{GH} can be calculated as:

$$F_{GH} = -\mu F_{GV} sign(V_f) \tag{4}$$

where F_{GV} is the vertical component of the ground reaction force, μ is empirically calculated friction coefficient and equals 0.15. Hence,

$$F_e = [F_{GH} \quad 0 \quad -F_{GV}]^T \tag{5}$$

The first two joints of the robot are driven by servo DC motors with amplifier gains $k_1 = 375\text{Nm/V}$ and $k_2 = 15\text{Nm/V}$, whereas, the knee joint in this case is assumed to be ideally driven by torque directly for convenience.

Most robotic systems have parametric uncertainty problems, the active prosthetic above-knee prosthesis is not an exception. Controller is considered due to its robust characteristic which is good at maintaining the performance in terms of stability, tracking errors, or other specifications despite parametric uncertainty, external disturbances or unmodelled dynamics present in the system. In this section, a robust passivity based controller is implemented to the 3-link robot system for trajectory tracking of these three joints.

In order to run the 3-link robot with the robust passivity based controller, the parameters of this manipulator are chosen as shown in Table 1. In addition, the uncertainty level is determined as 1:3 in this simulation, which means the value of parameters are selected arbitrarily in the range of 30% fluctuation from the nominal value, the dead zone of the controller is chosen as 1. The trajectory references for these two joints are sine waves, the amplitude, frequency, and phase angles are 1, 1, $\pi/2$, and 1, 1, 0 respectively. The controller is adjusted to give a better performance by tuning the controller gains L and K through trial and error. The system is simulated for 20 s.

Table 1. Simulation parameters for the three-link robot

Parameters	Values	Units
m_1	315.5	kg
m_2	43.28	kg
m_3	8.75	kg
m_0	2.33	kg
l_2	0.425	m
l_3	0.527	m
c_2	−0.339	m
c_3	0.32	m
I_{2z}	0.435	kg m^2
I_{3z}	0.062	kg m^2
J_m	0.000182	kg m^2
b_1	9.75	Nm s
b_2	1	Nm s
f	83.33	Ns/m

3   Results and Discussion

The Matlab Simulink simulation results are shown in Fig. 2: Diagrams show the tracking of the preset angular value in the knee and ankle joints. Simulated output values are in red while desired input values are given in blue.

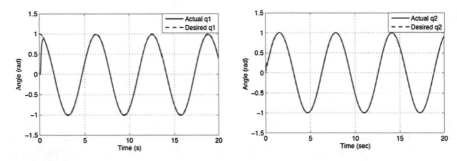

Fig. 2. Tracking in the knee (q1) and ankle (q2) joint angle

The results of the simulations show that the passivity-based robust regulator is able to trace the trajectory of the joints close to the desired values, although another 30% of the parametric uncertainty is present within the model. Parametric uncertainty, which is an irreducible element in the robot management system, is successfully solved by the implementation of the robust passivity-based control.

4   Conclusion

Most robotic systems have parametric uncertainties, robotic prosthetic devices are no exception. In order to overcome this problem, a robust regulator based on passivity has been implemented. Its benefits include that it is good in maintaining performance in terms of stability, error tracking and/or other specification in spite of parametric uncertainty, external disturbances or unmodelled dynamics within the system. In this paper, the robust controller is applied to a robotic system consisting of three links and showed good results regarding tracking of angle trajectories.

References

1. Jelačić, Z.: Impedance control in the rehabilitation robotics. In: Hadžikadić, M., Avdaković, S. (eds.) Advanced Technologies, Systems, and Applications II, IAT 2017. Lecture Notes in Networks and Systems, vol. 28. Springer, Heidelberg (2018)
2. Jelačić, Z.: Scattering problem in the rehabilitation robotics control design, Matematički institut SANU, Mini-simpozijum "Stohastičke oscilacije i zamor: Teorija iprimene", Beograd, 04 July 2017 (2018)
3. Rupar, M., Jelačić, Z., Dedić, R., Vučina, A.: Power and control system of knee and ankle powered above knee prosthesis. In: 4th International Conference "New Technologies NT-2018" Academy of Sciences and Arts of Bosnia and Herzegovina, Sarajevo, B&H, 28–30 June 2018 (accepted)

Influence of Additional Rotor Resistance and Reactance on the Induction Machine Speed at Field Weakening Operation for Electrical Vehicle Application

Martin Ćalasan[✉], Lazar Nikitović, and Milena Djukanovic

Faculty of Electrical Engineering, University of Montenegro, Dzordza
Vasingtona bb, 81000 Podgorica, Montenegro
{martinc,milenadj}@ac.me, lazar.nikitovic.
norsk@gmail.com

Abstract. This paper presents the influence of additional rotor resistance and reactance on the induction machine speed, at field weakening operation, for electrical vehicle application. For that reason, firstly, the usage and position of electrical machines in electrical vehicle are described. Also, this study shows how to calculate ratio of rotor resistance and reactance for desired maximum stator frequency (speed of field) and vice versa. It is validated that additional rotor resistance can increase induction machine (i.e. electrical vehicle) speed, while additional rotor reactance can decrease induction machine (i.e. electrical vehicle) speed.

Keywords: Induction motors · Speed · Flux weakening region

1 Introduction

Electrical vehicles represent our future. The usage of electrical vehicle, also called an electric drive vehicle, is in constant growth. It uses one or more electric motors or traction motors for propulsion [1–3]. It is the most used electrical vehicle motor. The principal position of electrical motor in electrical vehicle is shown in Fig. 1.

The main fields of research in electric vehicles are concentrating on achieving high speeds and high accelerations. Therefore, researches are also oriented towards high speeds of electric machines.

Induction motors in high performance applications can operate in wide ranges of the motor's mechanical speed [4]. However, in general, induction motor's design parameters are not possible to adjust (such as the winding turns distribution) without changing values of performance parameters (torque, efficiency, rating, etc.) [5]. For operation at higher than rated frequency, induction motor is fed with constant voltage which leads to the flux weakening operation [4]. For operation above rated speed the ac motors are flux weakened [6–8].

It is necessary to emphasize that the selection of the flux reference and the base speed is very important [9–13]. Furthermore, this problem is rarely analyzed together with current regulation [13]. Review and explanation of flux-weakening in high

© Springer Nature Switzerland AG 2019
S. Avdaković (Ed.): IAT 2018, LNNS 60, pp. 333–341, 2019.
https://doi.org/10.1007/978-3-030-02577-9_33

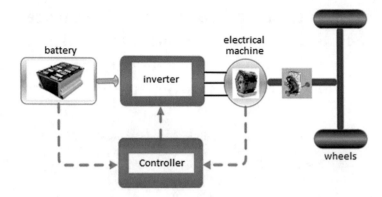

Fig. 1. Principal position of electrical motor in electrical vehicle

performance vector controlled induction motor drives is described in [14, 15]. The characteristic of torque in flux weakening region is described in [16–19], while the scalar control is presented in [19].

In this paper, the influence of rotor resistance on the maximum stator frequency of induction machine is analyzed. Therefore, this paper has similarities with [1]. However, in [1] the influence of machine designs parameters, mainly the mutual and leakage inductance, on the flux-weakening performance of induction machines (IMs) for electrical vehicle application is analyzed. On the other hand, in this paper, two equations for calculating maximum stator frequency (speed of stator field) for a given rotor resistance of induction motor and vice versa at field weakening operation will be presented. This paper, also, presents possibility of changing maximum speed of induction motor just by adding or reducing rotor resistance/reactance in flux weakening operation.

Paper is organized as follows. In Sect. 2 is described flux weakening region of induction machine and its importance for electrical vehicle. The calculation of maximum machine speed in flux weakening operation is presented in Sect. 3. Simulation results which describe paper proposals are presented in Sect. 4. In Conclusion, a short description of research results is given.

2   Electrical Vehicle and Induction Machine Torque-Speed Curve

There are two basic types of electrical vehicles: plug-in hybrid electric vehicles (PHEVs) and all-electric vehicles (AEVs).

PHEVs run on electricity for shorter ranges (6 to 40 miles), and then switch over to an internal combustion engine running on gasoline when the battery is depleted. However, AEVs run only on electricity. In addition to charging from the electrical grid, both types are charged in part by regenerative braking, which generates electricity from some of the energy normally lost when braking.

Block diagram of electrical vehicle is presented in Fig. 1. The main part of electrical vehicle is one or more electrical motors. Electric motors can provide high power-to-weight ratios, and need to operate in wide operation speed. The most popular electrical motors for electrical vehicle are induction machine and DC (Direct Current) machine. On the other hand, batteries can be designed to supply the large currents to support these motors. Although some electric vehicles have very small motors (for example 15 kW), many electric vehicle have large motors and brisk acceleration. Today, it is well known that Venturi Fetish, a two-seater electric sports car, can develop power of 220 kW and top speed of around 160 km/h. However, in order to develop that amount of speed, induction machine (IM) needs to operate in flux weakening mode of operation [6–8].

The operating speed range of the IM drive can be divided in three sub-regions:

- Constant torque region ($\omega < \omega_b$),
- Constant power region ($\omega_b \leq \omega < \omega_c$) and
- Constant slip frequency region ($\omega \geq \omega_c$), as shown in Fig. 1 (where ω_b and ω_c are base and critical stator angular speeds, respectively). The maximum torque of the induction motor is limited by the current and voltage ratings of the power inverter and thermal current limit of the IM. However, for high speed operation, flux needs to be weakened, as shown in Fig. 2. Also, during that operation a special attention should be paid to current limitation (see Constant power region in Fig. 2).

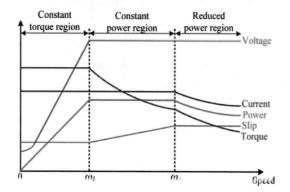

Fig. 2. Torque-speed characteristics of induction machine

3   Calculating Maximum Rotor Speed in Flux Weakening Operation

This section describes the way of calculating maximum rotor speed in flux weakening operation using rotor resistance, rotor reactance and rated slip.

If a is defined as ratio of rotor resistance and rotor reactance, and b as $\frac{P_j(1-s_r)s_r}{a^2+s_r^2}$, where P_j is normalized power and s_r is rated slip, then it is possible to calculate the slip of induction motor by following expression [3]:

$$s = \frac{1}{2\left(1 + b\omega_{sn}^2\right)} \pm \frac{1}{2}\sqrt{\frac{1}{\left(1 + b\omega_{sn}^2\right)^2} - \frac{4ba^2}{\left(1 + b\omega_{sn}^2\right)}}. \tag{1}$$

Knowing the fact that the slip cannot be a complex value, we can evaluate the maximum operating frequency for the motor drive:

$$\frac{1}{\left(1 + b\omega_{sn}^2\right)^2} \geq \frac{4ba^2}{\left(1 + b\omega_{sn}^2\right)}. \tag{2}$$

Normalized stator speed can be determined by solving the Eq. (2):

$$\omega_{sn} \leq \frac{\sqrt{1 - 4ba^2}}{2ab}. \tag{3}$$

Previous expression shows that the maximum normalized stator speed is obtained as:

$$\omega_{sn(max)} = \frac{\sqrt{1 - 4ba^2}}{2ab}. \tag{4}$$

The previous expression is derived in [4]. However, observing derived equations (i.e. involving Eq. (4) in Eq. (1)) it can be concluded that

$$s = 2ba^2. \tag{5}$$

Therefore, the maximum stator speed is obtained when slip is $2ba^2$, and the maximal rotor speed can be expressed as:

$$\omega_{r(max)} = \left(1 - 2ba^2\right)\frac{\sqrt{1 - 4ba^2}}{2ab}. \tag{6}$$

Using relation (6) one can easily calculate maximum rotor speed knowing the values of a, b, number of poles (p) and network frequency (f).

4   Calculating Active and Reactive Resistance Ratio for Desired Rotor Speed in Flux Weakening Operation

Equation (6), for different values of rotor resistance and reactance, calculates the maximum rotor speed which induction motor can achieve during flux weakening operation. But relation (6) has one weak point: It cannot calculate the ratio of rotor resistance during which induction motor can achieve desired maximum rotor speed in flux weakening operation. For that purpose this paper presents equation which calculates the ratio of rotor resistance and reactance for desired rotor speed, when the motor is in flux weakening region.

Expressing a from Eq. (6) one can calculate ratio of rotor resistance and reactance for given desired maximum rotor speed ω_1 as:

$$a = \sqrt{\frac{-c_1 + \sqrt{c_1^2 - 4c_1c_2}}{2}}, \tag{7}$$

Where coefficients c_1, c_2 are calculated as $c_1 = \frac{2s_r^2 - 4s_r^2\left(P_j(1-s_r)s_r\right) - 4\omega_s^2\left(P_j(1-s_r)s_r\right)^2}{1-4\left(P_j(1-s_r)s_r\right)}$, $c_2 = \frac{s_r^4}{1-4\left(P_j(1-s_r)s_r\right)}$ Expression (7) can be very useful, but it is important to be careful when choosing desired rotor speed. One must be aware of current limitation. If rotor resistance increases than the rotor current will drop below rated current, but the total loss will be bigger (efficiency of motor will decrease). On the other hand, decreasing rotor reactance, the current will become greater than rated. These current limitations are not presented in this paper.

5 Simulation Results

In order to verify the effectiveness of equations shown in this paper, a computer graphs are developed in MATLAB software and presented in this section. Parameters of examined induction motor are given in Table 1.

In Figs. 3 and 4 the impact of additional rotor reactance and resistance of the rotor speed, respectively, for different values of the base power, is presented. As it is evident,

Table 1. Parameters of examined 60 Hz induction motor

Parameter	Value	Parameter	Value
Mechanical power	4 kW	Rotor resistance	0.183 Ω
Rated speed	1767.1 rpm	Rotor reactance	0.841 Ω
Number of poles	4	Stator resistance	0.277 Ω
Number of phases	3	Stator reactance	0.554 Ω
Connection	3tai	Magnetizing reactance	20.3 Ω

for higher value of the reactance, the speed is lower. However, for higher value of the additional rotor resistance, the speed is higher.

The usage of derived Eq. 7 is presented in Fig. 5. As it can be seen, desired rotation speed can be easily calculated using the value of parameter a. Furthermore, as the higher speed can be obtained by using higher value of additional active resistance, then by using Eq. 7, the value of the additional resistance can easily calculated (see Fig. 6):

$$R_{add} = a \cdot X_{eq} - R_{eq}. \tag{8}$$

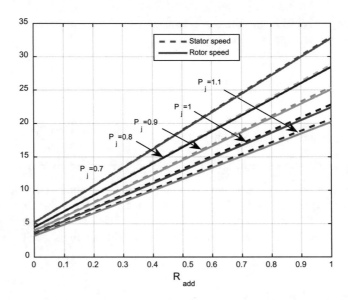

Fig. 3. Stator flux speed and rotor speed – additional rotor resistance characteristics for different value of the normalized power.

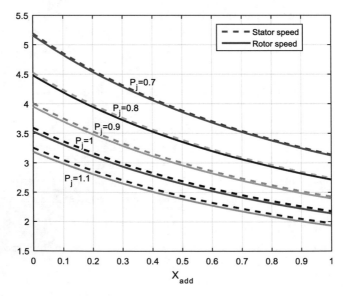

Fig. 4. Stator flux speed and rotor speed – additional rotor reactance characteristics for different value of the normalized power.

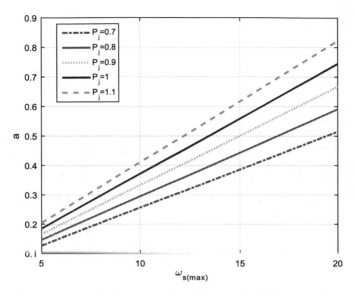

Fig. 5. Parameter a - stator flux speed, for different value of the normalized power.

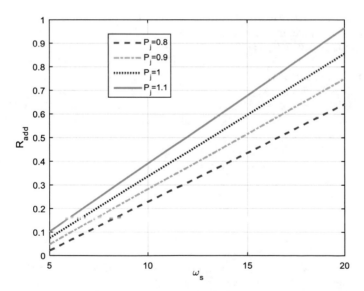

Fig. 6. Desired speed – additional resistance characteristics for different value of the normalized power.

6   Conclusion

In this paper, basic characteristics about electrical vehicle and position of electrical motors in its structure are presented. Also, it is noted that for high speed operation of electrical machines, as well as electrical vehicle, a field weakening operation should be considered.

Basic equation about impact of additional rotor resistance and reactance on maximum speed of stator field and rotor speed are presented. Also, the inverse equation, which gives the relation between desired stator field speed and ratio between additional rotor resistance and reactance, is presented. Corresponding simulation results are also given and discussed.

In the future work, the impact of additional rotor reactance and resistance of the machine current will be analyzed.

Acknowledgment. This paper is prepared within COST project - Action CA16222 "Wider Impacts and Scenario Evaluation of Autonomous and Connected Transport".

References

1. Guan, Y., Zhu, Z.Q., Afinowi, I., Mipo, J.C.: Influence of machine design parameters on flux-weakening performance of induction machine for electrical vehicle application. IET Electr. Syst. Transp. **5**(1), 43–52 (2015)
2. Vicatos, M.S., Tegopoulos, J.A.: A doubly-fed induction machine differential drive model for automobiles. IEEE Trans. Energy Convers. **18**(2), 225–230 (2003)
3. Farasat, M., Trzynadlowski, A.M., Fadali, M.S.: Efficiency improved sensorless control scheme for electric vehicle induction motors. IET Electr. Syst. Transp. **4**(4), 122–131 (2014)
4. Krishnan, R.: Electric Motor Drives: Modeling, Analysis, and Control. Prentice Hall, Upper Saddle River (2001)
5. Jiang, S.Z., Chau, K.T., Chan, C.C.: Performance analysis of a new dual-inverter pole-changing induction motor drive for electric vehicles. Electr. Power Compon. Syst. **30**(1), 11–29 (2002). ISSN 1532-5008
6. Sepulchre, R., Devos, T., Jadot, F., Malrait, F.: Antiwindup design for induction motor control in the field weakening domain. IEEE Trans. Control Syst. Technol. **21**(1), 52–66 (2013)
7. Levi, E., Wang, M.: A speed estimator for high performance sensorless control of induction motors in the field weakening region. IEEE Trans. Power Electron. **17**(3), 365–378 (2002)
8. Huang, M.S., Liaw, C.M.: Improved field-weakening control for IFO induction motor. IEEE Trans. Aerosp. Electron. Syst. **39**(2), 647–659 (2003)
9. Xu, X., Novotny, D.W.: Selection of the flux reference for induction machine drives in the field weakening region. IEEE Trans. Ind. Appl. **28**(6), 1353–1358 (1992)
10. Qingyi, W., Yang, L., Hui, L.: Optimal flux selection of induction machine in the field-weakening region. In: 2012 Asia-Pacific Power and Energy Engineering Conference (APPEEC), pp. 1–5 (2012). ISSN 2157-4839
11. Seok, J.K., Sul, S.K.: Optimal flux selection of an induction machine for maximum torque operation in flux-weakening region. IEEE Trans. Power Electron. **14**(4), 700–708 (1999)

12. Nguyen-Thac, K., Orlowska-Kowalska, T., Tarchala, G.: Comparative analysis of the chosen field-weakening methods for the direct rotor flux oriented control drive system. Arch. Electr. Eng. **61**(4), 443–454 (2012)
13. Briz, F., Diez, A., Degner, M.W., Lorenz, R.D.: Current and flux regulation in field-weakening operation of induction motors. IEEE Trans. Ind. Appl. **37**(1), 42–50 (2001)
14. Krishnan, R.: Review of flux-weakening in high performance vector controlled induction motor drives. In: Proceedings of the IEEE International Symposium on Industrial Electronics, ISIE 1996, pp. 917–922, June 1996
15. Sahoo, S.K., Bhattacharya, T.: Field weakening strategy for a vector-controlled induction motor drive near the six-step mode of operation. IEEE Trans. Power Electron. **31**(4), 3043–3051 (2016)
16. Nisha, G.K., Lakaparampil, Z.V., Ushakumari, S.: Torque capability improvement of sensorless FOC induction machine in field weakening for propulsion purposes. J. Electr. Syst. Inf. Technol. **4**(1), 173–184 (2017)
17. Kim, S.H., Sul, S.K.: Maximum torque control of an induction machine in the field weakening region. IEEE Trans. Ind. Appl. **31**(4), 787–794 (1995)
18. Tripathi, A., et al.: Dynamic control of torque in overmodulation and in the field weakening region. IEEE Trans. Power Electron. **21**(4), 1091–1098 (2006)
19. Smith, A., Gadoue, S., Armstrong, M., Finch, J.: Improved method for the scalar control of induction motor drives. IET Electr. Power Appl. **7**(6), 487–498 (2013)

Programming of the Robotic Arm/Plotter System

Milena Djukanovic[1]([⊠]), Rade Grujicic[2], Luka Radunovic[2],
and Vuk Boskovic[2]

[1] Faculty of Electrical Engineering, University of Montenegro,
DzordzaVasingtona bb, 81000 Podgorica, Montenegro
milenadj@ac.me
[2] Faculty of Mechanical Engineering, University of Montenegro,
DzordzaVasingtona bb, 81000 Podgorica, Montenegro

Abstract. Plotter, as a type of CNC machine, is a machine that requires great precision in work, so in its creation person must take into account all the details that can later result in the error in plotting a given vector. Error may occur due to defective electro-mechanical components, or poor calibration of the given mechanical elements, but the most common mistake is made in the programming process. In this regard, the aim of this paper is the entire mechatronic process of connecting electronic and machine components into a compact, portable and precise machine that would replace a human in performing some routine hand-executed functions, such as writing, drawing, and engraving. Realization of the given solution would have low production costs, while small dimensions and easy portability would allow a great application in everyday life.

Keywords: Plotter · Mechatronic process · Programming
Routine hand-executed functions

1 Introduction

Robotics is a multidisciplinary science that besides knowledge of electronics and mechanical engineering requires knowledge of programming languages, and it deals with the design, construction, operation and use of robots, as well as programming of their control and sensory feedback. As people were trying to do the everyday routine activities as quickly as possible, there was a desire to robotize them. Major advantages of robotization over humans are precision, speed and infinite number of same repetitions of a given task. They differ, from autonomous robotic hands which are used in industry for automation of all kinds of activities such as: cutting, welding, engraving, painting, to simpler mechatronic systems that work in automation. Until recently, robots were used only in large industrial plants, which have drastically changed in recent years. Nowadays, they have a big variety of usage, starting from small housework and finishing with complex tasks, such as precise drawing, cutting, welding and assembling parts. The intelligence that one robot possesses directly depends on the program itself and its adaptability to unpredictable situations [1–3]. The focus of this

© Springer Nature Switzerland AG 2019
S. Avdaković (Ed.): IAT 2018, LNNS 60, pp. 342–354, 2019.
https://doi.org/10.1007/978-3-030-02577-9_34

paper is connecting electro-mechanical components into a robotic system capable of replacing human presence in the function of signing, which would improve current technology and greatly accelerate the process itself.

2 Plotter

CNC plotter is a machine designed to plot a data vector sent by a computer. Plotters may use various head extensions, such as markers, lasers, drills, but most common ones are pen plotters. Main difference between usual dotted printers and plotters (Figs. 1 and 2) is that the plotter imitates human hand movement without lifting a pen from the paper until the vector is drawn, allowing the plotting of complex lines with high precision [4].

Fig. 1. Dotted plotter

Usual plotters consist of linear sliders that allow the movement of the plotter head in two mutually perpendicular directions. Movement is done by two threaded spindles, which, with the help of programmed stepper motors, guide the pen to its precise location. Microcontroller, as a control unit, represents a link between hardware and software, connecting program defined vector fields with electronics and motors. Sensors that are put in the process enable additional precision of the machine, lowering the possibility of error making.

Unlike usual plotters, the plotter developed in this work is a remote controlled machine that is working independent of users distance. Working process of the machine is real time data transaction between the input unit (the smart-phone device), and the output unit (the plotter itself), plotting the given vector in real time.

Usage of this system would be the verification and signing of the distant documents, and the very principle of plotting that signature would result in its authenticity and validity.

Smart-phone software would need to provide automatic data storing in database for reuse, as well as scaling of the entered values for various, variable document formats that are located on the working platform. The development of various plotters and realization of its management was a topical topic of numerous researches, such as [5, 6].

Fig. 2. Pen plotter [10]

The breakthrough made in this paper, compared to the previous solutions, is reflected in the specificity of the purpose of the device itself and the changes it brings with itself, such as the formation of the appropriate management base, the realization of management, communication between the machine and the user, and so on.

Conceptual design solution of the robotic arm/plotter, which is used as basis of this paper, is shown in Fig. 3, [7]. This solution is developed in SolidWorks software [8], and rendered in PhotoView360.

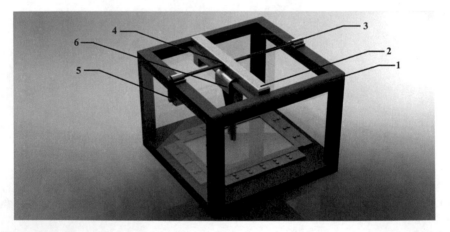

Fig. 3. Conceptual design of the plotter: 1-frame, 2-slider, 3-lead screw, 4-lead screw nut, 5-stepper motor, 6-plotter head

3   The Working Principle

The given system consists of multiple subsystems that are connected over Internet, system is shown in Fig. 4.

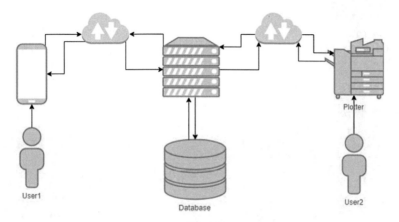

Fig. 4. Subsystem connections

(1) First subsystem is for data entry

Access to signing will be provided by a special application. User 1, as a main user in the system, interacts with the smart-phone by entering data (signature) in the application.

The task of this application is scaling and forwarding all data to the server, which are necessary for plotter to execute its function. Simple user interface would allow its users to use the application with ease, allowing them fast and efficient work.

(2) Second subsystem is a server

Application allows data storage of the input vectors if the user1 requests it, which would significantly affect the speed of the future process, by choosing the same data for signing the documents over and over instead of entering the new ones. If that option is checked, dataflow in the subsystem change its direction. Instead of directly transferring the vectors to the microcontroller and the plotter, database in which the data is directly stored is added in the flow, from which the control unit is reads them. Transferred data would be encrypted with special algorithm in order to prevent signature falsification and to ensure data security. Database system that would be used is relational database management system (RDBMS), which would allow easy access to its users, performing fast enough for this type of application.

(3) Last subsystem is a control unit

As an input element of this system, data directly transferred from user1 or data stored in the database is used, and for the output element, that vector is drawn on the document.

Input elements are modified by a specific algorithm [shown in the head] and are used for stepper motor rotation which enables translator movement of the plotter head. Programming of this system, not only that it requires knowledge of application and microcontroller programming, but also requires great knowledge of mechanics and electronics which are used throughout the whole system.

Figure 5 is representing schematic on which electronic components are seen, which are attached to the plotter machine directly. Microcontroller, as the heart of the process itself, is connecting all of the input and output units in one compact, semi-automatic process. User 2, who is shown in the Fig. 2 next to the plotter, may control client-client connection using the push-buttons, results are seen in real-time on LCD screen shown in Fig. 5.

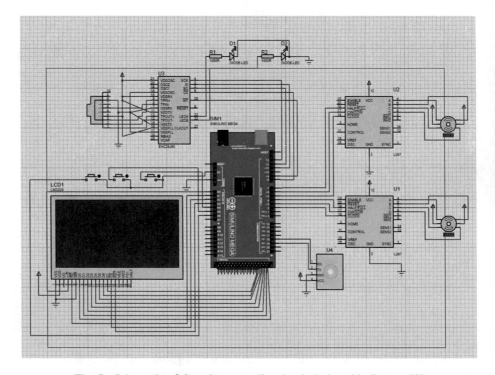

Fig. 5. Schematic of the microcontroller circuit designed in Proteus [9].

Digital pin-outs of the microcontroller, set as an output pins are indirectly across motor-drivers U1 and U2 controlling the stepper motors. Signals that are used for stepper motor control in this configuration are step signal, direction of rotation signal and half/full switch step signal. Ethernet module, shown as U3 on Fig. 5, enables the microcontroller to connect to the internet, and therefore allows dataflow from the camera to the application, and from the application to the motors.

Camera, shown as U4 on Fig. 5, as an input unit, has a primary and only goal to picture the working platform, which will be further forwarded to the application and user1.

3.1   Compatibility of Input and Output Devices

Since the input and output units are of a different format, a certain data scaling must be performed so that the data can be correctly plotted using the stepper motors. Data is also divided into input and output, so there are more variables in the process that are describing the same thing.

On the smart-phone application (input data) related to data scaling, there would be a two-dimensional vector that figures position of the pressed screen pixel at given moments, as well as two constants that figures the maximum value of screen pixel on both X and Y axis. Program functions for taking currently pressed pixel value are pre-embedded in every smart-phone device and they are quite fast in execution, so the processing of data itself would quickly be performed and it would enable the user to smoothly, without lag, perform the action.

Data that was loaded on the input unit must be scaled, and in order to do this, it is necessary to know some dimensions of the plotter. Using the given formulas (formulas 1 and 2), new two-dimensional vector is made out of input vector, but with scaled values.

$$X = \frac{curr_x}{sc_x} x_{\max}, \tag{1}$$

$$Y = \frac{curr_y}{sc_y} y_{\max}, \tag{2}$$

where X and Y are point coordinates on the plotters working platform, $curr_x$ and $curr_y$ – pixel coordinates on the smart-phone, sc_x and sc_y – smart- maximum phone resolution, $x_{\max}$ and $y_{\max}$ – dimensions of the platform (Fig. 6).

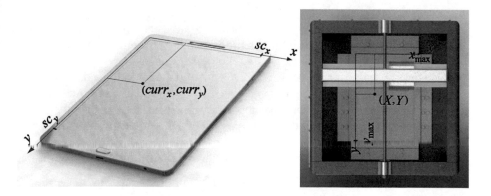

Fig. 6. Scaling the pixels

3.2   Solution for Motor Movement

When making the plotter, the biggest problem of the whole process is the movement of the pen across the paper of the given document. Since there is a large amount of data in vectors that needs to be plotted in a short time, a fast and simple software solution was needed. Solving of this problem consists of a couple of stages.

- First stage is the observation of two consecutive elements of the vector as a line that connects two dots (shown in Fig. 7). By subtracting X and Y values of new dots with old values, you get an increment in x axis written as dx, and increment in y axis written as dy. Ratio of dx and dy is defining further execution of the code.

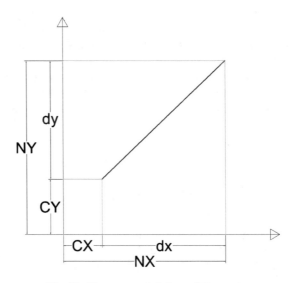

Fig. 7. Two connected dots of the vector

CX and CY are representing the current position of the pen tip, which is pressed on the paper/document. NX and NY are representing the positions on which we want the pen to come, performing its plotting along the way.

Difference of these values (NX-CX and NY-CY) is shown as dx and dy in Fig. 7.

- Second stage is observing the movement as a lot of elementary movements, which are in fact minimal movement that threaded spindle makes for one step of the stepper motor. Threaded rod defined in this paper has a stroke of 4 mm, and for 1 step of a motor, which is 1.8° in rotation, it makes translatory movement of 0.02 mm. Elementary movement is way less than the diameter of the tip of the pen.

Figure 8 shows the line on which the pencil should move (red line), and the real line of its movement (purple line). Taking into account the line size, which is approximately one fifth of a millimeter, and the size of elementary motion, which is 0.02 mm, it is clear that the pencil will move along the given path without any visible problems, leaving behind a visually straight line.

This stage is divided into three possible cases. Ratio of dx and dy is defining which case will be chosen. Each case has its code that controls the movement of the X and Y motor (Chart 1).

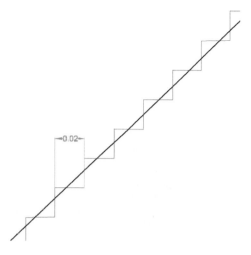

Fig. 8. Elementary movement

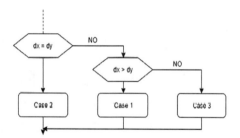

Chart 1. Control flow graph [10]

(1) dx > dy

Stepper motor that is rotating the X axis spindle will rotate dx/dy times more than the other motor. For one step, motors have pre-defined constant angle of rotation, and ratio dx/dy is usually not integer, so it should be split into two variables.

First variable is an integer type, representing floor division of dx/dy, which is shown in Fig. 9 as int P. Second variable is a float type and it represents remainder of the division of two operands (dx/dy), which is shown in Fig. 9 as float O. Stepper motor X is covering P steps in one loop, while motor Y covers 1, code is executed in the same order all until the remainder passes value of 1, then motor X gets one more step in that iteration. This kind of movement is creating an unbalanced pen path, but for the human eye, elementary movement of 0.02 mm is not visible, so the mistakes of that scale are negligible.

(2) dx = dy

In this case, the ratio of dx/dy is 1, with no remainder, so the working principle of stepper motors is in fact simultaneous, step by step, creating a line at an angle of 45°.

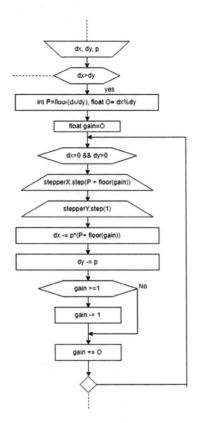

Fig. 9. Algorithm for motor control [11]

(3) dy > dx

It is same as in case (1). Ratio of the input data is with remainder, so two variables are created, creating an unbalanced rotation of the stepper motors.

3.3   Example

The moment from descent to lifting the pen is taken for one vector.

Vector consists of X and Y components, they define the positions of the dots that will be connected by writing. More dots in one vector, means a bigger precision.

Data of one vector are shown in Table 1, and they are graphically represented in Fig. 10. As in the real world, where pen would connect the dots with linear motion, so are they connected with straight lines on graph in Fig. 10. Units in Table 1 are in millimeters (Table 3).

For this example, these two dots shown in Table 2 were taken. First dot is with coordinates CX (12 mm) and CY (8.2 mm). Last dot has coordinates NX (12.2) and NY (8.4). In this example, dx and dy are same, so this is the second case where motors

Table 1. Data

X1	Y2
3.3	7.1
3.2	7
3	7.15
2.95	7.8
3.1	8.1
3.6	8.35
4	8.45
4.5	8.3
5	8
5.4	7.8
5.8	7.7
6.25	7.6
7	7.3
8	7.05
8.5	7
9	7
9.4	7.1
9.6	7.2
10	7.3
10.4	7.4
10.8	7.5
11	7.6
11.2	7.7
11.5	7.9
12	8.2
12.2	8.4

Table 2. Data of two dots

	X	Y
C	12	8.2
N	12.2	8.4

work simultaneous. RX and RY columns in Table 4 are representing number of steps that motor X and Y make in one iteration of the algorithm, and CX and CY are values of pen position after each iteration. Graphic representation of motor movement over paper is shown in Fig. 11.

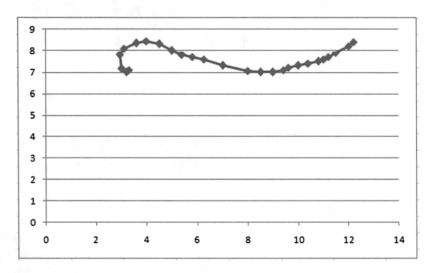

Fig. 10. Graphic representation of data in Table 1.

Table 3. Input data

dx	0.2
dy	0.2
p	0.02
P	1

Table 4. Number of steps

RX	RY	CX	CY
0	0	12	8.2
1	0	12.02	8.2
0	2	12.02	8.24
2	0	12.06	8.24
0	2	12.06	8.28
2	0	12.1	8.28
0	2	12.1	8.32
2	0	12.14	8.32
0	2	12.14	8.36
2	0	12.18	8.36
0	2	12.18	8.4
1	0	12.2	8.4

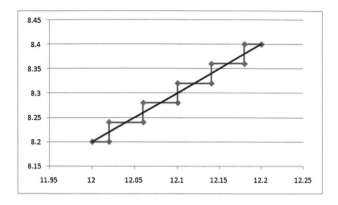

Fig. 11. Graphic representation of data in Table 4.

4   Conclusion

An algorithmic solution for robotic arm/plottermovement is presented in this paper.

This type of plotter would help speed up actions when a person is away from the document to sign, whether it is on a business trip, stuck in traffic, on vacation, etc.

Future elaboration would consider the realization of the presented machine, according to developed model, and selected components, materials and adopted dimensions.

Acknowledgment. This paper is prepared within two projects - "Montenegrin Wearable Robots (MWR)" supported by Ministry of Science of Montenegro and COST Action IC1403 CRYPT ACUS, supported by COST (European Cooperation in Science and Technology).

References

1. Wisskirchen, G., Biacabe, B.T., Bormann, U., Muntz, A., Niehaus, G., Soler, G.J., von Brauchitsch, B.: Artificial intelligence and robotics and their impact on the workplace. IBA Global Employment Institute (2017)
2. UK Essays, Machines vs Human Workers Business Essay (2013). https://www.ukessays. com/essays/business/machines-vs-human-workers-business-essay.php?cref=1
3. Soffar, H.: Advantages and disadvantages of using robots in our life (2016). https://www. online-sciences.com/robotics/advantages-and-disadvantages-of-using-robots-in-our-life/
4. Shivakumar, M., Stafford, M., Ankitha, T.H., Bhawana, C.K., Kavana, H., Kavya, R.: Robotic 2D plotter. Int. J. Eng. Innov. Technol. (IJEIT) **3**(10), 300–303 (2014)
5. Karthik, S., Reddy, P.T., Marimuthu, K.P.: Development of low-cost plotter for educational purposes using Arduino. In: IOP Conference Series: Materials Science and Engineering, vol. 225, no. 1 (2017)
6. Instructables. https://www.instructables.com/
7. Solidworks modeling software. http://www.solidworks.com/

8. Djukanovic, M., Grujicic, R., Radunovic, L., Boskovic, V.: Conceptual solution of the robotic arm/plotter. In: 4th International Conference "New Technologies NT-2018" Development and Application, 14–16 June 2018, Sarajevo, Bosnia and Herzegovina, paper accepted for publishing
9. Proteus design, simulation software. https://www.labcenter.com/
10. Online diagram software. https://www.draw.io/
11. AxiDraw machine. https://www.axidraw.com/

Effects and Optimization of Process Parameters on Seal Integrity for Terminally Sterilized Medical Devices Packaging

Redžo Ðuzelić[1]($\boxtimes$) and Mirza Hadžalić[2]

[1] University of Bihać, 77000 Bihać, Bosnia and Herzegovina
duzelicredzo@gmail.com
[2] Carefusion BH 335, Cazin, Bosnia and Herzegovina

Abstract. Plastics material can be applied in various industries. The reason for this are plastics properties (physical, mechanical, and technological) and low costs. Medical device packaging process is shown in this thesis, with analysis of input parameters for reliable performance. One of the greatest challenges in process development is determination of optimal process parameters. Determination of optimal process parameters for packaging process, can be conducted with help of statistical analysis software. Application of statistical software Minitab is conformal, since physical process of medical device packaging can be adequately described.

Keywords: Medical device packaging · Sealing · Temperature
Pressure · Optimization

1 Introduction

The application of plastic materials to the manufacture of the product is justified by the characteristics defined by this material and refer to a number of technical and economic advantages against other materials. From the current statistical data is known for many years the growth of the use of technology for the production of plastic products, as well as the problems of disposal and recycling of plastic waste. For this very reason, research in this field has taken place in recent years, with the aim of achieving various optimal solutions in the application of plastics processing technologies. The plastics can be applied in the packaging process of sterilized medical products because they allow the product to be sterilized for a long period of time. Packaging of medical products is performed in several phases, which require the analysis of the process parameters and their optimization.

In the paper is given an example of the application of the Minitab software package for the analysis and optimization of polyethylene foil and porous paper sealing parameters, which are the basic components for the packaging of medical products designed for sterilization.

S. Avdaković (Ed.): IAT 2018, LNNS 60, pp. 355–365, 2019.
https://doi.org/10.1007/978-3-030-02577-9_35

1.1 Significance of Plastic Materials on the Market

The influence of plastic materials on the 21st century trends is enormous. According to EcoWatch, plastics fall into the line of slow disintegrating materials. It takes 500 to 1000 years to completely disintegrate plastic materials, which is a potential hazard to the environment. Over the past 10 years, the total amount of processed plastic materials has been greater than the total amount of plastics processed until then. Today, 50% of the plastics produced is recycled. These facts imply the real need for a more rational management of plastic materials [1].

Global claims on plastics in 2005 amounted to 230 million tons, while in 2015 it amounted to 322 million tonnes worldwide, Fig. 1. According to research at Grand View Research, Inc., the expected share of plastics processing industry in 2020 will amount to 654 billion US dollars.

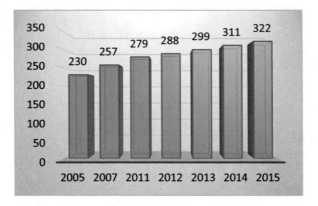

Fig. 1. Application of plastic materials in the world [1]

The application of plastics in different branches of industry is justified because their basic characteristics are low specific weight, low cost, do not require much energy for processing, can be easily combined with other materials etc.

The global medical device manufacturing industry's total share of the global market is about 400billion US dollars according to data published on the *Statista* portal. The global market share of medical device packaging worldwide is about $ 22 billion a year, or 5.5% of the total value of the medical device manufacturing industry. Taking into account the complexity and high cost of the medical devices themselves, this numbers show very high importance to the packaging of terminally sterilized medical devices [2].

2 Packaging for Terminally Sterilized Medical Devices

The packaging for terminally sterilized medical devices is carried out using a polymeric material material. The process of sealing materials of polymeric origin involves the bonding of two or more types of materials by applying heat and pressure. An example of the output of sealing process for polyethylene foil with porous paper is shown in Fig. 2.

Fig. 2. Seal area for sealing of polyethylene foil and porous paper

Input parameters of the sealing process for polyethylene foil and porous paper are [3]:

- sealing temperature T_s (°C),
- sealing time t_s(s) and
- sealing pressure p_s(MPa).

The output parameters of the process are measured/controlled respecting standard test methods. According to ISO 11607-1 and ISO 11607-2 there are a number of recommended test methods for the packaging of terminally sterilized medical devices testing, and for further analysis are essential the following two [4, 5]:

- seal strength F_s (N\15 mm) and
- burst pressure p_b(kPa).

2.1 Seal Strength

The test methodology as well as the criteria are described in the EN 868-5 standard. The same standard prescribes a minimum seal strength of 1.2 N on a length of 15 mm [6]. The seal strength of the sealed area is obtained by testing samples of a width of 15 mm on the dynamometer, according to Fig. 3a. Figure 3b shows a sample prepared for testing.

2.2 Burst Pressure

The test methodology is described in ASTM F2054/F2054M-13. The measurement of the pressure value used to separate the sealed area is performed on a device called the *Burst tester*, shown in Fig. 4.

The minimum burst pressure value is not determined by the standard because the size and packaging materials are different in almost every process, and therefore values

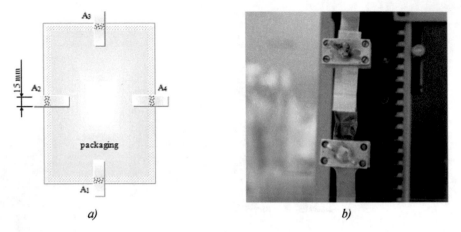

Fig. 3. Measuring seal strength (a) samples positisions (b) sample prepared for measurement

Fig. 4. Burst tester [7]

can not be standardized. The minimum allowable value is obtained by engineering research and analysis [6].

2.3   Application of Software in Analysis and Optimization of Process Parameters

The Minitab software, whose tools are based on statistics with the aim of improving quality, was used to analyze the sealing process parameters as part of packaging for terminally sterilized medical devices. These are essentially tools that use the collected statistical data to analyze the measurement results and help to make decisions related to a given problem [8].

2.3.1 Input Settings in the Minitab Software Operating Environment

For the purpose of analyzing and optimization of process parameters, the *Minitab* software offers several tools, of which is the Design of Experiments (DoE) most popular. The first step is to set the input settings, which are related to the definition of the research flow, Table 1.

Table 1. Input settings in the Minitab

Features	Setting
Type of design	Full factorial
Number of factors (k)	3
Number of blocks (b)	1
Number of replicates for corner points (n)	2
Number of central points per block (n_0)	6
Nr. Of runs (N)	22

Number of runs is determined by equation:

$$N = b \cdot \left(n \cdot 2^k + n_0 \right) = 1 \cdot \left(2 \cdot 2^3 + 6 \right) = 22 \tag{1.1}$$

The next step in defining settings is to determine the boundary values (min and max) of input process factors. Figure 5 shows the input parameter intervals used in the example.

Factor	Name	Type	Low	High
A	temperature	Numeric ▼	140	180
B	time	Numeric ▼	1,0	2,0
C	pressure	Numeric ▼	4,5	7,0

Fig. 5. Boundary values for factors

Sealing temperature (A) is given in (°C), sealing time (B) in (s), and sealing pressure (C) in 10^{-1} (MPa).

3 Measurement and Analysis of Output Parameters of the Sealing Process

In the process of measuring the value of the input parameters of sealing process, 22 experiments were performed. The preparation of the test samples was carried out on the packaging machine for packaging medical products (devices). All the samples were collected and stored in conditions that fit the conditions of the clean room, ISO 8 class, which controls conditions in terms of the amount of particles present, temperature, humidity and pressure inside the room.

3.1   Measurements and Results

Figure 6 shows the results for the mean values of the seal strength F_{avg}, expressed in N on a length of 15 mm. The mean seal strength values are shown depending on the sealing temperature (a) and sealing time (b) of the sealing of polyethylene foil and porous paper.

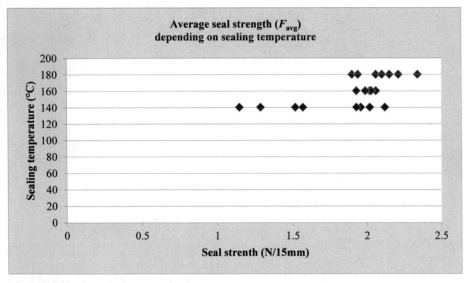

a)

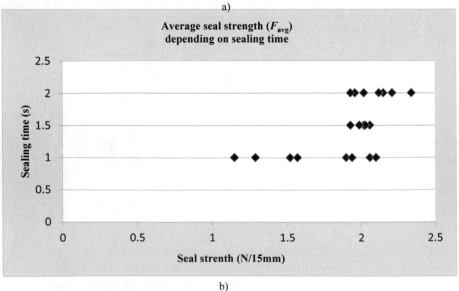

b)

Fig. 6. Seal strength depending on (a) sealing temperature and (b) sealing time

Figure 7 shows the results of the mean value for burst pressure measurement p_{avg}, expressed in (kPa). The mean pressure values are shown depending on the sealing temperature (a) and the time (b).

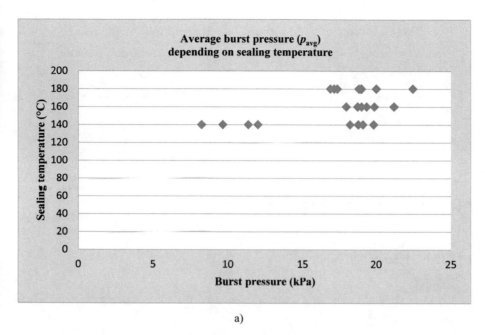

a)

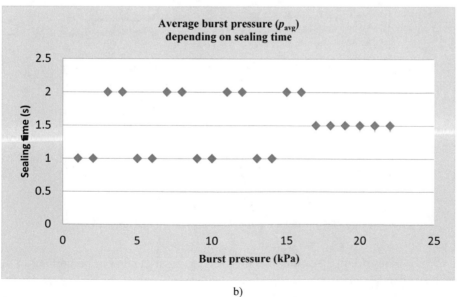

b)

Fig. 7. Burst pressure depending on (a) sealing temperature and (b) sealing time

3.2   Analysis of Measurement Results Using Minitab Software

The significance of the influence of the input parameters on the output sizes of the pressure and the seal strength for the packaging are given in Table 2.

Table 2. Impact of input parameters on seal strength and burst pressure

Significance of the input parameters to the defined output parameters for sealing		
Input parameters	Output parameters	
	Seal strength F_{avg} (N/15 mm)	Burst pressure p_{avg} (kPa)
Sealing temperature-A	B, A, C, AB	B, A, AB
Sealing time-B		
Sealing pressure-C		

According to the analysis (Fig. 8) on the value of the seal strength of the packaging, the influence of the input parameters was for:

- sealing temperature 20.5%,
- sealing time 20.5% and
- sealing pressure 9.85%.

Figure 8 shows a diagram of the influence of the input parameters on the seal strength of the packaging.

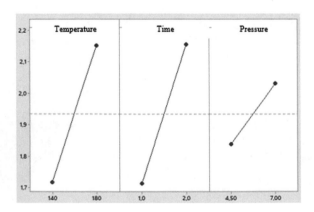

Fig. 8. Impact of the input parameters on seal strength

In the case of burst pressure, the effect of the input parameters, according to the analysis (Fig. 9), was:

- sealing temperature 24.96%,
- sealing time 27.47% and
- sealing pressure 0%.

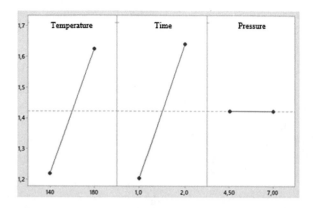

Fig. 9. Impact of the input parameters on burst pressure

Figure 9 shows a diagram of the influence of the input parameters on the burst pressure.

4   Sealing Process Parameters Optimization

Optimization of the process parameters of sealing polyethylene foil and porous paper was performed in Minitab. According to the analysis, the optimum input parameters of the polyethylene foil and porous paper sealing process in the packaging of terminally sterilized medical devices are:

- sealing temperature $T_{os} = 157\ °C$,
- sealing time $t_{os} = 2.0$ s and
- sealing pressure $p_{os} = 4.5 \cdot 10^{-1}$ MPa.

Optimization target for output parameters (responses) are given in Table 3.

Table 3. Optimization target

Output parameters of sealing process	Target
Seal strength (mean value $- F_{avg}$)	$F_{avg} \to$ max.
Burst pressure (mean value $- p_{avg}$)	$p_{avg} \to$ max.

Predicted values of responses, in case of optimal input process parameters are given below:

- seal strength: $F_s \to F_{o(max)} = 2.0694$ N/15 mm
- burst pressure: $p_b \to p_{b(max)} = 20.1659$ kPa.

Figure 10 shows optimal values and diagram of optimization of process parameters in the phase of sealing polyethylene foil and porous paper as part of the process of packaging terminally sterilized medical devices.

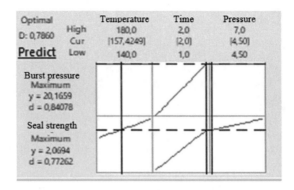

Fig. 10. Optimization diagram for sealing process parameters

By analyzing the optimal values of the input parameters it can be noticed that:

- Optimum sealing temperature is less than the maximum process temperature, $T_{os} = 157\ °C < 180\ °C$ (T_{max}), which saves energy consumption, higher tool utilization, etc.
- Optimum sealing time is equal to the maximum processing time $t_{os} = t_{smax} = 2$ s. This means that the significance of the impact of the sealing time as the input parameter on the output dependent variables of the seal strength F_s and the burst pressure p_b is confirmed. Also, with the optimal sealing time with the optimal values of other input parameters, a good integrity of the package sealed area can be achieved.
- Optimal value of the sealing pressure has a minimum value of the process pressure $p_{os} = p_{s(min)} = 4.5 \cdot 10^{-1}$ MPa. This confirms the lower significance of the sealing pressure to the integrity of the package sealed area in comparison to the temperature and the time of the sealing, as was obtained by the analysis in *Minitab*.

A deviation from the optimal values of the input parameters would lead to a poorer quality of the sealed area. The result would be:

- For values lower than optimal input parameters, the integrity of the packaging of medical products could be impaired, which could jeopardize legitimate shelf life duration.
- For values higher than optimal input parameters, greater seal strength/burst pressure would be achieved, but with poorer visual characteristics.

As a minimum force of 1, 2 N/15 mm is prescribed by the standard EN 868-5, predicted values obtained with the optimum process parameters are conformal.

5   Conclusion

Application of appropriate standards is of vital importance for packaging process for terminally sterilized medical devices. In addition, standards often do not include the all conditions in which the processes are performed, so it is necessary to perform experimental and other forms of analysis that will improve the quality of products, as well as the reliability and stability of the process. Application of *Minitab* software, as well as statistical processing of measurement results in the process of packaging for terminally sterilized medical devices, enables the achievement of optimal values of the process parameters, thus achieving the techno-economic feasibility of the process.

References

1. https://www.grandviewresearch.com/
2. https://www.statista.com/
3. AL. MA. Srl Packing and Packaging Machinery: Thermoforming and in-line Blister Packing Machines
4. ISO 11607-1 Packaging for terminally sterilized medical devices – part 1 requirements for materials, sterile barrier systems and packaging systems AMENDMENT 1 (2014)
5. ISO 11607-2 Packaging for terminally sterilized medical devices – part 2 validation requirements for forming, scaling and assembly processes AMENDMENT 1 (2014)
6. Franks, S.: Seal strength and package integrity – the basics of medical package testing. TM Electronics, Inc. (2006)
7. https://www.bfsv.de/en/tests/sterilepackages/
8. https://www.minitab.com/uploadedFiles/Documents/getting-started/Minitab17_GettingStarted-en.pdf

Control of Robot for Ventilation Duct Cleaning

Milos Bubanja[1(✉)], Milena Djukanovic[2], Marina Mijanovic-Markus[1],
and Mihailo Vujovic[1]

[1] Faculty of Mechanical Engineering, University of Montenegro,
DzordzaVasingtona bb, 81000 Podgorica, Montenegro
milosbubanja@hotmail.com, marinami@ac.me,
vujovic_mihailo@yahoo.com
[2] Faculty of Electrical Engineering, University of Montenegro,
DzordzaVasingtona bb, 81000 Podgorica, Montenegro
milenadj@ac.me

Abstract. In this paper we will present control algorithm of robot for cleaning and inspection of ventilation ducts. Importance of keeping ventilation ducts in good condition should not even be discussed; it is enough to know that most countries in the world have strict regulations and laws that govern this. Analysis is mostly focused on code that controls robot, which is written in LabVIEW, and results of testing performed on robot prototype. Tests were performed in different environments. Robot's movement was tested by putting different kinds of obstacles in its path to see its performance and simulate its operating environment. Because of its importance for functioning of robot, Wi-Fi signal strength was tested, and amount it drops over distance and through different materials. Time of robot's autonomy was also put to test.

Keywords: Robot · Duct · Cleaning · Inspecting · Ventilation system
Air conditioning system

1 Introduction

HVAC (heat, ventilation and air conditioning) technology (Fig. 1) is present almost everywhere nowadays and is integral part of all up-to-date residential buildings, houses, industrial and office buildings etc. Its main purpose is to provide air that is thermally stable and of good quality, enabling the user to control air temperature.

Increase in usage of HVAC technology also caused great increase in need for proper management and maintenance of HVAC systems [1]. One of the most important parts of maintenance of these systems is its cleaning from dust and other kinds of filth. It is very hard to complete this task by hand, and so due to immense technological advancements specialized robots are being made to take over this job.

In this paper detailed look at proposed robot for inspecting and cleaning of ventilation systems is provided as well as tests performed on it. Robot's movement capabilities were tested by placing different types of obstacles in its path to see how it handles it, these included slopes, trenches, big steps and small steps. Due to its importance in controlling the robot, strength of Wi-Fi signal and its drops over distance as well as through different types of materials are tested. Autonomy of robot's battery

© Springer Nature Switzerland AG 2019
S. Avdaković (Ed.): IAT 2018, LNNS 60, pp. 366–374, 2019.
https://doi.org/10.1007/978-3-030-02577-9_36

Fig. 1. Example of HVAC system

was also put to test. More detailed analysis of these tests and their results are provided further down the paper. All parts used in building of robot are connected with screw connection so it can be easily modified. That also allows users to place additional equipment on it should the need arise. Robot was built in Laboratory for mechatronics on Faculty of Mechanical Engineering, University of Montenegro.

2   Mobile Robot Design

The mobile robot developed for inspection and cleaning of ventilation ducts is shown on Figs. 2, 3 and 4. It is shown with camera for ducts inspection and without cleaning brushes. Robot has four driven wheels with differential control.

Fig. 2. Mobile robot for inspection and cleaning of ventilation ducts

Fig. 3. Mobile robot (side view)

Fig. 4. Mobile robot, front view

3   Robot Control

Microcontroller for control of this entire system is produced by National Instruments and has been programmed in LabVIEW. This software allows great flexibility and ease in coding.

In the coming part of the paper, codes which control movement, camera that sends information to user as well as code that allows robot to be controlled by joystick are shown.

Control system is developed for four independent driven wheels. Figure 5 shows LabVIEW code for robots movement. This part of the code controls all four motors which drive the robot. On the Fig. 5 is shown part of the code which allows robot to receive user's commands from joystick to control robot, and act upon those commands.

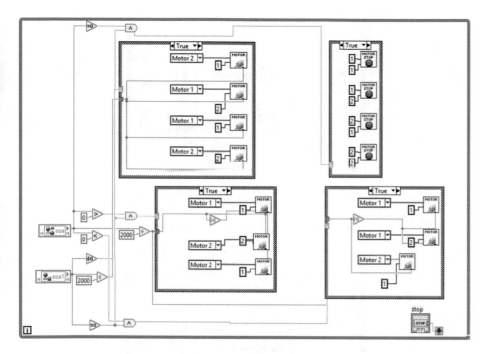

Fig. 5. Code for movement

Data received from joystick is stored in shared variable (*osa* and *osa1*). These variables, depending on the position of joypad stick, variate from −35000 to 35000 by linear function. These variables are divided by 2000 to achieve better sensitivity on joypad sticks. Data for movement forwards and backwards are stored in variable *osa1*. Movement is realized through *if* loop (Fig. 6).

Input parameters for motor*sub VI* (this subroutine in LabVIEW) are: data from joypad, position of DC motor controller in controller chain, position of DC motor on controller itself. On one DC motor controller only two motors can be connected. If variables are equal to zero motors are broken. Robot stops its movement fairly quickly because of existence of *if* loop dedicated to braking. When joypad stick is released, DC motor controller sends electric current that differs in polarity to previous movement.

In case of turning left or right, the *if* loops shown in Figs. 7 and 8 are used.

In case of turning, three motors are in use. If we want to turn right (Fig. 7), DC motor controllers sends current in required direction for left front wheel and left rear wheel to move forwards and right front wheel to move backwards. In this case, right rear wheel does not receive any current, and moves freely.

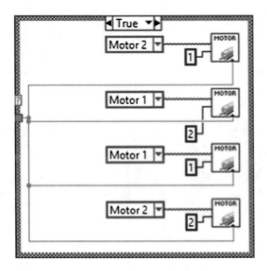

Fig. 6. *if* loop for forwards/backwards movement

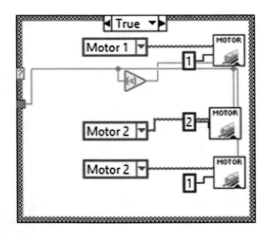

Fig. 7. *if* loop for turning right

In case of turning left (Fig. 8), DC motor controllers send current in required direction for right front wheel and right rear wheel to move forwards and left front wheel to move backwards. In this case, left rear wheel does not receive any current, and moves freely.

During the testing, it was concluded that this wheel control is best suited to the needs defined, because of very good maneuverability in small space and when wheel sliding is not important.

The code shown at the Fig. 9 allows camera to send information it captures in real-time. The code allows precise monitoring of robots position as well as inspecting the state of ventilation system. Example of what the user sees is given on the left side of the picture, while the code itself is presented on the right side.

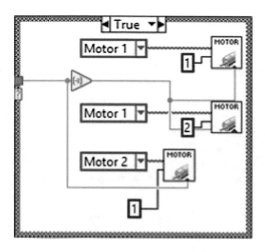

Fig. 8. *if* loop for turning left

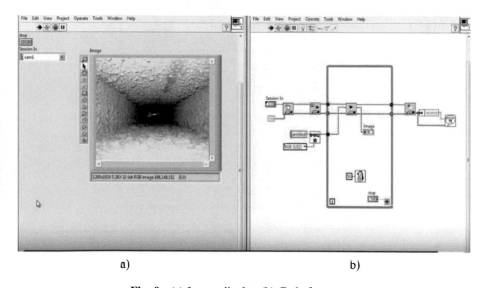

a)   b)

Fig. 9. (a) Image display (b) Code for camera

Code presented in Fig. 9b is created by using vision and motion tools in LabVIEW. Complete program for acquisition of video data can be made using this palette of tools. Firstly, the camera is initialized using IMAQdx configuration *grab.vi*. This function is located outside of *while* loop because initialization of camera should perform only once when the program starts. Inside of *while* loop there is IMAQdx*grab.vi* which is the main *VI* in charge of recording still images, and thanks to *while* loop this process in being repeated constantly every fifty milliseconds, thus creating video data.

On front panel indicator there is Image Display (Fig. 9a), through which operator can view inside duct. This is the only thing shown to operator, as all the other processes are being completed in background. Outside of *while* loop there is also image buffer which offers different options for customizing video operator receives. Camera is connected to myRIO controller via USB port.

In Fig. 10 is presented code that allows information to be sent from joystick to robot. Because operator can connect multiple joysticks to laptop, he needs to initialize used joystick firstly. This is realized through device *ID.vi*. Inside *while* loop acquired *inputs.vi* is located. With it operator acquires different types of data which then are stored in according types of variables.

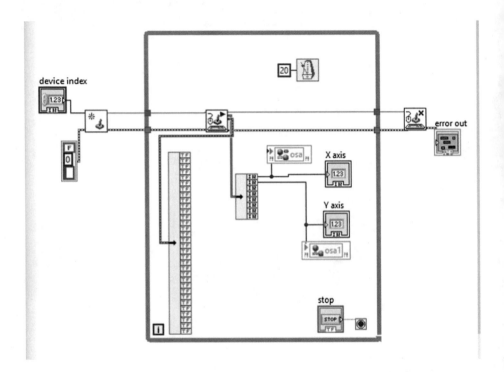

Fig. 10. Code for joystick

Most interesting is position of joystick whose values are stored in *osa* and *osa1* shared variables (Fig. 11). If there is need for reading other types of input data (button states for example) this can also be achieved with the use of this code.

Using delay function data transmission is being repeated in cycle of 20 ms. Shared variables which are being read in this interval are harmonized with myRIO micro-controller. If any error occurs program closes inputs automatically.

4 Testing of Mobile Robot

Testing was done on robot model shown on Figs. 2, 3 and 4. Main important robot characteristics that are tested are:

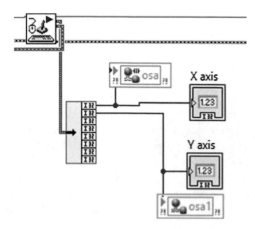

Fig. 11. Part of the code used for reading position of joystick stick

(1) Wi-Fi connection strength depending on distance and environment,
(2) Robots capability to cross obstacles,
(3) Autonomy of robot.

If the conditions were ideal, it could be simply applied inverse square law and determined the range of our Wi-Fi module. In this case signal loss over the distance of fifty meters was shown to be about 50%. Since this was not the case, there were additional losses caused by obstacles (walls, ventilation walls). Because of variation and complexity of environment signal strength may vary even if tests were done over the same distance but surrounding were different. The calculations for obtaining these data are not from the subject of this paper, but the results obtained experimentally fully satisfy the needs of the operator.

Different obstacles are created for the testing robot's ability, which were: trenches, small steps, slope, big steps (Fig. 12). Performance of the robot during these tests also proved to be satisfactory, as it managed to get over these obstacles with little to no difficulty.

Longevity of robots autonomy was calculated earlier in [2].

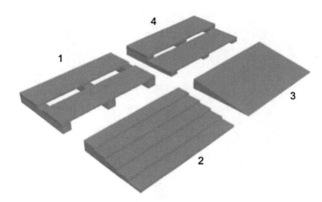

Fig. 12. Types of obstacles

5   Conclusion

In this paper is presented the control algorithm for proposed mobile robot for cleaning and inspection of ventilation ducts. Testing is performed in ventilation duct model. Robot was put to different kinds of tests, all of them meant to simulate real working conditions and check how it fares. Testing include robot's movement capabilities, maneuverability, strength of Wi-Fi control signal, and longevity of robot's battery autonomy. It managed to complete all the tests with satisfactory results.

It is also analyzed the solution for program made in LabVIEW, on which robot and all of its equipment runs.

All things considered, the robot is completely capable of completing its given task inspection and cleaning of ventilation ducts with minimal error range.

Future work includes testing of robot work with mounted brushes in real ventilation ducts.

Acknowledgment. This paper is prepared within two projects - "Montenegrin Wearable Robots (MWR)" supported by Ministry of Science of Montenegro and COST project - Action CA16116 "Wearable Robots for Augmentation, Assistance or Substitution of Human Motor Functions".

References

1. http://www.philcoaircontrol.com/wp-content/uploads/2016/07/ACCA-180.pdf
2. Bubanja, M., Markus, M.M., Djukanovic, M., Vujovic, M.: Robot for cleaning ventilation ducts. In: 4th International Conference "New Technologies NT-2018" Development and Application, 14–16 June 2018, Sarajevo, Bosnia and Herzegovina, paper accepted for publishing

Software for Assessment of Lipid Status

Edin Begic[1]($\boxtimes$), Mensur Mandzuka[2], Elvir Vehabovic[3],
and Zijo Begic[4]

[1] Department of Cardiology, General Hospital "Prim. Dr. Abdulah Nakas",
Sarajevo, Bosnia and Herzegovina
edinbegic90@gmail.com
[2] OSB AG, Munich, Germany
[3] Health Care Centre Maglaj, Maglaj, Bosnia and Herzegovina
[4] Pediatric Clinic, CCU Sarajevo, Sarajevo, Bosnia and Herzegovina

Abstract. Following indexes are used for the assessment of cardiovascular risk: Castelli Risk index I and II (CRI-I and II), Atherogenic Index of Plasma (AIP), atherogenic coefficient (AC) and CHOLIndex. It is important to emphasize that conventional lipid ratios give a clearer picture in the lipid status of the patient, although the conventional parameters are in physiological values. Aim of this article is development of software that could assist in the everyday work of both laboratory personnel and health workers (nurses, doctors). The developed software, is easily available, and represents a tool which will ease daily work in laboratory and will provide insight into the complete lipid status of the patient for the purpose of a high quality and comprehensive assessment of cardiovascular risk.

Keywords: Cardiovascular risk · Lipid status · Lipid indexes
Atherosclerosis · Software

1 Introduction

Lipids are widespread in all tissues and play a significant role in all life-cycle processes, help in digestion, provide energy conservation and serve as fuel in metabolism, represent structural and functional components of biological membranes, and as isolators provide nerve conduction and maintain body heat. In addition to very useful roles, they are linked to the pathology of lipoprotein metabolism and atherosclerosis (proaterogenic potential). Considering this potential, they play a major role in the cardiovascular system itself, or in cardiovascular risk assessment and in prevention of cardiovascular incident as such. Atherosclerosis makes about 80% of cardiovascular diseases and has a great health-social significance. The process of atherosclerosis begins with the emergence of endothelial dysfunction by the well-known risk factors for atherosclerosis. Lipid metabolism disorders have a fundamental importance in atherosclerosis process. The basic mechanism of arterial thrombosis is endothelial damage. At the very process of endothelial damage smoking, insulin resistance, hyperglycemia and hypercholesterolemia have an affect. Hypercholesterolemia works in a way to accelerate all stages of atherosclerosis, ranging from the initial stages of endothelial dysfunction due to reduction of the synthesis of oxides of nitrogen, further decreasing the vasodilatory and

© Springer Nature Switzerland AG 2019
S. Avdaković (Ed.): IAT 2018, LNNS 60, pp. 375–381, 2019.
https://doi.org/10.1007/978-3-030-02577-9_37

antithrombotic endothelial properties, increasing the endothelial potentiation and leukocyte adhesion, leading eventually to endothelial degeneration and arterial thrombosis [1, 2]. Cardiovascular diseases are detected by clinical assessment to which, along to basic procedures (medical history, physical examination, ECG, routine laboratory tests), should be added specific methods of diagnostics, non-invasive tests, and in which in addition to specific laboratory test should also be added values of cholesterol, triglycerides, and lipoprotein fractions. Based on the established cardiovascular risk assessment in primary prevention of cardiovascular diseases, we now have a system of estimation of individual risk of a ten-year fatal cardiovascular event. In clinical practice, prevention of cardiovascular disease helps those with low cardiovascular risk to retain at that level for life and in healthy individuals to maintain the characteristics of a healthy life, according to the target values of proven risk factors. The first risk assessment tables are derived from the Framingham study and based on the existence and severity of the major, conventional risk factors, the 10-year probability of developing coronary diseases can be estimated [3, 4]. Dyslipidemia, especially hypercholesterolemia, is the most important variable risk factor for coronary artery disease. There is clear evidence that the risk of vascular atherosclerosis is directly dependent on the level of cholesterol in plasma, which is why all recommendations for screening and treatment are related to total or LDL cholesterol. Screening of dyslipidemia is recommended for all men aged 35 and older and for women aged 45 and older. Earlier screening is recommended for all patients with family history of early coronary disease or familial dyslipidemia. The treatment of dyslipidemia is determined by the values of lipids, but also by the evaluation of cardiovascular risk [5, 6]. The beginning of therapy is very questionable. After cardiovascular incident, therapy with statins, drugs that affect cholesterol and triglyceride levels, is for a life, and especially after interventional coronary procedures. The main risk factors for cardiovascular diseases are elevated values of total (TC) and LDL cholesterol (LDL). It has been shown that concentrations higher than 1.5 mmol/L of HDL-cholesterol (HDL) have a protective effect. The values of HDL and LDL fraction are essential, i.e. the therapeutic regimen based on them is planned or revised. Because of the high probability of false positive results (primarily the influence of nutrition and the life style of patients), the current laboratory results are often not the best indicator of the lipid status of the patient. Cholesterol and triglyceride (TG) values are sometimes not the best predictor of the lipid status of the patient, and cannot clearly influence on the assessment of cardiovascular risk [7]. For this reason, clear markers of lipid status are developed as parameters, as one of the risk factors for increased cardiovascular risk [1–4]. For the assessment of cardiovascular risk the following indexes are used: Castelli Risk index I and II (CRI-I and II), Atherogenic Index of Plasma (AIP), atherogenic coefficient (AC) and CHOLIndex [8–11]. It is important to emphasize that conventional lipid ratios give clearer picture in the lipid status of the patient, although the conventional parameters are in physiological values. They give a clearer picture in setting the true rate of cardiovascular risk, but they are also useful in monitoring of the therapy. Laboratory findings, in addition to standard tests (TC, TG, HDL, LDL), in today's medical practice, give immediate insight into these indexes, which are of great help to general practitioners as well as specialists (endocrinologists, cardiologists). The calculation of the above mentioned indexes represents a small mathematical process, so development of software that would do this automatically would be a great help. The atherogenic index of plasma is calculated from serum, so for this reason (because of the laboratory procedure itself) it is excluded from processing in the development of this software. Data from big observational studies, including Framingham Study, the LRCP and the PROCAM, suggest that total/HDL cholesterol ratio is a better predictor of cardiovascular risk than self-interpreted total cholesterol, LDL cholesterol and HDL cholesterol [12]. Also, the relationship proved to be a good prediction of the carotid intima-media thickness, and is a better predictor than the isolated interpretation [8, 12]. LDL/HDL ratio is also powerful predictor of cardiovascular risk, as indicated by the Helsinki study [13]. Both ratios have a great clinical benefit also in the follow-up of medication treatment. Benefit of these two ratios was also demonstrated in Framingham Heart Study and Coronary Primary Prevention Trial in patients with vascular changes and after coronary intervention for the purpose of cardiovascular risk assessment [8]. The atherogenic coefficient places HDL in the primary view, which is essentially the goal of lipid metabolism disorders treatment. From all of this, the impression is that the evaluation of the mentioned indexes should be performed in routine laboratory examinations, even under basic, primary health care conditions, since they are essentially simple, free and easily accessible.

2   Aim

Development of software that could assist in the everyday work of both laboratory personnel and health workers (nurses, doctors).

3   Materials and Methods

For the purpose of doing the needed calculations a web application was developed which is hosted in the cloud. The latest web technologies were used to enable optimal user experience on both computer screens as well as mobile devices. Validation on user input is performed to check that all entered parameters are within their respective ranges. The programming language used in development is Python 2.7, coupled with the Flask web framework. To deliver optimal user experience and responsive design, Semantic UI library was utilized.

4   Software

The software, as first step offers a possibility to enter gender, values of TC, TG and HDL cholesterol value (while values are entered, the reference ranges of the mentioned variables are visible in the background, TC 0–6.8 mmol/L, TG 0–2.8 mmol/L, HDL 0.77–1.18 mmol/L for males and 0.77–2.28 mmol/L for females) (Fig. 1) (reference laboratory ranges are the once from the institution that software was tested first - reference laboratory ranges can be changed during use at any time).

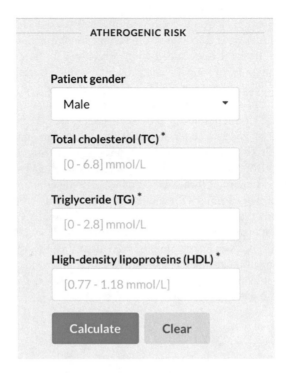

Fig. 1. First step in software startup

LDL cholesterol values are obtained from Friedewald's formula (1) (if TG values are higher than 4.52 mmol/L, the procedure is stopped).

$$LDL = TC - HDL - \frac{TG}{2.2} \qquad (1)$$

From the value of triglycerides (2), the value of very low-density lipoprotein (VLDL) cholesterol can be calculated (they are included in the composition of TG, and thus have a role in cardiovascular risk assessment).

$$VLDL = \frac{TG}{2.2} \qquad (2)$$

CRI-I (also known as cardiac risk ratio [CRR]) has a great importance because it reflects coronary plaques formation and the thickness of intima-media in the carotid arteries of young adults (the value is calculated based on TC/HDL). CRI- I shows the significance of HDL (normal value 2.4–7.1 (recommended 4.5) (3) [5].

$$CRI_I = \frac{TC}{HDL} \qquad (3)$$

CRI-II (atherosclerosis index) (4) represents LDL/HDL ratio (normal value is ranged 1.4–5.7), and literature suggests that it is a better indicator of cardiovascular risk in comparison to absolute concentrations [7, 8].

$$CRI_{II} = \frac{LDL}{HDL}$$ (4)

Atherogenic Coefficient (AC) (estimated as (TC-HDL)/HDL) estimates the total cholesterol concentration in all other cholesterol fractions compared to the HDL fraction (5). AC > 3.0 is an abnormal value [6, 9].

$$AC = \frac{TC - HDL}{HDL}$$ (5)

CHOLIndex is a relatively new simple index with proven use in prediction of likelihood of developing coronary arterial disease (CAD) with more accuracy than the other lipid ratios (6) (equation is for value of TG < 4.52 mmol/L) and CHOLIndex > 2.07 represents an abnormal value for assessment of cardiovascular risk [9, 11]. Software solution will give answer on LDL and VLDL values and mentioned indexes (Fig. 2).

$$CHOLIndex = LDL - HDL \qquad (TG < 4.52 mmol/L)$$ (6)

ATHEROGENIC RISK

3.75
Low density lipoproteins (LDL)

0.45
Very low density lipoproteins (VLDL)

RISK FACTORS

5.2
Castelli's risk index (CRI-I)

3.75
Castelli's risk index (CRI-II)

4.2
Atherogenic Coefficient (AC)

2.75
CHOLINDEX

Fig. 2. Software next step - values of LDL, VLDL, CRI-I, CRI-II, AC and CHOLIndex

5 Potential of the Application

The software solution was tested on fifty patients in the laboratory of Health Care Center Maglaj, Bosnia and Herzegovina. Solution was tested on laboratory results of total cholesterol, triglyceride, and cholesterol fractions. The fact is that the software will be acceptable by medical workers as a tool that can make daily work easier. It is deployed in the cloud, so it can be easily accessible from both smartphones and workstations using the following link: https://atherogenic-risk.herokuapp.com.

6 Conclusion

The developed software, is easily available, and represents a tool which will ease daily work in laboratory and will provide insight into the complete lipid status of the patient for the purpose of a high quality and comprehensive assessment of cardiovascular risk. Importance of this assessment will be in early diagnostic of cardiovascular pathology and prevention of cardiovascular incident. Also this solution is good option for monitoring of medical treatment.

References

1. Nigam, P.K.: Serum lipid profile: fasting or non-fasting? Indian J. Clin. Biochem. **26**(1), 96–97 (2011)
2. Gasevic, D., Frohlich, J., Mancini, G.J., et al.: Clinical usefulness of lipid ratios to identify men and women with metabolic syndrome: a cross-sectional study. Lipids Health Dis. **13**, 159 (2014)
3. Du, T., Yuan, G., Zhang, M., et al.: Clinical usefulness of lipid ratios, visceral adiposity indicators, and the triglycerides and glucose index as risk markers of insulin resistance. Cardiovasc. Diabetol. **13**, 146 (2014)
4. Milionis, H.J., Elisaf, M.S., Mikhailidis, D.P.: Lipid abnormalities and cardiovascular risk in the elderly. Curr. Med. Res. Opin. **24**, 653–657 (2008)
5. Nair, D., Carrigan, T.P., Curtin, R.J., et al.: Association of total cholesterol/high-density lipoprotein cholesterol ratio with proximal coronary atherosclerosis detected by multislice computed tomography. Prev. Cardiol. **12**, 19–26 (2009)
6. Olamoyegun, M.A., Oluyombo, R., Asaolu, S.O.: Evaluation of dyslipidemia, lipid ratios, and atherogenic index as cardiovascular risk factors among semi-urban dwellers in Nigeria. Ann. Afr. Med. **15**(4), 194–199 (2016)
7. Bhardwaj, S., Bhattacharjee, J., Bhatnagar, M.K., et al.: Atherogenic index of plasma, castelli risk index and atherogenic coefficient new parameters in assessing cardiovascular risk. Int. J. Pharm. Biol. Sci. **3**(3), 359–364 (2013)
8. Millán, J., Pintó, X., Muñoz, A., et al.: Lipoprotein ratios: physiological significance and clinical usefulness in cardiovascular prevention. Vasc. Health Risk Manag. **5**, 757–765 (2009)
9. Ogbera, A.O., Fasanmade, O.A., Chinenye, S., et al.: Characterization of lipid parameters in diabetes mellitus – a Nigerian report. Int. Arch. Med. **2**, 19 (2009)

10. Ogunleye, O.O., Ogundele, S.O., Akinyemi, J.O., et al.: Clustering of hypertension, diabetes mellitus and dyslipidemia in a Nigerian population: a cross sectional study. Afr. J. Med. Med. Sci. **41**, 191–195 (2012)
11. Akpınar, O., Bozkurt, A., Acartürk, E., et al.: A new index (CHOLINDEX) in detecting coronary artery disease risk. Anadolu Kardiyol Derg. **13**(4), 315–319 (2013)
12. Frontini, M.G., Srinivasan, S.R., Xu, J.H., et al.: Utility of non-high-density lipoprotein cholesterol versus other lipoprotein measures in detecting subclinical atherosclerosis in young adults (the bogalusa heart study). Am. J. Cardiol. **100**, 64–68 (2007)
13. Manninen, V., Tenkanen, L., Koskinen, P., et al.: Joint effects of serum triglycerides and LDL cholesterol and HDL cholesterol concentration on coronary heart disease risk in the Helsinki Heart Study: implications for treatment. Circulation **85**, 37–46 (1992)

Electrical Machines and Drives

Automated Data Acquisition Based Transformer Parameters Estimation

Elma Begic[1(✉)] and Tarik Hubana[2]

[1] Public Enterprise Elektroprivreda of Bosnia and Herzegovina, Konjic,
Bosnia and Herzegovina
elma.begic@epbih.ba
[2] Public Enterprise Elektroprivreda of Bosnia and Herzegovina, Mostar,
Bosnia and Herzegovina
t.hubana@epbih.ba

Abstract. The advancement of new technologies has brought many changes in traditional electric power systems, especially in the terms of the new monitoring systems, real time load flow calculations and advanced computer simulations and analysis of electric power systems. For each of these applications, the knowledge of the accurate system components parameters are crucial. However, the parameters of system components are not easily accessible, especially when it comes to transformers parameters. Since it is generally required to disconnect the transformer from the power system in order to measure and calculate the parameters, the transformer off-line time needs to be reduced as much as possible. This paper proposes a method for automated data acquisition based transformer parameters calculation, which reduces the transformer off-line time, and improves power quality. Results conducted on a real power transformer demonstrated that the developed hardware and user interface software are easy to use, with fast and accurate calculation. This paper makes a contribution to the existing body of knowledge by developing and testing an automated method for transformer parameters calculation, whose application represents an improvement when compared to the traditional process of calculating the transformer parameters.

1 Introduction

The power transformer represents a crucial element in any transmission or distribution network. Management and operation of the power systems have changed dramatically in recent years due to technological advancements and new regulatory requirements. Nowadays, electric power systems (EPS's) have integrated monitoring systems where the correct models of system elements are important. Accurate model of the certain EPS element gives better insight into the system operation, and more accurate load flow models and simulations. With the advancement of Smart Grid technologies [1, 2], a new supervision systems emerge, with real-time simulations, where the accurate models are even more required. The majority of the power transformers in EPS's, especially the distribution transformers are quite old, and parameters of this transformers are hard even impossible to obtain. The distribution system operator

© Springer Nature Switzerland AG 2019
S. Avdaković (Ed.): IAT 2018, LNNS 60, pp. 385–395, 2019.
https://doi.org/10.1007/978-3-030-02577-9_38

(DSO) companies have constant problems in this area when it comes to modelling different transformers in the process of integration of renewables [3], and power quality analysis [4]. Transformer parameters depend on many factors as transformer shape, windings placement and type etc. Because of that, the best approach is to estimate parameters by measuring, i.e. with the shot-circuit and open-circuit test. These tests require the transformer disconnection and quite long process of the testing and parameters calculation. This can significantly affect the consumers if there is no backup power supply, and worsens the SAIDI and SAIFI performance parameters of a whole system.

Precise estimation of transformer equivalent circuit parameters plays an important role in many aspects of power transformer condition monitoring, and fast and accurate parameters estimation is still a focus of many researchers. Authors in [5] and [6] have proposed a genetic algorithm based method for parameters estimation. Also, there are many approaches for on-line parameters estimation [7–9], but the activity of the grid still presents the essential source of problems [10].

Hence, the method that estimates the transformer parameters quickly and accurately, and provides short power interruptions is required. The advancement of electronic devices, with affordable prices enables their usage in many applications. Thus, the automated transformer parameters estimation system is proposed as an answer to all the previously mentioned problems.

2 Background

Transformers are widely used for different purposes in almost all areas of electrical engineering. They are used in electronic circuits for all kinds of current and voltage rectifiers that are used for control, regulation, signalization, protection and transmission of electrical impulses. Transformers that are used for this purpose have low power, and low voltage, but they can work in either very narrow or wide frequency range. Larger and more powerful transformers are in most cases constructed as three-phase transformers, even though the common practice in the USA is still to use 3 single-phase transformers. In case of three-phase transformers, that are common in Europe transmission and distribution power systems, every phase in a transformer has one separate winding. The phase windings are mutually connected, and a unique three-phase winding is created in that way. If the magnetic saturation of the windings is neglected, the transformer can be represented as shown in Fig. 1.

Parameters of the equivalent circuit are being determined in the process of the transformer design. However, this parameters can be determined by conducting two different operation tests: open circuit and short circuit test. The common practice is to conduct this test separately, by disconnecting the transformer from the grid, and afterwards calculate the parameters.

The open circuit test is an operating mode of transformer, where the low voltage (LV) circuit is open (there is no current trough LV windings). With neglected resistance R_1 and reactance $X_{1\sigma}$, since the $R_{Fe} \gg R_1$ and $X_\mu \gg X_{1\sigma}$, the equivalent circuit of the transformer in this case is shown in Fig. 2.

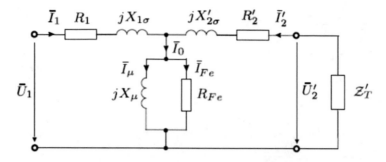

Fig. 1. Equivalent circuit of the transformer [11]

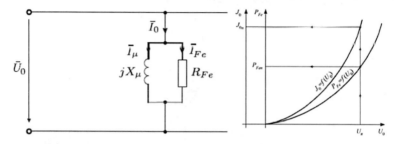

Fig. 2. Equivalent circuit of the transformer during the open-circuit test (left) [11] and the transformer characteristics during the open-circuit test [12]

During this test, the input current and power are measured and used for the parameters calculation, according to the following equations [11]:

$$\varphi = \cos^{-1} \frac{P_0}{U_0 I_0} \tag{1}$$

$$I_{fe} = I_0 \cos \varphi \tag{2}$$

$$I_{\mu} = I_0 \sin \psi \tag{3}$$

$$R_{fe} = \frac{U_0}{I_{fe}} \tag{4}$$

$$X_{\mu} = \frac{U_0}{I_{\mu}} \tag{5}$$

Opposed to the previous test, the short circuit test is conducted by short circuiting the LV side of the transformer and feeding the transformer from the high voltage (HV) side. During the short circuit test, because of the small impedance of the short circuited side, the current through the shunt branch can be neglected. Figure 3 shows the equivalent circuit of the transformer during the short-circuit test.

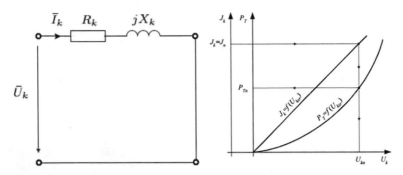

Fig. 3. Equivalent circuit of the transformer during the short-circuit test (left) [11] the transformer characteristics during the short-circuit test [12]

Transformer parameters are afterwards calculated from the following equations [11]:

$$\varphi = \cos^{-1} \frac{P_k}{U_k I_k} \tag{6}$$

$$Z_k = \frac{U_k}{I_k} \tag{7}$$

$$R_k = Z_k \cos \varphi \tag{8}$$

$$X_k = Z_k \sin \varphi \tag{9}$$

$$R_1 = R_2 = \frac{R_k}{2} \tag{10}$$

$$X_{1\sigma} = X_{2\sigma} = \frac{X_k}{2} \tag{11}$$

3   Automated System Design

In the following section the automated system design and components used for assembly will be presented. The system is designed with the following parts: Velleman K8055 and minilab 1008 data acquisition (DAQ) devices, current transformers (CT-s), mechanical relays, DC motor, AC/DC rectifier, LM317 voltage regulator, laptop and two USB cables.

The Velleman K8055 (VM110 for a pre-assembled board) is a low cost digital IO board [13]. The K8055 interface board has 5 digital input channels and 8 digital output channels. In addition, there are two analogue inputs and two analogue outputs with 8 bit resolution. [13] All communication routines are contained in a Dynamic Link Library (DLL), thus the custom Windows applications in Delphi, Visual Basic, C+

+ Builder or any other 32-bit Windows application development tool that supports calls to a DLL can be easily developed.

Figure 4 (right) shows the structure of the Vellaman board, where digital inputs are labelled with the number 1, analogue inputs with number 2, setting of the output voltage *A1* with number 5, setting of the output voltage *A2* with number 6, address choice with number 7, analogue inputs with number 8, digital outputs with number 9 and USB cable connection with number 10.

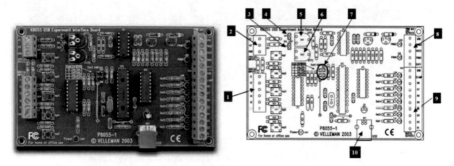

Fig. 4. Vellaman K8055 visual look (left) and the board components (right) [13]

Besides the Vellamann K8055 DAQ device, the miniLAB 1008, that offers a low-cost solution for multifunction measurement applications, is used for measurement. The miniLAB 1008 features eight 12-bit analogue input signal connections and 28 digital I/O connections. It is powered by the +5 V USB supply [14]. No external power is required. Two screw terminals rows provide connections for eight analogue inputs, two 10-bit analogue outputs, four bidirectional digital I/O lines, and one 32-bit external event counter. The analogue input connections can be configured with software as either four single-ended or eight differential channels. All analogue connections terminate at the screw terminals [14]. The miniLAB 1008 USB device is shown in Fig. 5.

As adjustable voltage regulator, a LM317 device is used [15]. LM317 is a linear voltage regulator with 1% output voltage tolerance. For this purpose it is used for the DC motor control. Input voltage of this regulator is changed in the range of 10-40 V, and the output voltage is described in the following equation:

$$U_{iz} = 1.25V\left(1 + \frac{R_2}{R_1}\right) \tag{12}$$

The resistors provide better accuracy and the additional stabilisation is improved because of the capacitance between the output end earth. This voltage regulator has an additional cooler to avoid excess heating of the device.

The DC motor is used for the autotransformer regulation. The motor is connected to the autotransformer over a timing belt, and in this manner the voltage can be regulated. The current is measured with the current transformer (CT) with the sensitivity of 250 mV/A. Rated power of the fuses used in this system is 20A. The mechanical relays

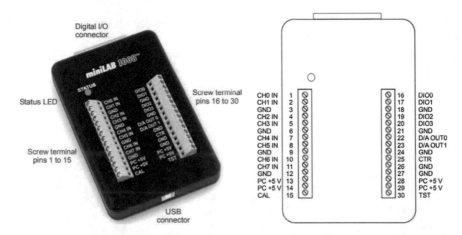

Fig. 5. Minilab 1008 external components and (right) and main connectors and pin outs (left) [14]

(24 V (DC), 6A, 250 V (AC)) are used, and they are controlled via AC/DC rectifier and voltage regulator LM317.

All these components provide the automatic measurement system that will automatically adjust the voltages and currents during the short and open circuit tests, and perform measurement when the conditions are met. The system is shown in Fig. 6.

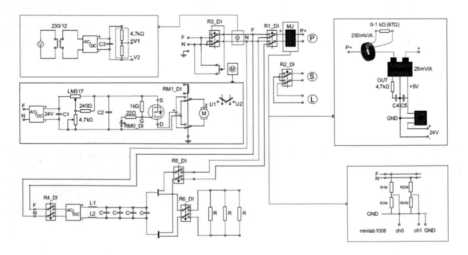

Fig. 6. Scheme of the automated measurement system

The designed hardware is coupled with appropriate user interface software developed in Microsoft Visual Studio 6.0 [16]. The software is used for control, measurement, data processing and finally for the calculation of parameters and visualisation of the transformers equivalent circuit.

4   Results and Discussion

This section will discuss the test system and the results of the proposed automated transformer parameters estimation method.

4.1   Test System

The proposed measurement system can work with the transformers with power up to several KVA. The open and short circuit tests are conducted via autotransformer and mechanical relays. Controlled via the Velleman K8055 DAQ board, the DC motor moves the autotransformer regulator, thus changing the voltage level. The Velleman K8055 DAQ board also controls the relays and changes the transformer LV side connection in either open or short circuit position. The measurement of the electric signals performed via minilab DAQ measurement board and CT's, altogether paired with the developed user interface software in Visual Studio 6.0. The test system is shown in Fig. 7.

Fig. 7. Designed automated measurement system

4.2   Developed User Interface

Figure 8 show the outline of the developed user interface software. The red numbers in the boxes show the main software functions. Function labelled with number1 shows the command *connect* that check the connection of the Velleman board with the computer, i.e. the software.

If the board is connected, the rated values of current and voltage need to be given as input in the text boxes labelled with number 3. Then it is possible to run one of the commands labelled with number 2. If the command labelled with number 4 is selected, the software will stop, and all the analogue and digital outputs will be shut down. Text

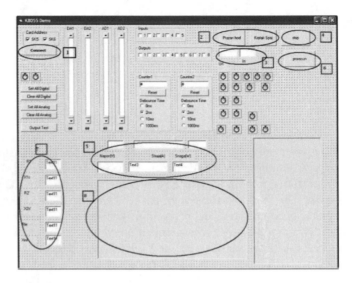

Fig. 8. Outline of the developed user interface software

boxes labelled with number 5 show the measured values of the voltage, current and power, respectively. After the one of the commands labelled with number 2 is selected, it is necessary to press the *proračun (calculate)* button, labelled with number 6. This command calculates the transformer parameters according to the Eqs. (1)–(11). After the calculations, the equivalent circuit of the transformer with all the parameters will be shown in the software window.

For the testing purposes, the mechanical relay R4 is connected between the LV transformer contacts. In the case of open circuit test, the mechanical relay is turned OFF, and there is no current flow between the contacts. In the case of short circuit test, the mechanical relay R4 is turned ON, and thus the LV side of the transformer is short circuited.

4.3 Automated Tests

In case of the open circuit test, the relay R3 need to be turned ON, and relay R4 OFF. In order to start the test, the command *Prazan hod (Open circuit test)* needs to be selected in the user interface software. As a result, the relays will be set in the appropriate setting, and the measurement will be started. Measurement takes values each 1 ms. The high voltage (HV) side voltage is continually measured during the test, and when it reaches the rated voltage value (with tolerance ±1 V), the measurement process stops.

In the case of short circuit test, the mechanical relay R3 need to be turned ON, and relay R4 OFF. The LV transformer side contacts are short circuited via relay R4. When the *Kratak spoj (Short Circuit)* command is selected, the HV side voltage is set to 0 V. Then the R3 and R4 relays are turned ON, and the measurement process starts. The voltages and currents are measured continuously until the current doesn't rise up to the rated value. When the current rises up to the rated current of the transformer (with tolerance of ±0.2 A), measurements are stopped, and the measured values are forwarded for the parameters calculation.

4.4   Transformer Parameters Calculation

The equations for the calculation of transformer parameters are programmed in the software. The calculation process proceeds according to the Eqs. (1) to (11). After both tests results, it is necessary to run the *Proračun (Calculate)* button, and as a result the transformer parameters will be calculated and the equivalent circuit of the transformer will be shown in the software, as shown in Fig. 9.

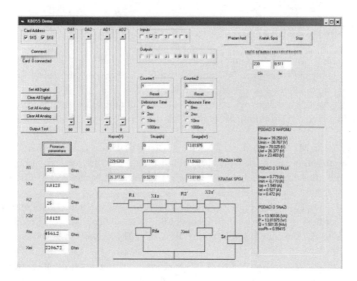

Fig. 9. Results of the measurement and the equivalent circuit of the transformer

5   Conclusions

In this paper an approach for automated transformer parameters calculation is presented. The system is completely developed and tested, both the hardware and software part, which is demonstrated in the paper.

The results demonstrated the efficiency of the system, and showed that this approach is simpler to use, much faster than the traditional process, doesn't require much circuit changes and the attention of the personnel while conducting the short and open circuit tests. Besides, no additional measurement devices are needed, since the measurements are carried out with the DAQ devices. The regulation system is galvanic isolated from the network voltage, resulting in the lower probability of failure. However, the automated system does not have fast dynamic response and requires additional 220 V AC. The calculation process is fast and thus improves the power quality and reduces the intentional outages, since the transformer offline time is reduces as much as possible.

The proposed system is tested and developed for the transformers up to several kVA, and is not suitable for the distribution transformers, but with few modifications, the system could be applicable to the higher power transformers. These modifications would in first place include voltage and current transformers, and an upgrade in rated power of other elements. Future research directions would be to adapt the proposed system for the 10(20)/0.4 kV distribution transformers, since there is a large number of transformers without exact parameters in the system. This paper presents a part of an ongoing research to improve the planning, operation and simulation process of the power distribution system.

References

1. Jadhav, V., Lokhande, S.S., Gohokar, V.N.: Monitoring of transformer parameters using Internet of Things in smart grid. In: International Conference on Computing Communication Control and automation (ICCUBEA), Pune (2016)
2. Hubana, T., Šarić, M., Avdaković, S.: Approach for identification and classification of HIFs in medium voltage distribution networks. IET Gener. Transm. Distrib. 12(5), 1145–1152 (2018)
3. Šemić, E., Šarić, M., Hubana, T.: Influence of solar PVDG on electrical energy losses in low voltage distribution network. In: Hadžikadić, M., Avdaković, S. (eds.) Advanced Technologies, Systems, and Applications II, IAT 2017. LNNS, vol 28. Springer, Cham (2018)
4. Hubana, T., Begić, E., Šarić, M.: Voltage sag propagation caused by faults in medium voltage distribution network. In: Hadžikadić, M., Avdaković, S. (eds.) Advanced Technologies, Systems, and Applications II, IAT 2017. LNNS, vol 28. Springer, Cham (2018)
5. Mossad, M.I., Azab, M., Abu-Siada, A.: Transformer parameters estimation from nameplate data using evolutionary programming techniques. IEEE Trans. Power Deliv. 29(5), 2118–2123 (2014)
6. Thilagar, S.H., Rao, G.S.: Parameter estimation of three-winding transformers using genetic algorithm. Eng. Appl. Artif. Intell. 15(5), 429–437 (2002)
7. Zjang, Y., Zhang, H., Mou, Q., Li, C., Wang, L., Zhang, B.: An improved method of transformer parameter identification based on measurement data. In: 5th International Conference on Electric Utility Deregulation and Restructuring and Power Technologies (DRPT), Changsha (2015)
8. Zhang, Z., Kang, N., Mousavi, M.J.: Real-time transformer parameter estimation using terminal measurements. In: IEEE Power & Energy Society General Meeting, Denver, 2015 (2015)
9. Bhowmic, D., Manna, M., Chowdhury, S.K.: Estimation of equivalent circuit parameters of transformer and induction motor from load data. IEEE Trans. Ind. Appl. PP(99), 1 (2018)
10. Staroszczyk, Z.: Problems with in-service (on-line) power transformer parameters determination - case study. In: 17th International Conference on Harmonics and Quality of Power (ICHQP), Belo Horizonte (2016)
11. Mašić, Š.: Električni strojevi. Elektrotehnički fakultet, Sarajevu (2006)
12. Mitraković, B.: Ispitivanje električnih mašina. Naučna knjiga, Belgrade (1991)

13. Velleman: Velleman (2018). https://www.velleman.eu/products/view/?id=351346. Accessed 11 Feb 2018
14. Measurement Computing: miniLAB 1008 USB-based Analog and Digital I/O Module - Users GUIDE. Measurement Computing Corporation, Norton (2006)
15. Texas Instruments: LM317 3-Terminal Adjustable Regulator, Dallas (2016)
16. Microsoft: Visual Studio 6.0 (2018). https://msdn.microsoft.com/en-us/library/ms950418.aspx. Accessed 13 Feb 2018

Evaluation of Losses in Power Transformer Using Artificial Neural Network

Edina Čerkezović[(✉)], Tatjana Konjić, and Majda Tešanović

Faculty of Electrical Engineering, University of Tuzla, Tuzla,
Bosnia and Herzegovina
edina.cerkezovic@untz.ba

Abstract. This paper presents an application of artificial intelligence for the analysis of total losses in power transformers. The method is based on a multilayer feed-forward neural network that uses the Levenberg-Marquard algorithm to adjust the network parameters. The analysis was carried out on a three-phase dry transformer 1000 kVA, 6000/400 V. The data used for developing the neural network were obtained experimentally by measuring on low voltage side of the transformer. The inputs to the developed neural network are: the mean value of the load current, the temperature and the losses in the copper, and the output is the total losses. The database contains 1441 samples obtained by changing the load every 30 s in the interval of 12 h. The network model was developed for a temperature of 25 °C, and then the same model was used to determine total losses at a temperature of 68 °C. Obtained results from the developed neural network were compared with the measured data. The low error value indicates that this neural network can be used for different load and temperature.

1 Introduction

Transformer is a static device that, on the principle of electromagnetic induction, converts electricity from one alternating system to another of the same frequency, but changes the voltage and current values [1]. Energy transformer is an important device of the electric power system because it enables electricity transmission at the appropriate voltage levels suitable for end-users. The losses that occur in the operation of power transformers are inevitable. An important requirement in transformer design is that transformer losses remain at a satisfactory level. The losses in the transformer are affected by various factors such as temperature, resistance, voltage, current, load, copper quality and many others. Due to load variation in an electric power system, it is very difficult to consider all these factors at the same time. Souza [2] applied the artificial neural network to overcome this problem, but he applied his work to a single-phase transformer and the model required several parameters for input in the neural network. Using an artificial neural network, it is possible very efficiently calculate losses in the transformer. In recent times, the use of neural networks for analyzing losses in the transformer is one of the most interesting methods [3–6]. Many researchers use artificial neural networks to predict transformers losses at the design stage [7, 8]. The advantages of the trained neural network are fast achievement of

© Springer Nature Switzerland AG 2019
S. Avdaković (Ed.): IAT 2018, LNNS 60, pp. 396–404, 2019.
https://doi.org/10.1007/978-3-030-02577-9_39

desired results and high accuracy in solving complicated problems. There are several types of artificial neural networks suitable for solving various problems including: RBF (Radial Basis Function) networks and multilayer feed-forward networks. In this paper, a multilayer feed-forward network with a back-propagation learning algorithm was used to calculate losses in a three-phase transformer. Data of load current and temperature were obtained by measuring while losses in the copper and the total losses were calculated based on the transformer data. These data were used for training and testing the neural networks. The paper is organized in the following way: in the second chapter, the losses in a transformer and their components are briefly explained, the third chapter describes developed neural networks, the fourth chapter presents the obtained results and at the end the conclusions drawn from this research and possible future research are noted.

2 Losses in Transformer

Losses in transformer are divided into three basic groups: losses in the core of the transformer, losses in conductors and stray losses [9]. Stray losses in power transformers are caused by a wasteful magnetic field in winding and connecting conductors. The losses in the core of the transformer or the iron losses (P_{Fe}) can be determined by open circuit test. They consist of losses due to eddy currents and losses due to hysteresis. They are dependent on change of frequency and magnetic induction. Copper losses (P_{Cu}) are obtained from a short circuit test and they consist of copper losses on the primary and copper losses on the secondary winding. We can calculate them using a relation:

$$P_{Cu} = qR_1I_1^2 + qR_2I_2^2 \tag{1}$$

where is:

P_{Cu} - losses in copper (W),
R_1 - resistance of primary winding (Ω),
R_2 - resistance of secondary winding (Ω),
I_1 - primary winding current (A),
I_2 - secondary winding current (A),
q - number of transformer phase.

The resistance of copper winding at different operating temperatures can be determined by the relationship:

$$R_t = R_h \frac{235 + \theta_t}{235 + \theta_h} \tag{2}$$

where is:

R_h - resistance of the copper coil in the stationary state at the temperature θ_h (Ω),
R_t - resistance of the copper coil in the stationary state at a temperature θ_t (Ω),
θ_h - transformer temperature in cold state (°C),

θ_t - transformer operating temperature (°C).

Total losses (P_u) are calculated as:

$$P_u = P_{Cu} + P_{Fe} \tag{3}$$

3 Proposed Neural Network

Artificial neural network (ANN) is a system consists of a large number of simple data processing elements. Such systems are capable to collect, memorize and use expert knowledge. They have an ability to learn from a limited number of examples. There are a large number of different neural networks that can be divided according to: number of layers, relationship between neurons, signal propagation path, way of training neural networks, type of data [10].

Analysis of total losses in a transformer using artificial neural network was carried out on a three-phase dry transformer of the following characteristics: S_n = 1000 (kVA), U_{n1} = 6000 (V), U_{n2} = 400 (V), I_{n1} = 96.34 (A), I_{n2}= 1445 (A), P_{Cun} = 7600 (W), P_{Fe} = 770 (W), R_1 = 0.273 (Ω), R_2 = 0.00012 (Ω). The measurement was carried out on the low voltage side of the transformer by power analyzer [11]. During 12 h of measurement a set of 1441 data for developing neural networks was collected at a temperature of 25 °C. The similar measurement was repeated for different working conditions (temperature of 68 °C). Those data will be use to confirm proposed network developed in the previous stage.

A multilayer feed-forward neural network was used to calculate total losses. The inputs to the network are: the mean value of load current I_{opt} (A), temperature T (°C) and copper losses P_{Cu} (W), and the output is the total loss P_u (W), as shown in Fig. 1.

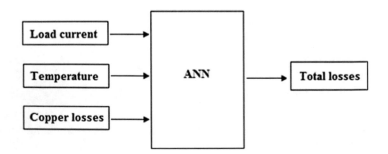

Fig. 1. Block diagram of proposed artificial neural network

One of the problems in the neural network development is to determine number of neurons in the hidden layer. Usually, the development of a neural network begins with one hidden layer of neurons. There are also no rigid rules for selecting the number of neurons in the hidden layer, but some of the following recommendations can be considered:

- The number of neurons in the hidden layer should be between the arithmetic values of the input and output parameters [12],
- Number of neurons = (input parameters + output parameters) * (2/3) [12],
- The number of neurons should be 70–90% of the arithmetic values of the parameters [12].

Number of neurons can also be determined using the sensitivity analysis. It begins with a minimum number of neurons and gradually increases until a satisfactory level of error is obtained. In procedure of developing neural networks it is also possible to select different activation functions. In this paper, on the base of the research in [3], a tangent-sigmoid transmission function in the hidden layer and a linear transfer function in the output layer was selected. For development of the neural networks, Neural Network Tool in the Matlab [13] was used. Total number of available data is randomly divided into three sets of data for training, testing and validation in the range of 70%, 15% and 15%, respectively.

Several different neural networks have been considered. However, in the purpose of the paper, feedforward neural networks with 2, 3 and 5 neurons in the hidden layer have been presented. The models were developed by three different algorithms: Levenberg-Marquardt, Bayesian Regularization and Scaled Conjugate Gradient [13].

4 Analysis Results

After development of several different neural networks for calculating the total losses in the observed transformer at a temperature of 25 °C, the results obtained are shown in Table 1. Change of number of neurons and training algorithms was made. Evaluation of different neural networks presented in Table 1 was based on the duration of the training process and Mean Squared Error (MSE).

From Table 1 it can be seen that in the case of using the Levenberg-Marquardt algorithm, increasing the number of neurons from 2 to 5, the training time increases slightly from 51 (s) to 54 (s), while the mean squared error increases from $3.56 \ 10^{-8}$ to $2.36 \ 10^{-8}$ (W). In the case of the Bayesian Regularization algorithm increasing the number of neurons from 2 to 5, the process lasts a little longer from 54 (s) to 57 (s), while the error is decreasing from $6.39 \ 10^{-9}$ to $2.02 \ 10^{-11}$ (W). The third algorithm gave the worst results. Increase number of neurons did not affect the duration of the training time - for each change of number of neurons (2, 3, 5) the training process lasts about 1 (s) and the error is approximately the same ($1.28–1.5 \ 10^{-2}$ W) that is very unsatisfactory compared to the results obtained by the previous two algorithms. Observing all the above facts, it can be concluded that the best responses were given by the neural network with 2 neurons in the hidden layer trained by Levenberg-Marquardt algorithm and the neural network with 5 neurons in the hidden layer trained by the Bayesian Regularization algorithm.

The developed neural networks were used to determine total losses at a new temperature of 68 °C. For the observed temperature, there was data of total losses obtained by calculation based on the measurements, which will be used to determine the error of the formed neural network. In the case that new input data related to a

Table 1. Characteristics of developed artificial neural networks

Temperature (°C)	Training algorithm	Number of neurons	Training time (s)	MSE (W)
25	**Levenberg-Marquardt**	**2**	**51**	**3.56 10^{-8}**
25	Levenberg-Marquardt	3	53	2.73 10^{-8}
25	Levenberg-Marquardt	5	54	2.36 10^{-8}
25	Bayesian Regularization	2	54	6.39 10^{-9}
25	Bayesian Regularization	3	56	3.71 10^{-9}
25	**Bayesian Regularization**	**5**	**57**	**2.02 10^{-11}**
25	Scaled Conjugate Gradient	2	1	1.5 10^{-2}
25	Scaled Conjugate Gradient	3	1	1.28 10^{-2}
25	Scaled Conjugate Gradient	5	1	1.45 10^{-2}

temperature of 68 °C passed through developed network with 2 neurons in the hidden layer trained by Levenberg-Marquardt algorithm, the mean square error was 9,728 10^{-5} (W). If the new date passed through other developed network (5 neurons in the hidden layer trained by Bayesian Regularization algorithm), the mean squared error was 1,224 10^{-2} (W). Comparing the values of the errors it can be concluded that the neural network with 2 neurons in hidden layer trained by Levenberg-Marquardt algorithm gave better response.

Figures 2 and 3 show the transformer losses obtained by measurements - the desired output ("izlaz", "izlaz1") and obtained by the proposed network with 2 neurons in hidden layer ("data", "data1") for temperatures of 25 °C and 68 °C, respectively. It is possible to notice minor variations between these two outputs, so it can be concluded that, based on the developed neural network at 25 °C, it is possible to determine the losses at any other temperature with an acceptable error.

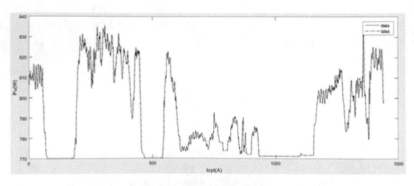

Fig. 2. Total transformer losses: red line ("izlaz") obtained by measurement; blue line ("data") obtained by the proposed neural network at temperature of 25 °C

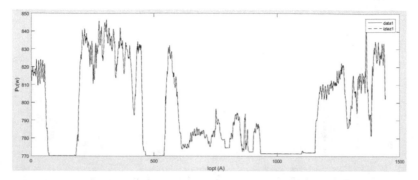

Fig. 3. Total transformer losses: dash red line ("izlaz1") obtained by measurement; blue line ("data1") obtained by the proposed neural network at temperature of 68 °C

In Fig. 4, a change of total losses depending on the load current at different temperatures was presented. By increasing the temperature, there is an increase of total losses.

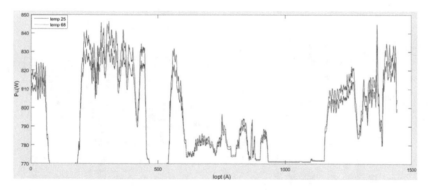

Fig. 4. Total losses obtained by proposed neural network at different temperature

Data were measured at a time interval of 12 h (14.00 h–02.00 h). Based on the obtained results and the neural network we can estimate the level of electricity consumption and losses that can occur the next day in the same time interval. With these data we can strive to reduce the total losses, thereby increasing the energy efficiency of the transformer, improving the cooling system and using better insulation materials.

In Figs. 5 and 6, dependence of total losses on copper losses at a temperature of 25 °C and 68 °C is shown, respectively. The increase in temperature increases the losses in the copper. Since core losses are constant for each load, total losses are increased by increasing copper losses. It can be noticed that in certain time interval, change of total losses depends on change in temperature and in load.

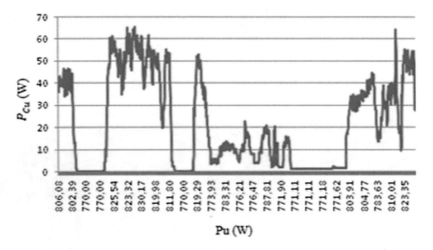

Fig. 5. Total and copper losses in the transformer at 25 °C

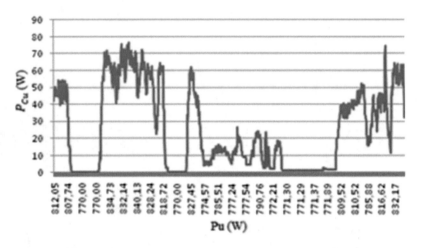

Fig. 6. Total and copper losses in the transformer at 68 °C

5 Conclusion

Analysis of total losses of a three-phase dry transformer, 1000 kVA, 6000/400 V is presented in this paper. Set of 1441 data was used to develop neural networks. The data collected by a power analyzer at temperatures of 25 °C and 68 °C. In the first stage, different neural networks were trained and tested on the base of data collected at temperature of 25 °C. After comparison of obtained networks characteristics, two networks were chosen for the next stage. In the second stage the proposed networks were used to obtained total losses in the transformer at other temperature of 68 °C. The performed analysis showed that it is possible with the relatively simple neural network (1 hidden layer and 2 neurons in the hidden layer) obtain total losses in the transformer.

The proposed network can be used to evaluate total losses at different temperatures for transformers of the same characteristics. Using modern methods of artificial intelligence, losses in the transformer as one of the most important parameters in phase of their design and construction can be much faster and reliably determined. Obtained results represent the base for further development and contribution to the analysis of losses in transformer. They can be used to reliably determine the nominal size in the design stage and to set the load limits of the transformer in different operating conditions.

Some additional researches could be done in the future. It would be interesting to test the developed network by data from different transformer's types. On the base of results that would be obtained, it would be possible to determine if the same developed model would be appropriate for losses evaluation for different types of transformers. During measurement phase power analyzer recorded different parameters (voltage per phase, current per phase, power per phase, voltage and current harmonics and flickers) that could be used for many analyses such as investigation of the influence of harmonic components or some other factors on the losses in transformer and the ways of mitigation them.

References

1. Kalić, Đ.: Transformers, Institute for Textbooks and Teaching Resources, Beograd (1991)
2. de Souza, A.N., da Silva, I.N., de Souza, C.F.L.N., Zago, M.G.: Using artificial neural networks for identification of electrical loses in transformers during the manufacturing phase. In: Proceedings of the 2002 International Joint Conference on Neural Networks, IJCNN 2002, 12–17 May 2002, vol. 2, pp. 1346–1350 (2002)
3. Suttisinthong, N., Pothisarn, C.: Analysis of electrical losses in transformers using artificial neural networks. In: Proceedings of the International MultiConference of Engineers and Computer Scientists 2014, IMECS 2014, 12–14 March 2014, vol. II, Hong Kong (2014)
4. Naresh, R., Sharma, V., Vashisth, M.: An integrated neural fuzzy approach for fault diagnosis of transformers. IEEE Trans. Power Deliv. 23(4), 2017–2024 (2008)
5. Leal, A.G., Jardini, J.A., Magrini, L.C., Ahn, S.U.: Distribution transformer losses evaluation: a new analytical methodology and artificial neural network approach. IEEE Trans. Power Syst. 24(2), 705–712 (2009)
6. Meng, K., Dong, Z.Y., Wang, D.H., Wong, K.P.: A self-adaptive RBF neural network classifier for transformer fault analysis. IEEE Trans. Power Syst. 25(3), 1350–1360 (2010)
7. Georgilakis, P.S., Hatziargyriou, N.D., Doulamis, N.D., Doulamis, A.D., Kollias, S.D.: Prediction of iron losses of wound core distribution transformers based on artificial neural networks. Neurocomputing 23, 15–29 (1998)
8. Yadav, A.K., Azeem, A., Singh, A., Malik, H., Rahi, O.P.: Application research based on artificial neural network (ANN) to predict no load loss for transformer's design. In: International Conference on Communication Systems and Network Technologies (2011)
9. Štrac, L.: Modeling electromagnetic properties of steel for calculating stray losses in power transformers, doctoral dissertation, FER Zagreb (2010)
10. Konjić, T., Švenda, G.: Making decision and optimization with application in the electricity system, 1st edn. Tuzla (2010)

11. Kupusović, A.: Impact of harmonics on losses and life time of transformers, Master's thesis, Faculty of Electrical Engineering, University of Tuzla (2017)
12. Jelušić, P.B.: Development of an automated calculation system using neural networks, Graduate thesis, University of Zagreb, Faculty of Graphic Arts, Zagreb (2016)
13. Matlab: Neural Network Toolbox, User's Guide. The Matworkss Inc. (2017)

Selection of the Optimal Micro Location
for Wind Energy Measuring in Urban Areas

Mekić Nusmir, Nukić Adis, and Kasumović Mensur[(✉)]

Department for Energy Conversion Systems, Faculty of Electrical Engineering,
University of Tuzla, Tuzla, Bosnia and Herzegovina
mensur.kasumovic@fet.ba

Abstract. Main objective of this paper is to analyze and propose solution for choosing optimal location for wind speed measurement in urban areas. When measuring wind speed on 12 months period it is crucial to find spot with best wind characteristics. In planning wind power plant there must be exact information about wind speed so it is highly important to have measurement in optimal location, where wind aggregate will be placed. This paper shows process of choosing optimal location using short-term measurement and Ansys software for wind simulation. Final goal is to have best output on long-term measurement and wind aggregate power.

Keywords: Wind speed measurement · Optimal location · Wind simulation
Wind power plant

1 Introduction

Distributed energy sources today are one of the most advanced areas of electrical engineering, and are increasingly used. Smart cities are especially topical, where, with the help of renewable distributed sources, households and all other elements of the smart city are being powered up (such as public lighting, common consumers…). In Bosnia and Herzegovina, unlike Europe, the process of switching to renewable sources and new technologies is running much slower. Currently, in Bosnia and Herzegovina, only small photovoltaic panels and solar panels are purchased as distributed energy sources, with the increasingly frequent presence of heat pumps in combination with pellet stoves or solar panels. The use of small wind farms as renewable energy sources has not yet been achieved, especially since a much higher investment is needed to acquire these devices than photovoltaic panels. Although these devices have time to return investments that resemble time investment returns for photovoltaic panels (that means the user will earn more money from the VE), their high initial investment is still a problem. Another problem is the relief of our area, which is generally unfavorable for the construction of wind farms because it is a mountainous area. Only the south of our country has suitable conditions for the construction of wind farms because of its geography and sea influence. Along the south, in the north, wind farms can also be operated, but with much less power than the south. This area is suitable for the construction of small wind farms, while the area of the central part of Bosnia and Herzegovina has a significantly lower wind potential. Nevertheless, it is possible to

© Springer Nature Switzerland AG 2019
S. Avdaković (Ed.): IAT 2018, LNNS 60, pp. 405–415, 2019.
https://doi.org/10.1007/978-3-030-02577-9_40

install wind farms with smaller power units that will meet household needs or cover their consumption. Especially in this are the farms and small businesses that often have high electricity bills and where it is necessary to reduce these costs. The project that is being processed here is for a small poultry farm, in the village of Oskova Banovići. The buyer's request is to install a wind power plant of 1 kW in his yard to cover the consumption of one part of the farm. Measurements and analysis of wind turbine conditions in the field were performed in order to perform the optimal location for setting up the wind turbine of this power. The basic parameters that are taken into account when optimizing the location are the roughness of the terrain, the position of the objects and the height at which the wind turbine is placed.

2 Wind Characteristics

The main characteristic of the wind is its medium speed, the extent and frequency of the wind speed change, and the maximum speed that has occurred in the last 5 years. The wind speed is measured using an anemometer. In addition to speed, it is also necessary to know the direction of the wind that blows so that must be an anemometer that has the ability to measure the direction. In this way, information is obtained which shows the degree of wind intensity in the appropriate direction. Wind measurements are carried out for a minimum of one year in order to get the accurate picture of the change in the wind and the expected strength of the wind turbine. Measurements are made every day throughout the year and values are taken at intervals of 10 min. In this way, a large number of measurements are obtained, as well as accurate data necessary for the construction of a wind farm. The wind speed depends largely on the location on which it is located, or the relief of that location. The concept of roughness of the surface, which is especially important for small wind turbines, is introduced, because they are at lower altitudes and the surrounding objects can interfere with the flow of wind. Because of this, the height at which the wind turbine is installed is important, since with height increases wind speed.

The most important parameter for a wind turbine is the power/energy it receives from the wind. This energy is further transformed into the kinetic energy of the wind turbine, and then into the electric energy by an electrical generator. The wind deliver its kinetic energy to the turbines by the appearance of aerodynamic forces on the blades of the turbine that make it rotate. This further implies that besides the wind speed on the amount of energy transferred, the shape of the turbine is also affecting energy conversion. The term for wind power, depending on wind speed and turbine design, is given as:

$$P = \frac{1}{2}\rho A v_o^3 C_p \qquad (2.1)$$

From previous equation it can be seen that the power of the wind, which is passed on to the aggregate, depends on the third degree of velocity and this applies to a wide range of winds with less oscillations. Wind power also depends on the turbine power factor, which represents turbine utilization. For modern wind turbines, power factor rate is ≈ 0.45, and for some turbines it goes up to 0.50. For small wind turbines this factor is smaller.

On the wind speed, great influence has height at which the wind farm is set up, as well as the roughness of the terrain that lies in front of the wind farm. With the height of the wind speed, the goal is to have such solutions for the pole of the wind aggregate that it can withstand the forces that occur and have the proper height for optimal wind speed. As far as small wind power plants are concerned, the goal is to achieve as little price as possible, and as the height increases the price increases evenly, so solution must be determined by the minimum height at which the wind turbine can run smoothly. The expression for relating the dependence of speed on height and roughness is [1]:

$$v_{(h)} = v_{(hr)} \frac{\left(\ln\left(\frac{h}{z_o}\right)\right)}{\left(\ln\left(\frac{h_r}{z_o}\right)\right)} \tag{2.2}$$

From this expression it can be determined the wind speed at the height h in relation to the reference wind velocity. Therefore, if we measure the speed, which is insufficient for the operation of the wind turbine, it is possible to increase the height of the pole in order to obtain the appropriate wind parameters necessary for normal operation.

The parameter z_o represents the roughness of the surface and it is recalculated for some of the characteristic environments in which the wind farm can be located. In addition to the equation above, a simplified equation for the calculation of the wind velocity is used, depending on the height given as [1]:

$$v_{(h)} = v_{(hr)} \left(\frac{h}{h_r}\right)^m \tag{2.3}$$

where parameter m is introduced, which depends on the roughness of the surface as given [1]:

$$m = 0.096 log_{10}(z_o) + 0.016((log_{10}(z_o)))^2 + 0.24 \tag{2.4}$$

The recalculated values for z and m are given in the table:

Table 1. Values of the surfaces roughness for the appropriate terrain [1]

Type of terrain	Z_o[mm]	m
Calm open see	0.2	0.104
Snow	3	0.100
Rough pasture	10	0.112
Crops	50	0.131
Scattered trees	100	0.160
Many trees	250	0.188
Forest	500	0.213
Suburbs	1500	0.257
City centers	3000	0.289

Especially important for small wind farms is wind turbulence. It is a part of the wind that does not move linearly but has a turning feature and creates additional forces on the turbine blades that are not at the same angle as the turbine-propelling force. In other words, these turbulences cause a decrease in turbine speed, jerking, and other similar effects, and the materials suffer additional efforts. This effect is particularly pronounced in rural and urban environments, where there are high buildings and where the edges of the roofs are under sharp or right angles. All this affects the creation of turbulences that create adverse effects for the operation of the turbine. Depending on the position of surrounding objects, there is a turbulence that occurs in wind blowing, so it is necessary to analyze the place where the wind turbine will be installed. It is necessary to choose a place with a constant wind speed and avoid spots with vortices.

According to all of the above mentioned wind characteristics, when choosing a site, the following things should be considered:

- The average annual wind speed must be at least 5 [m/s]
- The height of the aggregate must be at least 50% higher than the other facilities
- If mounted on the roof, unit should be set as close to the center of the roof as possible to take advantage of the roof effect.

3   Location Characteristics

The location of the property on which the wind power plant is built is located in the valley area, on two sides surrounded by hills on which there is a thick forest. Beside the estate, there is a river, along with a rare tree. The settlement that is located along the valley is rare, in front of the estate is located on one side several houses that are placed in front of this location. On the other hand, there is an agricultural parcel on which part of the house is located. Showing the plot with all the more important objects is in the following picture (Fig. 1):

Fig. 1. Satelit picture of location

This image was used to analyze the wind motion in order to determine the optimal location of the wind turbine setting. The user expressed his desire to install the turbine to its ancillary building where it would be firmly fixed on the top plate, thereby reducing the cost of the pole and the necessary foundation. In this case, the pillar would be less than 3 m, making it easier to make the foundation for a complete wind plant. The point of the first measurement was made on the spot marked with a red dot (P1). This is a 5 m high building, while an anemometer is placed at a height of 12 m, so this object has no effect on the wind characteristics. Since it is a valley area, the direction of the wind is almost always the same in the direction of SE-NW.

4   Location P1 Analysis

Location P1 represents the end user's desire and is optimal from the aspect of making the pole and foundations for the wind plant. The anemometer is set at a height of 12 m and is set to measure the wind speed at intervals of 5 min to justify the selection of this location for the wind plant. The measurements were made in a few days and the following results were obtained (Fig. 2):

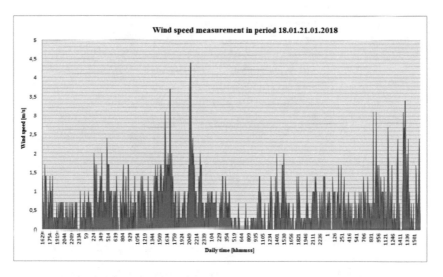

Fig. 2. Measuring results in period 18.01.-21.01.2018 on location P1

From this picture, it can be seen that the measured wind speed is very unfavorable for the installation of a wind turbine. The average wind speed in this case is 0.64 [m/s], and it is insufficient even for the period when it has the least wind during the year. In addition, there are very large changes in velocity, or many turbulence in wind speed for very short time intervals. It can also be seen that only at short time intervals, speed above is required 2 [m/s].

Considering that the results were pretty poor for wind speed measurement, the Ansys 17.0 software was used to further analyze the wind distribution at this location. Using Ansys 17.0, in 2D domain, calculation was made to analyze influence of nearby objects. First geometry was made for analyzing influence of height and then for analyzing influence of objects distribution. All objects that have influence on air flow were drawn and defined as wall in boundary conditions. Second boundary condition was inlet wind speed set on 2 [m/s] in direction given by anemometer. The opposite side is set as outlet. In background of Ansys the k-ω SST model was used, which describes turbulent motion. This model gives the most accurate results and is a combination of standard k-ε model which has best results for air flow away from the boundaries (wall) and k-ω model which has best results for air flow near wall. For using k-ω model it's necessary to have very frequent mesh around walls, otherwise it can cause destabilization of calculating method. As material air with constant Density (default value was set) and constant Viscosity (default value) were used. Solution methods that were used are:

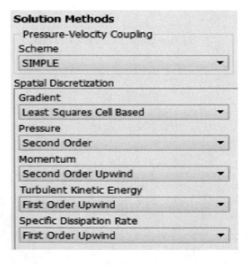

The k-ω model is used, which accurately describes turbulent motion. This model gives the most accurate results and a combination of standard k-ε and k-ω models. The anemometer is placed at the height of the house located in front of the location, and the house that is still in front will also be affected by the measurement. An analysis of this case was made and the following results were obtained:

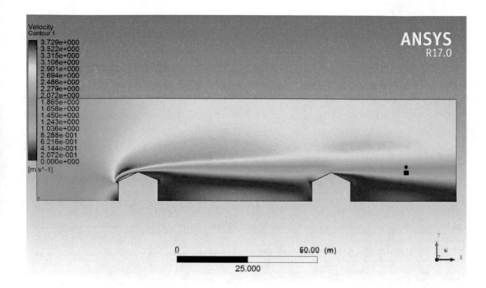

Fig. 3. Simulation of object impact on location P1 wind speed measurement

From Fig. 3 it can be argued that objects in front of point P1 have a significant impact on it. There is a pronounced effect of the roof, where wind gains additional speed due to roof inclination. It can also be seen that the anemometer is placed somewhere on the boundary of the impact of the closer house. By lifting the pole by an additional 3 m, the effect of the roof can be fully utilized, where significantly better wind speeds will be obtained. With the additional pole lift, the wind effect decreases, so optimization is needed in this case. Particularly important in this case is that the average speed obtained is insufficient, therefore it is not enough to increase it by 30% or 50%, but it is necessary to increase it at least 2–3 times in order to be able to analyze the location for setting up a small wind turbine.

5   Selection of Optimal Location Using Simulation

The originally analyzed location recommended by the user, due to the influence of surrounding objects has unfavorable wind characteristics. Using the Ansys 17.0 fluid analysis software, the entire plot and surrounding objects were analyzed to determine the optimal location for measuring the wind speed. The following results were obtained:

Figure 4 shows the detailed influence of the surrounding objects on the distribution of the wind energy field. The initial measurement was made at point P1 marked with red color. It is exactly the effect of the objects on this point where it is in a "green" area that has a much lower speed than the environment. Therefore, this point has an unfavorable location along the z axis (the analysis in Fig. 3) and the unfavorable location along the xy axis (the analysis in Fig. 4). By further analysis of the results obtained by the simulation, it was found that the optimal location for the setting of the wind turbine measurement point P2 is marked in black. From this site it can be seen that the highest wind speed is at the very top of the plot, however these locations are unfavorable due to the neighboring river at a lower angle of the plot. Often the earth's rush occurs due to the influence of the river, therefore, by placing a wind turbine on this site, it would damage its stability. With this area, the red color in the picture is also in the immediate vicinity of the house. This location is also not possible for the installation of a wind turbine because there is a summer garden where people spend their free time. In addition, the wind turbine would create noise during the night, which would have an impact on the sleep of the people who live there. The selected location is optimal from the aspect of object allocation, user needs and wind speed.

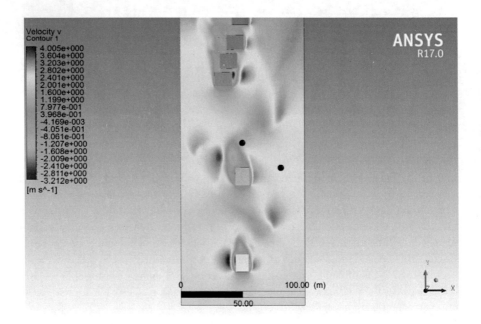

Fig. 4. Simulation of wind speed distribution on location

6   Location P2 Analysis

Using software analysis and measurement at P1 conclusion is that it is not an optimal location for measuring the wind speed for the wind turbine installation. Software analysis and analysis of the actual situation at the field determined the location of P2 as optimal. In order to measure and prove that it is optimal point, an anemometer is places at a height of 12 m. By measuring at intervals of 5 min, the following results were obtained:

Measurement shown in Fig. 5 is for the period 02.02.2018 until 08.02.2018. In this period, the weather was identical to weather on period 18.01-21.01.2018, therefore it is possible to compare results on these two locations. First of all, an average wind speed of 1.48 [m/s] was obtained on location P2, which is 2.3 times more than at P1, or

131%. Both locations are characterized by large changes in wind speed. It should be noted that the speed is generally about 1.5 [m/s], which is still insufficient for the long-term and optimal operation of the 1 kW wind turbine. Therefore, additional optimization is required in height, that is, to increase the pole height in order to obtain optimal conditions for the operation of the desired wind turbine. With the average and minimum wind speeds that are important for running and starting a wind turbine, it is also necessary to analyze the maximum wind speed. In the analyzed case it is 8.5 [m/s] which is quite high considering other more frequent speeds. This will be particularly noticeable when the height of the pole increases, that is, according to the increase in height, the maximum wind speed will be increased, which can be detrimental to the wind turbine.

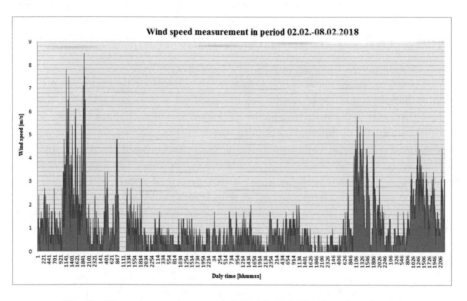

Fig. 5. Measuring results in period 02.02.-08.02.2018 on location P2

7 Conclusion

The aim of this paper is to describe process of optimal location selection for anemometers to determine the wind potential for future wind turbine installations. The initial location where the measurement was made is suggested by the user. Advantages of that location are additional height and foundation, because it would use the existing building and it's construction. Such location is acceptable when there is enough wind potential in the whole observed area. Therefore, the location that is least demanding from the aspect of installing and maintaining wind turbines is taken. The first measurements showed that recommended location do not have the potential to set up wind turbine, and a more detailed software analysis showed the big impact of surrounding objects on this location. All of these objects have a significant influence on the reduction and the occurrence of wind speed turbulence. Therefore, with the assistance of the software, it is possible to analyze the entire site and determine best location that has optimal wind characteristics and lowest impact on environment. The selected location P2 also has insufficient wind speed to set the desired wind turbine in this period of year and on this height. Therefore wind power analysis can not be made from first measurements, but data is taken throughout the whole year. With optimization on xy axes, it is needed to optimize height of pole. Using Eq. (5) and Table 1, it is possible to calculate the wind speed at altitudes above the measured 12 m without the need for re-measurement, based on the results obtained at this height. Optimizing by height, is comparison of the pole and foundations price with the additional amount of energy gained by increasing the height. This analysis is done after detailed measurements to ultimately determine the power that can be installed for the planned budget.

This paper also show that with the stochastic nature of the wind, a large influence on the generated energy from the wind turbine has the location of the aggregate. Unlike large wind farms set up in open fields, in small wind power plants that are being built in settlements must be analyzed great number of factors beside wind speed. Namely, every location on which the aggregate is set is different from the predecessor, so an analysis must be done for each. The first measurements to be obtained must be a condition indicator, but those results are not used for power and budget calculating. The reason for this is the stochastic nature of the wind, which changes with the weather conditions at frequent intervals but also the changes that occur in annual wind movements. All the great currents and air movements across the wide area also have influence on micro-locations, so measurements must be made throughout the year. At this location a significant increase in wind is expected in spring and autumn, where according to the user the winter period is with lowest wind speed intensity.

Also shown in this paper is the importance of applying new solutions such as location optimization with software. Though complex sites with many objects it may be a problem for all of them to be entered into software and it will increase processing time, even if it takes lot of time optimal location is most important information for small wind power plants. Using software huge facilitations can be made, and they represents in the fact that the software field analysis, values obtained with measurement, and field conditions can give expert all needed information for optimal location choosing.

References

1. Wood, D.: Small Wind Turbines Analysis, Design, and Application. University of Calgary University, Calgary (2011)
2. Saoke, C.O.: Analysis of wind speeds based on the Weibull model and data correlation for wind pattern description for a selected site in Juja, Kenya, Jomo Kenyatta University of Agriculture and Technology (2011)
3. Helgason, K.: Selecting optimum location and type of wind turbines in Iceland, Master of Science in Decision Engineering, School of Science and Engineering Reykjavík University (2012)

4. Manwell, J.F., McGowan, J.G.: Wind Energy Explained, Theory Design and Application, 2nd edn. University of Massachusetts, Boston (2009)
5. ANSYS Fluid Dynamics Verification Manual-ANSYS, Inc., Southpointe, 275 Technology Drive Canonsburg, PA 15317, October 2012

Computer Science

Quantifier Elimination in ACF and RCF

Mirna Udovicic[1](✉) and Dragana Kovacevic[2]

[1] Sarajevo School of Science and Technology, Sarajevo, Bosnia and Herzegovina
`mirna.udovicic@ssst.edu.ba`
[2] Catholic school center, Sarajevo, Bosnia and Herzegovina
`bosanska15@yahoo.com`

Abstract. In this paper we will explain some basic notions related to quantifier elimination in the first order theories. We will give a general algorithm for quantifier elimination for any theory. The examples of theories which admit QE are theory of dense linear order (DLO), theory of algebraically closed fields (ACF) and theory of real closed fields (RCF). At the end, we will show the applications of quantifier elimination in ACF and RCF. The interesting applications can be seen in geometry, biology and control theory.

The examples of theories which admit QE are theory of dense linear order (DLO), theory of algebraically closed fields (ACF) and theory of real closed fields (RCF). A general algorithm for quantifier elimination for any theory T is presented below. Also, we have presented a very interesting applications of quantifier elimination over the reals in biology. We were interested in the change of the qualitative behaviour of a parameterized system of non-linear differential equations as they occur in epidemiology. The equations present rational functions of the parameters. Our problem can be formulated as a first order formula over the reals and can be solved by QE method.

Keywords: Quantifiers · Elimination · Model · Field

1 Introduction

The first real quantifier elimination procedure was published by Tarski at the end of 1940s [1]. During the 1970s Collins developed the first elementary recursive real quantifier elimination procedure [2,3], which was based on cylindrical algebraic decomposition (CAD). An implementation by Arnon was available around 1980 [4]. CAD has undergone many improvements since then and establishes an active research area until today.

In this paper, we focus on geometric theorem proving. Related applications of real quantifier elimination methods include computational geometry [5] and solid modeling [6]. Theorems of elementary geometry have traditionally been considered an important test case for the scope of methods in automatic theorem proving. In particular, they have stimulated a variety of algebraic techniques for their solution.

© Springer Nature Switzerland AG 2019
S. Avdaković (Ed.): IAT 2018, LNNS 60, pp. 419–429, 2019.
https://doi.org/10.1007/978-3-030-02577-9_41

An original solution of a difficult geometry theorem is shown in this paper. Also, some other applications of quantifier elimination in geometry are presented. A given algorithm for quantifier elimination can be applied to any theory which admit QE. It is important to note that converting a formula to a prenex normal form is not a part of this algorithm.

The formula describing a general problem contains nine parameters, but it is possible to fix some of the parameters by biometrical arguments.

2   Quantifier Elimination

Let us show the example of a formula with quantifiers which is equivalent to a formula without quantifiers.

Suppose we are given a formula $\varphi(a, b, c)$ in a set of real numbers $\mathbf{R}$,

$$\exists x \left(ax^2 + bx + c = 0\right).$$

By the quadratic formula, we have the following equivalention:

$$\varphi(a, b, c) \leftrightarrow \left[(a \neq 0 \wedge b^2 - 4ac \geq 0) \vee (a = 0 \wedge (b \neq 0 \vee c = 0))\right],$$

so φ is equivalent to a quantifier free formula.

Now let us introduce some basic definitions which are of importance for quantifier elimination.

The language L is recursive if the set of codes for symbols from L is recursive. The first order theory T is recursive if the set of codes for axioms for T is recursive. An L-theory T is complete if for every sentence φ in language L the following holds:

$$T \vdash \varphi \ \text{ or } \ T \vdash \neg\varphi.$$

For each theory T arises question of its decidability, i.e. existence of algorithm which for given $\varphi \in Sent_L$ gives an answer whether $T \vdash \varphi$ or $T \nvdash \varphi$. In the case of recursive complete theory in a recursive language, the answer is affirmative.

Definition 1. *A theory T of language L admits quantifier elimination if for every formula $\phi(\overline{v}) \in For_L$ there exist a quantifier free formula $\psi(\overline{v})$ such that:*

$$T \vdash \forall \overline{v} \left(\phi(\overline{v}) \leftrightarrow \psi(\overline{v})\right).$$

Every logic formula is equivalent to its prenex normal form:

$$Q_1 x_1 \ldots Q_n x_n \varphi(x_1, \ldots, x_n, y_1, \ldots, y_m),$$

where $Q_i \in \{\forall, \exists\}$ and φ is a formula without quantifiers in DNF; formula of the form $\forall x \varphi$ is equivalent to $\neg \exists x \neg \varphi$; $\exists x (\varphi \vee \psi) \leftrightarrow \exists x \varphi \vee \exists x \psi$ is a valid formula. Using the previous we see that an $L-$theory T admits quantifier elimination if and only if for every $L-$formula of the form $\exists x \varphi(\overline{y}, x)$, where φ is a conjunction of atomic formulas and negations of atomic formulas, exists equivalent quantifier free formula $\psi(\overline{y})$.

The examples of theories which admit QE are theory of dense linear order (DLO), theory of algebraically closed fields (ACF) and theory of real closed fields (RCF). A general algorithm for quantifier elimination for any theory T is presented below.

Algorithm for QE

Input: formula φ in language L of T
Output: quantifier free formula ψ which is equivalent to φ

begin
    Convert φ to prenex normal form $Q_1 x_1 ... Q_n x_n \chi(x_1, ..., x_n, y_1, ..., y_m)$;
    $i = n$;
    while $i > 0$ do
        {if Q_i is $\forall$ replace $Q_i x_i \chi_i$ with $\neg \exists x_i \neg \chi_i$

        transform the matrix of a formula to DNF

        let the existential quantifier pass through disjunction

        eliminate quantifier $\exists$ using a specific algorithm for T
        $i = i - 1$
        }
end

Now we will present an original general algorithm for QE. It is important to note that converting a formula to a prenex normal form is not a part of this algorithm. The algorithm is recursive and convenient for implementation in Mathematica. Resolve is a function implemented in Mathematica which gives a solution for a formula which contains only existential quantifier. In the case that the input is a formula which contains quantifiers in a scope of quantifiers, we will apply the algorithm to its subformula first.

```
//a function Eliminacija returns a formula without quantifiers
//equivalentφ

Eliminacija(φ)
    begin
        if φ has a form (φ₁ EQUIVALENT φ₂)
        {
            φ = (φ₁ → φ₂) ∧ (φ₂ → φ₁);
            rez = Eliminacija(φ);
            return (rez)
        }
        if φ has a form (φ₁ OR φ₂)
        {
```

```
        r₁ = Eliminacija(φ₁);
        r₂ = Eliminacija(φ₂);
        rez = r₁ ∨ r₁;
        return (rez)
    }
    if φ has a form (φ₁ AND φ₂)
    {
        r₁ = Eliminacija(φ₁);
        r₂ = Eliminacija(φ₂);
        rez = r₁ ∧ r₁;
        return (rez)
    }
    if φ has a form (φ₁ IMPLIES φ₂)
    {
        r₁ = Eliminacija(φ₁);
        rez = ¬r₁;
        r₂ = Eliminacija(φ₂);
        rez = rez ∨ r₂;
        return (rez)
    }
    if φ has a form (NOT φ₁)
    {
        r₁ = Eliminacija(φ₁);
        rez = ¬r₁;
        return (rez)
    }
    if φ has a form (∀xφ₁(x))
    {
        r₁ = ¬φ₁;
        r₂ = ∃xr₁;
        rez = Eliminacija(r₂);
        rez = ¬rez;
        return (rez)
    }
    if φ has a form (∃xφ₁(x))
    {
        rez = Resolve(φ);
        return (rez)
    }
end
```

A function Eliminacija is implemented and tested in Mathematica, which is illustrated in the examples below.

Input: Implies[Exists[x,x+3>2],Exists[x,x^2<0]];
Output: False

Input: Or[Exists[x,x+3>2],Exists[x,x^2<0]];
Output: True

Input: Exists[x,x+3>2&&x+y>0];
Output: y∈Reals

3   Applications of QE in Geometry

The language of fields is $L = \{+, -, \cdot, 0, 1\}$, where $+$ and $\cdot$ are binary function symbols, $-$ is unary function symbol and 0 and 1 are constant symbols.

We could axiomatize the class of algebraically closed fields by adding, to the axioms of fields, for each $n \geq 1$, the axiom:

$$\forall a_0 \cdots \forall a_n \exists x \left(a_n x^n + \cdots + a_0 = 0 \right).$$

As example of ACF, we can take the field of complex numbers, which is the algebraic closure of the field of real numbers.

As we have noticed in the introduction, in order to obtain the algorithm for quantifier elimination in algebraically closed fields, it is sufficient to know how to eliminate the existential quantifier in the formula of the form:

$$\exists x \left(t_1 (x) = 0 \wedge \cdots \wedge t_m (x) = 0 \wedge t (x) \neq 0 \right),$$

where coefficients of t_i and t are polynomials from $\mathbf{Z} [y_1, \ldots, y_k]$, $y_i \neq x$. The crucial part of the QE algorithm is the polynomial pseudo-division algorithm.

We could axiomatize the class of real closed fields by adding, to the axioms of ordered fields, the axioms:

$$\forall x \exists y \left(x = y^2 \vee -x = y^2 \right)$$

$$\forall a_0 \cdots \forall a_{2n} \exists x \left(a_0 + a_1 x + \cdots + a_{2n} x^{2n} + x^{2n+1} = 0 \right), \text{ for each } n \geq 1.$$

As example of RCF, we can take the field of real numbers. Any formula in RCF without quantifiers is equivalent to a disjunctions of formulas of the following form:

$$t_1 = 0 \wedge \cdots \wedge t_m = 0 \wedge q_1 > 0 \wedge \cdots \wedge t_m > 0,$$

where t_i, q_j are polynoms with coefficients in $\mathbf{Z}$.

Now we will prove a very difficult geometry theorem using method of quantifier elimination in ACF. The example is given below.

Example 1. Suppose we are given a square $ABCD$. Let E be a point such that CE is parallel to a diagonal BD and it holds: $BE = BD$. A point F is the intersection point of BE and DC. Prove that the equality $DF = DE$ holds.

Let us denote given points by coordinates in the following way:

$$A\left(0,0\right),\ B\left(u,0\right),\ C\left(u,u\right),\ D\left(0,u\right),\ E\left(x_1,x_2\right)\ \text{and}\ F\left(x_{3,}u\right),$$

where u is a length of a side of square and a point A is a center of coordinate system.

Using the coordinate notation and some basic geometry calculation, it can be easily seen that a given problem is equivalent to the following logic formula:

$$\forall u \forall x_3 \forall x_2 \forall x_1 \left[(t_1 = 0 \land t_2 = 0 \land t_3 = 0) \rightarrow t = 0\right], \tag{1}$$

where $t_1 = 0$ is a formula

$$x_1^2 - 2x_1 u + x_2^2 - u^2 = 0,$$

$t_2 = 0$ is a formula

$$ux_1 + ux_2 + x_2^2 - 2u^2 = 0,$$

$t_3 = 0$ is a formula

$$x_2 x_3 - ux_2 - ux_1 + u^2 = 0,$$

and $t = 0$ is a formula

$$x_3^2 - x_2^2 + 2ux_2 - x_1^2 - u^2 = 0.$$

We will first consider the following subformula of a formula (1),

$$\forall x_1 \left[(t_1 = 0 \land t_2 = 0 \land t_3 = 0) \rightarrow t = 0\right] \tag{2}$$

A formula (2) is equivalent to:

$$\forall x_1 \left[\neg \left(t_1 = 0 \land t_2 = 0 \land t_3 = 0\right) \lor t = 0\right] \quad \text{equivalent}$$
$$\neg \exists x_1 \neg \left[\neg \left(t_1 = 0 \land t_2 = 0 \land t_3 = 0\right) \lor t = 0\right] \quad \text{equivalent}$$
$$\neg \exists x_1 \neg \left[\neg \left(t_1 = 0 \land t_2 = 0 \land t_3 = 0\right) \lor t = 0\right] \quad \text{equivalent}$$
$$\neg \exists x_1 \left(t_1 = 0 \land t_2 = 0 \land t_3 = 0 \land t \neq 0\right)$$

Since a negation of the previous formula

$$\exists x_1 \left(t_1 = 0 \land t_2 = 0 \land t_3 = 0 \land t \neq 0\right) \tag{3}$$

has an adequate form, we can apply the algorithm of QE on it.

Let us simplify a formula (3) first. Since from the formula $t_2 = 0$ directly follows the equality $x_2 = 2u - x_1$, we can substitute the previous equality into the formulas $t_1 = 0$ and $t_3 = 0$. It follows that a formula (3) is equivalent to the formula:

$$\exists x_1 \left(3u^2 - 6x_1 u + 2x_1^2 = 0 \land -u^2 + 2x_3 u - x_1 x_3 = 0 \land t \neq 0\right) \tag{4}$$

where $t \neq 0$ is a formula

$$x_3^2 - x_2^2 + 2ux_2 - x_1^2 - u^2 \neq 0.$$

We consider $3u^2 - 6x_1u + 2x_1^2$ and $-u^2 + 2x_3u - x_1x_3$ as polynoms by a variable x_1 and see that terms $2x_1^2$ and $-x_3x_1$ have the highest degree by x_1 equal 2 and 1, respectively. When applying the algorithm for QE we get the formula equivalent to a formula (4):

$$-x_3 \neq 0 \wedge \exists x_1 \left(-3x_3u + 2x_1x_3 + 2x_1u = 0 \wedge -u^2 + 2x_3u - x_1x_3 = 0 \wedge t \neq 0\right)$$

Now we combine the equalities in the previous formula in order to express a variable x_3 by u and x_1:

$$x_3 = 2u - 2x_1.$$

Also, we use the equality $x_2 = 2u - x_1$ that we have proved before. We will substitute these values for x_2 and x_3 in the formula $t \neq 0$. After some basic calculation, it can be seen that $t \neq 0$ is equivalent to:

$$- 6ux_1 + 2x_1^2 + 3u^2 \neq 0. \tag{5}$$

Since a formula (5) represents a negation of the subformula $3u^2 - 6x_1u + 2x_1^2 = 0$ of a formula (4), it is not possible that both of them are true and we have a contradiction.

So, our conclusion is that a formula (3) is false, which means that its negation is true. We have just proved the geometry property.

Example 2. Suppose we are given two curves of the highest degree equal two. We need to determine if the curves have the intersection point.

Let us denote formulas of the curves by $t_1 = 0$ and $t_2 = 0$:

$$t_1 = 0 \leftrightarrow A_1x^2 + 2B_1xy + C_1y^2 + 2D_1x + 2E_1y + F_1 = 0$$
$$t_2 = 0 \leftrightarrow A_2x^2 + 2B_2xy + C_2y^2 + 2D_2x + 2E_2y + F_2 = 0$$

We consider t_1 and t_2 as polynoms by a variable x and see that terms A_1x^2 and A_2x^2 have the highest degree by x equal 2 ($n_1 = 2$ and $n_2 = 2$).

So, our formula is:

$$\exists x \exists y \left(t_1 = 0 \wedge t_2 = 0 \wedge A_1 \cdot A_2 \neq 0\right) \tag{6}$$

Since a formula (6) represents a formula of a theory ACF, we can apply the algorithm of quantifier elimination for ACF.

Let us introduce the following notation:

$$t_1{}' = A_2 \cdot t_1 - A_1 \cdot x^{n_1 - n_2} \cdot t_2.$$

By algorithm of QE, a formula (6) is equivalent to:

$$A_2 \neq 0 \wedge \exists x \left(t_1{}' = 0 \wedge t_2 = 0 \wedge A_1A_2 \neq 0\right). \tag{7}$$

Since a subformula $t_1{}' = 0$ of a formula (7) has a degree by x equal 1, it represents a linear equation so we can calculate the value of x. After substitution of x into $t_2 = 0$, a value of y can be found.

Let us show the example with concrete values of A, B, C, D, E and F.

○ $x^2 + y^2 - 8x - 18y + 93 = 0$ and $x^2 + y^2 - 8x - 8y + 23 = 0$

Denote as $t_1 = x^2 + y^2 - 8x - 18y + 93$ and $t_2 = x^2 + y^2 - 8x - 8y + 23$. Since $A_1 = 1$ and $A_2 = 1$ hence condition $A_1 \cdot A_2 \neq 0$ is satisfied. Note that $n_1 = 2$ and $n_2 = 2$. Now we can make polynomial $t_1\prime = A_2 \cdot t_1 - A_1 \cdot x^{n_1 - n_2} \cdot t_2$. Solution of equation $t_1\prime = 0$ gives us $y = 7$. From $t_2 = 0$ we have $x = 4$. So we can see that there is intersection point for those two curves.

○ $x^2 + y^2 - 2x - 6y + 6 = 0$ and $x^2 + y^2 - 10x - 8y + 40 = 0$. Using same algorithm equation $t_1\prime = 0$ has no real solutions. Hence there is no intersection point for those two curves.

4   Applications of QE in Biology

Now we will show an interesting application of quantifier elimination over the reals in biology.

In epidemiology we are interested to get some information about the dynamics of different disease. One important notion is R_τ, the basic reproduction number, which represents the average number of secondary infections generated by one case. In this sense we have an expression which represents the dynamics of a disease: if $R_\tau < 1$ then the disease will die out, if $R_\tau > 1$ then we will get an epidemic. Our problem was to represent a formula for the total population depending on R_τ.

Let us introduce a certain epidemic model first. It is intended to model the epidemic of the AIDS disease. Consider a sexually active population which is divided into susceptibles X_1, X_2, infecteds Y_1, Y_2 and treated infecteds V_1, V_2. Let us assume that the first part of the population is the low active and larger group X_1, Y_2, V_1 and the second the high active and smaller group X_2, Y_2, V_2.

We will assume a constant influx into the suspectible population of $\mu N_0 \gamma_1$ and $\mu N_0 \gamma_2$, where N_0 is the stable population in absence of infection and $\frac{1}{\mu}$ is the average duration of sexual activity. γ_1 and γ_2 are the portion of the whole population going into the high and low part of the population, respectively $(\gamma_1 + \gamma_2 = 1)$. The susceptibles will be reduced through the per capita rate of natural mortality μ and through the force of infection $\rho_1 \lambda, \rho_2 \lambda$, where ρ_1, ρ_1 is the effective mean rate of sexual partner change per year in the low or high group, respectively. The infecteds through infection with the force of infection $\rho_1 \lambda$ or $\rho_2 \lambda$, depending on the respective group. They will be diminished through the per capita mortality rate μ, the AIDS induced per capita mortality rate ν, and through τ, the per capita rate of getting treated. The treated infecteds increase through τ. They will be diminished through mortality rate μ and through the AIDS induced, and through treatment reduced, mortality rate δ. (The time scale in our case will always be one year).

So the model can be described with the following non-linear differential equations:

$$\frac{dX_i}{dt} = \mu N_0 \gamma_i - (\rho_i \lambda + \mu) X_i$$

$$\frac{dY_i}{dt} = \rho_i \lambda X_i - (\nu + \mu + \tau) Y_i$$

$$\frac{dV_i}{dt} = \tau Y - (\delta + \mu) V_i,$$

where the subscript i denotes the homogenous behaviour with mean rate of partner change ρ_i. For the force of infection we will assume the proportionate mixing behaviour.

$$\lambda = h \frac{\sum_i \rho_i (Y_i + cV_i)}{\sum_i \rho_i (X_i + Y_i + cV_i)}, \quad \text{for } i = 1, 2.$$

where c represents the behaviour change through treatment $(0 < c < 1)$.

For that model it can be concluded that

$$R\tau = h \frac{\delta + \mu + c\tau}{(\tau + \gamma + \mu)(\delta + \mu)} \cdot \frac{\rho_1^2 \gamma_1 + \rho_2^2 \gamma_2}{\rho_1 \gamma_1 + \rho_2 \gamma_2}.$$

If we use the following substitutions:

$$r_i = \frac{\rho_i}{\mu}, v = \frac{\nu}{\mu}, d = \frac{\delta}{\mu}, t = \frac{\tau}{\mu},$$

we will get the formulas:

$$\frac{dX_i}{dt} = N_0 \gamma_i - (r_i \lambda + 1) X_i$$

$$\frac{dY_i}{dt} = r_i \lambda X_i - (v + t + 1) Y_i$$

$$\frac{dV_i}{dt} = t Y_i - (d + 1) V_i$$

Also, the values of λ and R_τ can be calculated.

After some basic calculation, the question can be transformed into a quantified formula:

$$\forall c \in (0, 1), \forall t > 0, \forall h > 0, \forall d > 0, \forall \gamma_2 < \frac{1}{2}, \forall v > 0, \forall r_1 > 0, \forall r_2 > 0, \forall p_1 :$$

$$r_1 < r_2 \wedge d < v \rightarrow (R\tau = 1 \wedge P(p_1) = 0 \rightarrow p_1 \leq 0),$$

where $\gamma_2 = 1 - \gamma_1$.

We can reduce a number of free variables and perform QE having different parameters fixed.

So, the problem described previously can be simplified with $\tau = 0$. For biometrical reasons it is possible to fix some of the given parameters $(h, \nu$ and $\mu)$ with certain values.

○ Let ρ_2, p_1 be free, then our formula looks like as follows ($\rho_1 = \frac{1}{2}, h = \frac{1}{5}, \nu = \frac{11}{120}, \mu = \frac{1}{30}, \gamma_2 = \frac{1}{5}$):

$$(\forall \rho_2)\,(\forall p_1)\,4\left(180 + 180\rho_2^2\right) = 75\left(12 + 6\rho_2\right) \wedge$$

$$\left(8 + 24p_1 + 20\rho_2 - 16\rho_2^2 p_1 + 198\rho_2^2 p_1^2 + 95p_1\rho_2 + 185\rho_2 p_1^2 - 32\rho_2^2 - 32p_1^2\right) p_1 = 0 \to p_1 \leq 0$$

Let us introduce the following notation: $t_1 = 4\left(180 + 180\rho_2^2\right) - 75\left(12 + 6\rho_2\right)$, $t_2 = \left(8 + 24p_1 + 20\rho_2 - 16\rho_2^2 p_1 + 198\rho_2^2 p_1^2 + 95p_1\rho_2 + 185\rho_2 p_1^2 - 32\rho_2^2 - 32p_1^2\right) p_1$.

We can transform a subformula of the previous one which contains only p_1 quantified into the following one:

$$(\forall p_1)\,(\neg\,(t_1 = 0 \wedge t_2 = 0) \vee p_1 \leq 0) \qquad \text{equivalent}$$
$$(\forall p_1)\,(t_2 \neq 0 \vee p_1 \leq 0) \vee t_1 \neq 0 \quad \text{equivalent}$$
$$(\forall p_1)\,(t_2 > 0 \vee t_2 < 0 \vee p_1 \leq 0) \vee t_1 \neq 0$$

In case $p_1 \leq 0$ the formula is always true. In case $p_1 > 0$, since it holds that a point $p_1 = 0$ is a zero of a polynom t_2, we know that in a theory of RCF the following formula is true:

$$(\forall p_1)\,(t_2 > 0 \vee t_2 < 0),$$

in an open interval around 0.

We have just proved that our formula must be always true.

The procedure will be the same if we use one more free parameter, in the formula:

$$(\forall \rho_1)\,(\forall \rho_2)\,(\forall p_1) : (\rho_2 > 0 \wedge \rho_1 > 0 \wedge \rho_2 > \rho_1) \Rightarrow$$
$$((4\left(720\rho_1^2 + 180\rho_2^2\right) = 75\left(24\rho_1 + 6\rho_2\right) \wedge$$
$$(-88\rho_1\rho_2^2 p_1^2 + 56\rho_1\rho_2^2 p_1 - 480\rho_1^2 p_1^2 \rho_2 - 335\rho_1 p_1 \rho_2 + 55\rho_2\rho_1 p_1^2 +$$
$$480\rho_1^2 p_1\rho_2 - 80\rho_1^2 + 128\rho_1^3 + 80\rho_1^2 p_1 - 20\rho_2^2 p_1 - 20\rho_2\rho_1 -$$
$$55\rho_2^2 p_1^2 - 256\rho_1^3 p_1 + 128\rho_1^3 p_1^2 + 32\rho_1\rho_2^2)p_1 = 0) \Rightarrow p_1 \leq 0).$$

Also, the equivalent formula that we get is a true formula.

5   Conclusion

The most interesting parts of this paper are the applications of quantifier elimination in geometry and biology. We have proved a very difficult geometry theorem using method of quantifier elimination in ACF. Using the coordinate notation and some basic geometry calculation, we have shown that a given theorem is equivalent to the logic formula which is always true. Also, we have presented an application of quantifier elimination over the reals in biology. We were interested in the change of the qualitative behaviour of a parameterized system of non-linear differential equations as they occur in epidemiology. The equations present rational functions of the parameters. Our problem was formulated as a first order formula over the reals and solved by QE method.

References

1. Tarski, A.: A Decision Method for Elementary Algebra and Geometry, 2nd edn. RAND, Santa Monica (1957)
2. Collins, G.E.: Quantifier elimination for real closed fields by cylindrical algebraic decomposition-preliminary report. ACM SIGSAM Bull. **8**(3), 80–90 (1974). Proceedings of EUROSAM 1974
3. Collins, G.E.: Quantifier elimination for the elementary theory of real closed fields by cylindrical algebraic decomposition. In: Automata Theory for Formal Languages, 2nd GI Conference. LNCS, vol. 33, pp. 134–183. Springer (1975)
4. Arnon, D.S.: Algorithms for the geometry and semi-algebraic sets. Technical report 436, Ph.D. thesis. Computer Science Department, University of Wisconsin-Madison (1981)
5. Sturm, T., Weispfenning, V.: Computational geometry problems in REDLOG. In: Automated Deduction in Geometry. LNAI, vol. 1360, pp. 58–86. Springer (1998)
6. Sturm, T.: An algebraic approach to offsettings and blending of solids. In: Proceedings of the CASC 2000, pp. 367–382. Springer (2000)

Constraint Satisfaction Problem: Generating a Schedule for a Company Excursion

Mirna Udovičić[(⊠)] and Nedžad Hafizović

Sarajevo School of Science and Technology, Sarajevo, Bosnia and Herzegovina
mirna.udovicic@ssst.edu.ba, nedzad.hafizovic@stu.ssst.edu.ba

Abstract. A large number of problems in AI and other areas of computer science can be viewed as special cases of the constraint satisfaction problem. A number of different approaches have been developed for solving these problems. Some of them use backtracking to directly search for possible solutions. Intelligent backtracking is used in this paper, but the algorithm is not standard. A specific problem of organizing an excursion for a company's employees is solved.

Keywords: Constraint · Chronological · Non-chronological backtracking · CSP

1 Introduction

Since the first formal statements of backtracking algorithms come from over 40 years ago [1,2], many techniques for improving the efficiency of a backtracking search algorithm have been suggested and evaluated. A fundamental insight in improving the performance of backtracking algorithms on CSPs is that local inconsistencies can lead to much trashing or unproductive search [3,4], which wastes time and computational power. A local inconsistency is an instantiation of some of the variables that satisfies the relevant constraints but cannot be extended to one or more additional variables and so cannot be part of any solution. Mackworth [4] defines a level of local consistency called arc consistency. Gaschnig [3] suggests maintaining arc consistency during backtracking search and gives the first explicit algorithm containing this idea.

Stallman and Sussman [6] were the first who informally proposed a non-chronological backtracking algorithm, called dependency directed backtracking, that discovered and maintained no goods in order to back jump. The first explicit back jumping algorithm was given by Gasching [7]. Gaschings back jumping algorithm (BJ) is similar to backtracking algorithm, except that it back jumps from dead ends. However, BJ only back jumps from a dead end node when all the branches out of the node are leaves; otherwise it chronologically backtracks. Prosser [8] proposes the conflict directed back jumping algorithm (CBJ), a generalization of (BJ) to also back jump from internal dead ends.

© Springer Nature Switzerland AG 2019
S. Avdaković (Ed.): IAT 2018, LNNS 60, pp. 430–438, 2019.
https://doi.org/10.1007/978-3-030-02577-9_42

In this paper, the original algorithm for solving one specific type of scheduling and organizing problem is presented. The problem was to make organize an excursion for all employees of a company, such that given constraints connected to a transportation and food are satisfied. Since we are given a condition that constraints are not binary in this example, a formulation of a problem is completely different and it is not possible to apply any of the methods mentioned above. Also, we assume that all variables are instantiated at the beginning of the algorithm. Precisely, we chose one faulty plan, which means that it does not satisfy the conditions required in the task. Our conclusion is that the approach to solving CSP in this paper is completely different from the previous ones.

2 Backtracking: A Method for Solving CSP

2.1 Preliminaries

A constraint satisfaction problem (CSP) is defined by a set of variables $X_1, X_2, ..., X_n$, and a set of constraints, $C_1, C_2, ..., C_m$. Each variable X_i has a nonempty domain D_i of possible values. Each constraint C_i involves some subset of the variables and specifies the allowable combinations of values for that subset. A state of the problem is defined by an **assignment** of values to some or all of the variables, $\{X_i = v_i, X_j = v_j, ...\}$. An assignment that does not violate any constraints is called a **consistent** or legal assignment. A complete assignment is one in which every variable is mentioned, and a solution to a CSP is a complete assignment that satisfies all the constraints.

In a literature, a discussion to CSPs is restricted to the discussion to problems in which each constraint is either unary or binary. It is possible to convert CSP with n-ary constraints to another equivalent binary CSP (Rossi, Petrie, and Dhar 1989). Binary CSP can be depicted by a constraint graph in which each node represents a variable, and each arc represents a constraint between variables represented by the end points of the arc.

2.2 Chronological Backtracking

CSP can be solved using generate and test paradigm. In this paradigm, each possible combination of the variables is systematically generated and then tested to see if it satisfies all the constraints. The first combination that satisfies all the constraints is the solution. A more efficient method uses the backtracking paradigm. In this method, variables are instantiated sequentially. As soon as all the variables relevant to a constraint are instantiated, the validity of the constraint is checked. If a partial instantiation violates any of the constraints, backtracking is performed to the most recently instantiated variable that still has alternatives available. Clearly, whenever a partial instantiation violates a constraint, backtracking is able to eliminate a subspace from the Cartesian product of all variable domains. Although backtracking is strictly better than the generate and test method, its run time complexity for most nontrivial problems is still exponential. The main reason for this is that when using BT we suffer from trashing.

In the Fig. 1, a fragment of the backtrack tree generated by the chronological BT algorithm for the 6-queens problem is shown. We see that, for example, a node labeled 25 consists of the set of assignments $\{x_1 = 2, x_2 = 5\}$. White dots denote nodes where all the constraints with no uninstantiated variables are satisfied (no pair of queens attack each other).

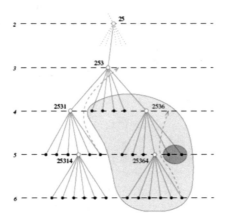

Fig. 1. A fragment of the backtrack tree for the 6-queens problem.

2.3   Non Chronological Backtracking

A no good is a set of assignments and branching constraints that is not consistent with any solution. Non-chronological backtracking algorithms can be described as a combination of a (1) strategy for a discovering and using no goods for back jumping and (2) a strategy for deleting no goods from the no good database. Non chronological backtracking algorithms are also called intelligent backtracking algorithms. A Back jumping algorithm (BJ) is similar to a chronological backtracking algorithm, except that it back jumps from dead ends. However, BJ only back jumps from a dead end node when all the branches out of that node are leaves: otherwise, it chronologically backtracks.

CBJ checks backwards from the current variable to the past variables. If partial instantiation of V_i is inconsistent with respect to some past variable V_g where $g < i$, then the index g is added to the conflict set CS_i of variable V_i. On reaching a dead end on V_i, CBJ jumps back to V_g where g is the largest value in CS_i. On jumping back to V_g the conflict set CS_i is updated such that it becomes the union of the conflict sets with index g removed. Conflict sets below V_g in a search tree are then annulled. If on jumping back to V_g there are no values left to be tried CBJ jumps back again to V_f, where f is the largest value in CS_g.

In the example shown in the Fig. 1, the light shaded part of a tree contain nodes that are skipped by Conflict Directed Back jumping (CBJ). A back jump is represented by a dashed arrow. In contrast to CBJ, BJ only back jumps from dead ends when all branches out of the dead end are leaves. The dark shaded part of a tree contains two nodes that are skipped by Back-jumping (BJ). Again, a back-jump is represented by a dashed arrow.

In this paper, intelligent backtracking is used but approach to a problem is completely different. A main difference can be seen from a formulation of a scheduling problem: given constraints are not binary so we can not use any of the methods mentioned above.

3   Example of Non-chronological Backtracking: Scheduling Problem

3.1   Description of a Problem

A result of this paper is a new algorithm for solving one complex scheduling problem. Intelligent backtracking is used but approach to a problem is completely different. The main difference can be seen directly from a formulation of a scheduling problem; given constraints are not binary, so none of the methods mentioned above can be used. Since a total number of variables is 200, this example is considered as a complex one. Detailed description of a problem is shown below.

In a company, the manager plans to organize an excursion for 200 employees. The manager needs to take care of a transportation and food. The possibilities for the transportation and food available are shown in a tables below.

Transport	Quality units	Cost
By bus	15	15 EUR
By bicycle	10	0 EUR
By foot	5	0 EUR

Food	Food units	Cost
Coffee and juice	5	0 EUR
Sandwich	10	0 EUR
Lunch	15	15 EUR

From the previous tables it can be seen that each transport choice is measured in so called quality units and each food choice in so called food units. The problem is how to make a choice for all employees having at least:

- k quality units
- m food units.

Expenses need to be limited to n EUR in total. Without loss of generality, let us assume that we are given the previous values: k = 1100, m = 1300, n = 1100 EUR. The starting schedule could be any schedule.

3.2   Solvable Algorithm

Now, it will be explained how the algorithm works. The main part of the algorithm is the *change* function which gives the solution if it exists. If that is not the case, the function returns 0.

It consists of three main parts. In the first part, it is checked whether the expenses are allowed. If that is not the case, a transport or food option of an employee is adequately replaced. After the change is made, there is a new schedule for which the function should check the validity. If the new schedule is the solution to the problem, the function returns 1. On the other hand, if the given schedule is not the solution, expenses are checked again. Once the expenses are allowed, the other conditions are checked in the other parts of the function. The procedure for the second and the third part is similar to the procedure of the first part which is explained above. The algorithm is given below.

```
changeTransport(employee, choice){ //changes transport for a given employee
        newChoice = bicycle;
    if (choice == bus){
        expenses -= 15;
        qualityUnits -= 5;
    }
    if (choice == foot){
        qualityUnits += 5;
        }
}
changeFood(employee, choice){ //changes food for a given employee
    newChoice = sandwich;
    if (choice == lunch){
        expenses -= 15;
        foodUnits -= 10;
    }
    if (choice == coffeJuice){
        foodUnits += 5;
    }
}

change (transport, food){ //main function consisting of three parts
    if (expenses are not allowed){
        for (i = 1 to 100){
                if (transport[i] == bus){
                changeTransport(transport[i], bus);
        }
        if (all conditions are allowed){
            return 1;
            }
        if (expenses are allowed break;
            if (food[i] == lunch){
                changeFood(food[i], lunch);
                if (all conditions are allowed){
                    return 1;
                }
                if (expenses are allowed) break;
            }
        }
    }
    if (there is not enough qualityUnits){
        for (i = 1 to 100){
            if (transport[i] == foot){
                changeTransport(transport[i], foot);
            }
```

```
          if (all conditions are allowed){
              return 1;
          }
          if (qualityUnits are allowed) break;
      }
  }
  if (there is not enough foodUnits){
      for (i = 1 to 100){
          if (food[i] == coffeeJuice){
              changeFood(food[i], coffeeJuice);
          }
          if (all conditions are allowed){
              return 1;
          }
          if (foodUnits are allowed) break;
      }
  }
  return 0;
}
```

The correctness of the algorithm will be discussed below. The main function is the only part that needs to be analyzed. The first condition in the main function is related to the expenses and it is most complex one, since both food and transport have cost as an attribute. It can be seen that a maximum number of changes of a schedule in the first part is 200. Without loos of a generality, it is supposed that the starting schedule in the Implementation part includes:

- bus as a transport option
- lunch as a food option

for the first 50 employees. Total amount of expenses for such schedule is 1500 EUR, which is more than allowed amount (1100 EUR). Since this plan is faulty, it needs to be changed in the if function that is related to expenses. The changes in a schedule are made until the expenses are allowed. After, if a current plan is not the solution, the algorithm checks for a quality units in almost identical procedure. While the adequate changes in a schedule are being made, the algorithm after every change checks whether a current schedule is a solution to the problem. In a case that the second part of the main function, the one regarding quality units, does not produce a solution, a condition related to food units is checked in the same way. Note that expenses stay the same when applying any of the changes related to a number of quality or food units. Hence, once expenses are allowed, they will not be checked again. Also, the same holds for the quality or food units, since these are also independent. From everything stated above, it can be concluded that the main *change* function is correct and that it provides a solution for any starting schedule, if the solution exists. Since the main function consists of three loops, each separated from the others and since there are no nested loops in the function, it can be said that this algorithm has a linear complexity, $O(n)$. This complexity is valid even for the worst case scenario. Worst case scenario means that the starting schedule is such, that the algorithm needs to change both transport and food option for the maximum number of employees. In each case, the algorithm loops through the schedule only once and then changes the options in linear time manner.

Analysis of setting and solving the given problem is performed with respect to the expenses constraint, since that constraint is the most complicated one. Let see the border case first:

- allowed expenses $\geq$ 3000 EUR.

In this case, any plan satisfies conditions for allowed expenses, and the analysis for quality and food units is to be done. If the problem is set in such way that the solution requires 1500 EUR or less quality units, and 2000 EUR food units or less, then the solution to the problem surely exist. However, if the quality units constraint is set to more than 1500 EUR and the food units constraint is set to more than 2000 EUR, the solution does not exist.

Let analyze the problem if:

- allowed expenses <3000.

As expected, the analysis is still dependent on other two constraints, quality and food units. The same expenses amount is taken as the amount that the algorithm was tested for (1100 EUR). If one more constraint is set, for example quality units (1100 EUR), a maximum number of food units, for which a solution exists, needs to be determined. If the transport option that has the most quality units and is free is chosen for all 100 employees, the total amount would be 1000 EUR, which is not enough. Consequently, some employees must have a bus as a transport option. It is not hard to calculate that the number of these employees is twenty. So, at least twenty employees have a bus as a transport, while other eighty have a bicycle. This case gives the number of quality units that satisfies the constraint (1100 EUR).

Since the cost of the transport was 300 EUR, allowed expenses for food are 800 EUR. A simple calculation can be used to conclude that 53 people have lunch as their food option. Next, to maximize the number of food units, sandwich option is chosen for other 47 employees. This produces 1060 + 470 food units which is 1530.

From the case described above, it can be concluded that the solution exists if the set number of food units is less or equal than 1530. It should be noted that the solution to the problem does not depend on the staring schedule. Accurately, if the problem is set in such way that the solution exists, for any starting schedule the solution will be a schedule that satisfies all set requirements.

3.3 Implementation

The algorithm was written in JavaScript programming language, using Visual-Studio Code software created by Microsoft. The starting schedule includes a bus as a transport option and a lunch as a food option for the first 50 employees. The other 50 employees are assigned a bicycle and a sandwich. All employees are stored to an array. Expenses are set to be at most 1100 EUR, while quality and food units are set to be at least 1100 and 1300, respectively. Expenses for such schedule are 1500 EUR, which is not allowed, so the plan is faulty. After the algorithm was run on an array containing 100 elements representing 100 employees, the algorithm returned 1, meaning that a solution was found. Solution included changed schedule for 13 employees that were planned to have a bus and a lunch for their transport and food options, respectively. These 13 employees were assigned a bicycle and a sandwich instead. Following these changes, the final schedule that is the solution to the problem includes:

- 37 employees having a bus and a lunch
- 63 employees having a bicycle and a sandwich.

The schedule that is mentioned above and is the solution to the problem, has total of 1180 quality units and total of 1370 food units. The total expenses are 1095 EUR. These numbers satisfy all set constraints.

This algorithm is applicable to any case, that is, to any constraint values. It will output whether the solution is found or not. For example if values of the constraints are changed to:

- expenses must be smaller than or equal to 1200 EUR
- quality units must be greater than or equal to 1200
- food units must be greater than or equal to 1350

the algorithm comes up with a solution with total expenses of 1200 EUR, total quality units of 1200, and total food units of 1400. The starting plan in this case was the same as in the previous.

3.3.1   Unsolvable Example

The algorithm described in this paper finds solution only if it is mathematically possible to find a plan that will satisfy all set constraints. For example, let's consider the following values of the constraints, given the same starting plan:

- expenses must be smaller than or equal to 900 EUR
- quality units must be greater than or equal to 1300
- food units must be greater than or equal to 1300.

If the constraints are set as above, the algorithm outputs that the solution cannot be found.

References

1. Davis, M., Logemann, G., Loveland, D.: A machine program for theorem-proving. Commun. ACM **5**, 394–397 (1962)
2. Golomb, S., Baumert, L.: Backtracking programming. J. ACM **12**, 516–524 (1965)
3. Gaschnig, J.: A Constraint satisfaction method for inference making. In: Twelfth Annual Allerton Conference on Circuit and System Theory, pp. 866-874, Monticello, Illinois (1974)

4. Mackworth, A.K.: Consistency in networks of relations. Artif. Intell. **8**, 99–118 (1977)
5. Mackworth, A.K.: On reading sketch maps. In: Fifth International Joint Conference on Artificial Intelligence, pp. 598-606, Cambridge, Mass (1977)
6. Stallman, R.M., Sussman, G.J.: Forward reasoning and dependency directed in a system for computer aided circuit analysis. Artif. Intell. **9**, 135–196 (1977)
7. Gaschnig, J.: Experimental case studies of backtrack vs. Waltz-type vs. new algorithms for satisfying assignment problems. In: Proceedings of the Second Canadian Conference on Artificial Intelligence, pp. 268–277, Toronto (1978)
8. Prosser, P.: Hybrid algorithms for the constraint satisfaction problems. Comput. Intell. **9**, 268–299 (1993)

Developing a Runner Video Game

Dalila Isanovic$^{(\boxtimes)}$

Sarajevo School of Science and Technology, Sarajevo, Bosnia and Herzegovina
dalila.isanovic@stu.ssst.edu.ba

Abstract. The video game industry has never been as powerful as it is today, and is getting bigger by the year, and so do the requirements when designing a new game. Even if a developer uses a game engine software, there are still challenges to face when creating a game that needs to be interesting and engaging, while running fast and without glitches. Games known as runner games have been popular in the recent years, especially on mobile devices. This paper aims to give an example of how such a game was implemented, including the basic game mechanics and all the other features working together to create a more engaging experience for the user. The possibility of adding non-player characters in the game is discussed, a feature not too commonly seen in this specific kind of video games. It also gives an overview of common techniques used for the creation of this type of games and discusses their use cases.

Keywords: Mobile game · Unity3d · Endless runner

1 Introduction

The gaming industry is getting bigger by the year. Developers now have to face more requirements when planning and developing a new game. It has to be challenging and interesting, while working smoothly even if it is run on handheld or mobile devices, which usually have less computing power. A certain kind of game, known as an infinite/endless runner game has been popular in the recent years, especially on mobile phones. Some of the best known examples are Temple Run or Subway Surf. However, while starting to create this kind of game, it was unclear as to what the best way to go about it was.

This paper aims to give an overview of common techniques used in the production of this kind of game, as well as ways to add new features that would enhance the gaming experience. It also describes how our game was developed.

1.1 Scope

The paper is concerned with developing the game in an optimal way regarding how various components of the code work together and how they organized, so as not to make the game performance-heavy. The actual graphical models used in the game and their purpose are described, however the details of their creation

© Springer Nature Switzerland AG 2019
S. Avdaković (Ed.): IAT 2018, LNNS 60, pp. 439–445, 2019.
https://doi.org/10.1007/978-3-030-02577-9_43

are out of scope of this paper, since we are now only concerned about the part of the game-mechanics code.

Also, as the game is implemented using the Unity engine, it uses libraries provided by that engine, therefore some parts of the implementation refer to the specific challenges and advantages that Unity offers.

2 Related Work

Although there is not much literature on this kind of game, there are some papers, such as [1], which identifies common patterns in designing this type of game and gives a guide on how to balance them out.

The topic of applying artificial intelligence in game development has been explored in works such as [2–4].

All of the functionality offered by the Unity engine was referenced in its technical documentation [5,6].

Although this paper's main focus is on programming game mechanics, [7] explains how noise functions, help in creating game objects, especially Perlin noise, which we used in the creation for this game's objects.

The topic of procedural content generation and how it could be applied to games is discussed in this paper. [8] gives a broad overview of applying this concept in video games.

For the development of non-player character (NPC) behavior, we used the concept of finite state machines (FMS), a concept that is explored in [9].

Finally, we discussed using goal-oriented action planning for our NPC behavior, which was also explored by [10], where GOAP was compared to FSM with the conclusion that GOAP was a better way for game AI.

3 Game Logic and Design

The game starts of as a typical running game. The character is pushed forward into the scene and keeps running in the same direction. On its way, the character stumbles upon objects they should avoid, objects that they can collect and get points for, as well as enemies.

The diagram in Fig. 1 presents the game flow visually.

The game also has two modes: attack and hide mode, which are scheduled to fire off randomly. The player cannot kill enemies if they are in hide mode, but has to pass quickly by the enemies without disturbing them. Obstacles and power-ups appear on the gamespace all the while the player is moving. As time passes, the player moves faster and faster, until it is too fast to handle. Along with the speed, the player's points increase.

Figure 2 shows a screenshot from the game.

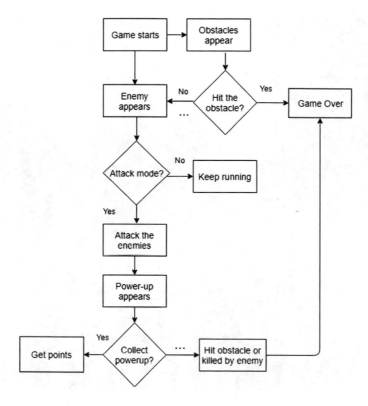

Fig. 1. Game flowchart

3.1 Game Objects

The specifics as to how the game models were created are out of scope for this paper, but it might be important to mention how they were created because it could have had an effect in the further development.

Figure 3 shows the mountain game object that appears on the sides of the player's view. Figure 4 shows the crystal game object that appears in the gamespace. Both models were create in Cinema4D, a graphics modelling software. They are saved as prefabs, meaning that they are static. However, if we wanted to generate these objects procedurally, thus making them unique as they are spawned each time, we would have to write special C# scripts for that. Nevertheless, the object being unique is not important for this kind of game, so we took the simpler approach.

4 Implementation

The game was developed in Unity, a game engine software, using the C# language. The models were created in Cinema4D, except for the placeholder character (shown in Fig. 2) which was imported from the Asset Store.

Fig. 2. The game environment

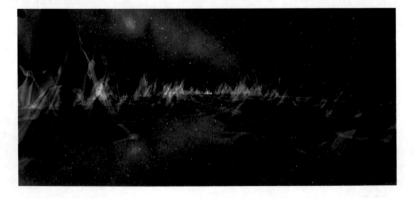

Fig. 3. The mountain object in the gamespace

The first step in the implementation was the main game component, the character running straight ahead, while the scenery appears alongside them as they go. There was number of issues to consider while making this component. Firstly, the scenery has to appear smoothly, so that it looks realistic. The appearing of the scenery can be quite a costly operation, especially considering that

Fig. 4. The crystal object in the gamespace

this process runs constantly in the game. Therefore, we had to find a way to make this operation as computationally low-cost as possible.

As in any game, it was also necessary to constrain the player in the game space. In the case of this game, that meant limiting the player so that they can up to a certain point to the left and the right side of the road. We could either put a plane under the character as a path and then use the *gravity* component in Unity, or use the *Clamp* function.

The next step was to spawn the game environment objects, in this case the mountains that appear alongside the character, and the obstacles on them. One of the options for the spawning the mountains is to have them procedurally generated using a noise function, such as Perlin noise. However, this was not the best option for this game since it was not deemed very important that the game scenery is unique at every frame. Also, it add a lot of additional complexity that is not really necessary.

On the other hand, we could instantiate the mountain game object and destroy it after it is out of view of the player. Destroying the game object would be necessary, otherwise it would clutter up memory rather fast. A better approach is caching the game object as a prefab an then calling it to reappear as we need. This is the preferred method, according to the Unity technical documentation [6].

Alternatively, we could have used a Navmesh component, a feature that the Unity framework offers. A navmesh is essentially a map that determines where an AI agent can move and where not. They can be precomputed, but can also be built at runtime [5]. Nevertheless, the only reason we did not choose to use a Navmesh in this instance is that the path our character can take is a straight road, so it was easy to determine where they can move.

After the core game mechanics was implemented, it was time to add additional functionality. As the character moves through the gamespace, the speed at which they moved would keep increasing, up to a point when it could be impossible for the user to keep up. The character needs to avoid obstacles on their way, and gets points for collecting items it stumbles upon on their way.

Now, all of the above mentioned functionality is commonly found in most endless runner games. At this point, the need arose for implementing other features of this game, that would make it more likely that it stands out on the market.

The objective was to have enemy non-player characters (NPCs) interacting with the player. Considering the nature of the game, those interactions should not last too long, and should work seamlessly with the player's movement. We decided to use a finite state machine (FSM) for this purpose. The character could be in the attacking, retreating, or idle state. If they are in attacking state and fight with the player, they lose points if the player shoots them. Once its health is low enough, they go into retreating state, moving away from the player. If the player is in hide mode, the NPC will be in idle state as long as the player does not collide with any of the NPCs.

A different approach would be to use goal oriented action planning (GOAP). Unlike FSMs, which has certain goals with actions to achieve them, GOAP finds a set of actions to be taken to fulfill a goal, usually using the A* algorithm [3,4,11]. GOAP tends to be the preferred method [10], but as we only had one agent in the first iteration for the production of our game, an FSM was good enough.

5   Discussion

Adding NPC behavior to runner games, especially mobile, is not very common, considering the most popular and best rated runner games. However, having the runner game pattern as the core of the game, while augmenting it with other features will transform it and make it more unique. Especially in the later iterations of development, the game might give a very rich experience for the players.

It is important to add that game aesthetics (concerning objects, colors and special effects) are a significant factor for success after the game is released. Game logic and mechanics, along with its visual representation are all requirements for a successful game.

6   Future Work

There is a number of possibilities how this project could develop in the future. Now that the game mechanics and main design are implemented, we can go through the next iteration of developing the game. The process could go in the direction of having a storyline behind the gameplay, which would unfold as the player progresses. The player could also have missions/challenges throughout the gameplay. Since virtual reality technology is progressing, seemingly exponentially, we could upgrade the game to it as well.

Nevertheless, the best way to make future strategies would be to have the users test out the game. Their opinions about the gameplay could be collected, so as to see in which direction the production should continue.

7 Conclusion

In summary, there are challenges, but also many possibilities for creating mini-runner games. On the one hand, we have to make sure that the game will run smoothly, especially if we are developing for mobile devices, which have limited resources. Game engines, such as Unity, can be a great tool, as they give an additional level of abstraction, which gives developers more time to focus on the actual gameplay from the user's perspective. Libraries that use artificial intelligence, whether through pathfinding, navmeshes, or finite state machines are of big help in this process.

On the other hand, while having the infinite runner mechanics as the starting point of the game, there are almost endless possibilities for adding new functionalities that would transform it to be a unique, interesting game. This can be done through adding new NPC behavior, generating content procedurally for unique, unexpected scenes, and many others.

References

1. Cao, D.: Game Design Patterns in Endless Mobile Minigames (2016)
2. Schwab, B.: AI Game Engine Programming (Game Development Series). Charles River Media Inc., Rockland (2004)
3. Millington, I., Funge, J.: Artificial Intelligence for Games, 2nd edn. Morgan Kaufmann Publishers Inc., San Francisco (2009)
4. Buckland, M.: Programming Game AI by Example. Wordware Game Developer's Library. Jones & Bartlett Learning (2005)
5. Unity Technologies. Unity - Manual: Navigation and Pathfinding (2018)
6. Unity Technologies. Unity - Manual: Object Spawning (2018)
7. Marschner, S., Shirley, P.: Fundamentals of Computer Graphics, 4th edn. A. K. Peters Ltd., Natick (2016)
8. Korn, O., Lee, N.: Game Dynamics: Best Practices in Procedural and Dynamic Game Content Generation, 1st edn. Springer, Heidelberg (2017)
9. Rehman, M., Jamil, A., AsadUllah, E.: An infinite runner game design using automata theory. Int. J. Comput. Sci. Softw. Eng. 5(7), 119–125 (2016)
10. Long, E.: Enhanced NPC behaviour using goal oriented action planning, January 2007
11. Hart, P.E., Nilsson, N.J., Raphael, B.: A formal basis for the heuristic determination of minimum cost paths. IEEE Trans. Syst. Sci. Cybern. 4(2), 100–107 (1968)

On-line Platform for Early Detection of Child Backlog in the Development

Alican Balik$^{(\boxtimes)}$ and Belma Ramic-Brkic

University Sarajevo School of Science and Technology,
Sarajevo, Bosnia and Herzegovina
alican.balik@stu.ssst.edu.ba, belma.ramic@ssst.edu.ba
http://www.ssst.edu.ba

Abstract. Early diagnosis and efficient interventions through educational methods may result in significant progress in early detection of autism spectrum disorder (ASD). The platform presented in this paper represents an on-line and web application for children, in-particular children with ASD. We hope it will help improve their self-esteem, increase their communication and leadership skills. The core concepts of this platform are to help children get rid of impairment of concentration and improve their self-confidence. Our contribution is reflected in the discussion session where we show the strengths and weaknesses of the current learning process and the advantage of here presented application software solution. This paper presents a work in progress and is done in collaboration with the professionals working at an NGO "EDUS Education for All", program devoted to evidence-based work with children with disabilities.

Keywords: Educational games · Education · Early detection · Autism

1 Introduction

Autism or Autism Spectrum Disorder (ASD) represent general terms for a neurobiological disorder that is detectable and cause discrepancies or differences in early age [10]. These disorders are characterized by difficulties in verbal or nonverbal communications, repetitive behaviours, lack of language development and socialization.

There are a number of factors that cause autism such as pollution, drugs, genetic barriers. A recent research published in 2012 shows that fathers are passing at least 80% of mutations leading to autism, because they give the DNA to a baby [13].

Parents often notice the autism in the first two to three years of their child's life. Research shows that most of the children with autism can make enormous improvements toward recovery by undergoing therapy in the early ages [10].

In this digital age, technology is no longer regarded as something alien or new, but as a normal environment. According to the latest trends, solely in US in

© Springer Nature Switzerland AG 2019
S. Avdaković (Ed.): IAT 2018, LNNS 60, pp. 446–456, 2019.
https://doi.org/10.1007/978-3-030-02577-9_44

2009 we had an increase of kid's ages 2–17 for 1.54%, while the same age gaming population growth was 12.68% [4]. Nowadays, children prefer playing games on modern devices such as smart-phones, tablets, laptops than going to therapists because they find it entertaining compared to the traditional activities that they do with therapists. They also get immediate positive feedback on the electronic devices, which is not always the case with other people either in kindergartens or elsewhere.

The work presented in this paper shows a novel application software developed in collaboration with Dr. Nirvana Pištoljević, education specialist, the concept and its development process as well as initial output and received corrective feedback.

2   Related Work

Recently, a number of researchers have tried to answer a question whether games are effective learning tools [3, 9]. They have looked into the aspect of Play and have identified a strong link between play and learning [8]. Furthermore, research has shown that PC-based interventions improve children's skills at a faster pace, compared to tarditional methods of teaching and learning [1, 2, 14, 16].

The trend of developing serious games is growing steadily. They cover a wide spectrum and are mostly created for English speaking regions. This represents a significant drawback for all other populations. And that is why we here present a tool, application software, primarily developed for the Balkan region with translation to other languages left possible.

There exists several platforms that help the children with autism improve concrete skills. One of them is called *Vithea*, known as Virtual Therapist for Aphasia Treatment for the children with autism [11]. It not only allows children to improve their skills, but also helps caregivers create a dynamic set of questions with images. The platform allows caregivers to create customized multiple choice exercises. Manually creating exercises and customization are key features to ensure that an application matches the characteristics of each child. They believe that creating exercises can be time consuming, so they implemented a module that enables the system to create an automated content that selects images and words from two prepared databases. The platform does not keep track of child's progress and nor child's respond. Collecting such data could be used for visualizing their progress. The content of the question can always be different, but the assessment method is always the same.

A platform that specifically focuses on children should have variety. The platform presented in this paper, not only allows caregivers to create dynamic content with different types of questions, but also stores children's progress to show the correct 3D games that kids can play, based on their progress.

3   The Platform Concept

The on-line platform presented in this paper is intended for children with ASD, teachers, therapists and parents. The goal of the platform is to discover autism in the early years and help those children improve their poor and low-developed skills. The flowchart shown in Fig. 1, shows the general interactive logic and structure of the presented platform.

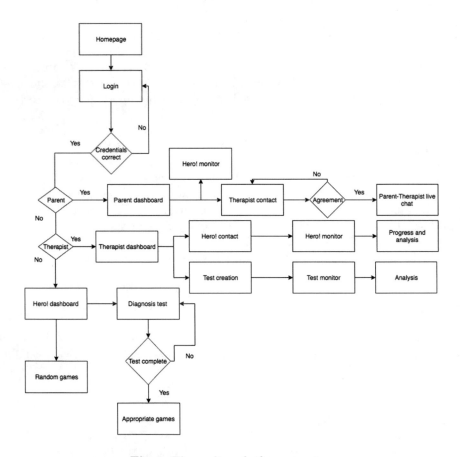

Fig. 1. The on-line platform structure

As can be seen from the Fig. 1, the on-line platform has three different dashboards. Parent account can be connected to a child's account while therapist's or teacher's account can be connected to many children accounts. Children are able to play games which are provided by their teacher through the platform itself, and take diagnostic tests intended for detecting the level of impairment of a particular skill. While playing the game or taking a test, parents and teachers/therapists can track and direct a child. Teachers and therapists are given a private communication channel with parents to discuss their children's progress. When a parent or teacher registers a child in a database, the child then needs to take a diagnostic test. Based on test results, the on-line platform offers several games specifically focussed on the identified poor child's skills.

3.1   Diagnostic Tests

Currently, in our database we have five different questions that are used for diagnostic purposes. Figure 2 shows the question type *drag and drop*. Users move the image to another part or replace the selected image between the two images to sort images by number. This type of question helps children learn the order of numbers and determine the level of the child's counting skills.

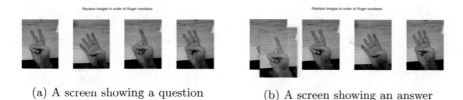

(a) A screen showing a question   (b) A screen showing an answer

Fig. 2. *Drag and Drop* type of a question

Figure 3 shows the question type *multiple choice*. A child is asked to count objects in the picture and select a correct answer from the given list.

Fig. 3. *Multiple Choice* type of a question

Figure 4 shows the question type *point to* using audio. This type of question is good training for auditory and speech. The title of the question is converted from text to speech and saved as MP3 in the database.

Figure 5 shows the question type *Supplement the string*. This type of question allows children to find the missing number and replace them with some of the offered numbers. These numbers are randomly given, but they always contain a correct answer.

Figure 6 shows the question type *puzzle*. This question type helps improve lack of visual, non-verbal thinking. Author of the test is allowed to determine dimensions, but the basic dimension is 3×3.

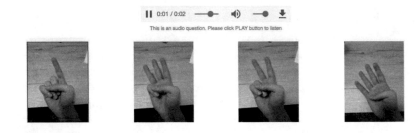

Fig. 4. *Point to* type of a question

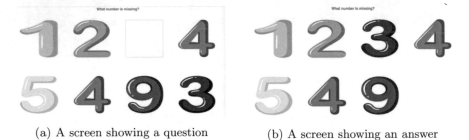

(a) A screen showing a question

(b) A screen showing an answer

Fig. 5. *Supplement the string* type of a question

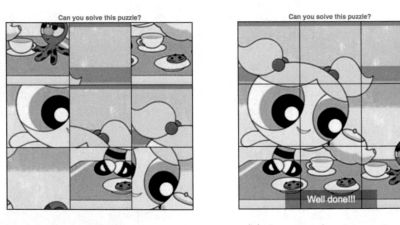

(a) A screen showing a question

(b) A screen showing an answer

Fig. 6. *Puzzle* type of a question

3.2 Teacher's Dashboard

A diagnostic test does not have question limit. It solely depends on the author of the test, which in this case is a teacher. The prototype of teachers dashboard is shown in Fig. 7. They are given an option for creating a test, adding children (here called "heroes") and seeing a complete visualization of child's progress over time.

Fig. 7. A prototype of teachers dashboard

Figure 8 shows the quiz creation page. The teacher must first select a quiz level. There may be more questions with different types. The response section is generated dynamically by the type of question. Each question and answer contents can have both an image and a text. Answers numbers are also dynamic and have validation according to the type of question.

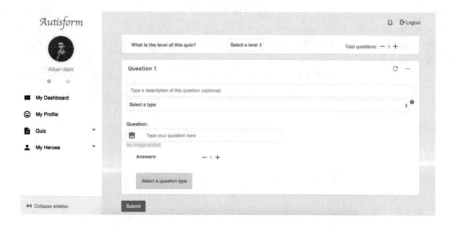

Fig. 8. Quiz creation

3.3 Hero's Dashboard

Hero's dashboard currently has test page and notification bar implemented (see Fig. 9). The platform forces all users to update their profile and verify their email addresses. This process is done by the parents of children because they are minors. Only then are they being able to use their dashboards. When a teacher adds a new hero, the added hero receives a notification which they can accept or reject.

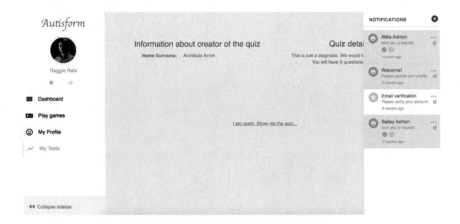

Fig. 9. Hero dashboard with notification bar

When a hero enters the test page, he is greeted with a welcome page with instructions related to the test. When the test begins, a time counter is triggered which helps the algorithm to calculate the time consumption for each test question. If a hero takes a test for the first time, he is allowed to pause the time counter. With a pause option, we wanted to encourage our little players to continue, although they may not know all the answers and may need some time to reach the goal.

Each hero has a level based on their progress. The platform has a level-up feature. Children can gain experience points when they are successful at repeated tests or while playing games. Figure 10 shows level badges from 1 to 22. New level badges will be created when the platform becomes available on-line.

Fig. 10. Level badges

4   The Development

The platform has both the back-end and the front-end component. The back-end represents a Maven project which is built in Java with Spring framework. Spring Framework is a Java based platform that provides comprehensive infrastructure support for developing Java applications.

We currently use an Apache HBase No SQL database. HBase is a column oriented database which is run in Hadoop for scalable storage which allows us to store large data sets. We use Apache Phoenix to use JDBC with power of standard SQL query commands [5–7].

SQuirreL SQL Client is a graphical Java program which allows us to monitor our records in the HBase [15]. The reason for using large database is that when a child plays a game or takes a test, the system monitors and records every action to create an accurate visualization of the child's progress.

The front-end is built in JavaScript framework called Vue.js [12] with ESLint 6. It is significantly lighter than other JavaScript frameworks such as Angular or React. It uses the same document object model like React and implements the Redux statement concept that allows you to run an application in another environment.

5 Results and Discussion

The work presented here is carried out in collaboration with Dr. Nirvana Pištoljević, education specialist working at an NGO "EDUS - Education for All". Initial prototype consisted of a teacher dashboard used for creating quizzes. The database consisted of five different question types with different question contents.

Initial feedback was overwhelming. During the interview, Dr. Pištoljević said that she liked the choice of tasks and that we chose something challenging "requiring kids to really focus. When someone is really interested in what they do, you can see it in his/her work." The main thing we should be careful about is the task determining the autism level. For that, teachers usually use long tests which children often find very boring. Therefore, she suggested making short assessments on different areas such as colours, numbers, and so on.

For example, learning counting can be composed of several tasks:

- Number recognition: Precomputed voice asks a question, a child listens and then answers;
- Dividing numbers into groups based on child's age level (0–10; 10–100; and so on);
- Different learning intentions: matching, point to, etc.

The assessment will be composed of pieces of games put in together as a sequence. Therefore, we first need to create a pool of games from which we will select questions and incorporate them into the assessment test. We believe that this method will allow teachers to upload prepared questions or they can mix and match and create a new test.

Children need to be offered a number of built-in levels that will allow them to progress within a particular skill, such as counting between 0 and 5. For example, a child might be told to count the number of children he/she sees in a picture, where the correct answer is two. A child is given an option of three possible answers: 10, 7 and 2. Since at this stage, they are learning numbers from 0 to 5, this is a relatively easy task for the child as they do not recognize numbers 7 and 10. The next level may include numbers that are close to each other, such as 2 and 3, requiring a child to think more deeply before selecting a correct answer. Furthermore, a number of options to choose from in order to give a correct answer depends on child's behavioural and intellectual level. Children at the beginning may be prompted with two or three choices, while older or more advanced children can be prompted with five or more options.

Commenting on currently implemented tasks and graphics, Dr. Pištoljević said that "the ones that you are showing are perfect and childish which it has to be if you want children to learn something from your games." Given tasks are called selection responses meaning we offer something and the child just selects what is right. However, in order to determine child's knowledge and understanding of numbers, we have to do the production, meaning a child has to write or say the correct number.

Writing and speaking features will be considered as part of the future work. We first need to gather data based on our assessment tests and monitor the child's progress. Currently, our platform supports a variety of tasks:

- matching;
- drag-drop;
- point-to;
- fill the missing;
- puzzle.

It is recommended that we now focus on a specific area, such as counting, and create requiring sub-levels. Focussing on a task such as *matching* is significantly more complex. For example, matching task, could be composed of several completely different sub-tasks, each reaching a different area:

- numbers to numbers;
- number in different colors;
- different shapes or different graphics that the child can understand the number. For instance: number 3 in any shape and/or any color;
- Question type *auditory point-to*: the computer says find three fingers and a child selects the picture that shows number 3.

Furthermore, it is recommended that the way of asking a particular question must vary so the child is not memorizing the pattern or starts to be bored. Another reason is that children with disabilities learn only one way. If we ask a same thing in different ways, they willl actually learn that it means the same. According to Dr. Pištoljević, "you teach them that you can say something in many different ways.". For example:

- Count how many cars.
- How many cars are there?
- Tell me, how many cars do you see?

6   Conclusion and Future Work

In this paper, we presented a prototype of a platform that can be used both, as an aid for early detection of ASD and for personalized learning as it supports individuals performance and personalisation capabilities. It can also serve parents as an accessory in education.

On the basis of the received corrective feedback, we plan to extend the platform in several directions, including taught concepts, complete speech support and new learning environments. The platform currently supports text-to-speech only for point-to question type. Furthermore, we plan to enable users from different countries to create their own dashboards in their own languages.

A more detailed analysis is needed to better understand the benefits of using the platform as opposed to traditional way of studying. That is why we plan to work with a larger number of children to verify the usefulness of a here proposed solution.

Furthermore, we will also consider the possibility of implementing drawing and writing features. For example, a child tries to create an image provided by a system such as a *stick men* and based on the drawing, we can calculate how accurate is it with machine learning environments.

References

1. Annetta, L.A., Murray, M.R., Laird, S.G., Bohr, S.C., Park, J.C.: Serious games: incorporating video games in the classroom. Educause Q. **29**(3), 16–22 (2006). https://www.learntechlib.org/p/103749
2. Bajraktarević, S., Ramić-Brkić, B.: Kockica: developing a serious game for alphabet learning and practising vocabulary. In: Hadžikadić, M., Avdaković, S. (eds.) Advanced Technologies, Systems, and Applications, pp. 349–358. Springer International Publishing, Cham (2017)
3. Connolly, T., Boyle, E., Boyle, J., Macarthur, E., Hainey, T.: A systematic literature review of empirical evidence on computer games and serious games. Comput. Educ. **59**(2), 661–686 (2012). https://doi.org/10.1016/j.compedu.2012.03.004
4. Forsans, E.: The video game industry is adding 2-17 year-old gamers at a rate higher than that age group's population growth. https://www.atjv.com/news/233_kids-and-gaming-2011.htm. Accessed 20 Mar 2018
5. The Apache Software Foundation: Apache Hadoop - Hadoop. http://hadoop.apache.org/. Accessed 20 Mar 2018
6. The Apache Software Foundation: Apache HBase - Hadoop. https://www.ibm.com/analytics/hadoop/hbase/. Accessed 20 Mar 2018
7. The Apache Software Foundation: Apache Phoenix - Hadoop. https://phoenix.apache.org/. Accessed 20 Mar 2018
8. de Freitas, S.: Education in Computer Generated Environments. Taylor and Francis, Routledge (2013)
9. de Freitas, S.: Are games effective learning tools? A reviw of educational games. Educ. Technol. Soc. **21**(2), 74–84 (2018)
10. Landa, R.J.: Diagnosis of autism spectrum disorders in the first 3 years of life. Nat. Clin. Pract. Neurol. **4**, 138–147 (2008)

11. Mendonça, V., Coheur, L., Sardinha, A.: VITHEA-kids: a platform for improving language skills of children with autism spectrum disorder. In: Proceedings of the 17th International ACM SIGACCESS Conference on Computers, Accessibility, ASSETS 2015, pp. 345–346. ACM, New York (2015). https://doi.org/10.1145/2700648.2811371
12. MIT: VueJS - The Progressive JavaScript Framework. https://vuejs.org/. Accessed 20 Mar 2018
13. O'Roak, B., Vives, L., Girirajan, S., Karakoc, E., Krumm, N., Coe, B., Levy, R., Ko, A., Lee, C., Smith, J., Turner, E., Stanaway, I., Vernot, B., Malig, M., Baker, C., Reilly, B., Akey, J., Borenstein, E., Rieder, M., Nickerson, D., Bernier, R., Shendure, J., Eichler, E.: Sporadic autism exomes reveal a highly interconnected protein network of de novo mutations. Nature 485(7397), 246–250 (2012)
14. Putnam, C., Chong, L.: Software and technologies designed for people with autism: what do users want? In: Proceedings of the 10th International ACM SIGACCESS Conference on Computers and Accessibility, Assets 2008, pp. 3–10. ACM, New York (2008). https://doi.org/10.1145/1414471.1414475
15. SQuirreL SQL - Sourceforge. http://squirrel-sql.sourceforge.net/. Accessed 20 Mar 2018
16. Tanaka, J., Wolf, J., Klaiman, C., Koeing, K., Cockburn, J., Heirlihy, L., Brown, C., Stahl, S., Kaiser, M., Schultz, R.: Using computerized games to teach face recognition skills to children with autism spectrum disorder: the Let's face it! Program. J. Child Psychol. Psychiatry 51(8), 944–952 (2010)

SPACE - Proprietary University and Gymnasium Information System

Emina Mekic$^{(\boxtimes)}$ and Emir Ganic

Sarajevo School of Science and Technology, Hrasnicka cesta 3a, 71000 Sarajevo,
Bosnia-Herzegovina
{emina.mekic,emir.ganic}@ssst.edu.ba

Abstract. The electronic information age and technology expansion are significantly and rapidly changing the world we live and work in. Consequently, one of the aspects greatly affected by the global modernisation is education, replacing piece by piece less effective traditional routines by modern, effective and technology-oriented educational procedures. Following this trend, as a groundbreaking university, Sarajevo School of Science and Technology (SSST) has created its own, specialised information system, SPACE, aiming to enrich and complete its quality and high-end educational services. SPACE has two instances: University and Gymnasium. Both instances also allow access through the use of the mobile version of SPACE, created currently only for iPhone users.

This paper presents the key ideas and functionalities of SPACE, motivation behind and benefits of developing custom-tailored information system opposed to purchasing ready-made software in terms of quality, suitability and user satisfaction. It also mentions the advantages of maintaining in-house development team which, in collaboration with other University services and facilities, immediately turns user feedback into useful SPACE integral parts, shaping SPACE according to the users' needs. This, coupled with constant SPACE customisation and numerous plans for future upgrades and improvements, demonstrates a very innovative approach to technology usage in education, especially in Bosnia and Herzegovina.

Keywords: University · Information system · Proprietary
Custom-tailored · ICT · Education · SSST

1 Introduction

1.1 Motivation

The performance of higher education is of immense importance for the competitiveness and advancement of nations [1]. Also, individual students have a great interest in the quality of their education, being aware that primarily high quality education can advantage them and turn them into assured, adaptable, innovative, entrepreneurial and employable competitors in a job market [2].

© Springer Nature Switzerland AG 2019
S. Avdaković (Ed.): IAT 2018, LNNS 60, pp. 457–467, 2019.
https://doi.org/10.1007/978-3-030-02577-9_45

Due to expansion of higher education and constantly increasing information technologies capability, new organisational structures are inevitable in order to catch up with and support new educational trends, as well as to fulfill a need for improved efficiency in administrative services [3].

As the university dedicated to providing high quality education, currently being positioned among the region's leading universities, Sarajevo School of Science and Technology (SSST)[1] has been constantly investing huge amount of effort to follow the most popular and efficient world trends, and to fulfill the highest standards in education.

Besides the fact that quality of education is closely related to the expertise and professionalism of teaching staff, there are other factors that support quality higher education. One of them is incorporating information technologies into educational flows, having been stated that performance of contemporary university heavily depends on its information system [4]. Along with the global trend of modernisation of educational processes and the rapid ongoing growth of the technology-driven education, where students need and use digital devices to maintain their portfolio, access news and information when they need it, get their grades, and manage their learning life, the Management of SSST decided to make a courageous step forward. They initiated the procedure of creating University's own information system, followed by the Gymnasium's shortly after.

The first release was delivered in March 2016, for University SPACE, and in January 2017, for Gymnasium SPACE. The information system was named SPACE, alluding to the virtual space as a piece of the system allocated for each user, containing all needed academic and study related information.

As already indicated, SPACE has two instances:

- *University SPACE* - developed first and with the intention of being used by SSST University staff and students [5]
- *Gymnasium SPACE* - developed upon the arrival of the SSST's new family member, Gymnasium, and intended to be used by Gymnasium staff, students and parents [6].

1.2   In-House vs. Application Packages Software

SPACE implementation was preceded by choosing the appropriate information system development strategy. Having choice of two major strategies, in-house development and the use of application packages, a comparison of their applicability to our particular case was made and is briefly presented below. Despite a growing trend toward the use of ready-made application software in information system development mainly because developing proprietary software is considered costly and risky, both in terms of development cost and the quality of resulting product [7], SSST chose to develop a proprietary information system. Taking into account information system's level of originality, as suggested concept underlying this choice [7], as well as the local context [8], the appropriateness of off-the-shelf solutions was clearly limited in this case. In a nutshell,

---

[1] http://ssst.edu.ba.

commercial ready-made software, available for use in a matter of minutes, but at the same time only partially adequate and suitable for the specific needs of SSST, was overpowered by more costly and time consuming, but absolutely adjusted, solution.

Even though still very young information system, due to the functionality richness as well as the extent to which it already serves the purpose, SPACE has managed to fulfill all initially set aims, justify high expectations and effort, and take the leading role in terms of managing and coordinating educational processes at SSST over.

The system is described in more detail in the following sections, from design and implementation process, and final products description, to future work and upgrade plans.

2   Design and Implementation

The introduction of the SSST information system has opened new possibilities to address some of the issues concerning the personal management of academic information and knowledge, main of them being the lack of integration, personalisation and customisation of existing academic information sources and systems [9]. Therefore, since it was crucial to the system suitability and success, and it will continually impact the further development, accurate investigations of the existing academic system had to be generated.

After weeks of preparation and many meetings with each department, administration and grading system researches, exploration of the overall academic structure as well as the departments and degrees, taking differences between departments, their schedules and specificities into account, many discussions and exchange of views, a clear idea of what the structure of the SPACE information system should look like, emerged. After that, the process of design and development started, whose significant steps and outcomes are briefly tackled in this and the following section.

SPACE can be described as intranet-based, integrated information system, whose major purpose is to serve central administrative needs of SSST, having all data stored in centralised database.

System design stage consisted of problem and requirements analysis, during which numerous diagrams and blueprints were produced. SPACE represents very complex system, composed of a number of differently functioning and interleaving modules, such as *Student Module*, *Professor Module*, *Student Services and Registry Module*, *Management Module* and others. Here we show selected user requirements for SPACE *Student Module*, being its main functionalities at the same time:

1. *Profile Management* retrieves student profile information, organises them and displays appropriate information on user profile, such as student name, surname, image, student ID number, phone number, address, city, and other, study-related information.

2. *Course Management* takes care of displaying a table with the list of all subjects a student is taking in the current semester, by selecting student's last enrolment and checking for the current year of study parallel to the current academic year and semester. By clicking on each subject, four tabs appear, with initially opened tab *General.*
3. *Course Information Management* obtains and displays general information about a specific course, including code, name, coursework percentage, final exam percentage, ECTS points and syllabus.
4. *Colleague Management* takes from the SPACE database all students taking a specific course in that semester. Name, surname, image and university email are information listed for each colleague.
5. *Grades and Attendance Management* take care of bringing and displaying student's grades or attendance records for a specific subject. Date, name, type and percentage of the exam, students' score and letter grade are shown after choosing *Grades* tab, while *Attendance* tab displays date and name of the lesson together with student's attendance record for that lesson.

A statechart UML diagram, describing flow of control from one to another state [10], for *Student Module* is shown in Fig. 1.

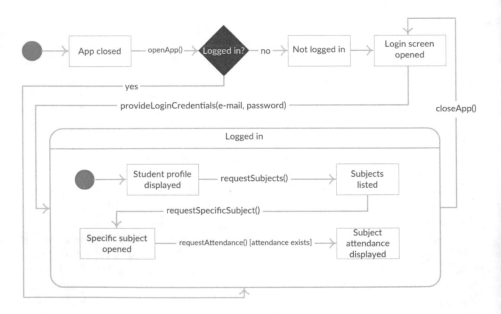

Fig. 1. Statechart diagram for SPACE - *Student Module*

Due to a heavy influence of appropriate system development approach on the quality of system being produced, object-oriented approach was chosen. The choice was mainly based on the fact that it exhibits the concept of system understandability, extensibility, resolvability, reusability, and maintainability [11].

SPACE is developed using popular PHP framework Laravel [12], and, as it represents the internal university and gymnasium system, a significant amount of effort and time during development was invested to incorporate reliable safety measures in order to ensure that security of the system is on adequate level. Additional tools used for development are NetBeans IDE and Bitbucket version control system. SPACE application uses MVC (Model View Controller) design pattern, which ensures clean and intuitive code management.

3   Results - Final Products

University SPACE. Since its first release, SPACE has been successfully serving a wide range of users, from students to various departments. All users are divided by roles, which control access to SPACE and direct them to the content they are allowed to see and intended to be used by their role (a student's dashboard is shown in Fig. 2). Having in mind that SPACE is equipped with a large number of diverse functionalities and that each role has different permissions and responsibilities regarding the academic process and requires specific approach, but is almost of the same importance from the aspect of SPACE user, explaining some of them will give clear overview of their privileges and manifest SPACE possibilities in the best way.

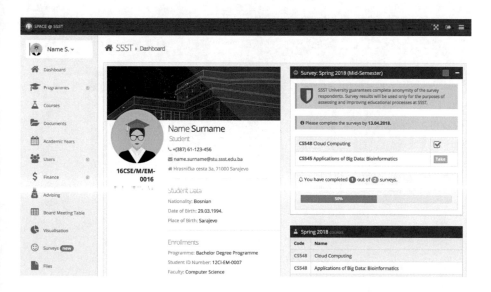

Fig. 2. Student's dashboard - *University SPACE*

It can be freely said that the most significant SPACE role is *Student*, for which this system is primarily created. SPACE allows students to see all subjects they are taking in the current semester (academic year, divided into semesters, is the central and key term in SPACE structure), checking grades and attendance for each subject. This provides students with effective, compact and always-available overview of and complete control over their academic progress, giving them a chance to be prevented from any unwanted situations if reacted on time, and involving student at the same time in the modern educational system flows. Interesting to mention is *Survey Module* (Fig. 2), incorporated into SPACE with the intention to simplify the process of rating course staff and their performance as well as to gain effective and precise information presentation, and transparency of results.

Professors represent another significant group of users, getting similar interface as students, but allowed to insert attendance, grades and course general information. Also, they can see visual representation of student's performance for their course (Fig. 3), which was developed under *Visualisation Module*. Furthermore, they can follow charts with a number of working hours displayed in different time intervals, with time of arrivals and departures being recorded using RFID cards by RFID reader in the entrance hall of the University. *Professor* can be at the same time *Dean* or *Advisor*, what gives him access to special types of summary tables and more privileges and responsibilities regarding supervising students' academic skills and progress.

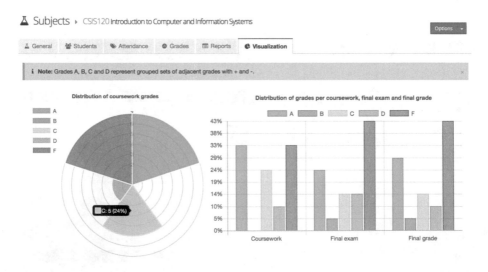

Fig. 3. *Visualisation Module - Professor's* profile

Similarly to the supervising role, *Management* role has a complete overview of all University instances and departments as well as academic processes and progress of each student.

Student Services, responsible for a wide range of administrative tasks, is also a notable group of users. Significant number of SPACE functionalities are dedicated to them, from creating and managing user profiles, subjects, associating subjects with faculties, creating and managing academic years and semesters inside them as well as managing each department by programme and academic schedule. Even though they are responsible for managing many SPACE records and maintaining many SPACE processes, they are exempt from associating every single student with subject they are taking in the current semester. The process of assigning subjects to students is absolutely automated in SPACE, where, by following the pre-determined procedure and triggered by the start of the specific semester under the current academic year, every student is automatically assigned appropriate subjects according to the previously inserted academic schedule. This represents one of the central and key processes in whole SPACE system. *Documents Module* allows them to easily (in a matter of seconds) generate various types of documents (statements, confirmations or academic transcript) on a student's request. SPACE automatically produces documents by collecting student's data from centralised database, efficiently replacing previous paper-based procedure and speeding up the process.

Finance Department role is equipped by the *Finance Module* functionalities for managing financial activities, such as invoice and warnings handling (both generating and sending). This role has insight into employees' working hours as well.

Administration role provides that group of users with needed functionalities, such as access to contact information of University staff and students.

There are more roles (*Teaching Assistant*, *Lab Demonstrator*, *External Examiner*, *Legal Department*, *IT Department*, *PR Department* and *Library*) and even more functionalities SPACE provides, but these represented previously are more than enough to get insight into principles of SPACE functioning, its purpose and significance.

Gymnasium SPACE. In accordance to the extension of the SSST family by the Gymnasium, a separate and adjusted SPACE branch has been created. Even though Gymnasium SPACE had to fulfill a number of strict and specific high-school related rules and requirements, it still heavily relies on the University SPACE, especially having in mind that it inherited a high percentage of its key ideas and functions.

Similarly to the University SPACE, Gymnasium SPACE provides students with complete overview of their academic records and progress, including their study information, grades, absences (excused and unexcused) and student conduct. A student's dashboard is very similar to Fig. 2 and shown in Fig. 4 (tables with subjects and absences are displayed).

Professors enter grades, absences and subject information on a daily basis, while *Form Tutors*, what is a new SPACE role, keep track of all students from their class by monitoring their academic progress, and handling their absences and student conducts. A *Director* role owns a right to supervise both students and professors, as well as to ensure all SPACE data are up-to-date.

Despite the uniqueness and innovativeness of the Gymnasium SPACE's general idea, what makes this information system even more characteristic and

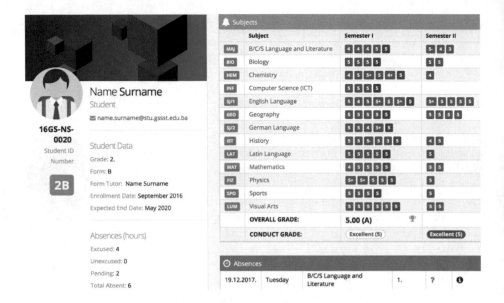

Fig. 4. Student's dashboard - *Gymnasium SPACE* (*Subjects* and *absences* tables)

valuable is the possibility for parents as well to follow their children's profiles and track their academic performance, including grades, absences and student conduct. Besides the timely awareness of the tiniest changes in one of these, they are enabled to promptly react on any unwanted change and act accordingly. In this way, one of the most important and usually most unreachable high school-education goals, to encourage parents' inclusion into educational process, is reached and ensured.

SPACE iOS Application. Finally, the mobile version of SPACE is developed, currently supported for iPhone users. Programming language used for iOS development is Swift [13], faithfully reflecting and supporting all significant SPACE functionalities for students. Figure 5 displays two SPACE iOS app's scenes, the *Login* and *Attendance* screens.

4   User Feedback and SPACE Maintenance

Even though SPACE has been tested throughout the entire process of development, a UAT (User Acceptance Testing) was conducted as well. A group consisted of ten SSST students - samples from different departments and years, representing real-life users and contributing test results with the most relevant feedback. They were asked to go through the entire process of using the system and test it, after what they were given 10-item usability questionnaire to express the results of testing, their feedback and experience with using SPACE. At the end, overall satisfaction score was more than satisfiable, since more than 90% of

Fig. 5. SPACE iOS app *Student Module*: *Login* screen (left) and *Attendance* tab (right)

users conveyed the positive experience, coupled with valuable comments on the possible future improvements. Moreover, this is not where SPACE testing ended, since every day of SPACE usage represents testing day at the same time, during which new ideas and suggestions are being collected, as well as the overall user satisfaction level examined.

After obtaining the user feedback, SPACE development team considers the request, communicates with appropriate University services or facilities if needed, prepares it for SPACE integration and implements it. The SPACE system is completely developed and managed by the University development team, which consists of developers, students and IT professionals. Internal development team structured like this, besides ensuring quality and highly appropriate design and implementation tasks performance, maintains SPACE primarily by following direct user feedback, constantly adjusting SPACE according to exactly what University and its users need. SPACE development is under continuous development, mainly having in mind numerous extensions and upgrades planned for the future.

5   Conclusions and Future Work

As it was highlighted through this paper, along with the global and revolutionary modernisation of education and educational processes and in order to simplify the process of tracking students' academic performance for both teaching staff and students themselves, the SSST University has created its own information system SPACE. Encouraged by extremely successful first release and according to the initial plan, there are many functionalities waiting to be implemented, for both University and Gymnasium instances, as well as in terms of mobile applications. As SPACE is planed to be large, comprehensive and fully-functional information system of the SSST, which means that it must become, by importance, function and role, a central warehouse and regulator of the whole educational system at SSST, the process of implementation may take a few years. That includes a number of improvements, upgrades, enhancements and revisions. Even though this list is getting bigger and bigger with every meaningful and innovative idea, some of them are: enriching intra-course collaboration possibilities in terms of communication and notifications, Office365, schedule and timetable integration, implementing new modules (library, restaurant and bookstore), implementing very modernised-education-related idea of recording students' attendance using RFID cards, and after the reasonable amount of time pass, even applying machine learning methods in order to analyze students' performance more deeply and thoroughly. Finally, creating Android mobile app would represent an excellent future step, since needs of almost entire student population will be satisfied.

Finally, according to the overall organizational structure and advantage of maintaining in-house development team, it is certain that SPACE will continue to grow and develop very quickly. It definitely promises to transform into huge and recognized information system, which will, besides obtaining much richer functionality spectrum and strengthening the name of a sensational, unique and innovative university and gymnasium information system at the same time, continue serving its purpose of cherishing education in the best way.

References

1. Ford, P., Goodyear, P., Heseltine, P., Lewis, R., Darby, J., Graves, J., Sartorius, P., Harwood, D., King, T.: Managing Change in Higher Education. SRHE & Open University Press, Buckingham (1996)
2. EU High Level Group: Modernisation of Higher Education (2015). http://ec.europa.eu/education/library/reports/. Accessed 2 Apr 2018
3. Allen, D.K., Fifield, N.: Re-engineering change in higher education. Inf. Res. 4 (1999)
4. Komka, A., Daunoravicius, J.: Information System of University: Goals and Problems, pp. 45–51 (2000)
5. SSST University (2016). SPACE@SSST: https://space.ssst.edu.ba
6. SSST Gymnasium (2017). SPACE@GSSST: https://space.gssst.edu.ba
7. Iivari, J.: Implementation of in-house developed vs application package based information systems. ACM SIGMIS Database **21**(1), 1–10 (1990)

8. Williams, J.O.: Appropriate information system development. Ph.D. dissertation, Radboud Univ. Nijmegen, Graz, Austria (2012)
9. Pienaar, H.: Design and development of an academic portal. Libri **53**, 118–129 (2003)
10. Burd, S.D., Satzinger, J.W., Jackson, R.: Introduction to Systems Analysis and Design: An Agile, Iterative Approach, 6th edn. Cengage Learning EMEA, Boston (2014)
11. Pan, J.: Requirements analysis for a student information system. M.Sc. dissertation, Univ. of Leeds, Leeds, UK (2014)
12. Otwell, T.: Laravel - The PHP framework for web artisans. Laravel.com (2015). http://laravel.com/docs/4.2. Accessed 2 Apr 2018
13. Apple Inc. (2014). The Swift Programming Language: https://swift.org/documentation/TheSwiftProgrammingLanguage(Swift4.1).epub

Change Detection of Hydrologic Networks Using Orthophoto Images in Bosnia and Herzegovina

Enes Hatibovic[✉] and Ajla Kulaglic

University Sarajevo School of Science and Technology,
Sarajevo, Bosnia and Herzegovina
enes.hatibovic@ssst.edu.ba

Abstract. A river is a natural water course, usually freshwater, flowing downhill from river source to river mouth. River course change is natural phenomenon which takes place during flood occurrence. Bosnia and Herzegovina has substantial water resources and water could in the future become one of the cornerstones of general economic development in many areas. Sustainable development in water management is only possible with the implementation of integrated water resource management principles, common problem solving in the main water management segments, especially in exploitation, water protection and water protection from harmful effects. In this paper, we focused on the river course detection of Bosna river in Bosnia and Herzegovina. In this paper, an approach for detecting changes in river flow using two ortophoto images is presented. The orthophoto images were taken in 2008 and 2012. A general and coherent increase in Bosna river flows has been detected.

Keywords: Bosna river · Orthophoto · Change detection
River flows · ISO Cluster Unsupervised Classification

1 Introduction

Floods, like droughts represent extreme hydrological phenomena affecting the population, social and ecological systems. The flood is defined as "temporary coverage of land water that is usually not covered by water and implies flooding of rivers, mountain streams, occasional watercourses in the Mediterranean and floods in the sea coastal areas, and can be excluded from the drainage of sewage systems". In other words, flooding is a natural phenomenon that marks an unusually high water level, which causes water from the basin to float across the shore and the blue surrounding. Waters (their precipitation), their spatial distribution, intensity and duration of precipitation are the main climatic causes of flooding, but in addition to these causes, flooding contributes to the capacity of the watercourse or watercourse network to receive and continue to convey water flow, condition in the whole basin, weather conditions prior to precipitation, ground cover and topography. Artificial impacts that can cause flooding may be the accentuation of water from accumulation and retention caused by

© Springer Nature Switzerland AG 2019
S. Avdaković (Ed.): IAT 2018, LNNS 60, pp. 468–475, 2019.
https://doi.org/10.1007/978-3-030-02577-9_46

the failure of dams or hydraulic engineering structures or their inadequate work and handling, then changes in the basin, river basins and inundation areas, etc. Special attention should be paid to changes in the basin, river basins and inundation areas that have arisen with anthropogenic influences, among which are the most important forests of forests, poor agricultural practices, inadequate water management, urbanization in areas of high flood risk and pressures that cause the population to their activities.

Talking about the causes of flooding in recent times certainly has to be pointed out to climate change, given that the flood itself, along with the droughts, falls into the effects most pointed to. Climate change as a cause of the flood should be further observed and analyzed.

In several years, several studies [1, 2] have been carried out on the impact of climate change on the water regime in the Sava river basin. Average season precipitation and temperature were analyzed, however, the results could not be used to analyze the impact of climate change on the flood. These data were the basis for the development of climate change forecasts on the likelihood of floods. The analysis also includes the results of the hydrological model of the river Bosna and flood data in May 2014.

The hydrological analysis was conducted by hydroplane model calibrated for flood in the fall 1974. The fall event was chosen for several reasons: the flood affected the entire basin since peak rainfall in the Drina river appeared two days after the peak in the Alps and at that time hydroelectric interventions in the basin have not yet been completed (dam on Piva, flood protection systems along the Sava river). The largest increase in precipitation due to climate change is expected in the fall period and high quality data for the whole basin. The Sava river basin is divided into 13 parts for the hydrological model. All the most important rivers are listed: Kupa, Una, Vrbas, Bosna i Drina.

Floods in the Sava River and Bosna river basin confirm the fact that floods with a longer return period can be expected at any time. The likelihood of flooding at the water station Doboj was analyzed in [3].

According to the study from the period 1961–1990, the flood in May 2014 has more than a thousand years of return, and according to a diagram that includes the period 1961–2014, the probability of flooding in 2014 is somewhat larger than the one-year return period. Due to the impacts of climate change, the probability of the first two prognosis periods increases considerably, and in the third period the diagram is almost the same for the second period and is not recognizable. The 2014 flood thus has a ten-year return period in the period 2011–2041. and a two-year return period from 2041 to 2100.

The characteristics of the Bosna River hydrological system will be affected by climate change impacts [3]. The impact on the karst valleys will be much smaller, and the Modrac reservoir and the floodplain of the Sprea river will significantly mitigate the increase in flood waves.

It is necessary to mention several studies conducted in the river basin of Bosnia. The first of them is "The impact of climate change on the floods in the river basin of Bosnia and Sava" [10]. A hydrological analysis was carried out for which the weather forecast for maximum daily precipitation and temperature was used for three periods in advance. In addition, during the study, the maximum flood-flow rate was calculated on the basis of daily precipitation with a return period of 20 and 100 years, and the flood was investigated from 2014 on the river of Bosnia.

The question of floods and their consequences and protection measures in the sixties and seventies of the 20th century was dealt with Vera Katz [4]. Katz treats concrete flooding events in individual river rivers in Bosnia and Herzegovina over the period considered, then provides numerous data on the financial aspect of the damage caused, the damage caused to settlements, infrastructure facilities and land, and the protection measures and defense mechanisms created to prevent damaging consequences in the future. It also discusses the general trend of industrial development and its consequences on river flows in Bosnia and Herzegovina at the end of the seventies and early eighties of the 20th century.

The paper is organized as follows. Section 2 describes the study area. There after, the data set used is explained in Sect. 3. Methods and results are discussed in the Sects. 4 and 5, while Sect. 6 concludes this paper and gives an outlook.

2   Study Area

The basin area of the river Bosna according to the data from "Hydrological studies of surface waters of Bosnia and Herzegovina, river basin of Bosna" (ZV and FHMZ, 2012), [5] amounts to 10.420 km², orographic. The river flows from the South to the North. At the top of the river flow are Dinarides, which exceed 2000 m, the central part of the basin occupies the surface of the central Bosnian hills, in the lower part flows through the floodplains of the floodplain of the Sava river. The geological structure is extremely diverse with typical karst regions. The total area of the basin on the right bank of Bosna is 6900 km² (about 65% of the total catchment area). In the Fig. 1, the main river flows with accent on Bosna River, highlighter in blue, in Bosnia and Herzegovina is presented.

Fig. 1. The river flows of Bosnia and Herzegovina, with emphasis on river Bosna

3   Data Set

In this study two optical orthophoto images were used. The ortophoto images were acquired in 2008 and 2012, respectively. Both images were acquired with UltraCam X, S/N UCX-SX-1-20915097. Pixel size on the ground for both images is 0.5 m. On the figure below (Fig. 2), two used images are presented. On the Fig. 2a) the ortophoto image from 2008 is presented, and on the right side, Fig. 2b) the ortophoto from 2012 is presented.

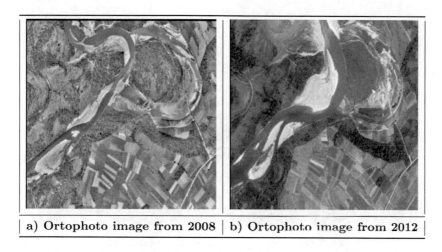

| a) Ortophoto image from 2008 | b) Ortophoto image from 2012 |

Fig. 2. The ortophoto images

4   Methods

In this study Geographical Information System (GIS) software ArcMap 10.3 was used to extract and compare changes in Bosna river flow from two different periods. The image classification is done to identify the changes in the river flow. The image classification refers to the extracting information classes from a multi band raster image [6]. The resulting classified raster from image classification was used to create polygon [7]. In this study the ISO Cluster Unsupervised Classification tool was used. This tool combines the functionalities of the ISO Cluster and Maximum Likelihood Classification tools. Unsupervised classification groups pixels with common characteristics and output a signature file and a classified image [7]. The ISO Cluster performs clustering of the multivariate data combined in a list of input bands. The resulting output, in this case signature file was used as an input for a classification tool, Maximum Likelihood Classification tool to classify the input image. From the classified images, river classification was done using raster obtained in previous step. Obtained two colors raster was converted to the polygon (vector), were polygons were used for constructing the shape file.

5   Results and Discussion

Using two ortophoto images obtained in 2008 and 2012, respectively, the river flow change detection of Bosna river was done. The original images are presented in the Fig. 3. For this purpose, ISO Cluster Unsupervised Classification, previously explained method was used. The input images were classified in the output raster with each class having equal probability weights attached to their signatures [8]. The classified image from 2008 (a) and 2012 (b) are shown in Fig. 3.

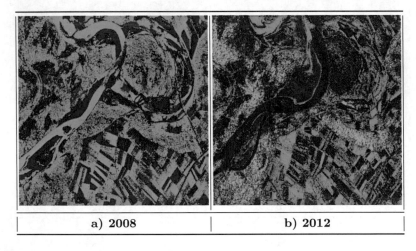

| a) 2008 | b) 2012 |

Fig. 3. The classified ortophoto images from 2008 (a) and 2012 (b) period using 4 classes

In this step, the number of classes used to classify the images was 8. The format of the file shown in the Table 1. The format of the file consisted of two columns, ID of the class and a prior probability of that class, respectively. From the table below it can be noticed that the classes 3 and 6 are missing.

Table 1. The classified ortophoto images from 2008 (a) and 2012 (b) period using 4 classes

1	0.3
2	0.1
4	0.0
5	0.15
7	0.05
8	0.2

Since the sum of probabilities presented is equal to 0.8, the remaining portion of the probabilities (0.2) was divided by the number of not specified classes (3 and 6) [20]. Therefore, classes 3 and 6 were assigned a probability of 0.1. Analyzing the probability values obtained, at the end, we used 4 classes in our classification method. For the second step, river extraction the raster with 4 classes was used. Decreasing the number of classes to two, where one class was river and another class was everything else, we obtained the raster with two colors. The output of this step is shown in the Fig. 4.

Using obtained shape file with two classes the conversion tool was used to convert our raster images to polygons. Using the polygons, the changes of Bosna river flows were detected (Fig. 5) [9].

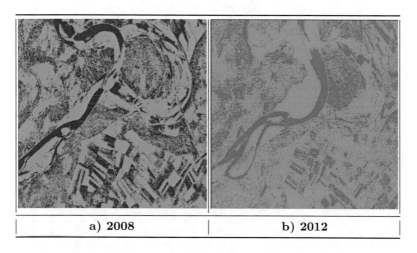

a) 2008   b) 2012

Fig. 4. The classified ortophoto images using 2 classes from 2008 (a) and 2012 (b) period

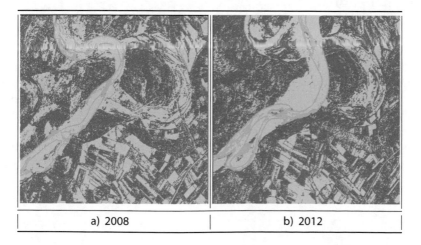

a) 2008   b) 2012

Fig. 5. The extracted river flow of Bosna river from 2008 (a) and 2012 (b) period

This raster clearly showed the flow of Bosna river. Analyzing the changes in course of Bosna river using only two images from period 2008 and 2012 noticeable changes were detected. In the first part of the Bosna river, the altitude varied 572 m while in the second part of the river the altitude changed for 277 m. The Fig. 6 represents the changes obtained during period from 2008 until 2012. From results it is evident that the river flow of Bosna river changed and in some parts even have disappeared.

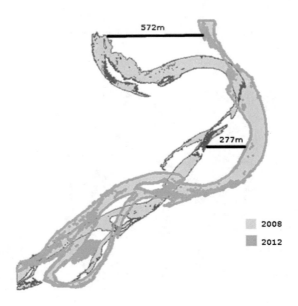

Fig. 6. The course change of Bosna river during analyzed period

6   Conclusion

The aim of this study was to investigate the river flow changes of Bosna river using orthophoto images. We focus only on two orthophoto images for river flow change detection. The analysis showed visible changes in the flow of Bosna river. In some parts of Bosna river even the disappearance of river flow is noticed. The study was carried out to demarcate the changed areas in the study for identifying possible river course change of Bosna river using GIS. The current approach was limited by the acquisition of orthophoto images as they were not obtained constantly. In order to benefit from high resolution images, the amount of acquired images should be increased. In the future work we will try to use frequently obtained high resolution images, such as Landsat remotely sensed images.

References

1. Babic-Mladenovic, M.: Transboundary flood risk management in the Sava river basin. In: Workshop on Transboundary Flood Risk Management, Geneva, April 2009
2. Das, J.D., Tutta, T., Saraf, A.K.: Remote sensing and GIS application in change detection of the Barak river channel, N.E. India. J. Indian Soc. Remote Sens. **35**, 301–312 (2007)
3. Upravljanje rizicima od poplava i ublažavanje njihovih štetnih posljedica, Akademija nauka i umjetnosti Bosne i Hercegovine, Sarajevo (2015)
4. Poplava, zemljotres, smog: Prilozi ekohistoriji Bosne i Hercegovine u 20. stoljecu. Zbornik radova. Udruženje za modern historiju/Udruga za modern povijest (UMHIS), Sarajevo (2017). Godina: https://ba.boell.org/sites/default/files/poplava_zemljotres_smog_-_kb.pdf
5. Federalni Hidrometeoroloski zavod, Klimatoloska analiza 2012. Godine. http://www.fhmzbih.gov.ba/podaci/klima/2012.pdf. Accessed 4 Mar 2018
6. Yuan, Z., Li, A., Yio, T.: Research on data collection and rapid map updating of forest resources based on 3S technology. In: The international Archives of the Photogrammetry, Remote Sensing and Information Sciences, Part B1, Beijing, vol. XXXVII. School of Surveying and Mapping, Henan Polytechnic University, Jiaozuo 454000, China (2008)
7. Iso Cluster Unsupervised Classification. http://desktop.arcgis.com/en/arcmap/10.3/tools/spatial-analyst-toolbox/iso-cluster-unsupervised-classification.htm. Accessed 6 Mar 2018
8. Iso Cluster. http://desktop.arcgis.com/en/arcmap/10.3/tools/spatial-analyst-toolbox/iso-cluster.htm. Accessed 6 Mar 2018
9. Measuring distances and areas. http://desktop.arcgis.com/en/arcmap/10.3/map/working-with-layers/measuring-distances-and-areas.htm. Accessed 6 Mar 2018
10. Imamovic, A., Trožic-Borovac, S.: Uticaj klimatskih promjena na vodni režim rijeke Bosne, Naše šume (2013)

The Role of Feature Selection in Machine Learning for Detection of Spam and Phishing Attacks

Ina Salihovic[(✉)], Haris Serdarevic, and Jasmin Kevric

International Burch University, Sarajevo 71000, Bosnia and Herzegovina
ina.salihovic@stu.ibu.edu.ba
https://www.ibu.edu.ba/

Abstract. With the increase in Internet use throughout the world, expansion in network security is indispensable since it decreases the chances of privacy spoofing, identity or information theft and bank frauds. Two of the most frequent network security breaches involve phishing and spam emails as they are an easy way to pass a virus or a malicious site, which can lead to extensive frauds. Despite the fact that there is an abundance of tools for detection and blocking of these types of messages and websites, society is still trying to combat and rise above said problem. The purpose of this paper was to exclude the human factor in security breaches executed in this manner with the use of various machine learning algorithms. For the purpose of training and testing of the most successful algorithms (Random Forest, k-Nearest Neighbor, Artificial Neural Network, Support Vector Machine, Logistic Regression, Naive Bayes) paper used two separate bases, UCIs Phishing Websites Data Set and Spam Emails Dataset together with Weka software, and found that the best results for both of them are achieved with the Random Forest algorithm. However, databases responded differently to feature selection algorithms, as the best result for phishing (97.33% accuracy) was accomplished through Ranker + Principal Components Optimization, and the best result for spam (94.24% accuracy) was accomplished through BestFirst + CfsSubsEval Optimization in Weka. These findings provide a base platform for future work towards a faster and more accurate online fraud detection.

Keywords: Phishing · Spam emails · Machine learning
Feature selection

1 Introduction

Equivalent to the 20th century technology that was marked by the rise of a microprocessor, 21st century's defining technology will be artificial intelligence. It is becoming more known with the rise of self-driving cars, and real time language translators, but it has even found its place in network security. This chapter will provide the definitions of cyber security, as well as the definitions and statistics regarding phishing and spam e-mails.

© Springer Nature Switzerland AG 2019
S. Avdaković (Ed.): IAT 2018, LNNS 60, pp. 476–483, 2019.
https://doi.org/10.1007/978-3-030-02577-9_47

Cyber Security and Attacks. According to Merriam-Webster dictionary cyber security is defined as the measures taken to protect a computer or computer system against unauthorized access or attack. This paper will focus on machine learning approach in detection of two types of cyber-attacks which usually come in pair spam e-mails and phishing [1].

Spam e-mails. Though there is no specific or unanimous definition of spam e-mail, many may agree that they can be depicted as unsolicited bulk messages whose sending, in contrary to junk mail, is more often than not paid by the receiver of the message, rather than the original sender [2,3]. Having an inbox cluttered with unwanted mail is insignificant compared to the danger these e-mails usually carry with them. According to Symantec 2017 Internet Security Threat Report (ISTR), 1 in 131 emails contained malware in 2016, which is the highest rate in 5 years. Furthermore, according to a different study done by Verizon, two-thirds of all malware was installed via email attachments in 2016 [4]. Besides the danger coming from installing malwares on one's computer, the receivers of this kind of email are susceptible to phishing scams and giving away the money to the sender, if they decide to respond to the message.

Phishing. By the definition from Merriam-Webster dictionary, phishing is a scam by which an Internet user is duped into revealing personal or confidential information which the scammer can use illicitly. The word itself was first used in 1996, when a group of hackers stole America Online (AOL) accounts by tricking AOL users into giving away their passwords [5]. Not much has changed in the past 20 years, and people are still falling into traps when it comes to this kind of scam [6]. Verizon's Data Breach Investigations Report, one of the biggest reports of its kind, says that two-thirds of electronic espionage cases can be traced back to phishing. Moreover, a study of 150,000 phishing emails by Verizon partners found that 23% of recipients open phishing messages, and 11% open attachments [7]. That means that one in ten people open an attachment not knowing what is inside.

To Sum Up. Considering the provided statistic data this paper aimed to exclude the human factor in the online frauds leaning on the spam and phishing. This meant that the fraud detection should be done automatically, and before it comes into contact with a potential casualty. Best platform was the use and potential accuracy improvement of machine learning algorithms which can be the base for development of better and faster tools in the future.

2   Related Work

Following chapter will review the accomplishments in using machine learning algorithms for the detection of spam emails and phishing websites thus far, as well as what this paper will try to accomplish with incorporating feature selection as an optimization technique.

2.1 Phishing

There have been various different approaches to its solving that can be sorted into three main types: (a) non-content-based approach which does not make use of the content of the website, (b) content-based approach which relies on the content of the webpage when classifying it as authentic or phishing, an (c) visual similarity-based approach that uses the visual similarity of known sites to recognize phishing [8].

This paper will use UCIs Phishing Websites Data Set, and an URL-based scheme for phishing detection with the help of machine learning algorithms priory optimized by feature selection. Phishing is a classical data mining classification problem, meaning that almost all classification algorithms can be applied in its solution. Most commonly used algorithms include Support Vector Machine (SVM), k-Nearest Neighbor (kNN), Artificial Neural Networks (ANN), Naive Bayes, Logistic Regression (LR), and Random Tree Forest (RTF).

Though there were many researchers who used SVM algorithm, Altaher proposed a hybrid SVM and KNN approach with 7 carefully selected features, with the claim that this duo can effectively be used for phishing website detection with low computational complexity in the training and detection stage. This hybrid had the accuracy of 90.04% and outperformed all of the other algorithms used, including Naive Bayes, kNN, SVM and ANN [9].

Another paper, done by Islam and Chowdhury compared different machine learning algorithms, with the goal of comparing their performance in terms of accuracy. Their research included Naive Bayes, Decision Trees, ANN, SVM and IBK lazy classifier, and a database with 11055 websites samples, each one consisting of 31 attributes. The best algorithm turned out to be Random Forest, with accuracy of 97.47%. However, it used all of the 31 attributes, which may have been unnecessary [10].

Another possible approach to the testing of these algorithms is Weka, data mining software programmed in Java. Weka contains tools for data preprocessing, classification, regression, clustering, association rules, and visualization and it is also well-suited for developing new machine learning schemes. The following project used it to compare Logistic Regression, Naive Bayes, k-nearest neighbors, Decision Trees, and Random Forest, and attribute subset selection was performed to further improve performance. It was tested on Phish-Tank, the largest malicious URL database, and before the subset selection was introduced, the biggest accuracy belonged to Naive Bayes algorithm (88.6%). However, Logistic Regression took the lead after the subset introduction, and it reached 89.9% [11].

2.2 Spam Emails

Different approaches have been used by different researches in the spam detection and the most popular one today is filtering (e.g. anti-spam filtering). Even though it is the most popular method, problems with filtering occur when the filter itself does not detect some email as spam. Most of nowadays commercial anti-spam filters lean on black-lists, white-lists and handmade patterns.

When it comes to machine learning in e-mail spam, each day becomes more promising. Many algorithms were implemented such as Random Forest, Naïve Bayesian, Support Vector Machine (SVM) and Neutral Network but out of all machine learning algorithms that have been used in spam e-mail filtering, Naïve Bayes one is most frequently used because of its simplicity, making it easy to implement with short training time and fast evaluation to filter email spam. Naïve Bayes classifier also can get highest precision that give highest percentage spam message manage to block if the dataset collects from single e-mail accounts [12].

The SVM approach in email classification problems showed great results, due to its robustness and ability to handle large feature spaces. Algorithm did not try to minimize the error rate, but rather separate the patterns in high dimensional space, resulting in quite insensitiveness to the relative size of each class. There occurred some disadvantages with this method such that calculation could be intensive while training the mode. Results showed that a properly trained SVM reaches a significantly higher classification accuracy than the Naive Bayes approach, showing an almost flawless accuracy of 98.6% [14].

Base of using K-nearest neighbor classifier was to look at the class of the messages that are closest to it first, to decide whether the message is spam or non-spam. This method of the k nearest neighbor algorithm was divided into two steps: step 1. Training and storing the training messages, step 2. Filtering. Given a message x, determine its k nearest Neighbors among the messages in the training set. If there are more spam's among these neighbors, classify given message as spam. Otherwise classify it as non-spam [13]. Some of the problems occurred within this algorithm were that there seems to be no parameter that could be tuned to reduce the number of false positives.

Obtaining results from many different approaches and techniques shows that the solution to spam could be on the long run. The combination of different filtering approaches such as collaborative filtering and machine learning will play the main role in these mixed filters. Many approaches showed that even combining of different learning algorithms gives some promising results, as different classifiers often make different errors [15].

While there are many papers which discuss the machine learning approach to the solution of the given problem, most of them use all features given by the database creator, treating them all as equals. Though a large number of different attributes describes the site in a better light, it has to be taken into account that each one included increases the cost of a system without possibly contributing to the classifier's performance. This is a good base and motivation for building a classifying system with smaller feature number. This will be attempted in this paper, in combination with already stated most commonly used algorithms.

3   Methods and Results

This chapter will introduce the algorithms, as well as the feature selection methods used in the spam and phishing detection. It will showcase the overall results and performance of the said algorithms and enhancement methods.

Algorithms were run in software WEKA, a collection of machine learning algorithms for data mining tasks where algorithms can either be applied directly to a dataset or called from Java code. The reason Weka was chosen is because it contains tools for data pre-processing, classification, regression, clustering, association rules, and visualization. It is also well-suited for developing new machine learning schemes [16].

For the purpose of solving the chosen problem, it was decided to use six of the most widely used algorithms for the databases similar to the two used in this paper (databases used for simple two-class classification): Artificial Neural Networks, Logistic Regression, Random Forrest, Support Vector Machine, k-Nearest Neighbor and Naive Bayes classifiers.

In the first case, the performance of these algorithms on unaltered databases was tested. The results for this case are shown in the Table 1 below.

Table 1. The accuracies of the algorithms tested on unalltered databases

Classifier	Spam	Phishing
Random Forest	95.48%	97.26%
k-Nearest Neighbor	90.76%	97.18%
Artificial Neural Network	91.05%	96.91%
Support Vector Machine	90.44%	94.80%
Logisitic Regerssion	92.41%	93.99%
Naive Bayes	79.29%	92.98%

The results showed that the Random Forrest Algorithm proved to be the best for both phishing and spam detection, with the accuracies of 97.26% and 95.48% respectively. On the other hand, during the first test, Naïve Bayes Classifier showcased the worst results with the error percentages of 7.02% and 21.71% respectively.

Since both databases had a relatively large number of attributes (31 for phishing, and 58 for spam), it was decided to try and use a feature selection optimization. WEKA has feature selection optimizers implemented in its package, and after running all of them, the only two options which suggested a decrease in relevance of certain attributes were BestFirst + CfsSubsEvaluation and Ranker + PrincipalComponents Optimization.

BestFirst searches the space of attribute subsets by greedy hillclimbing augmented with a backtracking facility. CfsSubsEvaluation evaluates the worth of a subset of attributes by considering the individual predictive ability of each feature along with the degree of redundancy between them [16]. BestFirst + CfsSubsEvaluation feature selection optimizer cut the number of attributes for phishing database from the original 31 to 10, and 58 features of the spam database were reduced to 16. After running the selected algorithms on now altered database, the results in the Table 2 below were obtained.

Table 2. The accuracies of the algorithms tested, with BestFirst + CfsSubsEval optimization in weka

Classifier	Spam	Phishing
Random Forest	94.24%	94.77%
k-Nearest Neighbor	91.11%	94.49%
Artificial Neural Network	92.07%	94.57%
Support Vector Machine	86.50%	93.29%
Logisitic Regerssion	92.11%	93.19%
Naive Bayes	78.88%	92.65%

In comparison to the test with original features, new results for the phishing database were decreased in accuracy with 1.53% on average (2.49% for Random Forrest, and 0.33% for Naive Bayes). As for the spam database, the accuracies of k-Nearest Neighbor and Artificial Neural Network algorithm were increased by 0.35% and 1.02% respectfully. Other algorithms experienced the decrease in accuracy of 1.47% on average.

Ranker ranks attributes by their individual evaluations, and PrincipalComponents performs a principal components analysis and transformation of the data. Dimensionality reduction is accomplished by choosing enough eigenvectors to account for some percentage of the variance in the original data [16].

Ranker + PrincipalComponents Optimization cut out only one of the phishing dataset attributes – Statistcal Report. The negligence of this feature brought the increase in the accuracies of Random Forrest and Support Vector Machine algorithm, and the average decrease in accuracy was 0.09%.

As for the impact of Ranker + PrincipalComponents optimization on the spam database, the number of attributes was reduced to 49 (out of 58), and the accuracies of all of the algorithms were reduced by 1.23% on average (1.72% for ANN, and 0.12% for Logistic Regression). The results can be seen in the Table 3 below.

Table 3. The accuracies of the algorithms tested, with Ranker + PrincipalComponents optimization in weka

Classifier	Spam	Phishing
Random Forest	94.13%	97.33%
k-Nearest Neighbor	89.94%	97.18%
Artificial Neural Network	89.33%	96.58%
Support Vector Machine	88.42%	93.89%
Logisitic Regerssion	91.35%	93.98%
Naive Bayes	78.87%	92.95%

4   Conclusion

Analysis of the results revealed that the best accuracies in detection of phishing websites and spam emails for both databases came from the Random Forest Algorithm, thus confirming the hypothesis that the same algorithm can be used for both problems effectively.

It was observed that the feature selection techniques used and tested in this paper improved the Random Forest algorithm accuracy by 0.05% compared to the unaltered phishing database, giving the overall accuracy of 97.33%. This increase in accuracy was accomplished by eliminating one of the features in the database – Statistic Report. An unexpected conclusion was that, even though one of the most widely used algorithms for spam detection is Naïve Bayes algorithm, its overall accuracy is significantly lower than any of the other algorithms.

Even though both databases used in this paper had carefully selected attributes which overall produced the best accuracies, the results showed that both of them have a good response to the decrease in the number of attributes. This allows an optional decrease in the complexity, and in the cost of the model, with a slight increase in accuracy for some of the algorithms, and almost negligible decrease for others. The results motivate future work to explore exclusion of additional variables of the data sets by possibly using different feature selection methods, which might improve the predictive accuracy of classifiers. Important point to add is that this paper used the chosen algorithms on completed databases, even though the detection of spam and email attacks should be done in real time, implying that the features should be extracted automatically, not manually.

The detection of spam emails and phishing websites remains one of the biggest threats to internet security, with every 1 in 10 people being a victim of network security breach induced fraud. Researches, scientists and media will have to deal with an ongoing problem on a larger scale, by protecting the society from the unexpected attacks, but also by raising the awareness about the dangers of the Internet. This paper is only one step on a long journey towards safer Internet.

References

1. Cyber Security Ventures. https://cybersecurityventures.com/hackerpocalypse-cybercrime-report-2016/lncs. Accessed 10 May 2018
2. Merriam Webester. https://www.merriam-webster.com/dictionary/spam/lncs. Accessed 09 May 2018
3. Jukic, S., et al.: Comparison of machine learning techniques in spam e-mail classification. SE Eur. J. Soft Comput. 4(1), 32–36 (2015)
4. Symantec Report. https://www.symantec.com/content/dam/symantec/docs/reports/istr-22-2017-en.pdf./lncs. Accessed 11 Mar 2018
5. Phising.org. http://www.phishing.org/history-of-phishing/lncs. Accessed 13 Mar 2018

6. Hodzic, A., et al.: Comparison of machine learning techniques in phishing website classification. In: Proceedings Book of International Conference on Economic and Social Studies, pp. 249–256. International Burch University, Sarajevo, BiH (2016)
7. Data Breach Investigations Report Executive Summary. https://www.verizonenterprise.com/verizon-insights-lab/dbir/lncs. Accessed 24 Feb 2018
8. Sahoo, B., et al.: Malicious URL detection using machine learning: a survey. arXiv.org/abs/1701.07179 [cs.LG] (2017)
9. Altyeb, A.: Phishing websites classification using hybrid SVM and KNN approach. Int. J. Adv. Comput. Sci. Appl. 8(6), 90–95 (2017)
10. Islam, M., Chowdhury, N.K.: Phishing websites detection using machine learning based classification techniques. In: 1st International Conference on Advanced Information and Communication Technology (2016)
11. Seelio by Keypath Education. https://seelio.com/w/32o4/phishing-website-detection-using-machine-learning-in-weka. Accessed 10 Mar 2018
12. Rusland, N.F., et al.: Analysis of Naïve Bayes algorithm for email spam filtering across multiple datasets. In: Materials Science and Engineering, vol. 226, IOP Publishing Ltd., Riga, Latvia (2017)
13. Saad, O., et al.: A survey of machine learning techniques for spam filtering. Int. J. Comput. Sci. Netw. Secur. 12(2), 66 (2012)
14. Bluszcz, J., et al.: Application of support vector machine algorithm in e-mail spam filtering (poster). Humboldt-Univerzität zu Berlin, Berlin (2016)
15. Sakkis, G., et al.: A memorybased approach to anti-spam filtering for mailing lists. Inf. Retrieval 6(1), 49–73 (2003)
16. Frank, E., Hall, M.A., Witten, I.A.: The WEKA Workbench. Online Appendix for "Data Mining: Practical Machine Learning Tools and Techniques", 4th edn. Morgan Kaufmann, Burlington (2016)

Challenges of Moving Database and Core IT Systems to Cloud with Focus on Bosnia and Herzegovina

Amar Svraka[1]([⊠]), Jasmina Nalic[2], and Almir Mutapcic[1]

[1] Sarajevo School of Science and Technology, Sarajevo, Bosnia and Herzegovina
amar.svraka@stu.ssst.edu.ba, almir.mutapcic@ssst.edu.ba
[2] Faculty of Electrical Engineering, J.J. Strossmayer Osijek, Osijek, Croatia
jasmina.nalic@gmail.com

Abstract. In the era of Big Data, a lot of companies are facing various challenges in processing large volumes of data and at the same time extracting and storing important information. For these reasons, the need for change has escalated, and database, as the core of every serious IT system, is one of the main aspects of it that has to be adapted. This paper analyses why the need for a change in database infrastructure as well as related work and research that has been done on similar topics so far. Following that background, the research in Bosnia and Herzegovina has been conducted to identify current situation in companies and institutions coming from three different business domains. Finally, the paper concludes with research findings and discussion about what can be done in the future to build on it and get companies more open to modern solutions and technologies regarding database infrastructure.

Keywords: Cloud computing · Cloud adoption · Cloud database

1 Introduction

For a long period of time, serious companies and corporations have relied on traditional Relational Database Management Systems (RDBMS). During the previous decade, with a huge rise of popularity of social networking sites, and many other web services and applications, we reached the era of Big Data with extremely high level of data throughput and transfer.

Digital economy, an economy based on Internet and social media is transforming the whole business world, since the core of such economy consists of web, mobile and Internet of Things concept. They are now the main ways of communication and interaction between companies and their users and customers. Companies are not only required to have safe and stable system within their organization, but they need to expand capabilities of their system to satisfy customer and user requirements in the most efficient way possible [1]. With these new trends in place, many companies intend to move from the traditional On-premise RDBMS infrastructure to NoSQL or Cloud database infrastructure.

© Springer Nature Switzerland AG 2019
S. Avdaković (Ed.): IAT 2018, LNNS 60, pp. 484–493, 2019.
https://doi.org/10.1007/978-3-030-02577-9_48

1.1   The Need for Change

Although, it is not simple to ascertain total quantity of the electronically stored data, IDC did a research in 2013 where they have roughly calculated it at a 4.4 zettabytes and with expectations of increase to 44 zettabytes (1021 bytes = one billion terabytes) by year 2020 [3]. Number of customers using online services is growing exponentially and therefore it is important that companies provide high performance, secure and 24/7 stable and available services, so they are competitive on the market.

Big Data is also expanding with users storing many different types of data which varies from unstructured to structured data. Traditional solutions were made long before the Cloud and Big Data, and were intended to run on a single server. NoSQL and Cloud databases emerged as technologies to cope with processing of the huge amounts of data [2]. It is important to acknowledge and understand that this data is not generated by humans only, but also by machines that include production of machine logs, RFID readers, retail transactions, vehicle GPS traces and various other data, all incorporated in Internet of Things concept. The following figure adopted from [2] shows the limitations of traditional database architecture (Fig. 1):

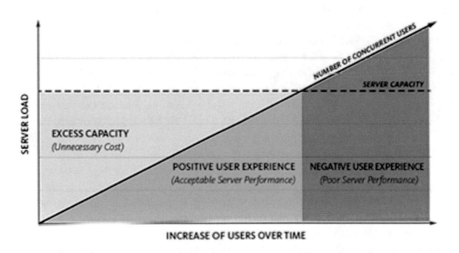

Fig. 1. Limitations of the traditional database architecture

However, the issue is not only about storage capacities, but also about rate at which these huge amounts of data can be accessed. Therefore, data writing and reading from multiple disks has been introduced by using RAID or HDFS in order to improve those rates. Furthermore, to improve analysis efficiency when it is needed to combine data there is a new programming model called MapReduce that transforms read and write operations into computation over sets of keys and values [3].

Basically, a lot of companies want to focus on development of their core business and customers without spending their resources in managing IT equipment and infrastructure. In order to gather, reach and support more customers, they have started to move their applications to Cloud, and with that, their databases as well.

2   Related Work

We have investigated various sources to identify research made on this topic and we found some papers and other research articles that dealt with similar issues but from different perspectives. One of the papers dealt with adoption of Cloud computing in Bosnia and Herzegovina [4]. They claimed that Cloud computing became popular with the beginning of global financial crisis when IT companies had to find more affordable solutions and reduce costs. Emphasis was put on network capability and reliability as key factors to usable, secure and efficient Cloud computing environment. It was written that European Commission had the strategy for Cloud computing potential liberation in Europe accepted while Bosnia and Herzegovina was yet to follow in the same footsteps. This thesis concluded with authors' opinion on what steps should be taken in order to raise Cloud services popularity and awareness in Bosnia and Herzegovina.

The main idea of the paper [5] was research regarding adoption of new IT technologies in areas where even the older ones have not been fully adopted and understood yet which is referred to as leapfrogging. Author described how IT technologies can help improve functioning of governance and market, and improve quality of life in communities. Next, author used leapfrogging to wireless ICT's and described benefits and risks of that case in particular. He presented a table with all factors and conditions that need to be satisfied for the leapfrogging to be successfully conducted. Finally, it was concluded that leapfrogging is a difficult and challenging process for developing countries, but if it is done correctly it can produce huge benefits and opportunities for the country and it can help them close the economic gap to developed countries.

Paper [6] discussed transition to Cloud from financial point of view where authors investigated level of profitability when acquiring Cloud resources. In the background section, authors defined and explained concept of NPV together with the tabular representation of different categories of quantifiable and less quantifiable costs when migrating to Cloud. Research methodology consisted of taking data from Amazon and Windows Azure, and calculating NPV for different Cloud service models and five different hosting options. In the final section, results of the research were shown in the form of line graphs, scatter plots and pie charts. For the future, authors intended to do similar investigation for the indirect costs.

In the article [7], author investigated challenges when choosing appropriate Cloud service model and type of it with regard to Law and requirements of the system that is to be deployed. Author investigated how law regulations in U.S changed over time, and how they are being continuously changed and adapted to support Cloud and encourage companies to use it. The paper was completed with a practical example of an application development on force.com which was to be used on Cloud. Conclusions were drawn that application's requirements always need to be carefully analyzed when choosing appropriate Cloud platform, and that Cloud vendors should work closely with government institutions to develop solutions that are legally acceptable.

2.1   Contribution of the Research

This research is unique among others that were presented in the previous section for various reasons. It provides information regarding current database infrastructure in companies from three different business domains in Bosnia and Herzegovina is presented as well as their readiness to adopt new, modern technologies. This information came from the actual investigation conducted in 2017 and not from data published by some external source. Furthermore, it describes in detail respondents' opinions on actual databases and their willingness and readiness to adopt Cloud services to support their business in any form. Finally, it concludes the findings of the research and analyzes its limitation. It is completed with recommendations for future work that should be conducted in order to raise awareness of Cloud computing services and all the benefits they provide.

3   Research in Bosnia and Herzegovina

In order to investigate situation in my country, we made a research that involved companies and their respective representatives from three different business domains where databases are highly incorporated and are very important in completing and satisfying business objectives.

3.1   Investigation Method

We chose questionnaire as an investigation method since we could design questions in such way so that we cover all aspects of the topic of interest and get information that is relevant and important for my research and work. In addition, questionnaires are cost-effective and do not take a lot of time for the participants to answer. Questionnaires reduce bias as everyone answers same questions and they allow participants to answer at the time convenient for them [8]. Finally, it is easy to analyze and interpret the gathered results.

Still it is important to understand and be familiar with disadvantages of questionnaires. Participant might have not understood certain question appropriately or participant might have been dishonest. Furthermore, it might have been difficult for the participants to express themselves the best way possible if they have found it harder to answer and explain certain questions in written form. Therefore, we made an effort to design the questions as clearly as possible in order to get information that we needed and meanwhile not consume a lot of time from the participants.

However, we were interested to investigate in more detail one particularly interesting participant whose company was one of the few that was actually using cloud services and we managed to schedule a Skype interview with their representative to get more feedback on their experiences after their transition to cloud.

3.2   Other Investigation Methods

In this section, we discuss other investigation methods that were available and could be used for my research:

- Interview: This method refers to one-on-one conversation where questions are asked in person, and respondents are given a chance to give broader answers and better express their opinions.
- Focus group: This concept refers to gathering certain group of participants to whom same conditions would apply and interview them in a group session [9].
- Collection of Institutional data: This method involves gathering data and producing information available through the Institutional Research [10].
- Collection of other (non-survey) data: This method was not applicable to my research since it involves measurement or performance recordings against certain systems and infrastructure without designing any questions [10].
- Interactive Workshops: This technique is similar to focus groups technique and it would involve personal participation of the potential respondents, which was not necessary since we have clearly set the focus of the research on a narrow area [10].

3.3   Research Description

My research involved representatives of 26 companies, from the following business domains: IT companies, Public Sector institutions and Financial Sector institutions. IT companies represent companies that are developing software and providing other computing services for various purposes. Public Sector refers to, mostly, government institutions while Financial Sector covers banks and micro-financing institutions. Some of the participants were sent questionnaire by e-mail with appropriate explanation, while some of them were given questionnaire in person. We avoided online surveys because of security concerns.

For the purposes of analysis, we tried to have equal number of participants in each business domain, and having in mind size of Bosnia and Herzegovina and amount of companies in it, we asked 45 companies, coming from these very important business domains, for their contribution to my research. However, we have obtained replies from 26 participants, while others were not interested to contribute to my research for various reasons. Still, we believe that we have gathered some useful data that showed certain trends among users regarding database infrastructure, as well as general thinking about modern database infrastructures and transition to cloud.

Questionnaire was formulated in such way so that it investigated current infrastructure in a company, level of satisfaction regarding various aspects of it as well as readiness of a company to switch to cloud or NoSQL solution and reasons for their answer. Participants were also able to express their thinking about cloud computing and service in a couple of sentences in the end.

3.4 Results Interpretation

We used data collected from the answers to produce simple visualizations in order to analyze results. The following figure is related to the question four and shows current database infrastructure that is used among companies that participated in the research (Fig. 2):

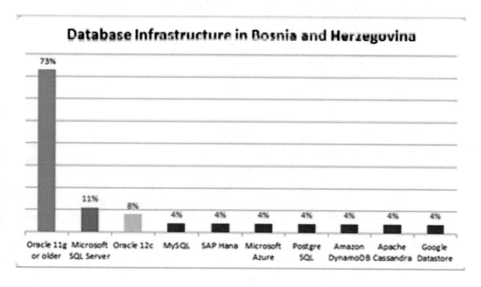

Fig. 2. Current database infrastructure in BH

It can be easily identified that majority of companies that we have investigated are using Oracle 11g or older database version. What is interesting is that ten out of ten participants from Financial Sector are all using this database. Second is Microsoft SQL Server and the third is Oracle 12c database. It is interesting that all other database infrastructures were also represented among participants except MongoDB. Majority of the participants were satisfied with performance of their database, while scalability and maintenance were the features that, in most of the cases, could be improved. Companies that have Oracle databases are mostly satisfied with its performance, stability and storage capabilities, and on the contrary, they mostly find scalability as the weakest point. What is interesting that similar answers had users who are working with Microsoft SQL Server.

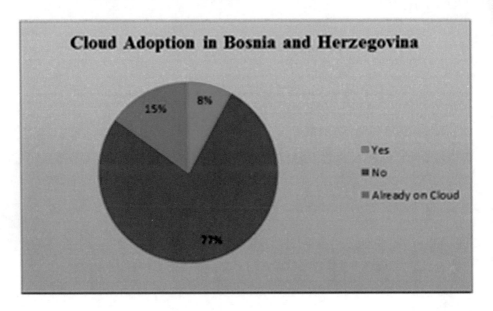

Fig. 3. Cloud adoption in BH

Regarding question number seven that deals with switching to cloud infrastructure, here are the results (Fig. 3):

Majority of the participants, 20 out of 26, were not interested into switching to cloud, while four were already in the cloud, and two were interested into switching to it. Reasons for not wanting to go to cloud are various. Three companies are not familiar with cloud at all, while some of them have just invested into current infrastructure and they are not thinking of anything else at the moment. In addition, for some of them there are legislation constraints, and concerns about security and Internet availability. Certain participants answered that their company is too big for going to the cloud.

Two participants answered that they are interested in cloud. One was microfinancing institution from Sarajevo that wants to acquire cloud infrastructure as a disaster recovery site, and at the same time test all capabilities of cloud. Other participant interested in cloud is an IT company that wants to offer their clients cloud services because of its affordability and high ROI rate. If we exclude companies from IT sector there is only one company that has database in the cloud, and one that has intention to adopt database cloud services (Fig. 4).

What is the most interesting part is that we found four companies that are actually using cloud infrastructure. Three of them are IT companies that are basing their back-end parts of applications and systems on databases in the cloud. What they find as advantages of cloud are that those infrastructures are easier to setup and that end-product is available in much shorter period of time than it is the case with On-premise installations. IT companies claim that cloud database solutions are much more financially affordable than the traditional

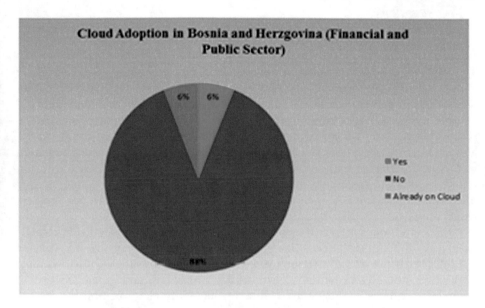

Fig. 4. Cloud adoption in BH (financial and public sector)

On-premise ones. On the contrary, they see small disadvantage in the fact that maintenance is partly done by vendor's company and that accessibility depends on Internet availability.

Finally, there is one micro-financing institution from Banja Luka that has acquired and implemented system on the public cloud infrastructure. Representative from that institution agreed for an additional Skype interview where we discussed their choice in more detail. They were not very satisfied with previous, on-premise, solution and high costs of its maintenance, thus they were attracted to cloud due to its affordability and the fact that system could be set up in a way so it is accessible from various locations. Currently they are very satisfied with their system, and they have not experienced any issues with security or availability of cloud services that are usually the biggest concerns regarding cloud computing.

It can be concluded from the research that majority of companies still believe only to the database infrastructure whose physical location they own, apart from the IT companies that must follow trends so they can be competitive on the global market. Without strong facts and proofs that cloud is absolutely safe and stable, not many of them would adopt it and decide to drop and abandon something that they already invested a lot of money in. Some of the respondents were not even aware of cloud computing and services that such infrastructure offers. Also, big companies and institutions believe only to the big players on the database market and open-source vendors still need to work hard in order to convince potential customers that their solutions are as advanced and secure as those of the big players.

4   Conclusion

In conclusion, it can be noted that companies in Bosnia and Herzegovina are approaching cloud based solutions and other modern technologies very carefully and warily. At the moment, most of the companies are sticking with established and secure On-premise database architectures and they are quite skeptical regarding new technologies with an exception of companies coming from the IT sector that had to embrace modern database solutions to be competitive on the global market. Legislation constraints are blocking larger government companies and banks in Bosnia and Herzegovina to move to cloud mostly since it is not allowed to store certain data outside of the country. In addition, larger companies believe that their systems are too big for cloud and they do not see how cloud would benefit their business.

Still, cloud computing as a modern solution brings a lot of benefits. It is easily manageable and upgradable which makes cloud databases very scalable. We see a lot of positives with it, both for the users and environment, and it is definitely part of IT and Big Data that is constantly improving and going forward. If companies and institutions start embracing not only cloud computing but all other modern IT solutions, we believe that benefits would be mutual for them and their customers since the idea of IT services use is to support the business, expand possibilities and enable company to grow in all aspects.

4.1   Research Limitations

There were some limitations regarding research that was conducted. It would have been much better if all of 45 participants that we initially wanted to have in my research agreed for participation and supported my research with their answers. That would have definitely led to a more detailed analysis and potentially different conclusions. Even more than these 45 participants would provide a far better sample, and provide even more variety in answers and maybe lead to new ideas. It should be investigated how to engage companies and community in Bosnia and Herzegovina to contribute to different research activities and support not only IT research, but research in other scientific areas as well.

4.2   Future Work

For the future work, we plan on including even more companies in my research in order to evaluate their knowledge and awareness of cloud computing and NoSQL technologies. Furthermore, we want to upgrade questionnaire with more questions like what would be the desired method to engage companies into thinking about cloud services. We also plan on including cloud vendors in this research and investigate their opinion on how to get companies more interested in cloud.

Furthermore, we believe that education is one of the key factors in order to get market in Bosnia and Herzegovina more open to cloud computing and new technologies in general. Companies that offer cloud infrastructure should constantly educate potential buyers and customers by using real life cases as examples and show how cloud infrastructure can benefit their company and business. They need to convince them and show that data and access to it will be safe, and that the vendor will do anything to ensure high availability rate and the best possible customer experience.

References

1. The database market is in play. https://www.mongodb.com/post/35653616737/the-database-market-is-in-play. Accessed 20 May 2017
2. Why NoSQL Database?. https://www.couchbase.com/resources/why-nosql. Accessed 25 June 2017
3. White, T.: Hadoop The Definitive Guide. O'Reily Media, Sebastopol (2015)
4. Škrbić, M., Huseinović, K., Nukić, L., Mušović, J., Kasumagić, L., Hebibović, A.: Cloud Computing in Bosnia and Herzegovina, TELFOR 2014, November 2014
5. Cvetkovic, S.: Development Study for Information Technology in Bosnia and Herzegovina. http://www.csiweb.org/pdf/bridges/bosnia/. Accessed 26 June 2017
6. Kwaku Kyem, P.A.: Failures in Technological Intervention and the Promise of ICT. http://www.iimahd.ernet.in/egov/ifip/feb2010/peter-kyem.htm. Accessed 24 May 2017
7. Louridas, P.: Up in the air: moving your applications to the cloud. IEEE Softw. **27**(4), 6–11 (2010)
8. Questionnaires - pros and cons. http://www.teach-ict.com/asa2ictnew/ocr/A2G063/331systemscycle/investigationmethods/miniweb/pg4.htm. Accessed 01 June 2017
9. Freitas, H., Oliveira, M., Jenkins, M., Popojoy, O.: The Focus Group, a Qualitative Research Method, February 1998. http://gianti.ea.ufrgs.br/files/artigos/1998/1998079ISRC.pdf. Accessed 26 May 2017
10. Alternatives to Surveys. http://www.swarthmore.edu/institutional-research/alternatives-to-surveys. Accessed 26 May 2017

A Survey on Big Data in Medical and Healthcare with a Review of the State in Bosnia and Herzegovina

Vedrana Neric[1]([✉]), Tatjana Konjic[2], Nermin Sarajlic[2], and Nermin Hodzic[2]

[1] ABC Software Development Ltd., Branch Office Tuzla,
Tuzla, Bosnia and Herzegovina
vedrana_neric@yahoo.com
[2] Faculty of Electrical Engineering, University of Tuzla,
Tuzla, Bosnia and Herzegovina
tatjana.konjic@etf.unsa.ba

Abstract. Healthcare systems are facing various challenges such as high healthcare costs, aging population, increased number of patients with chronic illnesses, dissatisfied patients, lack of medical specialists. Use of big data analytics and technologies is one of the ways to overcome problems and improve the current health systems. The main idea of this paper is to give a review of big data concept, summarize big data applications, and identify challenges in medical and healthcare with an overview of the current situation in Bosnia and Herzegovina.

Keywords: Big data · Big data analytics · Big data technologies
Medical · Healthcare

1 Introduction

Academicians define big data as huge size of unstructured data produced by a high-performance heterogeneous group of applications that span from social network to scientific computing applications. The data sets range from a few hundred gigabytes to zettabytes that it is beyond the capacity of existing data management tools to capture, store, manage and analyse [1–3]. The big data term often refers to the use of advanced methods, such as predictive analytic, to extract values from data. Depending on the content, the big data source can be human generated, machine generated, internal data sources, the web, and social media, transaction data, biometric data, etc. Further, the context data format may vary from being structured, unstructured, semi-structured, images, text, videos, audio, etc. [4,5]. A huge amount of data and information can be found almost in each area of work and life such as economy, industry, techniques, agriculture, etc. One of the areas with a huge amount of different data with high interest in work and research is medical and healthcare. In order to improve the quality of healthcare and reduce costs, it is necessary to analyse the generated data efficiently. Data analysis can help respond to new challenges and improve health services, such as providing patient-oriented treatment, early

© Springer Nature Switzerland AG 2019
S. Avdaković (Ed.): IAT 2018, LNNS 60, pp. 494–508, 2019.
https://doi.org/10.1007/978-3-030-02577-9_49

disease prediction, and so on. Big data analytics plays an important role in overcoming existing problems and achieving health improvement [6].

The amount of data in the field of healthcare is growing at a high speed, and this area is considered to be one of the fastest growing segments of the digital universe. According to the EMC report, health data are growing at 48% a year, and are projected to reach 2300 exabytes by 2020 [7].

The big data term not only refers to large amounts of data but also to new technologies for storing, searching, processing and interpreting data. Big data is usually described with 3Vs dimensions [1]:

- *Volume* - it collects and processes a large amount of data that is available for analysis,
- *Velocity* - the speed at which new data are collected is large and is higher than the processing speed,
- *Variety* - the data are diverse in terms of sources, format, and structure and are most often unstructured.

Some authors expand the big data description to 5Vs with dimensions [8]:

- *Veracity* - truthfulness, accuracy, reliability, predictability in terms of the quality of the collected data that can vary and thus affect the accuracy of the analysis,
- *Value* - there is always an interest in obtaining the right values based on the collected data,

and some authors expand it even on 8Vs with dimensions [9]:

- *Viscosity* - a speed element used to describe the delay time in the data relating to the event being described,
- *Variability* - the consistency of data in terms of availability or reporting interval,
- *Virality* - a data spread rate that describes how often data are collected and repeated by other users or events.

Due to a significant increase in the amount of data in the field of medical and healthcare, and the different possibilities for their applications, this paper provides a review of big data in medical and healthcare. The paper is organized as follows. After the introduction, Sect. 2 describes big data analytics in medical and healthcare. Big data technologies are described in Sect. 3 and several techniques for big data analysis in Sect. 4. Section 5 describes the results of research of the current state of big data in medical and healthcare in Bosnia and Herzegovina (BH). The final chapter provides concluding remarks and possible guidelines for future research.

2   Big Data Analytics in Medical and Healthcare

Big data analytics is the process of collecting, organizing and analysing large amounts of data using statistical models, data mining techniques and computer technologies, in order to discover patterns and other useful information.

Big data analytics helps in identifying and understanding the information contained in data and allowing organizations to make better decisions [10,11].

There are four types of the big data analytics [12]:

- *Diagnostic analytics* gives a look to the past in order to determine what has already happened and why,
- *Descriptive analytics* uses data aggregation and data mining to gain insight into what is now happening on the basis of input data,
- *Predictive analytics* uses statistical models and forecasting techniques to understand the future and give answers to the question of what might happen,
- *Prescriptive analytics* uses optimization and simulation algorithms for tips on possible results and answers the question of what should be done.

Big data analytics finds its application in all areas of life and business that generates a large amount of data, including the field of healthcare and medicine, where it is considered that in addition to information and communication technologies, the future of big data analytics lies.

In the field of medicine, there is a great potential for the application of big data analytics that is related to futuristic concepts such as the development of effective biological drugs, accurate disease prediction, control of the human genome, prolongation of the lifetime. In the field of medical and healthcare, large amounts of data are collected from different internal and external sources in different formats and forms. Some health data being collected are data from medical devices and sensors that measure vital signs of patients, data related to clinical trials, the human genome, drugs, adverse reactions, laboratory, epidemiological data and many others. Big data analytics has the ability to apply in medical and clinical research, where it can provide information that can contribute to the advancement of medicine. Special features of big data analytics are related to population data. Some population data that can be processed with the help of big data analytics are disease prevalence, epidemiological data, disease rate, mortality rate, a treatment used and duration of treatment. By analysing a large amount of patient data, it is possible to efficiently profile the patients that will lead to the application of adequate treatments and the improvement of patient care. Also, using big data analytics, it is possible to integrate and process complex data from a variety of sources, such as patient medical records, illness history of patients and their family members, laboratory results, diagnostic data, etc. [13,14].

3   Big Data Technologies

In order to be able to achieve improvements in the field of healthcare using a big data concept, the application of new technologies is necessary. Big data technologies are important for providing more accurate data analysis, better decision-making, reduction of business risks and costs. Healthcare services are no longer limited to being provided only at specific locations, such as hospitals and clinics, but are rather transmitted by communication technologies to patients on the move or at home. Healthcare and medical technology is used by healthcare professionals, doctors, engineers, patients, hospitals, clinics.

In order to respond to the challenges of big data concepts, new data management systems have been developed. The most widely used technology is the open source platform Hadoop. As an addition to this technology, there are other commercial technologies for big data processing such as Cassandra, Cloudera, MapR, and Hortonworks [15].

Many of these technologies do not have enough security tools, so there is a problem with safety and security of data [16]. One of the challenges for healthcare organizations is to make the right decision about which technology is best for them.

Big data capabilities and available primary technologies are summarized in Table 1 [1].

Table 1. Big data capabilities and their primary technologies [1].

Big data capability	Primary technology	Features
Storage and management capability	Hadoop Distributed File System (HDFS)	Open source distributed file system, Runs on high-performance commodity hardware, Highly scalable storage and automatic data replication
Database capability	Oracle NoSQL	Dynamic and flexible schema design, Highly scalable multi-node, multiple data center, fault tolerant, ACID operations, High-performance key-value pair database
	Apache HBase	Automatic failover support between Region servers, Automatic and configurable sharding of tables
	Apache Cassandra	Fault tolerance capability for every node, Column indexes with the performance of log-structured updates and built-in caching
	Apache Hive	Query execution via MapReduce, Uses SQL-like language HiveQL, Easy ETL process either from HDFS or Apache HBase
Processing capability	MapReduce	Distribution of data workloads across thousands of nodes, Breaks problem into smaller sub-problems
	Apache Hadoop	Highly customizable infrastructure, Highly scalable parallel batch processing, Fault tolerant
Data integration capability	Oracle big data connectors, Oracle data integrator	Exports MapReduce results to RDBMS, Hadoop, and other targets, Includes a Graphical User Interface
Statistical analysis capability	R and Oracle R Enterprise	Programming language for statistical analysis

Hadoop is a scalable, reliable, distributed computing platform that allows data storage, parallel access and distributed processing of large datasets. Thanks to an increasing number of solutions that are compatible, Hadoop is becoming a new operating system for business. Some solutions that can be performed over Hadoop with adequate performance are SQL, distributed MapReduce analytical platform, memory processing, flow processing, graphical analytics, etc. There are many reasons for healthcare organizations to use Hadoop. The main reason is the possibility of storage and rapid processing of large amounts of data. Hadoop is a free and flexible platform that can use hardware that organizations already own. Also, Hadoop is easy to use, does not require a large amount of administration and allows users to focus on tasks, without compromising on technical details of the environment [17].

Cassandra is one of the big data technologies that can also be used in medical and healthcare. It is an open source technology that has been developed to support continuous data growth and meet performance, efficiency, reliability, and scalability requirements. The main benefits of using Cassandra are horizontal scalability, replication, tolerance for failures, costs, performance, and operational management. Cassandra coordinates the existing infrastructure, so the stored data cannot be lost, as it is always available and includes all the functionalities that are needed by health organizations to manage patient data [18].

4   Techniques Applied in Big Data Analysis

There are numerous developed big data techniques, methods and algorithms, which are generally called big data techniques and can be divided into two groups: techniques for discovering new information and knowledge from data and predictive techniques. The most widely used big data techniques are derived from three areas: machine learning, branch of mathematics and databases. Some algorithms, such as neural networks, are taken from the field of machine learning, while decision trees and regression are taken from statistics. There are also several versions of classification and clustering algorithms that use database techniques and many new methods, algorithms and software have been developed [19].

Some techniques applied in big data analysis are:

- Neural Networks:
  Neural networks are a technique designed to imitate human brain work. Complex clinical big data can be processed and analysed using artificial neural networks that can help doctors in pattern recognition, classification, and making decisions [20]. Neural network analysis of the association between behavioural habits and chronic diseases can help in identifying risk factors for preventive healthcare [21]. This technique can contribute to precise and effective medical diagnosis of various diseases like cardiovascular diseases, cancer, and diabetes [22]. Artificial neural networks can also be used for prediction of kidney failure [23], for diagnosis of: kidney stones disease [24], skin diseases [25], diabetes disease [26], acute appendicitis [27], thyroid disease [28], coronary heart

disease [29], pancreatic cancer [30], lung cancer [31], breast cancer [32,33], in mammography interpretation [34], etc.

– Fuzzy Logic:
Fuzzy logic is in many ways close to human perception and is implemented as a simulation of human thinking and reasoning. Therefore this technique could be applied in the field of medical and healthcare for decision-making. Fuzzy algorithms for pattern recognition in medical diagnosis are very useful in early disease diagnosis and prognosis [35]. There are many fuzzy logic applications in medical diagnosis (tuberculosis, cancer, image and signal processing, aphasia, pharmacy, heart disease, asthma, diabetes, malaria, hypothyroidism, HIV, arthritis, anaesthesia, and meningioma) [36]. Fuzzy logic can also be used for: prioritizing service quality improvement in healthcare [37], evaluation of the risk of drug addiction [38], cancer risk analysis [39], evaluation of total antioxidant capacity that is important in preventing and managing a lot of diseases [40], etc.

– Nearest Neighbours Classification:
The nearest neighbour method is one of the oldest techniques used to classify data. The algorithm is to find the nearest neighbour, which is the data that has the most similar characteristics and the assumption is that it will behave similarly. There is an improved variant of this method, in the sense that one's behaviour is not observed, but a few neighbours and this is known as the K-nearest neighbours method. In this way, the behaviour and characteristics of a particular data can be predicted more accurately. K-nearest neighbours can be applied in medical data mining [41], predicting medical conditions [42], diagnosing heart disease patients [43], heart disease classification [44], to ECG data [45], for brain image retrieval [46], detecting and classifying MRI lung cancer images for large amount of data [47], etc.

– Decision Tree:
The decision tree is a technique used for classification and decision-making. This technique is based on the relationship between states and strategies, and results are predicted using a number of questions and rules for categorizing data. This technique finds its application in the field of medical and healthcare. Decision trees can be used to access cost-effectiveness in clinical research [48], for heart disease prediction [49,50], etc.

– Memory Based Reasoning:
The memory-based reasoning is a technique used to predict and classify. This technique is similar to neural networks, but with memory-based reasoning, similar data from the experience is required, but the correctness of data and patterns is not established. Case-based reasoning systems store information about situations in their memory. Case-based reasoning in the health sciences can be used for detection and prediction blood glucose control problems and help patients with diabetes on insulin pump therapy achieve and maintain good blood glucose control [51]. This technique can be used in medical domains [52] for different medical diagnosis systems [53].

– Rule Induction:
  Rule induction is a method of rules extraction from a set of observations to make easier decisions. The method is based on traversing database using logical functions on variables, taking into account the probability of occurrence of events and individual records in order to reveal hidden information from the data. Rule induction can be used to discover survival factors of patients after bone marrow transplantations [54], to elucidate co-occurrence patterns in microbial data [55], or as a technique of detecting severity of myocardial infarction [56].

Besides above mentioned techniques that are commonly applied in big data analysis there is also a great number of other methods, which can be used as well in the field of medical and healthcare. Some of them are:

– K-means clustering that can be used for solving problems of different health insurance claims [57], for pattern discovery in healthcare data [58], for health care analysis using clinical documents that contain a lot of medical information [59], in medical claims fraud and abuse detection [60],
– Genetic algorithms that can improve the clinical decision support systems, be used for rule extraction from patient's database and can help in medical diagnosis [61], for dimensionality reduction and improving accuracy in medical diagnosis [62], in predicting diabetes [63], for voxel-based medical image registration [64],
– Self-organizing maps that can be used for analysis of the impact of nutrition and lifestyle on health situation in the world [65], for automatic extraction of appendix from ultrasonography [66], for identifying regions of interest in medical images [67], for molecular subtyping of bladder cancer [68].

Also, hybrid methods are used to a large extent, which is a combination of several methods and use the best properties of each of them. Some hybrid methods refer to the following combinations:

– Particle swarm optimization and support vector machine that can be used for diagnosis of arrhythmia cordis [69], for cancer feature selection and classification [70], for disease detection in medical images processing [71], for ECG beat classification [72],
– Neural networks and the decision tree that can be used for eye diseases diagnosis [73], for migraine diagnosis [74],
– Genetic algorithms and fuzzy logic that can be used for developing systems to forecast outpatient visits in hospitals with high accuracy [75], for the breast cancer diagnosis problem [76], for designing diagnostic system for identifying disease of Chikungunya [77], heart disease prediction [78],
– Fuzzy and K-means that can be used for designing diagnostic systems for thyroid disease [79], for effective detection of Parkinson's disease [80], for medical image segmentation [81].

5   Big Data in Medical and Healthcare in Bosnia and Herzegovina

In Bosnia and Herzegovina there are Health insurance fund of Republika Srpska, Health insurance fund of the Brčko District, Health insurance and reinsurance institute of the Federation of Bosnia and Herzegovina and health insurance institutes at the cantonal level:

- Health insurance institute of Sarajevo Canton,
- Health insurance institute of Tuzla Canton,
- Health insurance institute of Herzegovina-Neretva Canton,
- Health insurance institute of Zenica-Doboj Canton,
- Health insurance institute of Una-Sana Canton,
- Health insurance institute of Canton 10,
- Health insurance institute of Posavina Canton,
- Health insurance institute of West Herzegovina Canton,
- Health insurance institute of Central Bosnia Canton,
- Health insurance institute of Bosnian-Podrinje Canton.

These health insurance institutes use different information systems for collecting and recording payments of health insurance contributions and other incomes as well as recording healthcare expenses for insured persons in primary, secondary and tertiary healthcare institutions and recording data about subjects that participate in the health insurance system processes. Collecting and recording this data enables obtaining good quality of data, reporting and making budget plans and other financial plans for these institutes.

The lack of these information systems that are reflected in data processing using different tools in different parts of the Federation of BH, nomenclature/code list system inconsistencies, poor or no data exchange, uneven and insufficient reports, different data processing at different levels within system, etc. are some of the main reasons for the development of the new information system HCFIS (Health Care Finance Information System).

HCFIS information system contains several logical units - modules:

Module for records,
- Module for data transmission between the organizational units,
- Module for reporting,
- Module for access control,
- Module for generating indicators,
- Module for automatic import of data on contribution payments,
- Module for data conversion from existing information systems,
- Module for backup and restore data.

Module for records performs the processing of the following data:

- Incomes records:
  - Health insurance contributions,
  - Donations,

- Payments from the budget,
- Cash and other payments,
- Payments from abroad.
- Expenses records:
  - Costs of treatment within one canton,
  - Costs of medicines within the canton,
  - Costs of treatment in health institutions belonging to other cantons,
  - Costs of treatment abroad,
  - Costs related to orthopaedic supplies,
  - Personal income tax refunds (sick leave),
  - Payment of travel expenses,
  - Payment of funeral expenses.
- Basic registers, nomenclatures and indicators records:
  - Taxpayer records,
  - Insured persons records,
  - Health institutions records,
  - Customer records,
  - Doctors and medical teams records,
  - Standard (unique) nomenclature/code list system
  - Indicators.

The system is based on the following code lists: states, cantons, municipalities, ZIP codes, activities, professions, points, health services, medicines, diagnosis-diseases, disease groups, orthopaedic supplies, the causes of disability, health institutions, banks, currencies, kinship, statuses of insured persons, types of taxpayers, types of health institutions, types of confirmation-approvals, types of expenses, types of incomes, budget organizations [82].

Using integrated healthcare information system could help citizens to achieve, and health workers to provide safer, more quality, more rational and better integrated healthcare. Also, it could enable participants in the health systems to have continuous and safe access to quality and up-to-date information. It could improve the efficiency of access to information, mutual communication, better quality monitoring and better planning of financial resources for healthcare.

Despite the plans that all healthcare institutes in the Federation of Bosnia and Herzegovina use the same HCFIS information system, only Health insurance institute of Tuzla Canton and Health insurance institute of Zenica-Doboj Canton use HCFIS, while other health institutes continue to use their old systems.

Beside Health insurance institute of Tuzla Canton, that uses HCFIS information system, in Tuzla Canton there is one of the leading health institutions in Bosnia and Herzegovina - Public health institution University clinical center Tuzla. This is a modern institution where highly specialized diagnostic procedures and new operational methods are introduced, developments in medicine followed with continuing staff education and high-quality research activities. University clinical center Tuzla uses Hospital information system consisting of Radiological information system and Laboratory information system. They also have system for processing, archiving and distribution of radiological images.

Primary healthcare institutions in Tuzla Canton use Ambulance information system. This system is used in all municipalities in this canton, but to varying degrees as manual records are still kept in addition to electronic medical records. The trend of using Ambulance information system in Tuzla Canton has increased in recent years and should continue to increase in the future.

Unlike some EU member states, in Bosnia and Herzegovina, big data in medical and healthcare has still not found real application. Given the fact that the data in the various information systems in some health institutions in BH, such as health institutions in Tuzla Canton, is based on the personal id number of insured persons, all this patient-related data can be linked and unified with tendency to apply big data applications that in future could significantly improve the quality of health services. These unified health data could be analysed using big data techniques and technologies and can help in making decisions about therapies, treating patients, and taking measures to prevent diseases. Processing of these unified health data can show trends in the occurrence of certain diseases and enable the management of diseases. Analysing records of sick leave can predict future diseases, detect epidemics of certain diseases, and see what are the most common causes of illness and in what direction should go to prevent these diseases. Using big data technology tools can improve epidemiological control by monitoring diseases that occur and result in mortality. Based on patient-related records in databases, patient profiling can be done. These data can also be used for various research and clinical studies. A special challenge is the development of predictive models specialized in the field of medical and healthcare due to the complexity of medicine and the large amount of data occurring in healthcare. Prediction of events reduces the uncertainty of complex things that surround us, increases profitability and success, so a predictive analysis, as a very important big data feature, can provides great opportunities in this area. A lot of money is spent in healthcare in an inadequate way. Big data applications can help in more efficient spending of resources and reduce healthcare costs with quality improvement.

In order to take advantages of big data applications in medical and healthcare, it is essential to collect all the necessary data, standardize, process and then analyse them. So, some efforts are needed to move from the domain of theory and begin to apply all big data benefits in practice on the territory of BH. With regard to the end result that could improve the overall health system in the country, increase efficiency in the treatment of patients and also increase the financial efficiency, it certainly has to be one of the priorities of the health insurance institutions.

6   Conclusion

Efficient integration of medical informatics, big data analytics, techniques and advanced technologies can influence the reduction of healthcare costs and the improvement of health outcomes through a good decision-making mechanism. A new approach to health and health system reform can provide an individual approach to treating patients, using the most effective therapy, better treatment outcomes, better use of drugs, and it generally improves the quality of healthcare. This paper describes the big data concept, the application of big data analytics in medical and healthcare, as well as some big data technologies and techniques that could contribute to more efficient use of data and improvement of the health system. Given that the application of these techniques is still largely in the domain of theory, and that practice at its beginnings, further research should be directed to their wider use in order to achieve as good results in the field of healthcare. There are also some big data research problems that still need to be solved such as real-time processing and visualization of multidimensional data, life-cycle of big data, efficient storage devices, big data computations, and social perspectives dimensions.

Big data application in Bosnia and Herzegovina is still at the very beginning of development. Healthcare institutions in BH collect a lot of different health data and there is a possibility to obtain values from these data. Patient-related data from different information systems in healthcare institutions should be standardized, linked and unified in order to be analysed using big data techniques and technologies. This could significantly improve the health system and provide great health benefits in the country.

The contribution of this paper is the state of the art in the field of big data applications in medical and healthcare with a review of the current situation in Bosnia and Herzegovina.

References

1. Bhadani, A.K., Jothimani, D.: Big data: challenges, opportunities, and realities. Chapter in an edited volume Effective big data management and opportunities for implementation (2016)
2. Cuzzocrea, A., Song, I.-Y., Davis, K.C.: Analytics over large-scale multidimensional data: the big data revolution!. In: Proceedings of the ACM 14th International workshop on Data Warehousing and OLAP (DOLAP 2011), pp. 101–104. ACM, New York (2011)
3. Agneeswaran, V.: Theoretical, engineering and analytics perspective. In: Srinivasa, S., Bhatnagar, V. (eds.) Big Data Analytics. Lecture Notes in Computer Science, vol. 7678, pp. 8–15. Springer, Heidelberg (2012)
4. Thomas, J., Sael, L.: Overview of integrative analysis methods for heterogeneous data. In: The 2015 International Conference on Big Data and Smart Computing (BigComp 2015), no. 1, pp. 266–270 (2015)
5. Thomas, J., Sael, L.: Maximizing information through multiple kernel-based heterogeneous data integration and applications to ovarian cancer. In: 6th International Conference on Emerging Databases (EDB), pp. 97–100. ACM Press (2016)
6. Anusha, S.K., Dhruti, H.G., Smitha, G.R.: Overview of big data's contribution in health care. Imp. J. Interdiscip. Res. **2**(11), 462 (2016)
7. EMC Digital Universe: Driving data growth in healthcare. Vertical industry brief report, p. 5 (2014)
8. Ganjir, V., Sarkar, B.K., Kumar, R.R.: Big data analytics for healthcare. Int. J. Res. Eng. Technol. Scie. **VI**, Special Issue, 2–5 (2016)

9. http://www.datasciencecentral.com/profiles/blogs/how-many-v-s-in-big-data-the-characteristics-that-define-big-data. Accessed June 2017
10. Zakir, J., Seymour, T., Berg, K.: Big data analytics. Issues Inf. Syst. **16**(II) (2015)
11. Neric, V., Sarajlic, N.: Big data: concept and technological solutions. In: 13th Conference International Council for Large Electric Systems BH K CIGRE, Neum (2017)
12. http://www.ingrammicroadvisor.com/data-center/four-types-of-big-data-analytics-and-examples-of-their-use. Accessed June 2017
13. Lazarevic, I.: Big data in pharmacy and medicine (part 2: big data analytics - business application). BB Inf. **243**, 29–30 (2015)
14. Vo, Q.D., Thomas, J., Cho, S., De, P., Choi, B.J., Sael, L.: Next generation business intelligence and analytics: a survey, arXiv:1704.03402 (2017)
15. Kune, R., Konugurthi, P.K., Agarwal, A., Chillarige, R.R., Buyya, R.: The anatomy of big data computing, Wiley Online Library (2015)
16. CSA: Big data security and privacy handbook. Cloud Security Alliance (2016)
17. Watson, H.J.: Big data analytics: concepts, technologies, and applications. Commun. Assoc. Inf. Syst. **34**, Article 65 (2014)
18. Sadhra-Rai, S.: Database technologies to manage big data in healthcare. Karos Health (2014)
19. Ali, A., Qadir, J., Raihan ur Rasool, Sathiaseelan, A., Zwitter, A.: Big data for development: applications and techniques. arXiv:1602.0781 (2016)
20. Kumar, S.S., Kumar, K.A.: Neural networks in medical and healthcare. Int. J. Innov. Res. Dev. **2**(8), 241–244 (2013)
21. Raghupathi, V., Raghupathi, W.: Preventive healthcare: a neural network analysis of behavioral habits and chronic diseases. MDPI J. Healthcare **5**(1), 1–13 (2017)
22. Amato, F., Lopez, A., Pena-Mendez, E.M., Vanhara, P., Hampl, A., Havel, J.: Artificial neural networks in medical diagnosis. J. Appl. Biomed. **11**, 47–58 (2013)
23. Osofisan, A.O., Adeyemo, O.O., Sawyerr, B.A., Eweje, O.: Prediction of kidney failure using artificial neural networks. Eur. J. Sci. Res. **61**(4), 487–492 (2011)
24. Kumar, K., Abhishek: Artificial neural networks for diagnosis of kidney stones disease. I. J. Inf. Technol. Comput. Sci. **7**, 20–25 (2012)
25. Bakpo, F.S., Kabari, L.G.: Diagnosing skin diseases using an artificial neural network. Intech (2011)
26. Soltani, Z., Jafarian, A.: A new artificial networks approach for diagnosing diabetes disease Type II. Int. J. Adv. Comput. Sci. Appl. (IJACSA) **7**(6), 89–94 (2016)
27. Park, S.Y., Kim, S.M.: Acute appendicitis diagnosis using artificial neural networks. Technology and Health Care IOS Press (2015)
28. Rastogi, A., Bhalla, M.: A study of neural network in diagnosis of thyroid disease. Int. J. Comput. Technol. Electron. Eng. (IJCTEE) **4**(3), 13–16 (2014)
29. Atkov, O.Y., Gorokhova, S.G., Sboev, A.G., Generozov, E.V., Muraseyeva, E.V., Moroshkina, S.Y., Cherniy, N.N.: Coronary heart disease diagnosis by artificial neural networks including genetic polymorphisms and clinical parameters. J. Cardiol. **59**(2), 190–194 (2012)
30. Sanoob, M.U., Madhu, A., Ajesh, K.R., Varghese, S.M.: Artificial neural network for diagnosis of pancreatic cancer. Int. J. Cybern. Inform. (IJCI) **5**(2), 41–49 (2016)
31. Gorynski, K., Safian, I., Gradzki, W., Marszatt, M.P., Krysinski, J., Gorynski, S., Bitner, A., Romaszko, J., Bucinski, A.: Artificial neural networks approach to early lung cancer detection. Cent. Eur. J. Med. **9**(5), 632–641 (2014)
32. Sandhu, I.K., Nair, M., Shukla, H., Sandhu, S.S.: Artificial neural network: as emerging diagnostic tool for breast cancer. Int. J. Pharm. Biol. Sci. (IJPBS) **5**(3), 29–41 (2015)

33. Utomo, C.P., Kardiana, A., Yuliwulandari, R.: Breast cancer diagnosis using artificial neural networks with extreme learning techniques. Int. J. Adv. Res. Artif. Intell. (IJARAI) **3**(7), 10–14 (2014)
34. Ayer, T., Chen, Q., Burnside, E.S.: Artificial neural networks in mammography interpretation and diagnostic decision making. Hindawi Publishing Corporation, Computational and Mathematical Methods in Medicine (2013)
35. Begum, S.A., Devi, O.M.: Fuzzy algorithms for pattern recognition in medical diagnosis. Assam Univ. J. Sci. Technol. **7**(II), 1–12 (2011)
36. Prasath, V., Lakshmi, N., Nathiya, M., Bharathan, N., Neetha, N.P.: A survey on the applications of fuzzy logic in medical diagnosis. Int. J. Sci. Eng. Res. **4**(4), 1199–1203 (2013)
37. Woldegebriel, S., Kitaw, D.: Application of fuzzy logic for prioritizing service quality improvement in healthcare a survey. Int. J. Sci. Knowl. **6**(1), 23–31 (2014)
38. Singh, M.K., Rakesh, L., Ranjan, A.: Evaluation of the risk of drug addiction with the help of fuzzy sets. J. Bioinform. Seq. Anal. **2**(4), 47–52 (2010)
39. Yilmaz, A., Ayan, K.: Cancer risk analysis by fuzzy logic approach and performance status of the model. Turkish J. Electr. Eng. Comput. Sci. **21**(3), 897–912 (2013)
40. Yilmaz, M.: Evaluation of total antioxidant capacity (TAS) by using fuzzy logic. Br. J. Math. Comput. Sci. **8**(6), 433–446 (2015)
41. Khamis, H.S., Cheruiyot, K.W., Kimani, S.: Application of k-nearest neighbour classification in medical data mining. Int. J. Inf. Commun. Technol. Res. **4**(4), 121–128 (2014)
42. Sun, J., Hall, K., Chang, A., Li, J., Song, C., Chauhan, A., Ferra, M., Sager, T., Tayeb, S.: Predicting medical conditions using k-nearest neighbors. University of Nevada Las Vegas (2017)
43. Shouman, M., Turner, T., Stocker, R.: Applying k-nearest neighbour in diagnosing heart disease patients. Int. J. Inf. Educ. Technol. **2**(3), 220–223 (2012)
44. Jabbar, M.A., Deekshatulu, B.L., Chandra, P.: Heart disease classification using nearest neighbor classifier with future subset selection. Ann. Comput. Sci. Ser. **XI**, 47–54 (2013)
45. Chitupe, A.R., Joshi, S.A.: Data classification algorithm using k-nearest neighbour method applied to ECG data. IOSR J. Comput. Eng. **14**(4), 13–21 (2013)
46. Charde, P.A., Lokhande, S.D.: Classification using k nearest neighbor for brain image retrieval. Int. J. Sci. Eng. Res. **4**(8), 760–765 (2013)
47. Thamilselvan, P., Sathiaseelan, J.G.R.: An enhanced k nearest neighbor method to detecting and classifying MRI lung cancer images for large amount data. Int. J. Appl. Eng. Res. **11**(6), 4223–4229 (2016)
48. Werner, E.F., Wheeler, S., Burd, I.: Creating decision trees to assess cost-effectiveness in clinical research. Biom. Biostat. **S7**(004), 1–3 (2012)
49. Pandey, A.K., Pandey, P., Jaiswal, K.L., Sen, A.K.: A heart disease prediction model using decision tree. IOSR J. Comput. Eng. **12**(6), 83–86 (2013)
50. Komal, G., Vekariya, V.: Novel approach for heart disease prediction using decision tree algorithm. Int. J. Innov. Res. Comput. Commun. Eng. **3**(11), 11544–11551 (2015)
51. Bichindaritz, I., Marling, C., Montani, S.: Case-based reasoning in the health sciences. In: Workshop at the Twenty-Third International Conference on Case-Based Reasoning (ICCBR) (2015)
52. Marling, C., Montani, S., Bichindaritz, I., Funk, P.: Synergistic case-based reasoning in medical domains. Expert Syst. Appl. **41**, 249–259 (2014)
53. Sharaf-El-Deen, D.A., Moawad, I.F., Khalifa, M.E.: A new hybrid case-based reasoning approach for medical diagnosis systems. J. Med. Syst. **38**(2), 1–11 (2014)

54. Sikora, M., Wrobel, L., Mielcarek, M., Kalwak, K.: Application of rule induction to discover survival factors of patients after bone marrow transplantation. J. Med. Inform. Technol. **22**, 35–53 (2013)
55. Thurimella, K.K.: Using Rule Induction to Elucidate Co-occurance Patterns in Microbial Data. University of Colorado, Boulder (2013)
56. Ridwan, R., Bony, M.I.S., Hasan, K.: Rule induction as a technique of detecting severity of myocardial infarction. In: International Conference on Bioinformatics and Biomedical Technology, vol. 29, pp. 34–39 (2012)
57. Fashoto, S.G., Adekoya, A., Gbadeyan, J.A., Sadiku, J.S.: Development of improved k-means clustering for health insurance claims. GESJ: Comput. Sci. Telecommun. **1**(47), 48–57 (2016)
58. Haraty, R.A., Dimishkieh, M., Masud, M.: An enhanced k-means clustering algorithm for pattern discovery in healthcare data. Int. J. Distrib. Sens. Netw. **2015**, 1–11 (2015)
59. Naaz, E., Sharma, D., Sirisha, D., Venkatesan, M.: Enhanced k-means clustering approach for health care analysis using clinical documents. Int. J. Pharm. Clin. Res. **8**(1), 60–64 (2016)
60. Wakoli, L.W., Orto, A., Mageto, S.: Application of the k-means clustering algorithm in medical claims fraud/abuse detection. Int. J. Appl. Innov. Eng. Manag. (IJAIEM) **3**(7), 142–151 (2014)
61. Thakur, N., Chauhan, R., Kumar, B.: Medical diagnosis using GA. Int. J. Electron. Comput. Sci. Eng. **1**(3), 1260–1266 (2012)
62. Singh, D.A.A.G., Leavline, E.J., Priyanka, R., Priya, P.P.: Dimensionality reduction using genetic algorithm for improving accuracy in medical diagnosis. Intell. Syst. Appl. **1**, 67–73 (2016)
63. Sapna, S., Tamilarasi, A., Kumar, M.P.: Implementation of genetic algorithm in predicting diabetes. Int. J. Comput. Sci. Issues **9**(2), 393–398 (2012)
64. Valsecchi, A., Damas, S., Santamaria, J., Marrakchi-Kacem, L.: Genetic algorithms for voxel-based medical image registration. IEEE (2013)
65. Mehmood, Y., Abbas, M., Chen, X., Honkela, T.: Self-organizing maps of nutrition, lifestyle and health situation in the world. In: 8th International Conference on Advances in Self-organizing Maps, pp. 160–167 (2011)
66. Kim, K.B., Song, D.H., Park, H.J.: Automatic extraction of appendix from ultrasonography with self-organizing map and shape-brightness pattern learning. BioMed Res. Int. **2016**, 1–10 (2016)
67. Teng, W.G., Chang, P.L.: Identifying regions of interest in medical images using self-organizing maps. J. Med. Syst. **36**(5), 2761–2768 (2012)
68. Borkowska, E.M., Kruk, A., Jedrzejczyk, A., et al.: Molecular subtyping of bladder cancer using Kohonen self-organizing maps. Cancer Med. **3**(5), 1225–1234 (2014)
69. Fei, S.W.: Diagnostic study on arrhythmia cordis based on particle swarm optimization-based support vector machine. Expert Syst. Appl. **37**, 6748–6752 (2010)
70. Maolong, X., Sun, J., Liu, L., Fangyun, F., Xiaojun, W.: Cancer feature selection and classification using a binary quantum-behaved particle swarm optimization and support vector machine. Comput. Math. Methods Med. **2016**, 1–9 (2016)
71. Huiyan, J., Lingbo, Z.: Improved PSO-SVM based disease detection in medical images processing. In: 6th International Conference on Computer Sciences and Convergence Information Technology (ICCIT) (2011)
72. Khazaee, A., Zadeh, A.E.: ECG beat classification using particle swarm optimization and support vector machine. Front. Comput. Sci. **8**(2), 217–231 (2014)

73. Kabari, L.G., Nwachukwu, E.O.: Neural networks and decision trees for eye diseases diagnosis, Chap. 3. Intech (2012)
74. Celik, U., Yurtay, N., Pamuk, Z.: Migraine diagnosis by using artificial neural networks and decision tree techniques. Online Acad. J. Inf. Technol. 5(14), 79–89 (2014)
75. Hadavandi, E., Shavandi, H., Ghanbari, A., Abbasian-Naghneh, S.: Developing a hybrid artificial intelligence model for outpatient visits forecasting in hospitals. Appl. Soft Comput. 12(2), 700–711 (2012)
76. Alharbi, A., Tchier, F.: A fuzzy-genetic algorithm method for the breast cancer diagnosis problem. In: The Ninth International Conference on Advanced Engineering Computing and Applications in Sciences (2015)
77. Mankad, K.B.: Design of genetic-fuzzy based diagnostic system to identify Chikungunya. Int. Res. J. Eng. Technol. (IRJET) 2(4), 153–161 (2015)
78. Santhanam, T., Ephzibah, E.P.: Heart disease prediction using hybrid genetic fuzzy model. Indian J. Sci. Technol. 8(9), 797–803 (2015)
79. Liu, D.Y., Chen, H.L., Yang, B., Lv, X.E., Li, N.L., Liu, J.: Design of an enhanced fuzzy k-nearest neighbor classifier based computer aided diagnostic system for thyroid disease. J. Med. Syst. 36(5), 3243–3254 (2012)
80. Zuoa, W.L., Wanga, Z.Y., Liua, T., Chenc, H.L.: Effective detection of Parkinson's disease using an adaptive fuzzy k-nearest neighbor approach. Biomed. Sig. Process. Control 8, 364–373 (2013)
81. Funmilola, A.A., Oke, O.A., Adedeji, T.O., Alade, O.M., Adewusi, E.A.: Fuzzy k-c-means clustering algorithm for medical image segmentation. J. Inf. Eng. Appl. 2(6) (2012)
82. http://www.fedzzo.com.ba/bs/clanak/informacioni-sistem/209. Accessed Feb 2018

Mechanical Engineering

Application of MSA as a Lean Six Sigma Tool in Working Conditions Automotive Firm from B&H

Ismar Alagić[1,2,3](✉)

[1] TRA Tešanj Development Agency, Tešanj, Bosnia and Herzegovina
ismar.alagic@gmail.com
[2] University of Zenica, Zenica, Bosnia and Herzegovina
[3] Engineering and Natural Science Faculty, International University of Sarajevo,
Sarajevo, Bosnia and Herzegovina

Abstract. Lean principles (value, current value, currents, withdrawal and perfection) have changed the behaviour of the participants in the work processes and created working environment in which the routing operation is performed on the process of labour, the elimination of losses in the process of work, continuous training of employees at all organizational levels in order to achieve shorter delivery times and cost competitive products. In this way, the companies have introduced the Lean concept in response to the activities known as the Japanese term "muda" that consumes resources, rather than creating value [1].

By using different tools and methods of Lean concept we aim to standardize and then constantly improve the quality of work processes in industrial systems under the basic motto of "standard process produces a standard product quality". For this reason, a growing number of companies worldwide invested significant funds in the development and implementation of tools of Lean concept that will allow them to have the ability of continuous and rapid adjustment of conditions that are constantly changing [2].

Six Sigma as a modern strategy management quality can help companies to achieve and maintain business success in the long period of time. This is one of the leading strategies that make the goals achievable through a structured and systematic project approach DMAIC (Define, Measure, Analyse, Improve, and Control). The main objective of Six Sigma strategy is oriented towards the improvement of the customer, in order to reach the company's business goals. Six Sigma is not just a method of improving the quality or process. This is the vision, philosophy, strategy and a set of tools, but the greatest savings and greatest revenue had companies that have implemented Six Sigma across the organization. Six Sigma philosophy must be understood by everyone within the organization and as well as all future steps of Six Sigma training program, where all employees are taught to understand the philosophy, cultural effects, the objectives of Six Sigma and the DMAIC projects approach [3].

This article provides proposal a set of approaches that are the basis for the development and application of the principles and MSA tool of Lean Six Sigma concepts. A special focus is given to the MSA method of applying Lean Six Sigma concepts in specific working conditions of domestic company from automotive sector.

© Springer Nature Switzerland AG 2019
S. Avdaković (Ed.): IAT 2018, LNNS 60, pp. 511–524, 2019.
https://doi.org/10.1007/978-3-030-02577-9_50

Keywords: Measurement system analysis (MSA) · Lean six sigma (LSS) DMAIC · Automotive

1   Introduction

Automotive industry is with the related industries the largest manufacturing sector in the world and occupies about 15% of the world's gross domestic product (GDP).[1] According to its size, the automotive industry is one of the world's largest employer therefore is directly or indirectly responsible for every ninth job in the developed countries.[2]

Development trends in the last 25 years that have introduced automotive manufacturers from the Far East (Japan, South Korea, etc.) have caused the shortening of new vehicle models development period from 7 to 3 years, which forced the final producers to access the redesign of the entire chain. Today's concept of supply chain management in the automotive industry is based on the vertical integration of all participants in the chain that delivered the final manufacturer 'integrated module' vehicles. Raising the efficiency of the production process and quality management in the automotive industry has historically been advocated using a number of concepts from Ford's assembly lines over the German concept The cycle-time and the Japanese concept of Kaizen, Kanban and Just-in-Time shipment to contemporary concept of "platform strategy", where vendors are responsible for the design and production of "integrated modules" for finished vehicles.

In this context, the last thirty years "Lean Production - Six Sigma" concept has become the dominant approach in world production, which leads to improvement in productivity and cost reduction in work processes. Powerful step of the automotive industry in Japan, compared to European and American competition was based on the principle of continuous improvement as one of the basic elements of improving production. The term "Kaizen" whose application management granted by Japanese company is an element of overall control of quality related to long-term continuous access to updates, with respect to human needs and quality [4].

At the operational level, Six Sigma uses a structured and systematic approach (DMAIC, DMADC cycle) to achieve operational excellence. The Six Sigma provides the tools to solve problems in various fields of business with statistical and non-statistical methods in a sustainable way. The goal of Six Sigma projects is to identify input factors that cause variations and deviations from the target in order to centre the process and minimize deviations.

Traditional methods such as "Six Sigma" are more focused on quality than on speed. Methods known as "Lean" are better to improve processes and speed, then quality. By combining these two types of methods is obtained method, Lean Six Sigma, which yields the best results in improving business organization.

Lean Six Sigma combines the two most important trends of development and improvement of operating results: improve the work (with the help of Six Sigma) and

---

[1] Mashilo (2015).

[2] Humphrey (2010).

increase the speed (through Lean principles). Lean Six Sigma is a business concept that focuses on increasing the zero point of profit and customer satisfaction.

Bosnia-Herzegovina's automotive production is no longer as it was once in the European market, partly because of political opportunities and partly because of the unfulfilled conditions that this market demands. Quality management and introduction of the concept of Lean Manufacturing - Six Sigma is in the hands of domestic companies and has not reached the level that is present primarily in the world automotive industry.

Over the past few decades dramatic changes have occurred in terms of rising awareness of quality, while access to quality management has been significantly improved in global frameworks. Today's development of the leading industry in the world increasingly imposes the immediate and closer cooperation of all participants in the production of end products. With the same aspect of the organizational chain vendor set-up, the appropriate co-operation of all the producers involved in the production of end products was ensured. Based on this, external stakeholders (suppliers and buyers) have to create the appropriate relationships that should enable immediate cooperation and participation in solving all the problems arising through the entire process of making the final product from development, procurement, technology, production, packaging and product distribution to the customer [5].

Quality has become a "phenomenon" of our time and the most important market factor in the international exchange of goods and services. To create a competitive product is a necessity for survival in today's general business climate. The market is becoming more selective in terms of quality. Therefore, the need to invest in quality becomes clear to all suppliers who strive to offer a quality product or service. Quality is a moving target that is defined by the customer. Only business organism, which is capable to track and hit a target that is constantly changing, can "produce" a quality product.

This article provides a set of approaches that are the basis for the development and application of the principles, methods and tools of Lean Six Sigma concepts. A special focus is given to the MSA method of applying Lean Six Sigma concepts in specific working conditions in BH companies, taking into account international experience in the study of this field.

2 Quality Management Standards in Automotive Sector

If we look at the quality in its general sense, we can say that the quality is "all around us" and that it presents a wide universal form in the living world. The term "quality" is known for all time of human civilization and from the beginning of mankind, and the actuality of interests has not ceased even now, on the contrary I think it has never been more topical as at the time of intensive development of sector approach to defining the concept of "quality" and "quality management". People have always wanted a quality of product, quality of service and ultimately quality of life.

Historically the development of quality philosophy can be divided into the following six phases:

(a) Quality control and inspection (Quality inspection - IQ);
(b) The technical aspect of quality control (Quality control - QC);

(c) Quality assurance (Quality assurance - QA);
(d) Quality Management (Quality management - QM);
(e) Total quality management (Total Quality Management-TQM), and the latest phase
(f) The quality of a man -Business Excellence (Quality of man -Business Excellence).

According to international sources of information[3] the advantages of an integrated supply chain are:

- 16%–18% improvement of characteristics of the shipment of goods;
- 25%–60% improvement in reduction of inventories;
- 30%–50% improvement in optimization of work cycle;
- 25%–80% improvement in the accuracy of prediction of the work actions;
- 10%–16% improvement in overall production;
- 25%–50% improvement in reducing costs in the supply chain;
- 20%–30% improvement in meeting labour standards and
- 10%–20% improvement in utilization of installed capacity.

Until the advent of harmonized technical specification ISO/TS 16949 automotive industry was the area with a number of schemes for assessment of the quality system based on sector specific requirements of individual manufacturers.

Standard QS 9000 harmonized requirements for manufacturers in the US and where "Big Three" (GM, Ford and Daimler-Chrysler) have their own suppliers. Until the first edition of ISO/TS 6949: 1999 the manufacturers of the automotive industry in the world had four different concepts for QMS according to the requirements of standards VDA 6.1, AVSQ, EAQF and QS 9000. Then in1996 the IATF was established. This working group IATF and technical committee ISO/TC 176 jointly made a technical specification that became the first edition of ISO/TS 16949: 1999. This was done in order to harmonize standards for the automotive industry, and therefore the drafting specified harmonized technical specifications. It can be said that the ISO/TS 16949: 1999 ("Quality systems - Suppliers in the automotive industry - specific requirements for the application of ISO 9001: 1994") took as its basis chapter 4 of the standard ISO 9001: 1994, and also included the requirements of standards VDA 6.1, AVSQ, EAQF and some new ones, agreed with international members.

In March 2002 the second edition of ISO/TS 16949: 2002 was published ("Quality management systems - Particular requirements for the application of ISO 9001: 2000 in mass production and manufacture of related service parts in the automotive industry"), which was developed in partnership between the IATF JAMA (Japan Automobile Manufacturer Association), and with the support of the technical committee ISO/TC 176, which is responsible for developing all the standards of the family ISO 9000. This second edition has 5 sections (modules) as well as ISO 9001, and the structure of sub-elements contains all the requirements of ISO 9001 with additions that improve the functioning of an important part of the automotive industry.

The third edition of ISO/TS 16949:2009 ("Quality management systems - Particular requirements for the application of ISO 9001:2008 in mass production and

---

[3] Pittiglio Rabin Todd & McGrath, SCM: Another Acronym to Help Broaden Enterprise Management, Lawrence S. Gould, www.autofieldguide.com.

manufacturing related service parts in the automotive industry") was issued in December 2009 with the full harmonization of the process approach and requirements contained in ISO 9001:2008 "Quality Management Systems - Requirements". Technical specification ISO/TS 16949:2009 was prepared by the International Automotive Task Force (IATF), with the help of ISO/TC 176, Quality management and quality assurance. This third edition of ISO/TS 16949 cancels and replaces the second edition (ISO/TS 16949:2002), which has been technically revised in accordance with ISO 9001:2008. A new edition ISO 9001:2015 (ISO 9001:2015 Quality management systems - Requirements) is currently under preparation.

The fourth edition bears the name the standard IATF 16949: 2016, which replaces ISO/ TS 16949: of 2009. This is how one of the most famous technical specifications evolves in the standard of world automotive industry. Although the standard is still in compliance with the latest version of ISO 9001: 2015, it is owned by IATF (International Automotive Task Force). One of the most widespread international standards of quality system has undergone new revision and in the fourth edition came with more than 65.000 companies in the world that have certified quality management system in accordance with the requirements of ISO/ TS 16949:2009. Standard IATF 16949: 2016 has a new structure that complies with the structure of control system according to the ISO 9001: 2015 requirements.

Employees in organizations that wish to introduce ISO/TS 16949 must be trained, in addition to the requirements of ISO/TS 16949, and the application of methods and procedures for the management of the project (development and/or the new product):

- APQP - planning product quality;
- FMEA - analysis of errors and their effects;
- SPC - statistical process control;
- MSA - measurement system analysis;
- PPAP - part of the approval process for the production;
- 8D - methodology for solving problems.

These methods and techniques that are defined in detail in the manuals published by AIAG (American Automobile Manufacturers Association) or the VDA (German Association), will be processed in the respective publication through concrete examples of application in real industrial conditions of BiH companies.

3 Measurement System Analysis (MSA)

Long time ago, lord Kelvin said: *"If you can't measure it, you can't fix, or improve it "*. This statement links very well to the mentioned statement of Deming: *"In God we trust, all others bring data"*.[4]

All this speaks in favour of the fact that measure system has an important role in the measurement phase of Lean Six Sigma project, because lacks of a measurement system

---

[4] "In God we trust, all others have to bring data", Deming.

can result in false and incorrect data and, thus, to wrong analysis and conclusions on where the problem is and what is the right solution.

Figure 1 gives a review that illustrates why data are basis for process improvement.

Looking at the Fig. 1, we can conclude that decisions are made based on feelings and personal experiences, and in most cases there is no proof that the process was

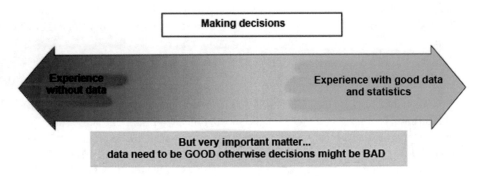

Fig. 1. Process of making decisions.

improved.

Namely, modern analysis of measurement system goes further than calibration of measuring devices.[5] Measuring devices can be well calibrated while measuring some characteristics, but the results can still be unacceptable. That is why it is necessary to test and "measure the measurement system" and quantify the deviation of measurement system. These deviations, if existing, can origin from wrong set ups of tools, measurement methods and mistakes of operators themselves. The measurement system must be checked: in its environment, by operators using it and by using proper samples.

When explaining what a measurement system is we have to start form measuring devices or gages, continue with operational definitions and procedures and, finally, end with people, operators who measure. Types of measurement system analysis are[6]

- Gage R&R – is a measurement system analysis for continuous data that can be measured in scale and be compared against each other;
- Attributive analysis – is analysis used for discrete data that have specifically determined minimum and maximum and are expressed in integer numbers and are usually used for visual measuring, inspections and subjective opinion (Fig. 2).

A general rule is that a gage must be at least 10 times more precise than the subject of measurement. Cause of such imprecision can be process variation that occurs in two forms:

[5] Kim Niles, "Characterizing the Measurement Process" – article http://www.isixsigma.com/index. php?option=com_k2&view=item&id=1306:characterizing-the-measurement-process&Itemid=207.

[6] Pyzdek, Keller, *"The Six Sigma Handbook: A Complete Guide for Green Belts, Black Belts, and Managers at All Levels"* (2010)- p. 289.

Fig. 2. Measuring tools must be 10 times more precise then the object.

- The first form of process variation can be that gage is precise, measurement correct, but still incorrect because of the use of wrong standard, pattern for measurement comparison. This kind of process variation is mostly caused by deviations inside gage, and test of repeatability is used for this kind of check.
- Another type of process variation can be that gage is correct, but imprecise, because different measurements show different values, i.e. imprecise values. This kind of process variation is mostly caused by operator's deviation and is checked by the test of reproducibility.

Basic features of measurement system are accuracy and precision. When discussing accuracy it is important to know the term of "bias". Bias is methodological system error in measuring, in the analysed case it would be "average" difference between "evaluator" and the correct value. In order to calculate bias, we need the "golden standard". Bias is equal to difference between average value of all measurements and "golden" standard which represents the reference value. In order to determine the bias, that is the "evaluator" accuracy, object of measurement is measured at least 10 times by the same "evaluator" (Fig. 3).

Another important term is precision and it represents "standard" deviation, declination, distribution of "evaluator" (the one performing measurement). Precision is checked by Gage R&R analysis, i.e. by the variation of measure device. Figure 4 shows review of interplay of (in)accuracy and (im)precision of measurement.

$$\text{(number of pieces)} \times \text{(number of evaluators)} > 15 \qquad (1)$$

Analysis of a certain measurement system, for example, lays down the following conditions for the measurement: 5 "products" (work pieces), 5 different settings of measure system, 3 inspectors/surveyors and 2 measurement repetitions.

While interpreting the results of measurement, we follow value percentage of Gage R&R that offers the following conclusions for obtained results:

% Gage R%R	Conclusions
<5%	Great measure system
≤ 10%	Measure system is satisfying (OK)
10%–30%	Acceptable, depending on costs, or how important measures are
>30%	Measure system needs to be fixed, it requires corrective actions

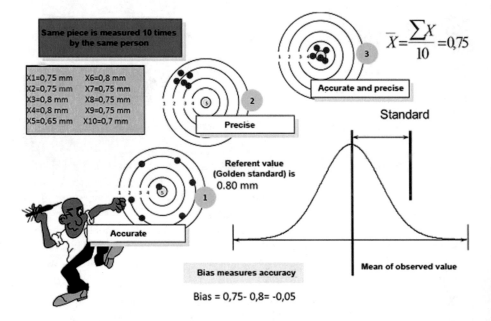

Fig. 3. Example of determination of "bias".

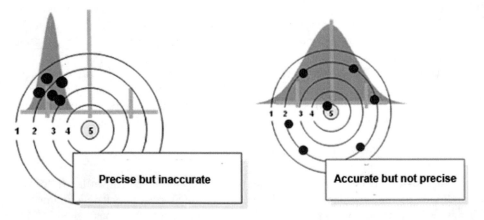

Fig. 4. Interplay of (in)accuracy and (im)precision of measurement.

It is visible, from the presented results, that, if the measure system is inadequate, that is, if the value of "Gage R&R" is higher than 30%, we have to correct the system by taking the following actions:

- Analyse possible cause of the problem;
- Check what the problem is: Repeatability and/or reproducibility?;
- Whether the problem is people, equipment, environment, procedure… (Ishikawa diagram).

Analysis of measurement system for attributive data is all about gaining data by the principle: "good - bad", "works – doesn't work" or "OK – NOK". Thereby, we need to test Repeatability and Reproducibility. To establish it we need to check each "piece" minimum twice by the same inspector. Since the status, or the result, can have only two states (OK - NOK; good - bad; tall - short; etc.) the result is not as precise as the numerically expressed analysis. However, attributive controls are often the only way of control, so it is important to do this analysis. In order to make this analysis it is necessary to take approximately minimum of 25–30 pieces/products, and the sample has to contain both, "good" and "bad" pieces. The next activity is marking of work pieces. An expert, i.e. a customer, marks (gives the final evaluation) using the principle "OK-NOK". After that, 2–3 inspectors randomly control pieces at least 2–3 times. The important thing, both in Lean Six Sigma project, as well as in use of any statistical method, is the procedure of collecting samples. The procedure is a special methodology for itself, but members of Lean Six Sigma team must be familiar with the basics, which is: the sample must be representative of the whole population and there cannot be systematic difference among data, why we take or reject certain samples. Also, it is necessary to take into consideration "Hawthorne effect"[7].

To be a successful Six Sigma team member it is necessary to know basic statistical terms. To lead Six Sigma project and use MSA method it is necessary to have advanced knowledge of statistical methods[8].

4 Example of MSA Application in Working Conditions of Domestic Firm

In the previous part of this article, was extensively demonstrated measurement system analysis - MSA, so below we will show patterns used for the application of the MSA from the companies in company Unis Tok from Kalesija.

In processing the MSA I have used the publication "Measurement Systems Analysis - MSA[9]", fourth edition, which is a binding frame for application of mentioned method. As is well known, measuring systems treats numerical analysis and attribute data, therefore we will pay attention to the next part of the presentation

---

[7] Hawthorne effect is theory which says that when we start measuring something, things change, operators pay more attention and work better. By that, the result of our measurement is not representative since it does not show real process picture, workers are relaxed and pay less attention while working. Namely, this theory got the name by "Hawthorne study" made from 1927-1932 in u Bell Western manufacture, where one of the goals was to research if better illumination influences operators' productivity. There was a conclusion that illumination has no influence, but productivity is increased due to knowledge somebody is monitoring them and measures their effects.

[8] For more information see: Pyzdek, Keller, *"The Six Sigma Handbook: A Complete Guide for Green Belts, Black Belts, and Managers at All Levels"* i Automotive Industry Action Group *"Measurement Systems Analysis"*.

[9] Measurement Systems Analysis-MSA, Fourth Edition, ISBN: 978-1-60-534211-5, Chrysler Group LLC, Ford Motor Company, General Motors Corporation (2010).

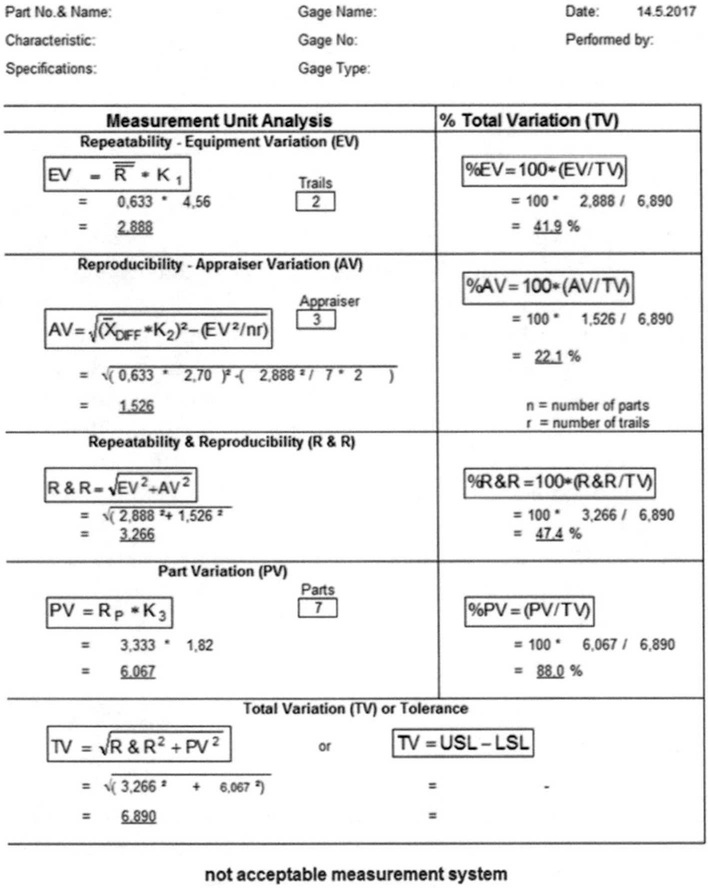

Part No.& Name:	Gage Name:	Date:	14.5.2017
Characteristic:	Gage No:	Performed by:	
Specifications:	Gage Type:		

Measurement Unit Analysis	**% Total Variation (TV)**
Repeatability - Equipment Variation (EV)	
$EV = \overline{\overline{R}} * K_1$ Trials $\boxed{2}$ $= 0.633 * 4.56$ $= \underline{2.888}$	$\boxed{\%EV = 100*(EV/TV)}$ $= 100 * 2.888 / 6.890$ $= \underline{41.9}\ \%$
Reproducibility - Appraiser Variation (AV)	
$AV = \sqrt{(\overline{X}_{DIFF} * K_2)^2 - (EV^2/nr)}$ Appraiser $\boxed{3}$ $= \sqrt{(0.633 * 2.70)^2 - (2.888^2 / 7 * 2)}$ $= \underline{1.526}$	$\boxed{\%AV = 100*(AV/TV)}$ $= 100 * 1.526 / 6.890$ $= \underline{22.1}\ \%$ n = number of parts r = number of trails
Repeatability & Reproducibility (R & R)	
$\boxed{R\&R = \sqrt{EV^2 + AV^2}}$ $= \sqrt{2.888^2 + 1.526^2}$ $= \underline{3.266}$	$\boxed{\%R\&R = 100*(R\&R/TV)}$ $= 100 * 3.266 / 6.890$ $= \underline{47.4}\ \%$
Part Variation (PV)	
$\boxed{PV = R_P * K_3}$ Parts $\boxed{7}$ $= 3.333 * 1.82$ $= \underline{6.067}$	$\boxed{\%PV = (PV/TV)}$ $= 100 * 6.067 / 6.890$ $= \underline{88.0}\ \%$
Total Variation (TV) or Tolerance	
$\boxed{TV = \sqrt{R\&R^2 + PV^2}}$ or $\boxed{TV = USL - LSL}$ $= \sqrt{3.266^2 + 6.067^2}$ $=$ $= \underline{6.890}$ $=$	

not acceptable measurement system

Fig. 5. Calculation of R&R for purposes of MSA analysis - ring bearing.

application of MSA in specific working conditions. When processing and calculations for MSA, as well as creating reports was used software application WinSTAT®. All the presented graphical representations are made using applications WinSTAT®. The procedure of calculation of R&R for purposes of MSA analysis - ring bearing is shown on Fig. 5.

From all the above it can be seen that the application of statistical techniques, methods and tools through Lean Six Sigma concept is unthinkable without strong software support embodied in specialized applications such as WinSTAT® in the above case (Figs. 6 and 7).

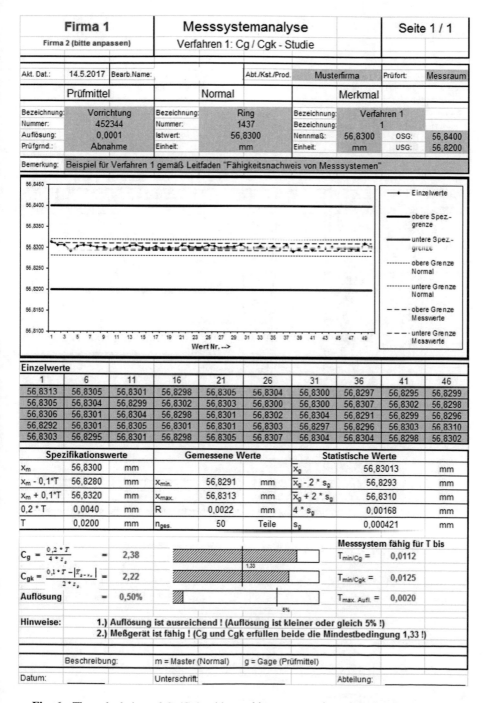

Fig. 6. The calculation of Cg/Cgk with graphic representation of the bearings product.

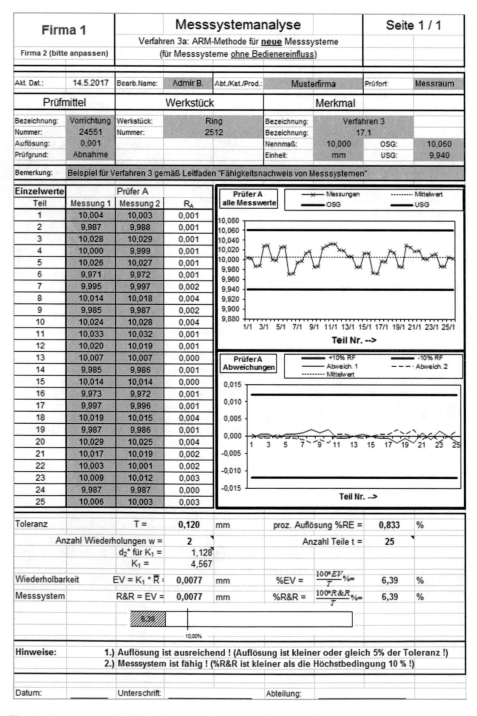

Fig. 7. An example of application of ARM method for a new measuring device for the bearing ring of the product.

5   Conclusion

The most common and the most often used tools within the second phase of Lean Six Sigma projects are:

- Process map;
- Ishikawa diagram;
- Cause-Effect matrix;
- FMEA (Failure Mode and Effect Analysis);
- Data Investigation Plan;
- MSA – Measure System Analysis;
- Visualization tools;
- Calculating process capacity (Cp, Cpk, Pp, Ppk, Sigma, PPM).

The importance of the MSA system is perhaps the best-painted in the statements of Lord Kelvin: "If something cannot be measured, you cannot fix/improve it" and Dr Deming, who said: "In God I trust, all others bring data". The data are the basis for the improvement of any process, in particular the production process[10].

The paper presents the application of MSA as a LSS tool in the concrete case study "Optimization of MSA analysis for ring bearing product in company Unis Tok Kalesija". The proposed intervention plan, company Unis Tok Kalesija, was implemented in practice after 6 months of entering the LSS program.

References

1. Alagić, I.: Upravljanje kvalitetom: Lean proizvodnja - Six Sigma, CIP - Katalogizacija u publikaciji, Nacionalna i univerzitetska biblioteka, 629.331:[658.5:005.6(075.8), COBISS. BH-ID 24216326. Štamparija-S, Tešanj (2017). ISBN 978-9958-074-08-0
2. Alagić, I.: Industrial Engineering and Maintenance: Lean Production - Six Sigma with application of tools and methods in specific working conditions, CIP - Katalogizacija u publikaciji, Nacionalna i univerzitetska biblioteka, 629.331:[658.5:005.6(075.8), COBISS. BH-ID 24232454, Štamparija-S, Tešanj (2017). ISBN 978-9958-074-09-7
3. Alagić, I., Božičković, R., Višekruna, V., Brkić, A.: Primjena Lean Six Sigma alata u konkretnim radnim uslovima firme iz automobilske industrije. Univerzitet u Istočnom Sarajevu, Mašinski fakultet, Festival kvaliteta 2017, Jahorina, BiH, 26–28 oktobar 2017
4. Alagić, I.: Razvoj modela sistema upravljanja kvalitetom u upravljanju lancem dobavljača automobilske industrije i primjena Lean Six Sigma alata u radnim uslovima firme iz autoindustrije. XIX Nacionalni, V Međunarodni naučno-stručni skup SISTEM KVALITETA USLOV ZA USPEŠNO POSLOVANJE I KONKURENTNOST, Hotel Kraljevi Čardaci SPA Kopaonik, Srbija, vol. 29, no. 11, 01 December 2017
5. Alagić, I.: Application of lean six sigma tools in order to eliminate bottlenecks in working conditions firm from B&H, vol. 19. međunarodni simpozij o kvaliteti, Plitvice, Republika Hrvatska, 21–23 March 2018

---

[10] Alagić Ismar, Prilog razvoju modela za ocjenjivanje sistema upravljanja kvalitetom u lancu dobavljača automobilske industrije, Doktorska disertacija, Univerzitet Džemal Bijedić, Mostar (2008).

6. Staudter, C., Meran, R.: Design for Six Sigma and Lean Toolset: Implementing Innovations Successfully. Springer, New York (2009)
7. El-Haik, B., Roy, D.M.: Service Design for Six Sigma: A Roadmap for Excellence. Wiley, Hoboken (2005)
8. Levine, D.M.: Statistics for Six Sigma Green Belts with Minitab and JMP. Pearson Education, London (2006)
9. Buehlmann, U.: Value stream mapping, leadership and skills development training. Bern University of Applied Sciences, Biel, Switzerland (2016)
10. Marić, B., Božičković,R.: Održavanje tehničkih sistema & Lean koncept. Univerzitet u Istočnom Sarajevu, Mašinski fakultet Istočno Sarajevo, Istočno Sarajevo (2014).ISBN 978-99976-623-09
11. Watson, G.H.: Six Sigma for Business Leaders-A Guide to Implementation. Methuen, GOAL/QPC (2004). ISBN 1-57681- 049-6

Resource Efficient and Cleaner Production – Case Study of Assessment and Improvement Plan in MADI Ltd. Tešanj

Ismar Alagić[1,2,3](✉)

[1] TRA Tešanj Development Agency, Tešanj, Bosnia and Herzegovina
ismar.alagic@gmail.com
[2] University of Zenica, Zenica, Bosnia and Herzegovina
[3] International University of Sarajevo, Sarajevo, Bosnia and Herzegovina

Abstract. Green Industry is industrial production and development that does not come at the expense of the health of natural systems or lead to adverse human health outcomes. It is an important pathway to achieving sustainable industrial development through sector strategy for the realization of Green Growth – Green Economy – Green Jobs in the manufacturing and related sectors.

Resource productivity is the quantity of good or service (outcome) that is obtained through the expenditure of unit resource. This can be expressed in monetary terms as the monetary yield per unit resource.

UNEP/UNIDO works to promote Resource Efficiency and sustainable consumption and production in both developed and developing countries. The focus is on achieving increased understanding and implementation by public and private decision makers of policies and actions for Resource Efficiency and sustainable consumption and production.

Resource Efficiency means:

- The careful selection of raw materials and energy resources;
- Minimization of waste, emissions, hazards and risks;
- Responsible management of material and energy flows during the production process;
- Attention to the use, recycling and disposal phases of the product life cycle.

This article is the overview of Resssource Efficient and Cleaner Production (RECP) assessment through audit done by author as UNIDO certified RECP expert.

This RECP report of MADI ltd. contained four sections: A detailed RECP assessment; Feasibility analysis; RECP options for following issues: materials and waste, Water and Effluents and energy efficiency and; Implementation improvement plan & monitoring.

This RECP assesses the RECP indicators in company MADI. The focus of this article also highlights the proposal for Point of Interventions (POI) within the pilot project in working conditions of company MADI.

© Springer Nature Switzerland AG 2019
S. Avdaković (Ed.): IAT 2018, LNNS 60, pp. 525–536, 2019.
https://doi.org/10.1007/978-3-030-02577-9_51

Keywords: Green industry · Resource efficient · Cleaner production Assessment · MADI

1   Introduction

Although the food sector is not a traditional pillar of BiH economy, it has a high share of GDP, well participation in BiH exports, but also a large deficit. The coverage of imports by exports is only 22.86%[1]. The absolute value of exports for this sector amounted to around 564 million BAM, while the value of imports is about 2.46 billion BAM. However, this industry has the potential for export of specific products in niche markets. There is the possibility of creating several clusters within this sector, which is characterized by a large number of sub-sectors, including: fruits and vegetables, meat and meat products, milk and dairy products, fish, honey and herbs, alcoholic and soft drinks. According to available data, the GDP share of this sector amounted to 11.1%, with the number of employees 27,952 in 1,793 companies distributed across the country that participated in the national exports to 7.36%[2].

Dairy products, as well as meat and meat products, still do not meet the safety regulations of the EU market. At the time of the conclusion of the text of this article came the news that the EU has approved the export of chicken meat products from BiH to the EU. Generally speaking, these products are mainly exported to countries of CEFTA, and certain unique market niches (e.g. Increased exports of beef to Turkey since 2015.), with a large concentration of the diaspora. The company Madi from the city of Tesanj has become the regional leader in chicken production in the Western Balkans, and is the fastest growing regional brand in this field [1].

Green Industry is industrial production and development that does not come at the expense of the health of natural systems or lead to adverse human health outcomes. It is an important pathway to achieving sustainable industrial development through sector strategy for the realization of Green Growth – Green Economy – Green Jobs in the manufacturing and related sectors. Having this as a initial input Enova company from Sarajevo through RECP national experts in Bosnia and Herzegovina have commenced the activities on more detailed and more scientific approach to given subject in order to explore resource efficient and cleaner production among ten domestic pilot-companies. These efforts were recognized by UN Mission in BiH and it resulted with awarding the project of "Resource Efficient and Cleaner Production in Bosnia and Herzegovina".

This article addresses in detail the situation in the most significant company from food processing sector in Bosnia and Herzegovina through the analysis of resource efficient and cleaner production (RECP) audit results. Findings of analysis this RECP Audit identify key obstacles in business operations and growth of companies concerning RECP requirements and provide recommendations for the required actions to eliminate the identified obstacles. The analysis of competitive advantage of company MADI was made on the basis of information gathered from company MADI and in

---

[1] Official data from Foreign Trade Chamber of Bosnia and Herzegovina, 2017.

[2] Source: I. Alagić: Exportbook-Customer Specific Requirements, pg. 28, Delegation of the Europeaan Union in Bosnia and Herzegovina, 2015.

accordance with the methodology developed by UNIDO [2] and of Porter's diamond [3], considering the structure and characteristics of production chain through the whole value chain with food processing sector. This article is the overview of Resource Efficient and Cleaner Production (RECP) assessment through audit done by author as UNIDO certified RECP expert.

2 Basic Facts About Food Processing Sector

Food processing industry in BiH in 1990, together with agricultural production, was seen as one production unit taking into account the interdependence of these two branches. Agricultural production participated with 9.5% in the formation of GDP of BiH in 1990, and was considered as an important branch of economy. Unlike agricultural production, the food processing industry was rather less developed and its share in the formation of GDP of BiH in the same period, together with beverage production, was 3.4%, and together with tobacco processing, this share was increased by additional 0.6%. Taking into account world standards applicable in the said period, BiH was an agrarian society where processing of agricultural products was a weak side[3].

Insufficient financial resources necessary to modernize the entire agricultural production (replacement of out-dated technology and equipment, improve the existing infrastructure of the agricultural sector, etc.) is one of the most important limiting factors of faster development of this sector. The lack of sufficiently strong, developed and recognizable local brands has resulted in import nearly 80 to 90% of the total demand for industrial processed food, while meeting the needs with its own primary agricultural production is much better. This shows the imbalance of the relationship between primary production of food and its processing. For the development of agriculture and thus food industry in BiH it is necessary to run projects related to:

- Shopping domestic industrial products;
- Developing awareness of the importance of regionalization in the production and distribution of its own food products;
- Development of rural areas and on the basis of improving rural living infrastructure;
- Sustainable technologies in agriculture and food production;
- Development of typical products especially products with geographical origin;
- Education of the importance of creating recognizable brands.

In accordance with the Strategy of Development of the Federation BiH by 2020 the following priorities for the sector of agriculture, food and rural development were set out:

---

[3] Source: Past agricultural policy and its future effect on development of food sector in the Federation of BiH, Faculty of Agricultural and Food Sciences of University in Sarajevo, 2011.

- Establishment of a functional institutional capacity for agriculture and rural development;
- Improving the competitiveness of the preparation, processing and trade, while raising quality level and safety of domestic products;
- Nature conservation and rational management of natural resources;
- Improvement of living conditions and income diversification in rural employment.

Similar priorities are in the strategic documents of the Republic of Srpska in the field of rural development and agriculture for the next mid-term period. The current points of weakness of the food industry in BiH are as follows:

- Lack of strong local brands;
- Low quality products;
- Old-fashioned design and packaging;
- Strong foreign import competition;
- Weak bargaining power to the domestic market;
- Limitation of the marketing of goods in the domestic distribution chains;
- Difficulties in exports and a small number of foreign customers;
- A large number of small producers (lack of readiness for higher capacities);
- Poor cooperation with agricultural and food universities.

Development of food processing industry in the territory of Tešanj in recent years ranks this Municipality amongst the leaders in the said sector in BiH. This is particularly evident in the field of water and mineral water production where a series of successful companies operate, including Tešanjska vrela (Tešanjski dijamant), Princess and Oaza (AS Grupacija). AS Grupacija currently represents one of the biggest industrial companies in BiH, which, besides the mentioned water bottling plant Oaza in the territory of Tešanj Municipality, includes a biscuit factory in Jelah that employs around 200 workers. In the field of meat processing, the company Madi also operates in Tešanj, which is one of the leaders today in the production of chicken meat in BiH. Besides the mentioned companies, there should be a mention of the company Subašić in the field of production of bakery products and pasta, company Jami that produces pre-cooked frozen meals, leaves of rolled-out dough, frozen dough and pastries, and company Biošamp – production of mushrooms, Dukat – mill and bakery, and Zlatna kap – cheese production. Company MADI was analyzed within the scope of this article.

3   Structure of Production and Export for Food Processing Sector – Case Study of Company MADI

Within the scope of market segment of production of chicken meat products in company MADI, the following market segments have been identified:

- Production of fresh and frozen chicken product;
- Hot dog production, and;
- Sausage production (Fig. 1).

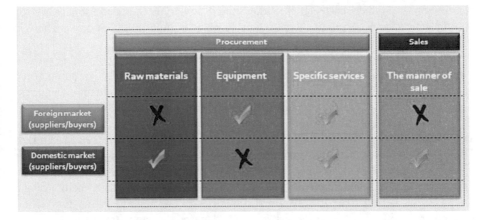

Fig. 1. General overview of production chain in market segment of production of chicken meat products.

Company MADI Ltd. Tešanj operates within the scope of this market segment. Major raw material used by this company in production process is chicken meat. Within the scope of its production facilities, MADI Ltd. Tešanj has a chicken slaughterhouse and chicken hatching technology installed in business premises of company Veterinarska stanica d.o.o. Tešanj, which has become the property of MADI Ltd. Tešanj through the acquisition. The process of provision of required raw material starts with chicks hatching, which are then transported for broiler rearing to subcontractors with whom MADI Ltd. Tešanj has a long-term cooperation and a partner relationship. After broiler rearing, MADI Ltd. Tešanj slaughter chicks and processes chicken meat in its own production facilities, and carries out final production of the aforesaid products. Considering that the aforesaid process of provision of raw materials is the result of a long-term partner relationship between the said actors, there are no major problems in regard to quantity, price, quality and delivery time of required raw materials. However, given that MADI Ltd. Tešanj is continuously developing and continuously increase the market share in the territory of BiH, there is a need for additional involvement of a number of subcontractors for broiler rearing.

In the production process of the final products MADI Ltd. Tešanj also uses different types of spices, packaging and cartons for packaging of final products. Major suppliers of spices are Ireks Aroma d.o.o. Zagreb and Wiberg Austria. Packaging for final products is procured from Serbia, from company Tipoplastika, while cartons are procured from company King Plus, also from Serbia. Major types of equipment used by MADI Ltd. Tešanj in the production process include slaughtering machines, chicken hair removal machines and chicks hatching equipment. All equipment is manufactured outside BiH and major manufacturers are companies Meyn from the Netherlands and Kilia from Germany. MADI Ltd. Tešanj possesses the most up-to-date equipment and has been continuously investing into updating of technology. Business services used by this company on a regular basis in the production process include services of broiler rearing. MADI Ltd. Tešanj possesses its own specialized transport vehicles for distribution of products on the BiH market, and it points out the need for services for optimization of transport routes with a view to reducing transportation costs. There is also a need for services of production process certification.

The sale of products on the BiH market is carried out through retail chains such as Bingo and Konzum, and through a large number of retail shops. The major competitors of company MADI Ltd. Tešanj on the BiH market in the same production range are Akova Impex d.o.o. Sarajevo, Perutnina Ptuj BH d.o.o. Breza and Bajra d.o.o. Travnik.

In the segment of production of chicken meat products, MADI Ltd. Tešanj appears as a vertical integrator in the production chain. The main line of business of MADI Ltd. Tešanj is the chicken meat processing and production of different final chicken meat products. Furthermore, it has been continuously engaged in market research, development of products and improvement of distribution network, which facilitates the access to the BiH market and market of countries in the region. In the production segment of products and semi-finished products used as input components in the production of chicken meat products, MADI Ltd. Tešanj has gradually been establishing a network of strategic partners and integrating a large number of companies from different sectors. MADI Ltd. Tešanj currently has a strategic partnership with company Veterinarska stanica/Veterinary Station d.o.o. Tešanj, which carries out chicks hatching for the needs of MADI Ltd. Tešanj. In addition, strategic partnership has been established with a large number of companies engaged in broiler rearing for the needs of MADI Ltd. Tešanj. In this manner, the production chain in the production sector of chicken meat products is being gradually completed from raw materials to final products, headed by company MADI Ltd. Tešanj. However, there is the potential for integration of increasing number of companies into the aforesaid system, primarily from the sector of chicken feed production, production of packaging and production of a variety of spices.

3.1   Analysis of Competitive Advantage of Company MADI According to Porter's Diamond

Findings of competitive advantage analysis of company MADI according to Porter's diamond are as follows:

(a) **Suppliers**:

- Broiler suppliers are farms for fattening broilers from Srbac, Derventa, Petrovo, Gracanica and Tešanj.
- They have cooperative contracts with about 100 farms.
- Payments are made per kg of broiler brooms reduced for invoiced raw materials delivered by farmers.
- They provide all raw materials and raw materials for these farmers.
- Fodder foods delivered by farmers for fattening broilers are imported from Serbia (VZ Subotica, Gebi Subotica).
- A strong negotiating position for MADI Ltd. because of the large quantities of fodder that are being procured.

- Additives and auxiliary raw materials for the production of meat products are purchased through intermediary firms from the EU (Raps, Ireks Aroma, Viberg).
- Technical assistance from suppliers only in the production of meat products by the technical staff of additives and subsidiary raw materials.
- Packaging is mainly imported from Serbia, especially for meat products. A part of the packaging for chicken packaging is purchased in BiH. The production equipment is mainly purchased from Germany.

(b) **Competitors**:

- The main competitors are: MI Akova Sarajevo and Perutnina Ptuj.
- Advantages of MADI Ltd. in relation to competitors are as follows: clear ownership structure, simple organizational setting, quick decision-making, closed production cycle (breeding flocks, one-day chicks, fattening, slaughtering, final processing).
- Disadvantages are as follows: insufficient branding-marketing, lower slaughtering capacity, lack of own production of animal feed.
- There is no significant scope of cooperation with competitors. Several times they had an attempt to reach agreement on the market and price policy, but without success. They have cooperation with certain companies in case of need for slaughter. They are not members of professional associations, etc.

(c) **Buyers**:

- Products Madi doo are mainly sold on the BiH market - 97%.
- Focus is on sale to smaller stores. They are less represented in retail chains in BiH. A significant buyer is Argeta BH, who buys chicken meat as a raw material for the production of chicken pie.
- Payment periods are up to 15 days.
- Export makes about 10% of the turnover (Serbia, Montenegro, FYRM, in some cases EU). Mainly chicken frozen meat is exported.
- There are no distributors/dealers on export markets. Specific customer requirements are reflected in the following: very short delivery times, product unloading at different customer locations

(d) **Increase a new competitions**:

- There is no indication in the chicken meat segment for new competitors (large investments are needed for new entrants).
- For companies from Croatia it is difficult to enter the market of BH with chicken meat (customs expenses, short shelf life of fresh meat).
- For companies from CEFTA countries, it is difficult to enter fresh chicken meat due to short shelf life, possibly presence in the frozen chicken meat segment.
- For the semi-long products of chicken meat (sausages, etc.) there is a big competition and there are continuously new participants.

(e) **Preparation of the substitute**:

- The meat is an essential food and must be present in the diet of people.
- Chicken meat is the cheapest meat, which is one of the reasons for great consumption.
- Due to religious rules, certain substitution is possible with beef, but it is more expensive than chicken meat.
- Substitution with semi-finished chicken meat products (sausages, etc.) may be beef and vegetable protein products (Soya).
- Changes in demand are visible due to the seasonal demand for chicken meat. The biggest challenge in integrated production is production planning.

(f) **Potential for improvements**:

- Expansion of the market in BH by entering large shopping centers.
- Branding the product.
- Expansion of export markets such as: Serbia, Montenegro, Macedonia, Kosovo, Albania, countries of the Middle East.
- Investing in new facility for slaughtering broilers (about 4 million BAM).
- Investments in new equipment for slaughtering broilers (about 6 million BAM). With these investments MADI Ltd. will have the largest capacity for slaughtering broilers in BiH.

(g) **Suport institution**:

- Good cooperation with institutions at the municipal level;
- Inadequate cooperation with institutions at entity level (Veterinary Office BH, Ministry of Agriculture of FBiH);
- The lack of a system for the adequate disposal of animal waste is one of the key issues that are not institutionally solved;
- In general, they did not use the support of institutions in business development.

(h) **Indicative conditions**:

- Customs tariff defining quotas for duty-free imports of frozen meat - more quotas negative impact/lower quota positive impact;
- The stimulation of exports has a very positive impact on the competitiveness of MADI Ltd.;
- The inability to export to the EU countries due to uncoordinated legislation in BH with the EU - a very negative impact;
- Inadequate application of environmental regulations - selective application of the obligation of wastewater treatment - very negative impact;
- Inadequate application of the law on the disposal of animal waste - a very negative impact.

4   Recp Assessment Case Study of Company MADI

This article provided a summary of the key factors related to the RECP assessment and intervention plan for improvements in case of company MADI carried out in 2017. In November 2017, the Author as certified UNIDO expert in cooperation with NCP for Bosnia and Herzegovina conducted a special survey of food-processing competitiveness capacities in Bosnia and Herzegovina which expressed its interest for participation in RECP programme. The survey was commissioned by and for the use of the UNIDO Programme which is carried out by National Contact Point (NCP) in Bosnia and Herzegovina.

During development of the methodology for this research, the Author have tried to fulfill the following requirements:

- A detailed RECP assessment; Feasibility analysis;
- RECP options for following issues: materials and waste, Water and Effluents and energy efficiency and;
- Implementation improvement plan & monitoring (Fig. 2).

Okolišni aspekt	😞	😐	😀
Kišna kanalizacija		X	
Otpadne vode		X	
Postupanje sa čvrstim otpadom		X	
Razdvajanje otpada		X	
Emisije u zrak		X	
Buka			X
Opasne materije		X	
Energija	X		
Okolišna politika			X

Fig. 2.   (a) MADI destination; (b) Fields of RECP assessment in MADI Ltd.

Some of input/output data for MADI Ltd. purposed for RECP Audit is shown in Fig. 3.

No.	Proizvodi/usluge	Instalirani kapacitet	Godišnja proizvodnja [kg]
1.	Svježe pileće meso	5.000 kom/h (17.472.000 kg/god)	14.514.822
2.	Proizvodi od mesa i mesa peradi (hrenovke, kobasice, oblikovani program, dimljeni program, slajsani program)	20 t/dan (6.240.000 kg/god)	2.966.477

Fig. 3.   List of input/output data in case of RECP Audit.

Type of wastes after pre-fabrication of chicken meat are shown in Fig. 4.

No.	Vrsta otpada	Godišnja količina	Jedinica mjere	Način zbrinjavanja u krugu kompanije	Način zbrinjavanja izvan kompanije	Trošak zbrinjavanja
1.	Perje	1.895.400	kg	Kontejner	Deponija za animalni otpad	Gorivo za transport i ljudski rad na zbrinjavanju
2.	Iznutrice glavice	1.701.000	kg	Kontejner	Deponija za animalni otpad	Gorivo za transport i ljudski rad na zbrinjavanju
3.	Krv	972.000	kg	Cisterna	Deponija za animalni otpad	Gorivo za transport i ljudski rad na zbrinjavanju
4.	Nogice	725.760	Kg	Kontejner	Deponija za animalni otpad	Gorivo za transport i ljudski rad na zbrinjavanju
5.	Kosti	1.301.400	kg	Kontejner	Deponija za animalni otpad	Gorivo za transport i ljudski rad na zbrinjavanju
6.	Masti sa uređajia za prečišćavanje otpadnih voda	1.895.400	Kg	Kontejner	Deponija za animalni otpad	Gorivo za transport i ljudski rad na zbrinjavanju
7.	Mulj sa uređaja za prečišćavanje otpadnih voda	312.000	kg	Kontejner	Deponija za animalni otpad	Gorivo za transport i ljudski rad na zbrinjavanju
8.	Uginuli i škartni pilići	62.242	kg	Kontejner	Deponija za animalni otpad	Gorivo za transport i ljudski rad na zbrinjavanju
						Ukupno: 30.000 KM/god

Fig. 4. Type of waste after pre-fabrication of chicken meat.

When processing and calculations for MSA, as well as creating reports was used software application GHG calculator®. All the presented graphical representations are made using applications GHG calculator®. Figure 5 shows calculative form of "absolute RECP indicators" from company MADI.

ABSOLUTE RECP INDICATORS					
Indicator	Unit	Baseline (B) (Before RECP intervention)	Year 1 A (After RECP implementation)	Change (C) C=100*(A-B)/B [%]	Difference Between A and B
Resource use					
Energy Use	[MJ/yr]	20.177.326,80	16.030.067,04	-20,55	-4.147.259,76
Materials Use	[ton/yr]	22.169,50	22.168,50	0,00	-1,00
Water Use	[m3/yr]	55.378,00	33.490,00	-39,52	-21.888,00
Pollution					
Carbon dioxide	[ton CO₂-eq/yr]	4.657,69	4.217,61	-9,45	-440,08
Waste-Water	[m3/yr]	55.378,00	33.490,00	-39,52	-21.888,00
Waste	[ton/yr]	8.865,20	8.864,20	-0,01	-1,00
Product Output					
Product Output: P	[ton/yr]	17.481,30	17.481,30	0,00	0,00

Fig. 5. RECP indicators after RECP audit in case of company MADI.

5   Conclusion

This article is the overview of Resource Efficient and Cleaner Production (RECP) assessment through audit done by author as UNIDO certified RECP expert. This RECP report of MADI ltd. contained four sections: A detailed RECP assessment; Feasibility analysis; RECP options for following issues: materials and waste, Water and Effluents and energy efficiency and; Implementation improvement plan & monitoring.

| RECP mjere | Koristi | | | | |
| | Ekonomske | | | Resursne | Okolinske |
	Investicija [KM]	Ušteda [KM/god]	Povrat investicije	Smanjenje potrošnje energije, sirovina i vode na godišnjem nivou	Smanjenje proizvedenog otpada, otpadnih voda i emisija u zrak
Upravljanje energijom 1. Upravljanje obračunatom snagom *Izrada detaljne studije i projekta za kontrolu vršnog opterećenja* *Realizacija optimalne metode smanjenja vršnog opterećenja (ugradnja hibridnog sistema)*	100.000 KM	100.000 KM	1 godina	Sistemsko rješenje plaćanja obračunate snage Smanjena potrošnja el. energije iz mreže za cca. 10%	Smanjenje emisija u zrak za 440 tCO2/god, odnosno za 9,45%

Fig. 6. Intervention Improvement plan in case of Energy management.

This RECP assesses the RECP indicators in company MADI. The focus of this article also highlights the proposal for Point of Interventions (POI) within the pilot project in working conditions of company MADI. The most important findings of above mentioned Assessment were as follows [2]:

- There are some problems concerning RECP requirements in case of company MADI, such as:
  - The company has no energy consumption monitoring system;
  - The company does not have a busy engineer;
  - Annual electricity costs 526,000 km;
  - cca. 39% of this amount is deductible on your strength;
  - There is only one electricity meter;
  - Firewood exposed to atmospheric influences and accepts excessive moisture;
  - There is no system for purification and reuse of wastewater;
- Intervention Plan for company MADI consisted of following measures:
  - Power Management - Energy Management (see Fig. 6);
  - Construction of a roof covering for wood - Material management (see Fig. 7);
  - Reuse of waste water - Water management (see Fig. 8);
- There is no ideal theory in the efficient use of resources and cleaner production, there are successful models that depend on us;
- The resource is in people who want and know;
- The RECP report is just one of the steps to success;
- The RECP report must go to RECP management (control point 2018).

RECP mjere	Koristi				
	Ekonomske			Resursne	Okolinske
	Investicija [KM]	Ušteda [KM/god]	Povrat investicije	Smanjenje potrošnje energije, sirovina i vode na godišnjem nivou	Smanjenje proizvedenog otpada, otpadnih voda i emisija u zrak
Upravljanje materijalom 2. Izgradnja nadstrešnice za pokrivanje drveta *Sprečavanje izloženosti drveta atmosferskim uticajima i prihvatu prekomjerne vlage*	15.000 KM	22.962 KM	8 mjeseci	Smanjenje potrošnje drveta za 42,52% u odnosu na trenutnu potrošnju za slučaj dobivanja iste količine toplote	-

Fig. 7. Intervention improvement plan in case of Material management.

RECP mjere	Koristi				
	Ekonomske			Resursne	Okolinske
	Investicija [KM]	Ušteda [KM/god]	Povrat investicije	Smanjenje potrošnje energije, sirovina i vode na godišnjem nivou	Smanjenje proizvedenog otpada, otpadnih voda i emisija u zrak
Upravljanje vodom 3.Ponovno korištenje otpadne vode *Instalacija uređaja za skidanje masti* *Prečišćavanje vode i njeno vraćanje za čišćenje nečistog kruga* *Izgradnja cjevovoda, rezervoara i nabavka pumpe, priključenje na postojeći sistem za pranje*	40.000 KM	69.452 KM	7 mjeseci	Smanjenje potrošnje vode iz gradskog vodovoda za 21.888 m3/god Smanjen utrošak hemikalija za pogon uređaja za 1 t/god	Smanjenje otpadnih voda za 21.888 m3/god, odnosno za 39,5% Smanjenje utrošenih hemikalija za 1 t/god, odnosno smanjenje ukupne količine otpada za 1%

Fig. 8. Intervention improvement plan in case of Water management.

References

1. Alagić, I.: A competitiveness analysis of the Tešanj municipality food processing industry, Internal report, TRA Tešanj Development Agency (2017)
2. Alagić, I.: Assessment report about possibility of RECP and improvement plan in company MADI, Tešanj, UNIDO (2016)
3. Porter, M.E.: The five competitive forces that shape strategy. Harvard Business Review, January 2008

The Effect of Test Temperature on Lap Shear Test Results of Two-Component Epoxy/Metal Adhesive-Bonded Aluminum

Amila Bjelopoljak[1]([⊠]), Petar Tasić[2], Murčo Obućina[2], and Ismar Hajro[2]

[1] Student of Faculty of Mechanical Engineering of University of Sarajevo, Vilsonovo šetalište 9, 71 000 Sarajevo, Bosnia and Herzegovina
amilabjelopoljak@yahoo.com
[2] Faculty of Mechanical Engineering of University of Sarajevo, Sarajevo, Bosnia and Herzegovina

Abstract. This paper presents results of experimental investigation of shear strength and breaking elongation of adhesive single-lap joints using aluminum specimens. It was carried out in order to understand effect of temperature on the joint strength, and to compare adhesive performance with manufacturer's specification. The adhered material used in tests was aluminum alloy 5754 in the form of thin sheets, bonded with two-component epoxy/metal adhesive intended for use with metals. This is easily weldable aluminum alloy, and usually that is the way of joining it. Therefore it is interesting to observe its behavior in case it is bonded. This is particularly interesting in cases where bonding is more viable solution than welding. Lap shear test was performed at room and elevated temperatures. It is shown that by increasing temperature shear strength significantly drops, while breaking elongation has almost the same value for all elevated temperatures.

Keywords: Single lap joints · Lap shear test · Two-component epoxy adhesive Temperature test

1 Introduction

Nowadays constructions, assemblies and parts in various industries require excellent properties regarding strength, toughness, corrosion and heat resistance. Moreover, it is required to have low carbon footprint, to be energy efficient and have acceptable price. Such requirements led to increased complexity of product geometry and usage of different materials to achieve single part. Such combinations include polymers or glass combined with metals, as well combinations of different metals. Used materials often cannot be joined by using conventional processes (e.g. welding). Also, welding usually cannot be used in case two metals have significantly different melting temperature. Still, it is required to have simple and economically viable ways to join and connect parts. Adhesive bonding is often used in such cases, thanks to advances in research and development of many kind of adhesives, technology and technique [1, 2].

© Springer Nature Switzerland AG 2019
S. Avdaković (Ed.): IAT 2018, LNNS 60, pp. 537–543, 2019.
https://doi.org/10.1007/978-3-030-02577-9_52

Adhesive bonding is very old joining technique, using natural adhesives. Industrial applications have been developed during late 18th century. Rapid increase of usage have begun after World War II, when synthetic adhesives were introduced. Today, adhesive bonding is widely used in automotive, aerospace and process industry. It is used not only for metals, but for materials that cannot (or hardly can) be welded, brazed or soldered (e.g. wood, stone, ceramics, glass, polymers). Adhesive joints can easily achieve required aesthetical and aerodynamic properties, and joining process is easy to automate or robotize. There are several other important advantages, including even stress distribution, excellent absorption of vibrations and, in most cases, low price [6–8].

Adhesive bonding is commonly done at low temperatures, rarely exceeding 100 °C. Metallurgically viewed, that is the most important advantage. Lack of heat input causes no change in microstructure and, consequently, properties of welded metals. That is why it found its way to be one of major joining techniques for aluminum alloys. Some of them can be problematic for welding (due to crack sensitivity), can require specific filler materials, pre-heating or post-weld heat treatment. It is also used in case of steel/aluminum combination, where alternative is solid state welding (e.g. friction stir or explosive welding) or mechanical joining (e.g. riveting) [7, 8].

However, bonded joints, as well as bonding process, have several disadvantages. Key ones are low operating temperature (usually not above 200 °C, due to significant drop of mechanical properties [1]), brittle behavior at low temperatures and reduced dynamic strength if adhesive layer is too thick. Usually, bonding requires excellent surface preparation, including degreasing by chemicals and mechanical roughening. Also, problem can be prolonged by time required for adhesive to be solidified [9–11].

2 Experimental Setup

Several examinations can be conducted over bonded joints, and several parameters can be considered for evaluation of bond properties. This paper covers basic one, lap shear test, and covers shear strength and strain of joint as parameters.

To evaluate bond properties, an experiment has been conducted. Specimens of aluminum 5754 (AlMg3, chemical composition given in Table 1) have been prepared and bonded with two-component epoxy/metal adhesive as adhesive single lap joints. This adhesive has been chosen among others because it is intended to be used with metals, specifically aluminum.

Table 1. Chemical composition of Al 5754 [3]

wt%								
Mg	Mn+Cr	Mn	Si	Fe	Cr	Zn	Ti	Cu
2.60–3.60	0.10–0.60	0.50	0.40	0.40	0.30	0.20	0.15	0.10

Specimens were cut from sheets to dimensions required by EN 1465:2010, as shown in Fig. 1.

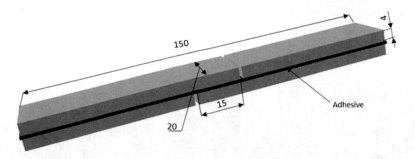

Fig. 1. Geometry and dimension of specimen [4]

Used adhesive is intended to be used for fast repairs of metallic parts and components (e.g. steel, cast iron and aluminum). It also can be successfully used for stone, house ceramics, wood, concrete and certain kinds of polymers (e.g. polyester, polyethylene and polypropylene). Examples of application includes cold-working tools, car parts, metal fences, housewares and radiators. After solidification, joint has metal color, resistance to (salt) water and chemicals, can withstand mild dynamic forces and vibrations and can be ground and painted. Most important, working temperature ranges from −40 °C up to +100 °C [5].

Adhesive manufacturer's specifications are given in Table 2.

Table 2. Adhesive specifications [5]

Tehnical properties	Tehnical specifications
- Extreme strength: up to 220 kg/cm^2	**Chemical base**: Resin: epoxy resin Hardener: modified amide
- Metal coloured	**Colour:** Grey
- Resistant to harsh conditions	**Viscosity:** Pasty
- Resistant to temperatures between −40 °C and +100 °C	**Solid contents**: approx. 100%find
- Filling	**Density:** approx. 1.2 g/cm^3
- (Sea) water resistant	
- Very good chemical resistance	
- Can be sanded, filed, drilled and painted after curing	

Prior to application of adhesive, specimens were mechanically cleaned, rinsed in warm water, cleaned with trichloroethylene, and carefully dried after one more rinse in warm water. No surface roughening has been applied.

Adhesive has been manually applied and evenly distributed over contact surfaces. Each joint has been made by using 1.0 g of bonding mixture made from two components in accordance to manufacturer's instructions. After that, each sample has been pressed by 300 N for 24 h to solidify.

Afterwards, examinations were done using Zwick tensile test device, as shown in Fig. 2. Test speed was 10 mm/min. Figure 2 also shows furnace used for heating up bonded specimens to testing temperature, i.e. 18 °C, 50 °C, 100 °C and 150 °C.

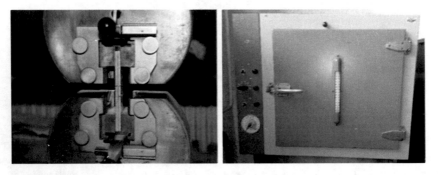

Fig. 2. Specimen in jaws (left) and furnace for heating up specimens (right)

Temperature was regulated by thermostat at furnace, and additionally controlled by thermocouples. Real temperatures achieved in furnace were:

- 48–56 °C for examination at 50 °C,
- 97–110 °C for examination 100 °C,
- 152–159 °C for examination at 150 °C.

Time between taking specimen out from furnace and starting test has been kept under 5 s. Tests were usually done in under 20 s. For every temperature, five specimens were tested.

Two parameters were observed during testing: shear strength and breaking elongation.

3   Results and Discussion

After experiments, it was possible to analyze results. Figure 3 shows change of shear strength as a function of temperature.

As expected, highest value of shear strength bond exhibits at room temperature. Manufacturer claims that this adhesive maintain its strength up to 100 °C. It is possible to see that is not the case, since it is almost halved at that temperature. However, it is interesting that bond still maintain relatively high strength even at 150 °C, temperature significantly above recommendations.

Figure 4 shows breaking elongation as a function of temperature. It is possible to note drop from 0.85% at room temperature to 0.55% at 50 °C. After that drop, breaking elongation has almost constant value. Although initial drop could be expected, it is unusual that further increase in temperature does not decrease breaking elongation.

Finally, a comparison between shear strength and breaking elongation at different temperatures has been made. This shows Fig. 5.

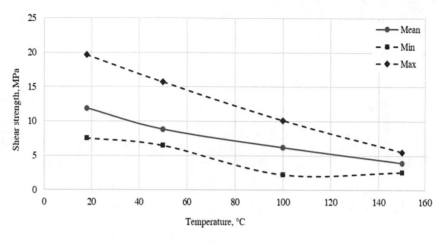

Fig. 3. Shear strength as function of temperature

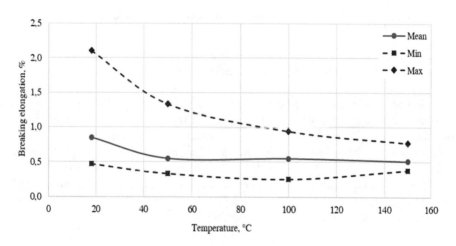

Fig. 4. Breaking elongation as function of temperature

As possible to see, this diagram shows combination of what those from Figs. 3 and 4 show. Increase in temperature leads to significant loss of strength, while breaking elongation stays almost the same at elevated temperatures.

4 Conclusion

One of widely used joining techniques for aluminum and its alloys is adhesive bonding. In comparison with other, it has numerous advantages, while still having just a few disadvantages. This paper presents results of lap shear test of specimens made of aluminum alloy 5754 bonded with adhesive, at room and elevated temperatures. This is generally easily weldable aluminum alloy, and therefore it is interesting to observe its

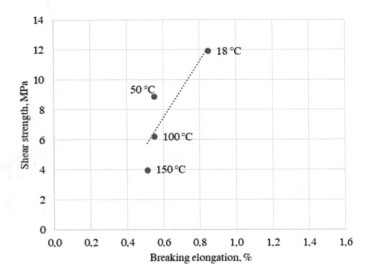

Fig. 5. Shear strength and breaking elongation as function of temperature

behavior in case it is bonded. For this purpose, two-component metal-epoxy adhesive has been used. It is claimed that it keeps key properties at elevated temperature (up to 100 °C).

Based on results presented here, it can be concluded that this adhesive does not perform strictly in accordance with manufacturer's specification. It has significant drop (two-fold) in strength at 100 °C. However, its breaking elongation has almost the same value of 0.5% at all elevated temperatures, after initial drop from 0.85% at room temperature.

Moreover, it is possible to conclude that this is adhesive usable at even 150 °C. Shear strength drops to 4 MPa (three-fold of that at room temperature), but breaking elongation stays the same as at 50 °C and 100 °C.

References

1. Kinloch, A.J.: Durability of Structural Adhesives. Applied Science Publishers, London (1983)
2. Petrie, E.M.: Adhesive bonding of aluminum alloys. Met. Finish. **105**(9), 49–56 (2007)
3. AlcoTec: AlcoTec Aluminum Technical Guide, Identification code ALC-10029B (2015)
4. International Organization for Standardization: Adhesives. Determination of tensile lap-shear strength of bonded assemblies. EN 1465 (2010)
5. Bison Epoxy Metal – Adhesive Specifications. https://www.bison.net/en/product.2267. Accessed 12 Apr 2018
6. Brockmann, W., Hennemann, O.-D., Kollek, H.: Surface properties and adhesion in bonding aluminium alloys by adhesives. Int. J. Adhes. Adhes. **2**, 33–40 (1982)
7. Johnsen, B.B.: Adhesive bonding of aluminium. Norwegian University of Science and Technology (2004)

8. Kinloch, A.J., Smart, N.R.: Bonding and failure mechanisms in aluminium alloy adhesive joints. J. Adhes. **12**, 23–35 (1981)
9. Müller, M., Valášek, P., Ruggiero, A., D'Amato, R.: Research on influence of loading speed of structural two-component epoxy adhesives on adhesive bond strength. Proc. Eng. **149**, 340–345 (2016)
10. Phung, L.H., Kleinert, H., Jansen, I., Häßler, R., Jähne, E.: Improvement in strength of the aluminium/epoxy bonding joint by modification of the interphase. React. Polym. **2003**(210), 349–358 (2004)
11. Bishopp, J.A., Sim, E.K., Thompson, G.E., Wood, G.C.: Important features in the adhesive bonding of aluminium. Int. J. Surf. Eng. Coat. **66**, 127–132 (1988)

Influence of Different Parameters on Mechanical Characteristics of Wood Welded Assemblies

Izet Horman[✉], Ibrahim Busuladžić, Senad Burak, and Ninoslav Beljak

Faculty of Mechanical Engineering, University of Sarajevo, Sarajevo, Bosnia and Herzegovina
horman@mef.unsa.ba

Abstract. The goal of this study was to test strength of joints for two lamellas, made of solid massive wood, rotary welded with dowels, respecting the change in direction of side penetration into the model, as well as the change of the tightness. Under the influence of friction and tightness the dowel wear occurs, where changing their geometry ultimately affect the strength.

The experiment was set to test the strength of welds in three series of samples prepared by models in which the two cases welding was done with constant tightness following the weld depth but with the change of the penetration direction into the elements, and one model that was subject to tightness change grading the bore of the weld depth.

Variable parameters such as; direction of dowel penetration through the elements in conjunction, tightness, direction of dowel penetration with respect to the line-grained (as incidental factor), species (only at the stage of selecting the material for the experiment) and wood humidity during testing (relative humidity of $50 \pm 5\%$) were adopted in the experiment.

1 Introduction

Welding wood was developed as a new way of joining wooden parts without the use of adhesives, with the help of heat generated by friction and pressure on the compound. Friction may be indicated by vibration or rotation. Welding wood is a process in which occur chemical-physical reactions. During the process, the surface layer of the wood elements in contact is melted, and it is initiated due to the influence of pressure and heat, which is usually generated by mutual friction under the influence of the elements being connected. In the interface area there is a complete destruction of the cellular structure of wood, and in the area of welding occurs a significant increase in bulk density of wood because the wood cells are completely destroyed. The walls of the cells burst and collapse due to the influence of heat and pressure that force a chemical reaction during the cooling of timber.

Based on previous research, optimal welding parameters were adopted for the experiment: Tightness 2 mm for dowels having diameter 10 mm [1, 2], the welding

© Springer Nature Switzerland AG 2019
S. Avdaković (Ed.): IAT 2018, LNNS 60, pp. 544–554, 2019.
https://doi.org/10.1007/978-3-030-02577-9_53

time is 3 to 4 s [3, 4] frequency, in the range of previous researches, which is conditioned by the parameters of drilling and pressure welding within 1.3 MPa [5].

2 The Relationship Between Bond Strength and Moisture Content

In order to select optimal models of welding, testing modes and type of woods, 13 samples are made and tested from Larch (Larix), Douglas-fir (Pseudotsuga) and exotics "Marblewood" (Marmaroxylon racemosum), using the dowel of beech (Fagus sylvatica). In this study, the special reception head is designed and manufactured for montage of samples on a universal testing machine.

Specimens have been made joining two elements of above-mentioned woods, having dimensions 200 × 35 × 21 mm, with four rotationally welded beech dowels. Cylindrical beech dowels are commercially procured, having following features: material beech class 1, diameter 10 mm, length 100 mm, grain direction parallel to the length.

To test the strength of the weld, three models of test samples are made:

- Model M1, having pre-drilled holes with a diameter of 8 mm and injecting dowels only from one side (Fig. 1(a)),
- Model M2, having pre-drilled holes with a diameter of 8 mm on the first transmission element and 6.7 mm at the second coupling means, with one-sided embossing of the dowel (Fig. 1(b)),
- Model M3, having pre-drilled holes with a diameter of 8 mm and with two-sided embossing of the dowel, two on each side of the sample (alternately) (Fig. 1(c)).

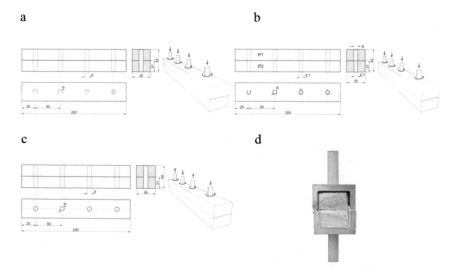

Fig. 1. (a) Model M1, (b) Model M2, (c) Model M3, (d) Receiving head

For model fixture in the receiving head according to EN 319:1996, it was necessary to develop special metal plates to replace the test blocks. Special backing head was designed and manufactured to place the test specimens in the receiving head Fig. 1(d).

Before testing, measurement of moisture content in wood was performed for each sample, using digital hygrometer. Results of measurement of moisture content for all samples have ranged from 9.0 to 14.1%. Measurements of force and displacement were carried out with the help of computer. Testing was performed under displacement of 5 mm/min.

The main goal of the experiment was to test the bond strength of welded rotational wooden dowels into the substrate by applying the change in the direction of penetration into the substrate and changing the profile of holes. As the accompanying results of experimental research, the data on the content of moisture during the test of pulling force are recorded. Before testing, for each sample humidity was measured. Despite the fact that after the welding, samples are conditioned under certain climatic conditions (temperature 20° ± 2°C, relative humidity of 50 ± 5%), the results of moisture content measurements in the wood samples were ranged from 9.0 to 14.1%. Reason for occurrence of dispersion for humidity results may be requested in the tolerance of the relative humidity during time past under the conditions of relative humidity of 50 ± 5%, and as well as in the manner of the "group" deposit.

Unlike, vibratory welding where increasing humidity wood has a negative impact on the bond strength, the results of this research showed that at the rotary welding, humidity positively affect the bond strength. Tables 1, 2, 3 and 4 shows the comparative arithmetic means ratio of measurement of pulling force and strength to the layering, grouped by grade of 1% of the difference in wood humidity.

Table 1. The results of the impact of wood humidity on pulling force and strength (model M1)

Model 1 Humidity		Number of samples	Pulling force (kN)	Difference of force	Shear strength (MPa)	Difference of strength
9.5–10.4	%	8	3.9205	6.52%	0.94255555	3.95%
10.5–11.4	%	9	3.3963	19.02%	0.78860934	19.64%
11.5–12.4	%	8	3.9241	6.43%	0.90161874	8.12%
12.5–13.7	%	5	4.1938		0.98132577	

Table 2. The results of the impact of wood humidity on pulling force and strength (model M2)

Model 2 Humidity		Number of samples	Pulling force (kN)	Difference of force	Shear strength (MPa)	Difference of strength
9.0–10.0	%	6	4.8767	12.62%	0.126868282	8.98%
10.1–11.0	%	7	5.0513	9.49%	0.130505166	6.37%
11.1–12.0	%	7	5.3001	5.03%	0.135961158	2.45%
12.1–13.0	%	7	5.5238	1.02%	0.140301132	−0.66%
13.1–14.1	%	3	5.5810		0.139377844	

Table 3. The results of the impact of wood humidity on pulling force and strength (model M3)

Model 3 Humidity		Number of samples	Pulling force (kN)	Difference of force	Shear strength (MPa)	Difference of strength
9.4–10.3	%	6	3.9115	18.18%	0.90830098	19.36%
10.4–11.3	%	7	4.2384	11.34%	0.97980018	13.01%
11.4–12.3	%	11	4.2957	10.14%	0.96630087	14.21%
12.4–13.3	%	6	4.7803		1.12632190	

Table 4. The results of the impact of wood humidity on pulling force and strength (all models)

All models Humidity		Number of samples	Pulling force (kN)	Difference of force	Shear strength (MPa)	Difference of strength
9.0–10.0	%	16	4.2646	11.47%	1.05233956	9.45%
10.0–11.0	%	23	4.2148	12.51%	1.01454784	12.70%
11.0–12.0	%	24	4.6205	4.08%	1.10583681	4.85%
12.0-13.0	%	19	4.5219	6.13%	1.09063430	6.15%
13.0–14.0	%	8	4.8173		1.16216267	

According to the survey, almost all compounds that are realized in wood with a higher percentage of humidity, there was a noticeable increase in pulling force. For example, for a given model M1 (Table 1), the results obtained for samples having humidity class from 9.5% to 10.4 are 3.95% lower compared to the results obtained in the humidity class from 12.5 to 13.7%. Only for model M1was recorded a drop of pulling forces increasing humidity between 10.5 and 11.4%, which is reflected in the review and grouped for all models (Table 4). The reason for the appearance of these oscillations in results is not considered during the experiment, but it may be due to poor weld of some dowels or anatomical defects at the position of the weld.

It is important to notice that it is much more appropriate to compare the results for the shear strength than for the pulling forces, for the reason that the surface of the weld for model M2 is lower than for the M1 and M3 model. This is due to the different radius of holes for the model M2 and taking in account that the surface in contact between dowels and holes is necessary to calculate the shear strength. After confirming an overview of results for the class of humidity from 12.1 to 13.0% (Table 2), pulling force is 1.02% lower than the average for the sample with humidity from 13.1 to 14.1%, while the strength of stratification is increased by 0.66%.

The reason for increasing the bond strength with the increase of humidity can be reflected in the appearance of swelling and shrinkage, whereby a dimensional distortion of the anatomical parts of the wood, which is reflected in the expansion of wood fibre, cell walls and lumens.

Theoretically, in the case of rotation of welded dowels, increase of humidity causes the appearance of shrinkage (decrease) of hole in which the dowel is welded, on the other hand cross-section of dowels increases (swelling). Since the wood is anisotropic material, it reflects different behaviours and characteristics in different directions. This is reflected in the occurrence of swelling which results in different values observed in the longitudinal, tangential and the radial/cross direction. The Table 5 shows values of the coefficient of swelling of conifers and deciduous following the directions.

Table 5. Average values of the coefficient of swelling for humidity change of 1%

	Radial (αr)	Tangential (αt)	Longitudinal (αl)
Coniferous	0.12%	0.24%	0.01%
Deciduous	0.20%	0.40%	0.01%

Considering the different values of coefficients of swelling following different orientation of fibers in dowels and lamellas causes increase the bond strength with increase of moisture content.

3   Strength of Welded Joint

Comparing the results with the help of the measured pulling forces for model M2 with the transition profile holes from 8 to 6.7 mm through the base one can see that the pulling forces is 27.27% lower for model M1 and 17.88% in relation to the model M3 (Table 6), but it is not fully justified considering that the welding dowels reduce volume and surface welds.

Table 6. Comparative ratio of the pulling force arithmetic mean for M1, M2 and M3

Model	Pulling force (kN)	Ratio to M2
M1	3.8098	27.27% < M2
M2	5.2385	
M3	4.3024	17.88% < M2

From the scientific point of view, it is much more appropriate to compare this to the welded joint strength of the layering. The average area of one dowel welded to a depth of 42 mm through the bore of 8 mm is 1076 mm^2, and dowels with changing bore profile with equal depth have smaller surface and is equal to 975 mm^2. Surface of welded joints, through the models M1 and M3, was 9.4% higher compared to the model M2 and as strength of welded joints is ratio of force per unit area, it follows that for the same values of pulling force strength of welded joints is growing in favour of welding. The average amount of strength of the welded joint with four dowels for model M1 is 0.8919 MPa, for model M3 is 0.9899 MPa and for the model M2 is 1.3427 MPa (Table 7).

Table 7. The comparative strength of the delamination for M1, M2 and M3

Model	Strength of the delamination (MPa)	Ratio to M2
M1	0.8919	33.57% < M2
M2	1.3427	
M3	0.9899	26.28% < M2

Comparing the results of the analysis for pulling forces and strength of the delamination, it can be concluded that the delamination strength shows 6.3% and 8.4% higher values of welded model M2 with respect to M1 and M3.

4   Statistical Analysis of the Results of Pulling Force and Strength - Normality Test of Data Distribution

As a first and basic step of any statistical data processing, it was carried out to check whether the distribution of the results obtained in the experimental section corresponds to the normal distribution.

Normality checking was done on the basis of:

- subjective estimates, based on the histogram graph (Figs. 3 and 4) and through visual checks how many results follow a normal distribution "normal - plot" (Fig. 2)
- Check it with the help of K-S and SW tests.

We analysed the results of pulling force and the results of the strength calculation of the stratification. For statistical analysis and data processing we used SPSS Statistic and MedCalc.

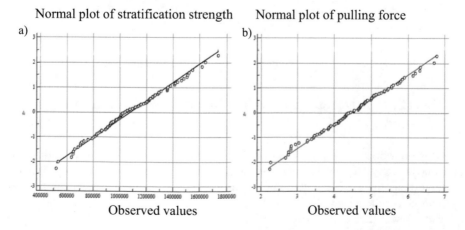

Normal plot of stratification strength   Normal plot of pulling force

a)   b)

Observed values   Observed values

Fig. 2. Normal-plot diagrams for the results of the stratification strength (a) and the pulling force (b)

- KS - Kolmogorov-Smirnov or K-S test; statistical test normality distribution
- SW - Shapiro-Wilk test; statistical test normality distribution
- SPSS Statistic - programming software for statistical analysis of data, IBM Corporation.

As a test to verify the normality of distribution take over the K-S and SW tests, which essentially represent a preliminary analysis of the data for the final selection of the type of analysis to be used in further work. The goal of this analysis is to determine the behaviour of distribution, because if the distribution turns out to be normal then it will be applied parametric statistics and if it proves that it is not normal it will apply nonparametric statistics.

Table on Fig. 3 represents the analysis of the results for force with the help of the program MedCalc, conducted under the KS test, which shows the acceptance of the normal distribution, actually the fact that the results do not differ significantly from normal. Also in Fig. 3 the histogram graph is shown and results obtained show that distribution of the results do not differ significantly and it follows a Gaussian curve of the interval estimation confidence.

Analogous to Figs. 3 and 4 shows the data analysis according to the calculated strength of the layering.

Variable	SILA_M123 SILA_M123
Sample size	90
Lowest value	2,2370
Highest value	6,7860
Arithmetic mean	4,4502
95% CI for the mean	4,2401 to 4,6604
Median	4,4110
95% CI for the median	4,2222 to 4,7130
Variance	1,0070
Standard deviation	1,0035
Relative standard deviation	0,2255 (22,55%)
Standard error of the mean	0,1058
Coefficient of Skewness	0,02604 (P=0,9155)
Coefficient of Kurtosis	-0,2708 (P=0,6539)
Kolmogorov-Smirnov test[a] for Normal distribution	D=0,0358 accept Normality (P>0.10)

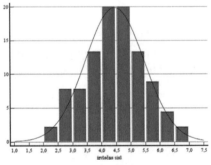

Fig. 3. Analysis of the distribution behavior of pulling force data

5 Direction of Dowel Penetration Relative to Orientation of Wood Fibers

Wood is orthotropic material. As in this case reflects the fact that the strength in one direction substantially greater than the strength in other directions. Tensile strength is the largest parallel to fibres but in this direction the shear strength is less than the shear strength perpendicular to the fibers. Tensile strength perpendicular to the fiber is significantly lower than that of parallel fibres [6].

Variable	ČVRSTOĆA_M123
	ČVRSTOĆA M123
Sample size	90
Lowest value	518712.4200
Highest value	1735756.0100
Arithmetic mean	1074794.1080
95% CI for the mean	1018540.8457 to 1131047.3703
Median	1037641.7600
95% CI for the median	992562.9800 to 1139230.1727
Variance	72135909821.1111
Standard deviation	268581.2909
Relative standard deviation	0.2499 (24.99%)
Standard error of the mean	28310.9539
Coefficient of Skewness	0.2039 (P=0.4096)
Coefficient of Kurtosis	-0.4598 (P=0.3159)
Kolmogorov-Smirnov test[a] for Normal distribution	D=0.0699
	accept Normality (P>0.10)

Fig. 4. Analysis of the distribution behaviour of stratification strength data

If the load acts at an angle to the direction of propagation of the fibres, the properties of wood can be determined by the general Hankinson's formula [7]:

$$N = \frac{PQ}{P\sin^n\vartheta + Q\cos^n\vartheta} \quad (MPa) \tag{1}$$

Where N is the strength at an angle ϑ to the direction of the fibers, Q strength across the P strength parallel to the direction of providing fiber, "n" is an empirically determined value [7].

The differences in the anatomical and physical structure of wood certainly affect the wood welding process, but it can be assumed that the chemical, anatomical and physical properties of the same type of wood influence the strength of welded joints. Within the same ring width there are variations in the proportion of early and late wood.

The same ring width and a higher proportion of late wood are a precondition for increasing the wood density, but it can cause an increase in strength of welded joints. According to investigations [8], with an increase in the wood density embedded force has increased. In this research, beech dowels are welded in Douglas fir wood perpendicular to the wood fibres (radial-tangential texture), and at different angles of penetration to the lines of rings (Fig. 5).

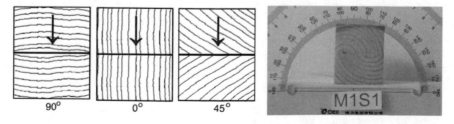

Fig. 5. Directions of dowel penetration to the lines of rings

6 Comparison of the Data in Relation to the Influence of Humidity Content and Angle of Penetration of the Dowel According to the Orientation of Growth Rings

Guided by the fact that the results of the research showed that the adhesion effect has made a comparative analysis of the behavioural results of two factors; on the one hand the direction of dowel penetration into the wood elements according to the orientation of growth rings and on the other hand the percentage of moisture in the wood. Table 8 gives an exact number of samples analyzed and grouped by grade humidity in the range of 1% and the corner of penetration and Fig. 6 shows a block diagram of the results obtained by statistical analysis of data.

Table 8. Number of test samples grouped by the angle of penetration and the wood moisture

Humidity class	Penetration angle			Total
	0°	45°	90°	
9–10%	6	5	5	16
10–11%	5	6	12	23
11–12%	3	11	10	24
12–13%	7	4	8	19
13–14%	1	3	4	8
Total	22	29	39	90

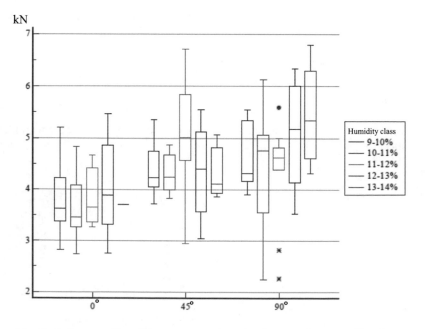

Fig. 6. Influence of humidity content and angle of penetration on pulling force

The diagram shows a noticeable increase of the pulling force value increasing the angle of penetration to the orientation of growth rings, but also a tendency to increase the force with the increase of moisture in the wood considering subgroups (class) of humidity. Since these are different numbers of samples in groups and subgroups, which are derived from a variety of models, analysis of experimental results can be regarded only such as observation analysis.

Also on the graph it is possible to notice that the wood fibers are not only broken off from the structure of wood and mixed with the melt, but also come up with their partial separation and blending in line welds binding so wood structure with melted material.

7 Conclusion

Depletion of cap by passing through the binding elements is changed and its geometry with the transition from the cylinder to the final shape like a truncated cone. It is much more appropriate to compare the strength of welded joints according to the delamination. It means that the compounds provided by the models M2 to delamination are more solid for 33.57% compared to M1, respectively, 26.28% compared to M3. Percentage differences that occur between the comparisons based on the force and on the strength, as a result of cap depletion, which results in reduced welding surface.

The angle of the dowel penetration by rotary welding in relation to the lines of rings and the direction of the wood strips is also one of the influential factors on the ultimate strength.

By treating the data, regarding the angle of penetration of the dowel versus the line of rings and dividing all experimental groups into three of them, results were obtained that the angular contact penetration of 45° yields that the compounds are 14.36% firmer in relation to the intersection angle of 90° or 3.5% in relation to the angle of penetration of 0°. With the use of the same kind of wood, this can be explained by the different percentages of early and late wood, i.e. the width of the rings.

The influence of humidity has also a significant impact on rotational strength of welded joints. Experimental results show that adhesion increases with increasing wood moisture content, so that for the observed interval from 9.0 to 14% moisture content, samples with higher moisture content achieve 12.7% higher bond strength.

The reason for increasing the bond strength, with the increase in humidity, may be sought in phenomena of swelling whereupon occur dimensional distortions of anatomical parts of wood. In this case, shrinkage of bore of its elements and swelling cap reinforces bracing.

To define the optimal set of welding parameters, anatomical characteristics of wood or at least for the default samples, including their density, the width of the ring and the cross angle should be taken into account.

References

1. Župčić, I., Mihulja, G., Govorčin, S., Bogner, A., Grbac, I.: Zavarivanje termički modificirane grabovine. Drvna Industrija **60**(3), 161–166 (2009)
2. Stamm, B.: Friction Welding of Wood. https://infoscience.epfl.ch/record/166109/files/150_c_01_COST_FP0904_2010_12_03.pdf. Accessed 15 June 2015
3. Ganne-Chedéville, C.: Soudage linéaire du bois: étude et compréhension des modifications physico-chimiques et développement d'une technologie d'assemblage innovante. Ph.D. dissertation, Faculté des Sciences et Techniques Nancy, Epinal, France (2008)
4. Ganne-Chédeville, C.: Predicting the thermal behaviour of wood during linear welding using the finite element method. J. Adhes. Sci. Technol. **22**, 1209–1221 (2008)
5. Rhême, M.: Strength and fracture characterization of welded wood joints - effects of moisture and mixed mode loadings. Ph.D. dissertation, École Polytechnique Fédérale de Lausanne, Lausanne, Suisse (2014)
6. Grubišić, I.: Konstrukcije malih brodova, Fakultet strojarstva i brodogradnje, Sveučilište u Zagrebu; poglavlje 222 Tehničke osobine drva (2005). https://www.fsb.unizg.hr/kmb/
7. Wood Handbook Wood as an Engineering Material. General Technical reports, FPL–GTR–113. U.S. Department of Agriculture, Madison, WI (1999). http://www.fpl.fs.fed.us/documnts/fplgtr/fplgtr113/fplgtr113.pdf
8. Pizzi, A.: Linear and high speed rotational wood welding-wood furniture and wood structures. In: Practical Solutions for Furniture and Structural Bonding, International Workshop Larnaka – Cyprus, 22–23 March 2007, pp. 25–38 (2007)

Modeling and Remodeling of PC Steam Boiler Furnace on the Basis of Working and Simulated Operating Parameters

Midhat Osmić[1], Izudin Delić[1(✉)], Amel Mešić[2],
and Nedim Ganibegović[2]

[1] Faculty of Mechanical Engineering, University of Tuzla,
Tuzla, Bosnia and Herzegovina
izudin.delic@untz.ba
[2] PE Elektroprivreda B&H, Sarajevo, Bosnia and Herzegovina

Abstract. Analysis of the operation of boiler plants showed significant deviations of the characteristic operating indicators obtained by the exploitation measurements compared to their design values. In order to reduce the variation of process parameters during the exploitation, in relation to their design values, a selection of relevant process parameters was made, and their impact on the constructional characteristics of the furnace and vice versa were analyzed. It is important to note that a large number of plants never reach the project defined operational parameters in exploitation, so in order to determine the interconnections between the constructional and energy characteristics of the steam boiler furnace, the design and simulation (off design) models were generated. The "design" of the simulation model shows the changes of the selected constructional parameters of the steam boiler furnace depending on the energy-process parameters, while the "off-design" of the simulation model shows a change energy-process parameters. By establishing these relationships, assumptions would be made for the use of the obtained data when modeling new and, to a certain extent, the reconstruction of existing plants.

Keywords: Steam boiler · Geometrical characteristics of the furnace
Combustion · Modeling and simulation

1 Introduction

When analyzing the process in the steam boiler furnace, a number of assumptions are introduced which facilitate the mathematical description of these processes. However, practice has shown that such an approach also leads to the appearance of certain deviations between the parameters obtained by these calculations and those parameters obtained in the exploitation itself [1, 2].

The steam boiler analysis approach is carried out in such a way that the plant monitoring does not merely present the measured data, but also selecting them in the set of important information. The purpose of power plant monitoring with parameter selection is in fact a measurement procedure with the possibility of reducing the amount of information using the appropriate system models, as indicated in the

© Springer Nature Switzerland AG 2019
S. Avdaković (Ed.): IAT 2018, LNNS 60, pp. 555–563, 2019.
https://doi.org/10.1007/978-3-030-02577-9_54

literature [3–5]. This issue represents the subject of interest of scientific institutions and experts in the field of thermoenergetics. Special attention is paid to the analysis of combustion inside the steam boiler using mathematical models and other methods and techniques. Nowadays, CFD as a tool for analyzing and simulating the process in the boiler, is applied a lot. It was used to simulate the effects of different tangential arrangements of burners on the performance of a 600 MW utility boiler to find the best arrangement of the burners [6]. Also, it was applied in the case of evaluation of a co-firing concept with the use of pre-dried lignite as supporting fuel, especially under relatively low thermal loads [7]. Also, tangentially fired furnaces with three different burner arrangements have been numerical simulated and analyzed with regard to their gas flow, temperature profile, and heat transfer [6]. Another approach is known as the zone method. The zone method is one of the most accurate methods in the simulation of heat transfer by radiation within industrial furnaces, but it can not be applied to all furnaces. Complex geometries of real furnaces should be replaced by simpler forms that better suit the zone method. This method needs the high power of numerical calculations and finding the best size for zones is one of the major criteria in this method [8]. On the other hand, Gate Cycle as the software includes detailed design and off-design analysis routines, to apply the program through various stages of plant design and analysis such as advanced systems studies, preliminary screening analyses, initial conceptual design and detailed engineering design [3]. Therefore, the geometry of the steam boiler is defined during the design of the plant and cannot be influenced during its exploitation. Through this work, it was attempted to show the connection of the energy and constructional parameters of the steam boiler, in order to indicate significant deviations of the exploitation parameters in relation to their expected and projected values. The steam boiler type OP 650 was taken as a research basis. [1, 9]. It is important to emphasize that access to analysis and monitoring of steam boiler operation is carried out so that it is not only excluded from displaying measured data, but also their grouping according to the degree of dominance. Of course, the purpose of such a grouping is to isolate and reduce the amount of relevant information used in the modeling process, as stated in the literature itself [3–5].

1.1   System Under Consideration – Boiler OP 650

The steam boiler OP 650 in TE Tuzla is a radiant with natural circulation of water. The boiler is made of two flow lines and it consists of: the chamber (part one) and the convective part of the boiler (part two). In the boiler, water heater, evaporative system, four primary steam preheating stage, two secondary steam preheating stage, primary and secondary steam collector system are located. Reconstruction of the thermo-block resulted with a boiler, which has a different combustion concept. In the reconstruction, the existing 8 mills were retained and instead of the classic system, a modern system was established in which mills provide coal dust to low NOx jet burners with a significantly higher number of metering equipment and extended burning along the furnace. The burners are arranged tangentially on the circumference of an oscillating fire chamber, resulting in the formation of a concentric O flame. Steam boiler OP650 burns pulverized coal mixture of lignite and brown coal (in the ratio of 70:30) with guaranteed quality (a = 24%; w = 40%; s = 0.8%; Hd = 900 kcal/kg).

2   Creating Simulation Model

The experimental data used to set simulation models are partly taken from the normative tests of the steam boiler system, and a part of the data is taken from the measuring equipment installed on the steam boilers, a detailed data classification is carried out in [1] and [9]. Simulation models of the combustion process in the defined steam boiler furnace are created in the commercial software package Gate Cycle. Gate Cycle is used both for designing and performance evaluation of thermal power systems in the project, as well as in the exploitation phase.

A GateCycle model represents a specific plant or equipment configuration. In design mode, the user specifies the performance attributes required. The software then calculates ("sizes") the equipment to match these performance criteria. In off-design mode, the software works in the other direction: the user defines operational conditions, and the GateCycle application calculates the corresponding "as-built" performance. GateCycle models are flexible, allowing an indefinite number of calculation cases to cover variations in design parameters as well as plant performance under "off-design" conditions. When the model is executed, a sequential modular algorithm solves the mass and energy balances for both the overall system as well as all individual components, and detailed reports for each are generated automatically. GateCycle software uses detailed analytical models of thermodynamic processes in thermal power plants, as well as processes of heat transfer and flow processes. To determine the convergence of the calculation, the software uses the following criteria: the calculated output quantities from each model element must correspond to the values from the previous iteration; mass and energy balance must be satisfied both for each element of the model and for the whole system; the output data from a particular model element must be equal to the input data in the following element; the standard model set in Gate Cycle usually converts after performing from 2 to 40 iterations [3].

Figure 1 shows an example of the generated steam boiler plant model in the user interface of the Gate Cycle software, and according to [3] and [9]. The created model has a basic, project version as well as a set of exploitation versions. The project version of the model defines physical quantities, and "non-project" versions of the model are used to analyze the performance of the plant in the event of a change in working conditions. As a basis for modeling and verifying simulation models, data obtained by the normative tests of the steam boiler plant were used. Normative testing is a procedure that verifies the warranty values which contracting authority provides from the equipment manufacturer. By carrying out the tests, the actual state of the equipment as well as all operating parameters is determined. Normative tests are performed with measuring instruments that are more accurate than those used for standard monitoring of the steam boiler plant. On the basis of these normative values, the simulation "design model" was set up, and according to [1, 3, 9]. The maximum error of the simulation model for the observed case was 9.2%, which was considered accurate enough for the model to be verified and used as such in further research [1, 9, 10].

After the model verification was carried out, the modeling of the process with the data obtained by measuring, i.e. creating the "design" model. After creating the "design" model, "off design" models were created. The normative simulation model of the

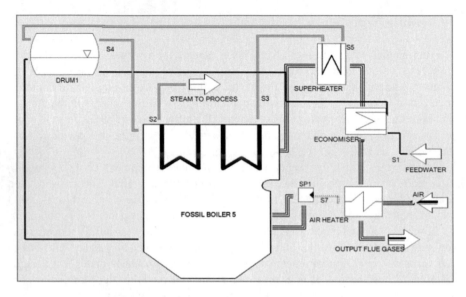

Fig. 1. Display of the design model of the boiler

steam boiler furnace was used as the basis for the creation of the "off design" model. Here, the geometry is fixed so that on the basis of the change of the energy-process parameters, the influence of the aforementioned on the constructional values would be determined.

3 Selection of the Influential Operating Parameters on the Steam Boiler Furnace

In order to isolate the most influential exploitation parameters on the processes that take place in the steam boiler furnace, an analysis of all relevant process parameters was performed. Using the simulation model, specially designed for this analysis, it was noted that two parameters have the main influence on the process in the steam boiler furnace: the coal amount in the combustion process and the flue gases temperature at the end of the furnace. Table 1 shows the values of these parameters as well as a range of their change.

Table 1. The interval of the observed parameters

Parameter	Minimum value	Maximum value	Interval
B_g	35 [kg/s]	55 [kg/s]	5 [kg/s]
T_{dp}	800 [°C]	1200 [°C]	100 [°C]

The value of the coal amount for combustion was set as variable in the existing simulation models in the range of 35–55 kg/s, while the other parameters were constant without changes. In this way, 150 simulation models were generated.

The next step performed was the analysis of the impact of the furnace shape factor on the constructional parameters. The furnace shape factor (F) is the ratio of the height and depth of the steam boiler furnace. Constructional parameters whose change was analyzed in the case of the process parameters change represent the furnace volume, the surface of the furnace walls, and the surface of the furnace cross-section.

3.1   The Influence of the Flue Gases Temperature

The chart in Fig. 2 shows the change in the volume of the furnace, depending on the change in the flue gas temperature, and for different values of the furnace shape factor. The character of the furnace volume change is exponential. By increasing the flue gas temperature, as well as by increasing the ratio of the height and depth of the furnace, the apparently required flue volume decreases. The deviation in the simulated cases is greater at lower flue gas temperatures, while at the flue gases temperature above 1200 [°C] this deviation is reduced to a minimum.

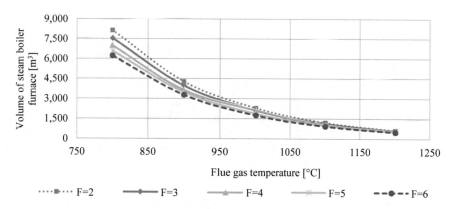

Fig. 2. The change in the volume of the steam boiler furnace in the function of the flue gases temperature for different values of the furnace shape factor.

Figure 3 shows the change in the furnace cross-section surface in the function of the flue gases temperature for different values of the furnace shape factor. The diagram of the change was created for the same temperature range. The deviation between the curves indicating the nature of the furnace area change in the function of the furnace temperature is more expressed than in the previous case. It can be concluded that the temperature influence on the furnace cross-section area is greater than on the volume of the steam boiler furnace.

The character of furnace walls surface change in the function of the flue gases temperature is shown in Fig. 4. It is obvious that the surface of the furnace walls has almost no changes, regardless the size of the furnace shape factor.

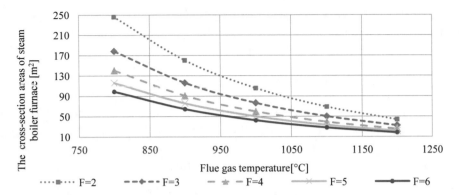

Fig. 3. The change in the cross-section area of the steam boiler furnace in the function of the flue gases temperature for different values of the furnace shape factor.

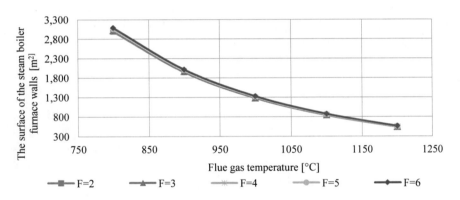

Fig. 4. The change in the surface of the steam boiler furnace walls in the function of the flue gases temperature for different values of the furnace shape factor.

The change in the size of the furnace walls surface is solely for the purpose of changing the flue gases temperature at the outlet from the furnace, which can be seen from the relation, that with the rise in the flue gases temperature the walls surface occupied the value of approximately 3000 m^2 at 800 °C, i.e. up to 500 m^2 at 1200 °C.

3.2   The Influence of the Coal Amount

As previously mentioned, another parameter whose impact will be analyzed is the coal amount coal that is fed into the boiler furnace. Figure 5 shows the character of the change in the furnace volume in the function of the coal amount at the steam boiler inlet for different factors of the steam boiler furnace. The observed range of change in the coal amount is 35–55 kg/s. According to the change character, it is concluded that this is a linear change in the constructional parameters of the furnace in the function of the coal amount at the inlet of the steam boiler. Also, there is little change in the furnace volume in the function of the furnace shape factor. It can be noticed that with the increase in the coal amount also increases the influence of the furnace factor.

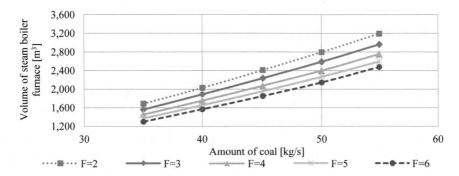

Fig. 5. The change in the volume of the steam boiler furnace in the function of the coal amount at the inlet of the steal boiler for different values of the furnace shape factor.

A similar situation is with the furnace cross-section area whose change in relation to the coal amount is shown in Fig. 6. It is a linear change, where the increase in the furnace shape factor increases the cross-section area.

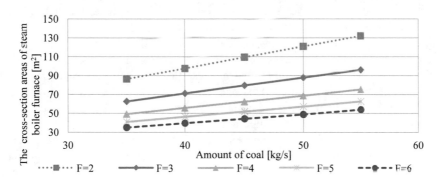

Fig. 6. The change in the cross-section area of the furnace in the function of the coal amount at the inlet to the steam boiler for different values of the furnace shape factor.

Thus, for the ratio 2, the cross-sectional area, in the case of a change in the coal amount from 35 to 55 kg/s, corrects by ~ 44 m^2, while, for example, for the ratio 6 and the same interval of change in the coal amount, the cross-section area is corrected by ~ 19 m^2.

The furnace factor influence on the surface of the steam boiler furnace walls at the increase of the coal amount is negligible. This is shown in Fig. 7. It can be concluded that the influence of the furnace shape factor in this case can be ignored.

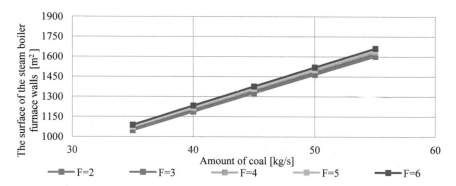

Fig. 7. The change in the surface of the steam boiler furnace walls in the function of the coal amount at the inlet to the steam boiler for different values of the furnace shape factor

4 Conclusion

The conducted research showed that the flue gas temperature at the end of the furnace as well as the coal amount at the steam boiler inlet have the greatest influence on the constructional parameters of the steam boiler. The basic geometric characteristics of the furnaces analyzed in this case are: the volume of the furnace, the surface of the furnace walls and the area of its cross-section. The previous analysis showed that the flue gases temperature at the end of the furnace is a very sensitive parameter that has a very significant influence on the constructional characteristics. The values of the flue gas temperature that are higher than the normatively declared given more precise results, and the values of the constructional parameters were close to the project values. However, for flue gas temperatures that are less than normative, a sharp rise in the parameters monitored is recorded. This phenomenon is not expressed when analyzing the influence of coal. In this case, the observed parameters had a linear change, but with a minimum and maximum value range that was significantly lower.

From the aspect of the flue gas temperature limit (AFT), it can be concluded that the furnace factor has almost no effect on the volume of the furnace. The furnace geometry is designed for a certain quality of coal and defined operating modes. For the reliability and efficiency of the boiler plant for coal, it is necessary to keep the input fuel parameters (coal) within the limits of the recommended values. In order to transform the input energy of fuel into heat through the combustion process and usefully used in the boiler, it is necessary to have a harmonized boiler furnace construction with a chemical composition of coal. Changing the quality of coal is a problem in the exploitation of the plant.

It is important to emphasize that results, provided by the Gate Cycle Software, are based on simplified combustion models assessed at stehiometric conditions and due to the fact that furnace has complex geometry it is recommended that this software is used in combination with CFD, for achieving the clearer picture of processes in the same furnace.

Therefore, it can be concluded that the results of this analysis can be used when designing new ones, and to a certain extent in the revitalization/reconstruction of existing plants, taking into account the above limitations.

References

1. Osmić, M.: Influence of the constructional characteristics of steam boiler combustion chamber on the combustion process of coal dust. Doctoral thesis, Mechanical Faculty of Tuzla (2016)
2. Gay, R.R., Palmer, C.A., Erbes, M.R.: Power Plant Performance Monitoring. R-Squared Publishing, Woodland (2004)
3. GE Energy: GateCycle^{TM}, Getting started & installation guide, SAD (2006)
4. Gulič, M., Brkić, L., Perunović, P.: Parni kotlovi. Mašinski fakultet Beograd (1988)
5. Brkić, L., Živanović, T.: Parnikotlovi. Mašinskifakultet Beograd (1997)
6. Yanqing, N., Xing, L., Yiming, Z., Houzhang, T., Xuebin, W.: Combustion characterisation of a four-wall tangential firing pulverized coal furnace. Appl. Therm. Eng. **90**, 471–477 (2015)
7. Khaldi, N., Chouari, Y., Mhiri, H., Bournot, P.: CFD investigation on the flow and combustion in a 300 MWe tangentially fired pulverized-coal furnace. Heat Mass Transf. **52** (9), 1881–1890 (2016)
8. Kakaras, E., Koumanakos, A., Doukelis, A., Giannakopoulos, D., Vorrias, I.: Oxy fuel boiler design in a lignite-fired power plant. In: The 6th European Conference on Coal Research and its Applications, vol. 86, no. 14, pp. 2144–2150. Elsevier (2007)
9. Osmić, M., Delalić, S., Buljubašić, I. Analysis of the impact of steam boiler furnace operation parameters on its geometric characteristics. Bosanskohercegovačka elektrotehnika J. (11) (2017). ISSN 1512-5483
10. Osmić, M., Delalić, S., Kudumović, D., Buljubašić, I.: Influence input parameters of combustion process on Output performance of steam boilers. TTEM-Technics Technol. Educ. Manag. J. **9**(2), 386–390 (2014)

Pulse Combustion Burner As Tool
For Increasing The Energy Efficiency

N. Hodžic$^{(\boxtimes)}$, S. Metovic, and S. Delic

Faculty of Mechanical Engineering Sarajevo, University of Sarajevo,
Vilsonovo Setaliste 9, 71000 Sarajevo, Bosnia and Herzegovina
hodzic@mef.unsa.ba

Abstract. In the revitalization and reconstruction of existing, and in particular in the design and construction of new industrial and thermopower boilers, the ever increasing demands on achieving higher energy efficiency, operating and time readiness, and the reduction of the emission of polluting components of flue gas into the environment have to be considered. These goals are far more difficult to reach for boilers in which as the primary fuel low-value, weakly reactive coal prone to fouling boiler heating surfaces. In connection with this, at the Faculty of Mechanical Engineering in Sarajevo, as part of the Pulse Combustion Laboratory, extensive research on the characteristics of combustion of gaseous fuel using pulse combustion burner, was carried out. Here is considered the application of a simple and robust water-cooled burner for pulsating combustion of a modular type, with aerodynamic valves and no moving parts. The burner can be used as a basic or auxiliary burner in a boiler combustion chamber to increase the turbulization of the combustion atmosphere and thereby to increase the efficiency of the chemical energy conversion from the fuel to the heat. In addition, the burner can also be used as a device for generating pressure waves and corresponding sound energy in the combustion chamber zones and/or flue gas ducts with the aim to remove ash and slag deposits from boiler heating surfaces - such application would contribute to more intense heat transfer in the boiler, decreasing fuel consumption, increasing plant operational and time readiness and reducing the emission of polluting components of flue gas to the environment. The paper presents the results of the laboratory research of the dependence of some of the operational performances of the geometry to the conditions of the burner. An analysis of the possible application of this burner for the cleaning of the heating surfaces was carried out on the basis of the results obtained on the boiler model. In this sense, the generation and propagation of the pressure waves and the accompanying sound energy of the broad spectrum in the boiler model give very encouraging results.

Keywords: Pulse combustion · Pressure waves · Sound energy

S. Delic—Master program student: Energy Department.

S. Avdaković (Ed.): IAT 2018, LNNS 60, pp. 564–571, 2019.
https://doi.org/10.1007/978-3-030-02577-9_55

1   Introduction

Conversion of primary energy from fossil fuels, in the majority of cases, also today begins with the combustion process. These processes, besides being accompanied by a series of problems of technical and technological nature, are often the most important pollutant producers. Problems are particularly pronounced in the case of burning low-value, weakly reactive coal prone to fouling, as most of coals in Bosnia and Herzegovina. Current research on coal combustion in the world is focused on further knowledge of slagging, fouling and corrosion during combustion, and research on emissions of CO_2, NO_x, SO_2 and trace elements, but always including the analysis of energy efficiency of the conversion across all segments of the final energy supply chain. It is known that gas fired boilers and combustion chambers, in terms of emissions into the atmosphere, are significantly more favorable than boilers combusting coals. This fact continues to provide a strong incentive to research coal combustion technology with a more favorable environmental impact and, in particular, to achieve greater energy efficiency compared to conventional technologies. In addition to the above technical aspects of these boilers, economical aspects are equally important and, because the future use of these power plants depends on these aspects and hence their impact on the environment. Because of the above mentioned reasons it is necessary to develop new and to improve already existing or insufficiently explored methods of high efficiency and low-waste combustion. Significant possibilities in this regard are provided by pulse combustion which, by its performance, belongs to group of low-waste and high-efficiency processes of chemical energy conversion contained in the fuel. In addition, the pulse combustion phenomenon can also be applied as auxiliary system to existing industrial and power boilers (and also high power), as device for cleaning the external side of the fouled surfaces, Fig. 1-left, as well as in the combustion chamber of these boilers in order to raise the level of turbulence of the combustion atmosphere under certain conditions and thereby improve the reaction of reagents, i.e. combustion efficiency. The basic reasons for the still insufficient use of the application of the process are intense noise, which is almost unavoidably generated during pulse combustion.

2   Laboratory Facility and Working Principle

At the Faculty of Mechanical Engineering in Sarajevo at the beginning of the 1990s (air-cooled burners, [1]) and at the beginning of this century (water-cooled burners [2]), laboratory tests of pulse combustion of gaseous fuels were performed, Fig. 1 - in the middle right. Compared to classical combustion methods, pulse combustion means a qualitative step in the following [1, 2]:

- more efficient combustion,
- pulsations have a suction and thrust effect,
- more efficient heat transfer,
- reduction of nitrogen oxide emissions,
- more flexible filling,
- self-cleaning of heating surfaces by flue gas.

Fig. 1. Example of fouled boiler surfaces (left); Water-cooled burner connected to flue pipe (in the middle) and to boiler model - furnace (right); [2]

As a forerunner of this combustion technique at the Mechanical engineering faculty in Sarajevo, a domestic method has been developed to prevent the formation of and to remove already formed deposits from boiler heating surfaces by detonation-wave cleaning[1] - see [3–8].

Working Principle: Basically, the burner for pulse combustion consists of: combustion chamber, air inlet and resonant tube. During the burner operation, an asymmetrical flow of media along the burner was provided thanks to the mechanical or aerodynamic valves mounted on the suction side. Namely, during the expansion of flue gas, during and after combustion which is necessarily interrupted - intermittently, the products are mainly expanding through the resonant tube further into the system or the environment, Fig. 2.

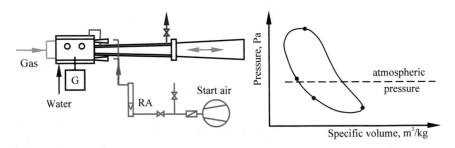

Fig. 2. Schematic representation of the water-cooled pulse combustion burner (left) and the theoretical working cycle in the p-v diagram, [2].

In addition to the combustion characteristics of the "free" in the lab-space positioned burner, also the characteristics of the burner mounted on the boiler-model were also investigated - the burner was mounted by two directed adapters: I - side one, simulating application application of the burner in the combustion chamber zone, and II - upper one: simulating the use of the burner in the convective zone of the boiler. In this

---

[1] This detonation-wave cleaning technique for cleaning boiler heating surfaces is patented: Smajević, I., Hanjalić, K.: Paten number: P1728/88 i P1756/88.

case, using various geometric forms of the burner and various heat loads, the intensity of the pressure pulses in the free space of the boiler-model and inside of the pipe-register, as well as the influence of the pulsating flow of the flue gas and the sound energy on the boiler walls, are observed and measured, Fig. 3 - [9, 10].

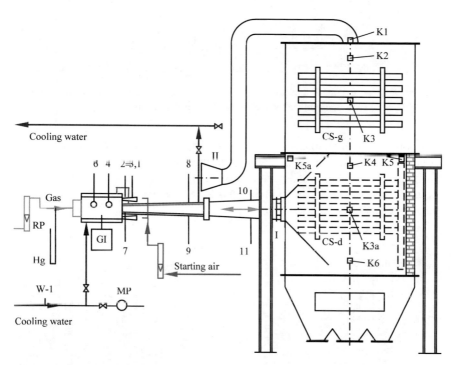

Fig. 3. Sheme of the pulse combustion burner application on the boiler-model in the combustion chamber zone with the indicated measuring points in the interior of the model - position I, upper pipe register (CS-g), [2]

3 Results and Discussion

Example of Burner Performance: As an example of the pulse combustion burner performance, the diagrams in Fig. 4 are given. The results are always related to the medium combustion chamber to which short or long air intakes (KUV, DUV), basic, short, medium or long resonant tube (T, K, S, D) are mounted. It is obvious that the mean temperature in the chamber is always less than 1200 °C, which is one of the essential preconditions for lower NO_x emissions. The total deflection of the pressure and the frequency of the process depend significantly on the combustion air intake length - both of the shown dependencies were obtained for the burner power of about 150 kW and the burner cooling conditions defined with cooling water flow of 0.4 l/s. In this case, the maximum deflection of the flue gas pressure is 80 kPa (long air intake valves and long resonant pipe) and the frequency range is wide, in this case from 90 to

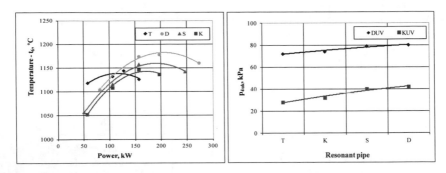

Fig. 4. Performance for pulse combustion burner: mean temperature of combustion products in the chamber (left); total pressure deflection in the combustion chamber (right)

142 Hz - otherwise, the frequency can go up to 180 Hz. Generated noise on the burner goes over 100 dB - the sound intensity measurement was performed on two sources: on the air intake valves and on the open end of the resonant tube (during both measurements the burner was freely set up in the laboratory - i.e. the resonant pipe was outside the flue gas duct) [2] - Fig. 5.

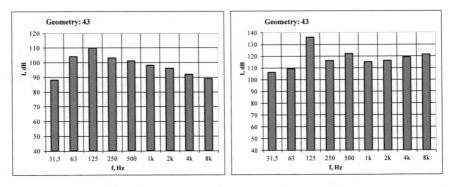

Fig. 5. Sound spectrum recorded from the air intake valves (left) and from the end of the resonant tube (right), with the fuel consumption of $3.68m_n^3/h$ LPG

In these places, very strong sound is produced, mostly in the region of audio frequencies, although it can be said that the spectrum, in the lower frequency range, is practically interfering with the infrared spectrum as well. In Fig. 5, the results of the sound intensity measurement from the air intake valves and from the open end of the resonant tube are presented, [2].

Weakening of the flue gas pressure pulses in the boiler model on experimental line is given in Fig. 6. The results refer to the burner geometry with code 43 - see [2], two thermal power regimes of the burner, 150 kW and 245 kW, and two positions of the pipe register, upper (CS-g), and lower (CS-d). Significant decrease in the intensity of the pressure pulses in the interior of the boiler model is noticeable in relation to the values

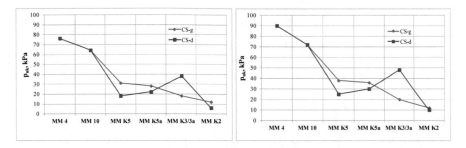

Fig. 6. Intensity of pressure pulsations inside the boiler model, burner power: 150 kW (left), 245 kW (right)

recorded at the corresponding measuring points on the burner. There is also an evident reduction in pressure pulses in the interior of the boiler model during pulse propagation through the pipe register - see the values for measuring points K3 and K2 in Fig. 6. In case of lowering the tube register to the lower position (inside the combustion chamber of the model boiler) it is possible to investigate the change of parameters, in the first place pressure, and in the case of longitudinal flow of pulsating flue gases over the pipes of boiler element, and in this way to investigate the possibility of using the burner as a device for preventing the formation of deposits and cleaning of already formed deposits on boiler heating surfaces - [2, 10]. In Fig. 6, for different heat loads of the burner, the results of the measurement for such a pipe register setting inside the model boiler are shown. The diagrams show two characteristics of the presented results: the lower values of total pressure deflection at measuring points K5, K5a and K2 in relation to the values at the same measuring points when the pipe register is in the upper position - this can be explained by the significant decrease of the pulse intensity inside of the pipe register; the maximum values of total pressure deflection in the boiler model appear in the middle of the pipe register, measuring point K3a - here it should be emphasized that this measuring point is on the axis of the resonant pipe at a distance of about 1.2 m from its end. Relatively high values of the total deflection of the pressure pulses within the pipe registers of the boiler elements is a desirable result with respect to the intent of applying a pulse combustion burner as a boiler surface cleaning device - [2, 10]

4 Conclusion

From the results of laboratory research it can be concluded:

- with the increase of the burner cooling intensity the pressure pulse frequency decreases as the pressure oscillation amplitude increases; during the experiment the frequency value ranged from about 80 to about 140 Hz and the total pressure deflection generated in the combustion chamber were over 90 kPa.
- intensity of the water-cooling of the burner does not significantly affect the pulse combustion burner operating flexibility.
- the average process temperature is relatively low and depending on the selected burner geometric form and thermal load, reached a maximum of 1175 °C;

- the sound intensity, as an unavoidable accompanying phenomenon during the burner operation, was also over 130 dB, with very wide sound spectrum.
- the values of high-frequency pressure pulses in surroundings of elements inside the boiler model, but also within them, are significant. Having this in mind, combined with multiple reflection of pulse flow and generated sound energy, it is recommended to perform tests of the pulse combustion burner using gas fuel under real conditions as a cleaning device of already formed deposits and to prevent formation of new deposits, primarily on the convective heating surfaces of industrial and power boilers, as well as high power boilers, in particular those in which use fuel with a higher tendency to fouling heating surfaces.

Improving fuel combustion efficiency as well as intensifying heat transfer in industrial and power boilers, as well as high power boilers, at the same time contributes to an increase in energy efficiency of the plant as a whole, significant reduction in fossil fuel consumption and reduction of adverse environmental impacts through emission of pollutant components of flue gases but also through the deposition of solid combustion products, slag and ash. All this results in a more favorable financial result of the plant and in an increase of the competitiveness on the market. Therefore, during the operation of boilers - especially those that use low-value, low-reactive, high-ash, and fouling prone coals - it is necessary to have efficient techniques or tools that reduce to a minimum and even minimize the mentioned problems. High-frequency exhaust gas pressure pulsations of up to 100 kPa and the associated generating sound energy of about 130 dB can be utilized in industrial and power boilers. Improving fuel combustion efficiency as well as intensifying heat transfer in industrial and power boilers, as well as high power boilers, at the same time contributes to an increase in energy efficiency of the plant as a whole, significant reduction in fossil fuel consumption and reduction of adverse environmental impacts through emission of pollutant components of flue gases but also through the deposition of solid combustion products, slag and ash. All this results in a more favorable financial result of the plant and in an increase of the competitiveness on the market. Therefore, during the operation of boilers - especially those that use low-value, low-reactive, high-ash, and fouling prone coals - it is necessary to have efficient techniques or tools that reduce to a minimum and even minimize the mentioned problems. High-frequency exhaust gas pressure pulsations of up to 100 kPa and the associated generating sound energy of about 130 dB can be utilized in industrial and power boilers. Pulse combustion burners are recommended for both purposes: in the combustion chamber as an additional boost for the turbulization of the combustion atmosphere and the stabilization of combustion and thus increasing the combustion efficiency, and as a tool for preventing of deposits and cleaning of the already fouled boiler heating surfaces, thus resulting in intensification of heat transfer. It has been shown that propagation of flue gas pulsation through the furnace and convective part of the boiler is significant and is estimated to be additional prerequisites for preventing or at least reducing intensity of deposition. These results represent a strong additional motivation to investigate the feasibility of using the pulse combustion burner under real conditions - on boilers of various purpose and power, as it will potentially make them more energy efficient, smaller in size and cheaper. In addition, under these conditions, the energy and time availability of the boiler will be greater and the negative impact on the environment will be smaller.

References

1. Smajević, I.: Istraživanje pulzirajućeg sagorijevanja gasovitog goriva sa analizom mogućnosti primjene, doktorska disertacija, Mašinski fakultet Sarajevo, Sarajevo (1991)
2. Hodžić, N.: Laboratorijsko istraživanje mogućnosti primjene gorionika za pulzirajuće sagorijevanje na kotlovima velike snage, Magistarski rad, Mašinski fakultet Sarajevo, Sarajevo (2007)
3. Hanjalić, K., Smajević, I.: Razvoj metoda za sprječavanje formiranja i otklanjanje naslaga sa kotlovskih ogrjevnih površina detonaciono-impulsnom tehnikom, (istraživački projekat: P-162-12, sufinansijer EP BiH i SIZ za nauku BiH), Završni izvještaj, Mašinski fakultet Sarajevo - IPES, Sarajevo (1985)
4. Smajević, I.: Doprinos razvoju detonaciono-impulsnog postupka za sprječavanje stvaranja naslaga na kotlovskim ogrjevnim površinama, Magistarski rad, Mašinski fakultet Univerziteta u Sarajevu, Sarajevo (1984)
5. Smajević, I., Hanjalić, K.: Zwanzig Jahre erfolgreiche Anwendung der Stosswelen-Reinigungstechnik in einem mit Kohle befeurtem Kraftwerk, VGB PowerTech. Int. J. Electric. Heat Gener. 71–75 (2004)
6. Hodžić, N., Metović, S., Smajević, I.: Water-cooled pulse combustor performances. In: 11th International Research/Expert Conference Trends in the Development of Machinery and Associated Technology, TMT 2007, Hammamet, Tunisia, 5–9 September 2007
7. Smajević, I.: Experimental study and computational modelling of gas-fired puls combustion. Int. J. Autom. Mech. Eng. (IJAME) 1, 1–12 (2010). ISSN 1985–9325 (Print); ISSN 2180-1606
8. Hanjalić, K., Smajević, I.: Detonation-Wave Technique for On-Load Deposit Removal In Coal Fired Boilers, VIII All-Russian Conference with international participation: Combustion of Solid Fuel/ VIII Всероссийская конференция с международным участием: Горение твердого топлива, The Ministry of Education and Science of RF, Russian Federation, Novosibirsk IT SB RUS, 13–16 November 2012
9. Hodžić, N., Metović, S., Smajević, I.: Pulse combustion burner as cleaning device of boiler heating surfaces. In: Experimental Investigation – 13th International Research/Expert Conference Trends in the Development of Machinery and Associated Technology TMT 2009, Hammamet, Tunisia, 16–21 October 2009
10. Smajević, I., Hodžić, N., Hanjalić, K.: Aerovalved gas pulse combustor for enhancement of heat transfer in large scale solid-fuel boilers. In: 10th International Conference Turbulence, Heat and Mass Transfer 8, 20–25 September 2015. © 2015 Begell House, Inc., Sarajevo (2015)
11. Putnam, A.A., Belles, F.E., Kentfield, J.A.C.: Pulse combustion. Prog. Energy Combust. Sci. 12(1), 43–79 (1986)
12. Mullen, J.J.: The pulse combustion furnace-how it is changing an industry. ASHRAE 26(7), 28–33 (1984)

Case Study on Small, Modular and Renewable District Heating System in Municipality of Visoko

Anes Kazagić$^{(\boxtimes)}$, Ajla Merzić, Elma Redžić, and Dino Trešnjo

Strategic Development Department, JP Elektroprivreda BiH d.d.-Sarajevo,
Sarajevo, Bosnia and Herzegovina
a.kazagic@epbih.ba

Abstract. Small modular district heating (DH) system has greater advantages than the traditional heating system in many aspects and it shows great development potential and broad market prospect in the future. Such DH grids can be fed by different heat sources, including solar collectors, biomass systems and surplus heat sources (e.g. heat from industrial processes or biogas plants that is not yet used). Especially the combination of solar heating and biomass heating is a very promising strategy for smaller rural communities due to its contribution to security of supply, price stability, local economic development, local employment, etc. The objective of the CoolHeating project, funded by the EU's Horizon2020 programme, is to support the implementation of "small modular renewable heating and cooling grids" for communities in South-Eastern Europe. This paper elaborates the developed concept of the renewable district heating system in Municipality of Visoko, as well as the corresponding function of each component.

Keywords: Heating · Cooling · District heating grid · Seasonal heat storage
Heat pumps · Solar thermal

1 Introduction

The heating and cooling demand in Europe accounts for around half of the EU's final energy consumption. During the urbanization process, the city-size increases, but the corresponding building energy consumption stays at a high level. The heating energy consumed is the most important part of building energy consumption, however, it is also the most wasted and has the most energy saving potential. With the increase of the heat user's number and heating network scale, the traditional heating system could not satisfy the current urban heating demand, and the shortcomings become prominent. The increasing degree of deregulation of the energy market and the increasing focus on energy efficient buildings place a demand on district heating systems to be more efficient than ever. The development of the technology provides opportunity and change for district heating, many new concepts are presented and applied to realize the "city smart network", aiming at energy saving and improvement of the comfort degree [1].

The future DH system should fulfill the goals like supplying low-temperature district heating for space heating and domestic hot water, integrating renewable heat

© Springer Nature Switzerland AG 2019
S. Avdaković (Ed.): IAT 2018, LNNS 60, pp. 572–581, 2019.
https://doi.org/10.1007/978-3-030-02577-9_56

sources, and operation planning. Therefore, it is important to support and promote renewable heating and cooling concepts, the core aim of the CoolHeating project (Fig. 1).

Fig. 1. CoolHeating logo and countries involved in the CoolHeating project and target municipalities (red dots)

2 The CoolHeating Project

The objective of the CoolHeating project, funded by the EU's Horizon2020 programme, is to support the implementation of "small modular renewable heating and cooling grids" for communities in South-Eastern Europe. This is achieved through knowledge transfer and mutual activities of partners in countries where renewable district heating and cooling examples exist (Austria, Denmark, Germany) and in countries which have less development (Croatia, Slovenia, Macedonia, Serbia, Bosnia-Herzegovina) (Fig. 1). Core activities, besides techno-economical assessments, include measures to stimulate the interest of communities and citizens to set-up renewable district heating systems as well as the capacity building about financing and business models. The outcome is the initiation of new small renewable district heating and cooling grids in 5 target communities up to the investment stage. These lighthouse projects will have a long-term impact on the development of "small modular renewable heating and cooling grids" at the national levels in the target countries. CoolHeating activities will have impact at tree levels: on national level, on target community level, and on "follower" community level. Core actions will be focussed on the target communities. Follower communities have the opportunity to learn from the project activities in the target communities [2].

An important instrument of the CoolHeating project is a handbook (Fig. 2) which was elaborated by the project partners: "Small Modular Renewable Heating and Cooling Grids - A Handbook". Although various information materials on technologies for small modular renewable heating and cooling systems exist, there was a need to

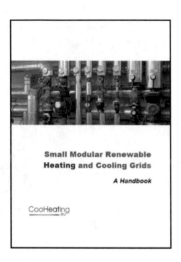

Fig. 2. CoolHeating handbook

create this up-to-date handbook that is accessible for free (Download here: http://www. coolheating.eu/images/downloads/D4.1_Handbook_EN.pdf) in various languages (English, Bosnian, Croatian, Macedonian, Serbian, Slovenian, German). In many of the CoolHeating target countries, there is lack of such information in national language. The handbook provides an overview of both, technical and non-technical aspects. The main characteristics of different heat sources from solar, biomass, geothermal and excess heat are described and the opportunities of their combination in small modular RE district heating and cooling system are presented. Seasonal and diurnal storage systems are included, as well as the use of heat pumps. Specific aspects of heating and cooling in smaller grids are shown.

3   Small Renewable Modular Heating and Cooling Grids

Small modular district heating/cooling grids are local concepts to supply households and/or small and medium industries with renewable heat and/or cooling. In some cases, they may be combined with large-scale district heating (DH) grids, but the general concept is to have an individual piping grid which connects a relatively small number of consumers. Often, these concepts are implemented for villages or towns. They can be fed by different heat sources, including solar collectors, biomass systems and surplus heat sources (e.g. heat from industrial processes or biogas plants that is not yet used) [3] (Fig. 3). A scheme of these grids is presented in Fig. 4.

Especially the combination of solar heating and biomass heating is a very promising strategy for smaller rural communities due to its contribution to security of supply, price stability, local economic development, local employment, etc. On the one hand, solar heating requires no fuel and on the other hand biomass heating can store energy and release it during winter when there is less solar heat available. Thereby, heat storage (buffer tanks for short-term storage and seasonal tanks/basins for long-term

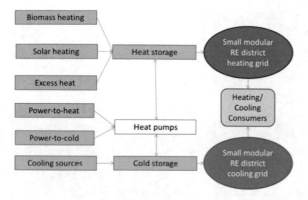

Fig. 3. Concept of small modular renewable heating & cooling grids [2]

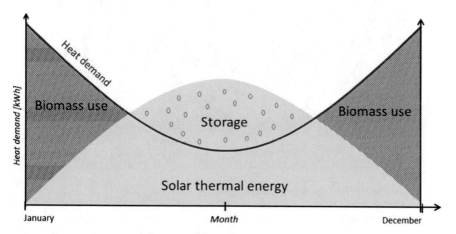

Fig. 4. Scheme of the seasonal heat demand and supply from solar and biomass sources in Europe [2]

storage) needs to be integrated. A scheme of a typical seasonal demand and supply of a combined small heating grid is presented in Fig. 5. The main advantages of a biomass/solar heating concept are:

- Reduced demand for biomass
- Reduced heat storage capacity
- Lower maintenance needs of biomass boilers.

With increasing shares of fluctuating renewable electricity production (PV, wind), the Power-to-Heat conversion through heat pumps can furthermore help to balance the power grid.

If the planning process is done in a sustainable way, small modular district heating/cooling grids have the advantage, that at the beginning only one part of the system can be realised and additional heat sources and consumers can be added later.

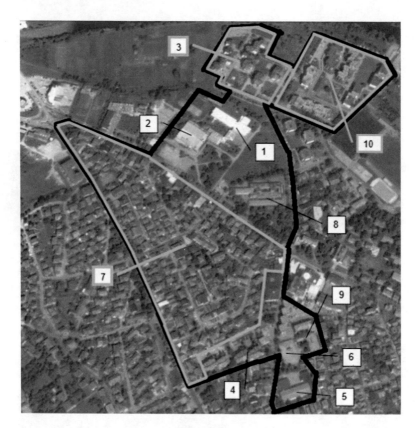

Fig. 5. Project area: the North-Eastern part of Visoko (1-secondary school "Hazim Šabanović"; 2-sports center 'Mladost'; 3-collective housing facilities – Luke; 4-collective housing facilities - center of the city; 5-primary school 'Safvet beg Bašagić' + music school; 6-Crèche; 7-individual housing facilities; 8-medical centre; 9-Social work centre; 10-collective housing facilities - Luke 2)

This modularity requires well planning and appropriate dimensioning of the equipment (e.g. pipes). It reduces the initial demand for investment and can grow steadily.

Besides small district heating, also small district cooling is an important technology with multiple benefits. With increased temperatures due to global warming, the demand for cooling gets higher, especially in southern Europe in which the target countries are located. In contrast to energy demanding conventional air conditioners, district cooling is a good and sustainable alternative, especially for larger building complexes. However, experiences and technologies are much less applied than for district heating. The CoolHeating includes both, heating and cooling in its planning process.

The CoolHeating target countries in south-eastern Europe have high solar irradiation which can be used both for heating and cooling. The combination of small district heating and cooling in the same planning step saves cost and efforts, even if some consumers will demand only either heating or cooling. Thereby, also technical synergies are created (piping, the use of heat pumps). CoolHeating will develop business plans for the target communities with the following characteristics:

- Seasonal storage
- Diurnal storage
- Renewable heating (e.g. with a solar thermal plant and a biomass boiler)
- Thermal cooling
- Utilization of the waste heat from thermal cooling for heating (e.g. hot water supply).

Small modular district heating/cooling grids have several benefits. They contribute to increase the local economy due to local value chains of local biomass supply. Local employment is enhanced as well as security of supply. The comfort for the connected household is higher as only the heat exchanger is needed in the basement of the buildings and no fuel purchase has to be organised. Due to all these benefits, the objective of the CoolHeating project is to support the implementation of small modular renewable heating and cooling grids for communities (municipalities and smaller cities) in South-Eastern Europe [2].

4   Case Study – Municipality of Visoko

Municipality of Visoko is one of the target communities within the CoolHeating project. The fundamental idea of district heating is 'to use local fuel or heat resources that would otherwise be wasted, in order to satisfy local customer demands for heating, by using a heat distribution network of pipes as a local market place.

District heating systems were well developed in towns and cities before the war. During the war, many systems fell into disrepair, and after the war could not recover customers due to a fall in the purchasing power of the population. The maintenance and investment in the remaining functioning district heating systems has been low, leading to obsolete technologies, as well as low efficiency and large heat losses on the network. District heating and cooling concept based on renewable energy sources would help meet rising urban energy needs, improve efficiency, reduce emissions and improve local air quality in Municipality of Visoko. Air quality especially badly suffers during the heating season due to heavy use of coal for heating. Existing heating systems are mainly individual and currently dominated by coal as the cheapest energy source on the market, therefore they should be upgraded or new networks created, using solid biofuel and solar and geothermal energy technologies. Depending on local conditions, renewable-based DHC would bring a range of benefits, including increased energy security, improved health and reduced climate impact.

Heating/cooling demand for the concept and initial situation
By reviewing statistical data compiled through the survey, a great presence of wood as a fuel for the heating system has been noticed, especially in family houses with the heating system that includes individual hand-firing solid-fuel furnaces, but also in a great number of categories of collective housing buildings. The DHC system concept is planned to cover the central area of the town. The zone includes different types of buildings. The main focus is on public buildings which are in the jurisdiction of the Municipality, which represent the biggest consumers in the town. On the other side,

there is private and collective housing outside of the jurisdiction of the Municipality, whose connection to the centralized heating system would contribute to the reduction of air pollution and more rational use of energy resources. In order to determine the consumer affordability of the heating costs, it is necessary to review the main socio-economic parameters of the population and energy consumption of households in Visoko.

Map: Potential buildings to be connected to the DH grid
The buildings to be considered by the DH concept are shown in Fig. 1 and marked by consecutive numbers. Amongst the buildings considered, Medical centre is the largest polluter on the municipality using coal as the primary energy source. Primary and Music school use fuel oil of questionable quality, unorderly supplied by relevant cantonal ministry. Therefore, it happens that school sometimes runs out of fuel, or has to keep sparing the fuel lowering the inner temperature with pupils often sitting in their jackets. Individual and collective housing use mainly coal obtained on black market, then firewood, gas, electricity, or the combination thereof. Collective housing as well as 25% of individual houses have individual central heating in use.

Development process for solution and concept
The DH system is planned so that the heat production would be achieved with different production units: heat pumps (water-water), solar collectors and existing peak load gas boilers which would start automatically if the heating output of the renewable energy sources DHC system is insufficient to cover the demand. All of these units would located in a central location in the northwest of the city and would be connected to a seasonal heat storage (Fig. 6). Long-term (seasonal) storage would mean storing heat for months and even from summer to winter. But a large thermal store would do more than just store solar heat from summer to winter. Introducing the large-capacity thermal store in the district heating system gives extremely good conditions for a very flexible production from combined production units. It is also planned that photovoltaics would be installed on the roofs of public institutions, which would contribute to the justification and sustainability of such project. Taking into account all the collected data, conducted surveys, performed energy audits, the heat demand required for this area is approximately 19 GWh per year, which also meets the requirements of the CoolHeating project.

Determining the capacities of the production units, and optimizing the operating mode itself, was done in the specialized software energyPRO (Fig. 7). EnergyPRO is a modelling software used primarily in relation to district heating projects to carry out an integrated detailed technical and financial analysis of both existing and new energy projects. The software was used to plan the optimal production for the energy plant for a whole year. We indicated the period for the optimization down to minutes with a detailed production plan. Inputs for the optimization are typically parameters such as content of stored energy at the beginning of the optimization period, expected energy demands within the period as well as expected fuel and electricity prices [4]. After all carried out calculations based on the inputs for all units separately, taking into account the prices of all energy sources, energy efficiency class of the facilities, estimated

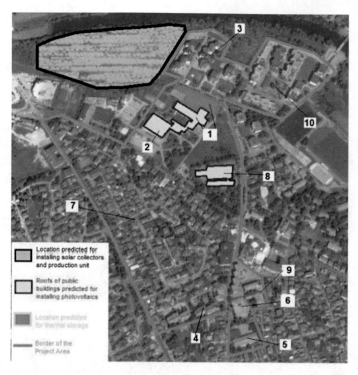

Fig. 6. Map of potential locations for installation solar collectors, thermal storages and heating plants

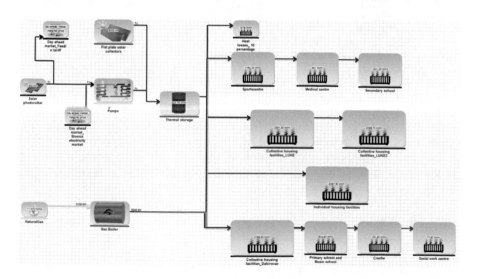

Fig. 7. EnergyPRO interface

operating time per day and other data, the total installed power for all production units is 9.4 MW$_{th}$ and area of 5000 m^2 for solar collectors.

After the elaborated technical concept, it was necessary to find the most optimal solution from an economic point of view. All financial calculations were made in an economic tool (Fig. 8) which is one of the deliverables of the CoolHeating project [5]. The basic pricing principles for district heating have been either market-based or cost-oriented. Private owners prefer prices close to prices for the competitive heat supply alternatives in order to capture the full benefits of district heating. Municipal owners have had a tradition of applying cost-oriented prices in order to share the benefits of district heating with the final customers [6, 7].

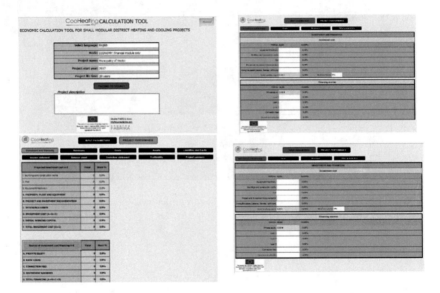

Fig. 8. Economic calculation tool for small modular district heating and cooling projects [5]

A large number of calculations were performed by the iterative process until the best economic parameters of the considered concept were obtained. Each iteration optimized individually all the capacities of production units and using the real investment costs of such technologies on the market and the corresponding fuel prices, operation and maintenance costs, district heating grid and land purchase costs. An overview of the obtained capacities of all production units for the most optimal variant in the techno-economic sense is as follows.

5 Conclusions

With a district heating network, substantial environmental benefits can be reached in line with a sustainable urban development. The individual technologies required in the energy systems of the future already exist, but the challenge lies in combining the right technologies in the right partnerships and eco systems. While many of the tools needed to create value from data are both available and – compared to basic metering for billing requirements – affordable today, they have the potential to be developed much further together with utilities and research institutes.

The analysis of the DH network in the Municipality of Visoko shows a positive technical potential of establishment and expansion. The future context foresees promising possibilities for district heating, but strong efforts are required in order to realise them. One important effort is to enhance the current district heating technology to align with future conditions associated to renewables and these new conditions should be met by introduction of the fourth generation of district heating technology.

In the coming half a year, the CoolHeating will work on a key challenge to make the energy supply in Europe more sustainable: it will support local actions in the target countries in south-eastern Europe in order to develop and establish concepts for small renewable modular heating and cooling systems.

Acknowledgements. The authors would like to thank the colleagues of WIP Renewable Energies as well as the CoolHeating partners for their contributions in the project: The authors would like to thank the European Commission and the Innovation and Networks Executive Agency (INEA) for the support of the CoolHeating project. CoolHeating has received funding from the European Union's Horizon 2020 research and innovation programme under grant agreement No. 691679. The project duration is January 2016 to December 2018. DisclaimerThe sole responsibility for the content of this paper lies with the authors. It does not necessarily reflect the opinion of the European Union. Neither the INEA nor the European Commission are responsible for any use that may be made of the information contained therein.

References

1. Lin, G., Xuyang, C., Jiaxin, N., Wanning, L., Tao, H., Chao, B., JunHong, Y.: Technologies in smart district heating system. In: 9th International Conference on Applied Energy, ICAE2017, 21–24 August 2017, Cardiff, UK (2017)
2. Rutz, D., Doczekal, C., Zweiler, R., Hofmeister, M., Laurberg, J.L.: Small modular renewable heating and cooling grids – handbook, WIP Renewable Energies, Munich, Germany (2017)
3. Lund, H., Werner, S., Wiltshire, R., et al.: 4th generation district heating (4GDH). Energy **68**, 1–11 (2014)
4. EnergyPRO User Guide. EMD International A/S, Aalborg
5. https://www.coolheating.eu/en/
6. Rutz, D., et al., Mergner, R., Janssen, R., Soerensen, A., Jensen, L.L., Doczekal, C., Zweiler, R., Puksec, T., Duic, N., Doracic, B., Sunko, R., Sunko, B., Markovska, N., Karanfilovska, M., Rajkovic, N., Bjelic, I.B., Kazagic, A., Redzic, E., Smajevic, I., Jerotic, S., Mladenović, S., Fejzovic, E., Babić, A., Petrovic, M., Kolbl, M.: The combination of biomass with solar thermal energy and other renewables for small heating grids. The CoolHeating Project, 25th European Biomass Conference & Exhibition, 12–15 June 2017, Stockholm, Sweden (2017)
7. Rutz, D., et al., Mergner, R., Duic, N., Puksec, T., Doracic, B., Sunko, R., Sunko, B., Markovska, N., Karanfilovska, M., Rajkovic, N., Bjelic, I.B., Kazagic, A., Smajevic, I., Jerotic, S., Materadzija, M., Fejzovic, E., Kolbl, M., Redzic, E., Bozhikaliev, V., Doczekal, C., Zweiler, R., Hofmeister, M., Jensen, L.L.: Small, modular and renewable heating grids in South-Eastern Europe. In: 11th International Conference Sustainable Development of Energy, Water and Environment Systems – Dubrovnik 2017, Dubrovnik, Croatia, October 2017

A Small-Scale Solar System with Combined Sensible- and Latent-Heat Thermal Energy Storage

Nijaz Delalić, Rejhana Blažević, Mirela Alispahić,
and Muris Torlak[✉]

Mechanical Engineering Faculty, University of Sarajevo, Vilsonovo šetalište 9,
71000 Sarajevo, Bosnia and Herzegovina
torlak@mef.unsa.ba

Abstract. This paper describes experimental investigation of storage of the heat obtained from the solar thermal collectors. The storage unit contains both the sensible-heat medium (water) and the latent-heat medium (sodium-acetate-trihydrate as phase-change material, PCM). The measured temperature variations are compared with the case without the PCM. The effect of PCM in the storage, keeping the temperatures in the tank at nearly the same level, is observed.

Keywords: Renewably energy sources · Solar thermal energy
Energy storage · Phase-change materials (PCM)

1 Introduction

Due to variable-in-time and unpredictable solar irradiance, the overall efficiency of solar thermal systems depends considerably on the efficiency of applied thermal-energy storage. In most cases, solar thermal energy is stored in the form of sensible heat, typically as hot water. Alternatively, latent heat can also be used applying a suitable phase-change material (PCM) [1, 2]. The latter allows larger storage capacities per unit volume or weight of the material used, as well as better storage stability. Use of latent-heat storage is expected to increase the efficiency of solar-to-heat systems, aiming at balancing availability of solar thermal energy and demands for hot water. This paper describes the investigation of the solar thermal systems in urban environment with limited space for energy storage, such as appearing in domestic use, but not restricted to this.

The use of a solar system for sanitary water heating and room heating is very widespread. Water tanks are used to balance hot water consumption and thermal energy supply from sun. The volume of these containers is proportional to the need for hot water, also when there is no sun. However, in urban areas there is less and less space available for storing the heat.

This is a good reason to use materials that can absorb higher thermal energy in the same volume. During the energy-charging day-time period, a PCM material absorbs the heat while it changes the phase from solid to liquid. Thus, if a PCM is added to the hot water, the heat absorbed by the storage tank is considerably higher. On the other hand,

S. Avdaković (Ed.): IAT 2018, LNNS 60, pp. 582–588, 2019.
https://doi.org/10.1007/978-3-030-02577-9_57

in the periods of reduced energy supply (night or cloudy weather or fog, for example) the heat required by the consumer can be obtained from the PCM, depending on the realized temperature differences. Thus, the latent heat in the PCM maintains the water temperature in the tank for a delayed time period, and also pre-heats the cold intake water if there is no sun heat. The PCM-based heat storage technology also stabilizes the temperatures and reduces water temperature fluctuations in the solar system.

Depending on the desired operating temperature ranges, appropriate PCM materials have to be selected. Their costs are also an important decision criterion. Mostly, non-toxic, non-flammable, and cost-effective PCMs are required. For applications in the small-scale, domestic solar systems their phase change can be in the temperature range between 50 °C and 70 °C.

2   Experimental Setup and Measurement Techniques

A small-scale solar system with integrated water (sensible-heat) and PCM (latent-heat) energy storage unit has been built and tested.

It includes the heat source consisting of eight solar collectors, whose dimensions are 600 mm × 1800 mm (total area of 8 m^2), which are mounted on the laboratory roof, see Fig. 1. The pipes connecting the heat-source and the storage device are filled with a water-glycol mixture (volume ratio 1.5:1) to prevent freezing in the winter time, and thermally insulated to minimize energy losses. Also a control unit is installed including a circulation pump, a safety valve, a non-return valve, and an expansion vessel. A heat exchanger made out of copper in the form of a helically coiled pipe is placed in the interior of the storage, as a part of the source/solar circuit, as shown in Fig. 2.

Fig. 1. The solar collectors installed on the laboratory roof: a photograph (left) and a corresponding thermo-graphic recording (right).

The energy storage unit consists of two cylindrical tanks. The primary one contains 230 l water (limited by the PCM-container design choice). The secondary tank, whose effective, available volume is 30 l, is filled with PCM. It is installed co-axially in the interior of the primary one. Its outer wall is made out of copper for better heat transfer from water to PCM and vice versa. An auxiliary 2 kW electric heater is installed in the primary water tank for the use in case of insufficient solar irradiance (in the tests

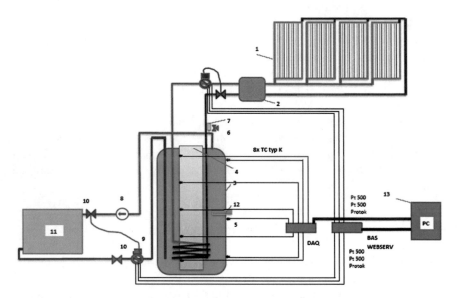

Fig. 2. A small-scale solar system with integrated water and PCM thermal energy storage unit: 1-Solar collectors, 2-Control unit with pump, 3-Water storage, 4-PCM storage, 5-Heat flow meter, 6-Safety valve, 7-Air release valve, 8-Circulation pump, 9-Heat flow meter, 10-Valve 11-Fan coil, 12-Electrical heater, 13-PC with data logger.

presented here it was however not activated). The outer walls of the storage unit are thermally insulated. Similar configuration was also tested in works [3, 4], where the PCM was placed in four containers located in the upper part of the water tank, but the amount of PCM was about 5 times smaller.

The consumer/heating circuit consists of a fan coil with 10 kW thermal power at three different fan speeds, circulation pump, safety valve, air release valve, and expansion vessel.

The experimental setup includes measuring the heat flow rate from the solar collectors during the energy charging stage (heat supplied to the storage), using ultrasonic flow meter and two PT-500 temperature sensors. The temperature sensors are installed in the supply and the return line of the solar circuit. The integration unit calculates the heat delivered to the storage. In addition to that, the temperatures are measured at four points vertically distributed along the storage tank using the type-K thermocouples (chromel-alumel). During the energy discharge from the tank, the heat released by the fan coil is measured by the ultrasonic flow meter and the two PT-500 temperature sensors on the supply and the return line of the fan coil/consumer circuit. The integration unit calculates the heat delivered by the fan coil. All measured values are collected by the data logger and stored on a computer, with the time step of one minute or less.

In this work, sodium-acetate-trihydrate is used as PCM. Typical physical properties of the sodium-acetate-trihydrate are shown in Table 1.

The estimated energy which is stored in sensible form in a typical 300 l water tank at the difference of 30 C between the initial and the final temperature (e.g. from 40 °C

Table 1. Basic material properties of sodium-acetate-trihydrate

Sodium-acetate-trihydrate	
Starting melting point	58 °C
Latent heat	226–270 kJ/kg
Density of solid phase at 20 °C	1450 kg/m^3
Density of solid phase in powder form	≈900 kg/m^3
Density of liquid phase	1280 kg/m^3
Specific heat capacity of solid phase at 25 °C	2790 J/kg K
Specific heat capacity of liquid phase	3000 J/kg K
Thermal conductivity of solid phase	0.7 W/K m
Thermal conductivity of liquid phase	0.4 W/K m

to 70 °C) is about 10.3 kWh. Using the data given in Table 1 for a combination such as employed in the experimental setup, consisting of 230 l water and 26 kg PCM (obtained by filling 30 l of the PCM in the solid, powder form in its container), one may expect the total energy of about 11.15 kWh stored at the same temperature range (7.9 kWh is the sensible heat in the water and 3.25 kWh is the sensible and the latent heat in the PCM). This is about 8% more storage capacity at the same occupied space.

A reasonably small change of the initial tank temperature would not involve considerable change of the stored energy, since the density and the specific heat capacity of the PCM do not differ significantly in the solid and the liquid phase.

Note that the capacity of combined storage with water and PCM would be larger with increasing fraction of the PCM, or if the difference between the initial and the final temperature is smaller. Clearly, an increase of the PCM fraction in the storage tank, however, would lead to higher costs. Hence, the optimum amount of the PCM has to be found taking the economic benefits into account.

3   Experimental Results

First, the tank containing 230 l water and no PCM is tested. The experiment was done on May 29th, 2018 in Sarajevo, Bosnia-Herzegovina (a place with a typical moderate continental climate in Southeast Europe). Figure 3 shows the temperature history at four measurement locations: T1 and T2 in the primary, solar circuit, and T5 and T6 in the storage tank at the distance 90 cm and 20 cm beneath the top side, respectively. Evidently, in the early afternoon, the temperature in the supply line from the solar collectors to the tank reaches the value of nearly 95 °C, while the temperatures in the tank approach the value of 80 °C. In the period between 12:00 and 14:30, the water temperature in the tank varies from about 62 °C to 80 °C, delivering thus the average heat supply rate of about 2 kW (limited by the weather conditions). About 14:30, the pump in the secondary, heater circuit is started, causing heat release by the fan coil. Consequently, the temperatures in the tank, T5 and T6, decrease. Also, a steep decrease of the temperatures T1 and T2 in the primary, solar circuit is seen, due to the weather condition. After the temperature on the top of the solar collectors or in the solar circuit

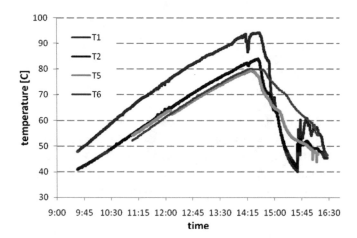

Fig. 3. Temperature history in the storage tank filled with 230 l water only (without PCM): T1 – temperature at the supply line, T2 – temperature at the return line, T5 – temperature measured 90 cm beneath the tank top side, T6 – temperature measured 20 cm beneath the tank top side.

falls below the tank temperature, the pump in the solar circuit is automatically stopped (shortly before 15:00). For that reason, the temperatures in the supply and the return line T1 and T2 are the same. Some time later, another, not very strong, heating process was automatically started, caused by increased solar irradiation.

The temperature history of the storage in the tank with 230 l water and 30 l PCM is shown in Fig. 4. The experiment was done on June 1st, 2018. The values are measured at four points, with slightly modified positions of T5 and T6 in the tank. They are located 20 cm and 80 cm from the tank top side, respectively. Similar trend is observed

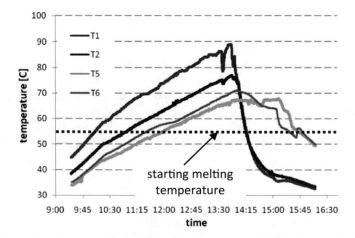

Fig. 4. Temperature history in the storage tank filled with 230 l water and 30 l PCM: T1 – temperature at the supply line, T2 – temperature at the return line, T5 – temperature measured 80 cm beneath the tank top, T6 – temperature measured 20 cm beneath the tank top.

in temperature variation in the primary, solar circuit as in the previous case. The temperature in the supply line is slightly below 90 °C shortly before 14:00, when the weather conditions suddenly change. Also the difference between the supply and the return line is similar. In the first case, the maximum difference is about 11.5 °C (in a wide period between 12:00 and 14:15) while in the latter case it is about 12.5 °C (at 13:50). The flow rate of the water-glycol mixture is the same in the both cases. However, despite the similar incoming heat flow, the tank temperature in the case with water and PCM is considerably lower. It reaches the maximum value about 70 °C, at about 14:00 while in the case with water only the same temperature value was reached at about 13:00. This effect is addressed to the latent heat stored in the PCM. Although, the heat source stopped about 14:00 (the pump in the solar circuit automatically switched off due to the weather change and the temperature in the solar circuit fell down significantly), the temperature in the tank is maintained at roughly the same level about 10 °C above the PCM starting melting temperature, for a further period between 60 min and 90 min. In the case without PCM, the tank temperatures decreased monotonically after the heat source switch-off. Here, with PCM, the tank temperatures start to decrease when the fan coil (including the fan, too) is started at 15:00 releasing the heat from the tank. It may also be seen that the tank temperature is more-or-less uniform during the charging/heating stage (the values of T5 and T6 are relatively close). However, energy charging process in the molten phase of the PCM is accompanied by the buoyancy effects. Additionally, during the energy discharge from the tank which causes solidification, inhomogeneity in the PCM develops, as it was also observed in an earlier work [5]. These two phenomena are assumed to be the cause of the larger differences between the T5 and T6 values during and after the melting, implying a vertical temperature gradient in the tank. The vertical temperature differences are clearly seen in the tank with water only, Fig. 3, where the buoyancy in the liquid is expected.

4   Conclusions

The installed experimental setup allows series of investigations on the solar thermal energy use and its storage striving for optimum flexibility and the balance of energy supply and demand.

The temperature levels obtained in the primary, solar circuit, as well as in the tank, both with water and with water and PCM, are satisfactory. Also a time delay in the tank-temperature variation was recorded, as expected, yielding a flattened character of the tank temperature history curve. Inhomogeneity in the PCM (entrapped air, buoyancy) requires additional tank design improvements.

Acknowledgment. This study is conducted within the project financially supported by the Ministry of Education, Science and Youth of Sarajevo Canton, Bosnia-Herzegovina, which is gratefully acknowledged.

References

1. Mehling, H., Cabeza, L.F.: Heat and Cold Storage with PCM. Springer, Heidelberg (2008)
2. Dincer, I., Rosen, M.A.: Thermal Energy Storage – Systems and Applications, 2nd edn. Wiley, Hoboken (2011)
3. Solé, C., Medrano, M., Nogués, M., Roca, J., Cabeza, L.F.: Economic, energetic and exergetic study of a water tank including PCM modules inside, Dept. d'Informàtica i Enginyeria Industrial, Universitat de Lleida, Pere de Cabrera s/n, 25001 – Lleida (Spain)
4. Solé, C., Medrano, M., Comellas, M., Nogués, M., Cabeza, L.F.: Water and PCM-water stores for domestic hot water and space heating applications, Centre GREA Innovació ConcurrentEdifici CREA, Universitat de Lleida, Pere de Cabrera s/n, 25001 – Lleida (Spain)
5. Torlak, M., Delalić, N.: Latent-heat thermal-energy storage in heat exchanger with plain and finned tube. In: Proceedings of the 3rd World Congress on Mechanical, Chemical, and Material Engineering, MCM 2017, paper no. HTFF-160, Rome, Italy, 9–10 June 2017

Author Index

© Springer Nature Switzerland AG 2019
S. Avdaković (Ed.): IAT 2018, LNNS 60, pp. 589–590, 2019.
https://doi.org/10.1007/978-3-030-02577-9

Printed in the United States
By Bookmasters